Electronic States of Inorganic Compounds:
New Experimental Techniques

NATO ADVANCED STUDY INSTITUTES SERIES

*Proceedings of the Advanced Study Institute Programme, which aims
at the dissemination of advanced knowledge and
the formation of contacts among scientists from different countries*

The series is published by an international board of publishers in conjunction
with NATO Scientific Affairs Division

A	Life Sciences	Plenum Publishing Corporation
B	Physics	London and New York
C	Mathematical and Physical Sciences	D. Reidel Publishing Company Dordrecht and Boston
D	Behavioral and Social Sciences	Sijthoff International Publishing Company Leiden
E	Applied Sciences	Noordhoff International Publishing Leiden

Series C – Mathematical and Physical Sciences

*Volume 20 – Electronic States of Inorganic Compounds:
New Experimental Techniques*

Electronic States of Inorganic Compounds: New Experimental Techniques

*Lectures Presented at the NATO Advanced Study Institute
held at the Inorganic Chemistry Laboratory and St. John's College,
Oxford, 8–18 September 1974*

edited by

P. DAY

St. John's College, Oxford, England

D. Reidel Publishing Company

Dordrecht-Holland / Boston-U.S.A.

Published in cooperation with NATO Scientific Affairs Division

Library of Congress Cataloging in Publication Data

NATO Advanced Study Institute, Oxford, 1974.
    Electronic states of inorganic compounds.

    (NATO advanced study institutes series : C, mathematical and physical sciences ; 20)
    Bibliography: p.
    1. Chemistry, Inorganic—Congresses.   2.   Excited state chemistry—Congresses.
I. Day, Peter R., 1928–   II.   North Atlantic Treaty Organization. Division of Scientific Affairs.
III.   Title.   IV.   Series: NATO advanced study institutes series : Series C, mathematical and physical
sciences ; 20.
QD475.N37   1974   547'.1'3   75–17752
ISBN-13: 978-94-010-1862-3   e-ISBN-13: 978-94-010-1860-9
DOI: 10.1007/ 978-94-010-1860-9

Published by D. Reidel Publishing Company
P.O. Box 17, Dordrecht, Holland

Sold and distributed in the U.S.A., Canada, and Mexico
by D. Reidel Publishing Company, Inc.
306 Dartmouth Street, Boston, Mass. 02116, U.S.A.

CONTENTS

PREFACE

In the last few years a surprisingly large number of new
experimental techniques have been devised to probe, often with
great subtlety, into the electronic structures of inorganic
substances.   Thus in favourable cases one now has the opportunity
of locating and assigning electronically excited states over a
vast energy range stretching from tens of cm^{-1} above the ground
state up to some 10^6 cm^{-1}.   The techniques are extremely dis-
parate in background, involving (among others) linearly and
circularly polarised electromagnetic radiation, electron kinetic
energy analysis and neutron scattering.   Furthermore, practition-
ers of many of the techniques may not be aware of how the
information which they are obtaining overlaps and complements
that obtained by other techniques.   The time therefore seemed
ripe to bring together a group of experts to survey, for an
audience of inorganic chemists, the basic theories and experimental procedures relevant to the different techniques, and the
relations between them.   In pursuing this aim we were fortunate
in having the very generous financial backing of N.A.T.O.,
through their Advanced Study Institutes programme, and the
present volume records the substance of lectures given at the
Institute which took place at the Inorganic Chemistry Laboratory
and St. John's College, Oxford, from 8-18 September 1974.
However, the volume is much more than a set of 'Proceedings';
by its completeness it comprises, rather, a textbook of the
new methods which should prove a useful guide for all inorganic
chemists interested in the electronic structures, both of
molecules and continuous lattice solids.

As the Director of the Institute I am deeply conscious of,
and extremely grateful for, the efforts of the other members of
the Organizing Committee, Mr. A.F. Orchard, Professor C. Furlani
and Professor L. Oleari, for the care taken by the lecturers in
preparing their material for publication, and finally to my wife
for typing the entire manuscript.   Everyone concerned in the
Institute will also be grateful to the staff of the Inorganic

Chemistry Laboratory, to the equipment manufacturers for their illuminating demonstrations, to the President and Fellows of St. John's College for their permission to use the College premises and to the N.A.T.O. Scientific Affairs Division for their financial assistance.

P. DAY
Oxford, March 1975

BASIC THEORY OF THE ELECTRONIC STATES OF MOLECULES

C.J. Ballhausen

Chemical Laboratory IV, H.C. Ørsted Institute,
Universitetsparken 5, Copenhagen Ø, Denmark

ELECTRONIC AND NUCLEAR COORDINATES

The wave-functions for a molecule are dependent both on the nuclear and electronic coordinates. The first task is, therefore, to try to separate the electronic and nuclear motions. We designate the coordinates and masses of the nucleus μ (Q_μ, M_μ) and the coordinates and masses for the electrons (r_i, m), where all coordinates $\underline{r}$ and $\underline{Q}$ are referred to a laboratory fixed coordinate system. The total non-relativistic Hamiltonian for a molecular system of $\underline{N}$ nuclei and $\underline{n}$ electrons is

$$\mathcal{H} = -\sum_\mu \frac{\hbar^2}{2M_\mu} \nabla_\mu^2 - \sum_i \frac{\hbar^2}{2m} \nabla_i^2 + V(r,Q) \tag{1}$$

$V(r,Q)$ is the potential energy term equal to

$$V(r,Q) = -\sum_{\mu i} \frac{Z_\mu e^2}{|Q_\mu - r_i|} + \sum_{\mu \nu} \frac{Z_\mu Z_\nu e^2}{|Q_\mu - Q_\nu|} + \sum_{i<j} \frac{e^2}{|r_i - r_j|} \tag{2}$$

Let us now suppose that all the nuclear masses are infinite. Clearly, this will quench the nuclear motions, leading to an 'electronic' Hamiltonian of the form

$$\mathcal{H}_E = -\sum_i \frac{\hbar^2}{2m} \nabla_i^2 + V(r,Q) \tag{3}$$

P. Day (ed.), Electronic States of Inorganic Compounds. 1–25. All Rights Reserved.
Copyright © 1975 by D. Reidel Publishing Company, Dordrecht-Holland.

The 'electronic' Schrodinger equation is then defined as

$$\mathcal{H}_E \Psi_n (r,Q) = W_n (Q) \Psi_n (r,Q) \tag{4}$$

Both $W_n(Q)$ and $\Psi_n(r,Q)$ are seen to contain the nuclear coordinates Q as parameters, and are indeed continuous functions of the 3N nuclear coordinates. In principle, we could solve (4) for all values of the 3N nuclear coordinates, and for each nuclear arrangement we will obtain a complete set of electronic wave-functions $\Psi_n(r,Q)$ and eigenvalues $W_n(Q)$. In order to treat the motions of the nuclei we introduce the so-called normal co-ordinates. This involves changing from the external coordinates Q to an internal molecular coordinate system ξ.

Let us assume that we can find at least one point, Q^o, in the 3N dimensional coordinate system for which all the first derivatives of $W_n(Q)$, with respect to Q, are equal to zero. Thus

$$\sum_\mu \left(\frac{\partial W(Q)}{\partial Q_{\mu x}}\right)_{Q^o} = \sum_\mu \left(\frac{\partial W(Q)}{\partial Q_{\mu y}}\right)_{Q^o} = \sum_\mu \left(\frac{\partial W(Q)}{\partial Q_{\mu z}}\right)_{Q^o} = 0 \tag{5}$$

Obviously the function $W_n(Q)$ will have an extremum at the point Q^o. Whether Q^o represents a minimum or a maximum depends, of course, upon the values of the second derivatives of $W_n(Q)$.

In the case where (5) cannot be fulfilled, and at least one of the 3N derivatives is different from zero for all values of the nuclear coordinate, this simply means that the potential hypersurface cannot have a minimum with respect to this nuclear coordinate, and that the corresponding molecular state is unstable. However, if for some point Q^o equation (5) can indeed be fulfilled, Q^o will be taken to represent a stable configuration of the molecule in the state $\Psi_n(r,Q)$.

We now introduce a set of 3N mass-weighted displacement co-ordinates $S_1 \ldots \ldots S_{3N}$ defined as

$$
\begin{aligned}
S_1 &= \sqrt{M}_1 (Q_{1x} - Q^o_{1x}) \\
S_2 &= \sqrt{M}_1 (Q_{1y} - Q^o_{1y}) \\
S_3 &= \sqrt{M}_1 (Q_{1z} - Q^o_{1z}) \\
&\vdots \\
&\vdots \\
S_{3N} &= \sqrt{M}_N (Q_{Nz} - Q^o_{Nz})
\end{aligned}
\tag{6}
$$

This transformation amounts to a translation of the Q-coordinate system followed by a change in scale on the axes. The first and second derivatives with respect to these displacement coordinates are seen to be

$$\frac{\partial}{\partial S_1} = \frac{1}{\sqrt{M_1}} \frac{\partial}{\partial Q_{1x}} \qquad \frac{\partial^2}{\partial S_1^2} = \frac{1}{M_1} \frac{\partial^2}{\partial Q_{1x}^2}$$

$$\vdots \qquad\qquad\qquad \text{and} \qquad\qquad \vdots$$

$$\frac{\partial}{\partial S_{3N}} = \frac{1}{\sqrt{M_N}} \frac{\partial}{\partial Q_{Nz}} \qquad \frac{\partial^2}{\partial S_{3N}^2} = \frac{1}{M_N} \frac{\partial^2}{\partial Q_{NZ}^2}$$

In this coordinate system, the point S°, for which all the first derivatives of the function $W_n(S)$ vanish, obviously lies at the origin $S^\circ = (0,0, \dots 0)$, and we can expand $W_n(S)$ in a Taylor series retaining terms up to second order only:

$$W_n(S) = W_n^\circ + \frac{1}{2} \sum_{k,l} \left(\frac{\partial^2 W_n(S)}{\partial S_k \, \partial S_l}\right)_\circ S_k S_l + \dots \tag{7}$$

where W_n° is the electronic energy calculated at the origin (or equivalently at Q°) and where all terms containing the first derivatives of $W_n(S)$ are absent by virtue of (5). The second derivatives of $W_n(S)$ are to be evaluated at the origin as indicated by the subscript 0.

The potential energy term in (7) is a so-called quadratic form, containing all of the cross terms $S_k \cdot S_l$. Introducing a linear combination of all the displacement coordinates (6) we can, however, reduce the potential energy terms to a form in which all of the cross terms have disappeared. Taking

$$\xi_t = \sum_k b_{kt} S_k \tag{8}$$

we can write for the potential energy (7)

$$\mathcal{V}_n(\xi) = \mathcal{V}_n^\circ + \frac{1}{2} \sum_{t=1}^{3N} k_t^n \xi_t^2 \tag{9}$$

The coordinates ξ_t represent all of the possible movements of the nuclei. Three of them will, therefore, describe the translations of the molecule as a whole, and three of them will characterize the rotations of the molecule. The translations and the rotations cannot however depend upon the intermolecular distances. The potential energy terms in (9) associated with the translations and the rotations, depending as they do upon the internuclear distances, must therefore be zero. This implies that for non-linear molecules the coefficients k_1^n to k_6^n are zero. (For linear molecules only k_1^n to k_5^n are zero.)

The 3N-6 linear combinations, ξ_t, of the cartesian nuclear displacement coordinates are called the normal coordinates of the molecule. They are of such a nature as to leave the centre of the mass of the nuclei unaltered, and the principal axes of inertia are likewise left unchanged. It is therefore natural to use, in the description of the molecule, a cartesian coordinate system with origin at the centre of mass, and with the coordinate axes directed along the principal axes of inertia for the nuclei in the equilibrium position. Hence, in the so-called <u>crude adiabatic approximation</u>, we take for the total wavefunction

$$\Psi_{mv}(r,\xi) = \chi_{mv}(\xi)\Psi_m^o(r) \tag{10}$$

where $\Psi_m^o(r)$ is the static electronic wavefunction at the equilibrium position of the nuclei, given by

$$\mathcal{H}_E(Q^o)\psi_m^o(r) = \varepsilon_m^o\psi_m^o(r) \tag{11}$$

and the vibrational part $\chi_{mv}(\xi)$ is solved using the potential surface for molecule, given by

$$\mathcal{U}_n(\xi) = \tfrac{1}{2}\sum_{t=1}^{3N-6} k_t^n \xi_t^2 + \mathcal{U}_n^o \tag{12}$$

The totality of all the static electronic wavefunctions span what has been called a <u>Longuet-Higgins</u> space. The completeness of this space implies that we can obtain a dynamical electronic wavefunction as a superposition of static electronic wavefunctions, viz.

$$\psi_k(r,\xi) = \sum_m \psi_m^o(r)c_{mk}(\xi) \tag{13}$$

Such an expansion is encountered in the so-called <u>Herzberg-Teller</u> <u>vibronic scheme</u>.

The transformation which generates the normal coordinates, ξ_t, from the displacement coordinates, S_k, depends very much upon the electronic state. Calling the electronic state <u>i</u> we have

$$\mathcal{U}_i = \mathcal{U}_o + \tfrac{1}{2}\sum_{k,l} S_k S_l V_{k,l}^{ii} \tag{14}$$

with

$$V_{kl}^{ii} = \int \psi_i^o(r) \left[\frac{\partial^2 V}{\partial S_k \partial S_l} \right]_o \psi_i^o(r) \, dr \tag{15}$$

Hence the transformation which generates the normal coordinates, ξ_t, from the displacement coordinates, S, depends crucially upon the electronic matrix element V_{kl}^{ii}. A particular transformation is therefore unique for the state i in the reduction of the quadratic form of the potential energy (7). The normal coordinates of, say, an excited state will in general correspond to a translation and rotation in the space spanned by the ground state normal coordinates. This 'effect' was first stated by Duschinky, and it is referred to as the Duschinky effect. The presence of the Duschinky effect may be detected by comparing the absorption and emission spectra of a molecule, since the 'mirror symmetry' of these two properties are destroyed.

Suppose now that a molecule possesses a certain number of identical atoms. This means that the Hamiltonian, $\mathcal{H}$, must be totally symmetric under all permutations of the identical nuclei. Let $\hat{P}$ be any such permutation operator, then

$$\hat{P}\mathcal{H} = \mathcal{H} \tag{16}$$

The electronic Schrödinger equation is

$$\mathcal{H}\psi = W\psi \tag{17}$$

Operating with P on equation (17) yields

$$\hat{P}(\mathcal{H}\psi) = \mathcal{H}(\hat{P}\psi) = (\hat{P}W)(\hat{P}\psi) \tag{18}$$

It is always possible to construct our wavefunctions ψ to be eigenfunctions of the permutation operator, $\hat{P}$. Hence

$$\hat{P}\psi = c\psi \tag{19}$$

Substituting (19) into (18) gives after division with the constant c

$$\mathcal{H}\psi = (\hat{P}W)\psi \tag{20}$$

A comparison with (17) shows that

$$W = \hat{P}W \tag{21}$$

In other words, the potential surface must be invariant under all permutations of identical nuclei.

The molecular point group to which a molecule belongs is a sub-group of the permutation group. Let $\hat{S}$ be a symmetry operator in the molecular point group. The potential energy surface is then for a non-linear molecule given by $W(\xi_1, \xi_2 \ldots \xi_{3N-6})$, where $\xi_1, \xi_2 \ldots \xi_{3N-6}$ are the vibrational symmetry coordinates.

Then using (21)

$$W = \hat{S}W(\xi_1, \xi_2 \ldots \xi_{3N-6}) = W(\hat{S}\xi_1, \hat{S}\xi_2 \ldots \hat{S}\xi_{3N-6}) \qquad (22)$$

Degenerate electronic wavefunctions: Jahn-Teller theorem

Consider now the case where we have two or more degenerate solutions to the electronic Schrödinger equation. The degeneracy which occurs for a certain nuclear arrangement, Q^o, is therefore symmetry dependent. In order to calculate the potential surfaces we expand $\mathcal{H}$ around Q^o and use degenerate perturbation theory. As basic functions we take the crude adiabatic degenerate electronic functions $\psi_1^o(r)$, $\psi_2^o(r) \ldots \psi_n^o(r)$. To second order in ξ the perturbing Hamiltonian is

$$\mathcal{H}^{(1)} = \mathcal{H} - \mathcal{H}^o = \sum_i \left(\frac{\partial \mathcal{H}}{\partial \xi_i}\right)_o \xi_i + \frac{1}{2} \sum_{i,j} \left(\frac{\partial^2 \mathcal{H}}{\partial \xi_i \partial \xi_j}\right)_o \xi_i \xi_j \qquad (23)$$

A displacement of the nuclei which destroys the degeneracy of the electronic functions will lead to a splitting of the state. To first order in the nuclear displacement coordinates this splitting is determined by the value of the matrix elements

$$\sum_i \xi_i \int \psi_m^o(r) \left(\frac{\partial \mathcal{H}}{\partial \xi_i}\right)_o \psi_n^o(r) \, dr \qquad (24)$$

A totally symmetric vibration cannot in first order destroy the molecular symmetry. Concentrating therefore on the non totally symmetric vibrations we observe that the matrix elements (24) will be different from zero provided the symmetric product* of the representation which is spanned by $\psi_1^o(r) \ldots \psi_n^o(r)$ contains a representation spanned by $\sum_i \left(\frac{\partial \mathcal{H}}{\partial \xi_i}\right)_o$. Now $\left(\frac{\partial \mathcal{H}}{\partial \xi_i}\right)_o$ must on the other hand transform like the normal vibrational coordinate ξ_i in the point group of the molecule, since otherwise the terms in the Hamiltonian $\sum_i \left(\frac{\partial \mathcal{H}}{\partial \xi_i}\right)_o \xi_i$ cannot transform like a totally symmetric function. In a systematic investigation Jahn and Teller showed that for all molecules, except linear systems, a nuclear arrange-

---

* If the wavefunctions span a double group representation the antisymmetric product should be used.

ment which will lead to electronic degeneracies will also
automatically have at least one non totally symmetric vibrational
symmetry coordinate present, such that the matrix elements (24)
are different from zero.

A symmetry determined degeneracy will therefore always be
removed by interactions with the vibrations of the molecule.
The further treatment of the effect split up into two parts.
The investigation of the potential surfaces leads to the concept
of the static Jahn-Teller effect. Solving for the vibrational
motions of the nuclei on the potential surfaces, thereby characterising the energy of the system, concerns the dynamic Jahn-Teller
effect. These questions are taken up in greater detail in
Professor Stephens' chapter.

SYMMETRY

In classical mechanics there are constants of motion such as,
e.g. the energy, the linear momentum p and others. In quantum
mechanics any quantity represented by A which commutes with $\mathcal{H}$ will
be a constant of the motion. This we write

$$[\mathcal{H} A] = \mathcal{H}A - A\mathcal{H} = 0 \tag{1}$$

Hence

$$A(\mathcal{H}\Psi) = \mathcal{H}(A\Psi)$$
$$\|$$
$$A(W\Psi) = W(A\Psi) \tag{2}$$

Notice that $(A\Psi)$ and Ψ are both solutions to the Schrödinger
equations with the same eigenvalue W.

Consider for instance a square planar molecule (Figure 1).
The potential energy term in the Hamiltonian is given by

$$-\mathcal{V} = \frac{Q_1 e^2}{\sqrt{(x-a)^2+(y-a)^2+z^2}} + \frac{Q_2 e^2}{\sqrt{(x-a)^2+(y+a)^2+z^2}}$$
$$+ \frac{Q_3 e^2}{\sqrt{(x+a)^2+(y+a)^2+z^2}} + \frac{Q_4 e^2}{\sqrt{(x+a)^2+(y-a)^2+z^2}} \tag{3}$$

Rotating the molecule by an angle π around an axis going
through nuclei 1 and 3 will interchange x and y and cause z to
become $-z$. This change of coordinates will leave $\mathcal{V}$ unaltered
provided $Q_2 = Q_4$. Alternatively we could, of course, just as
well interchange the identical nuclei 2 and 4. Hence we observe
that allowed transformations of the electronic coordinates equal

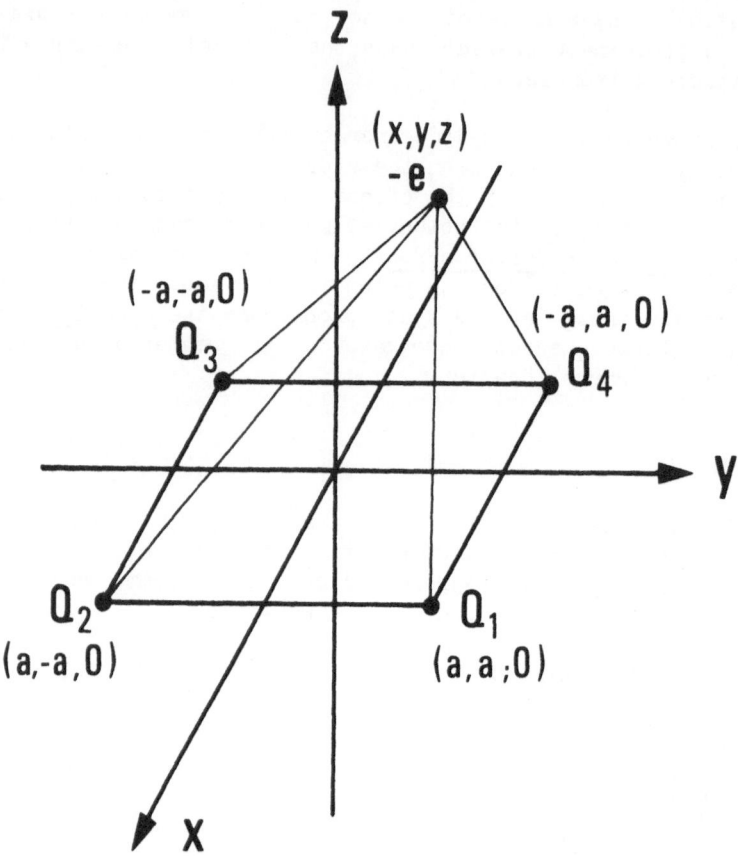

Figure 1.   Electronic and nuclear coordinates of a square planar
molecule.

allowed transformations of the nuclei.

Let us now consider the case $Q_1 = Q_2 = Q_3 = Q_4$.   In that
case a four-fold rotation, $\hat{C}_4$, of the molecule will commute with
$\mathcal{V}$.   Hence for a one-fold degenerate solution to the potential (3)
we must have acording to (2)

$$\hat{C}_4 \psi = c \psi \tag{4}$$

since if both $(\hat{C}_4 \psi)$ and ψ are to be solution to the same Schrödinger equation having the same energy, they can only differ by a
constant.   Applying the $\hat{C}_4$ operator four times we must, however,
find that the molecule has been rotated back to its original
position.   Therefore

$$\psi = (\hat{C}_4)^4 \psi = (\hat{C}_4)^3 c\psi = (\hat{C}_4)^2 c^2 = (\hat{C}_4)c^3\psi = c^4\psi \tag{5}$$

from which $c^4 = 1$ or $c = \pm 1$. A one-fold electronically degenerate wavefunction appropriate to a molecule having a four-fold rotational symmetry axis must therefore go into either plus or minus itself when rotated 90°. For a molecule possessing a three-fold rotational symmetry axis, we would have had

$$\psi = (\hat{C}_3)^3\psi = b^3\psi \tag{6}$$

or $b^3 = 1$; $b = 1$. We observe that a wavefunction can be characterized by its behaviour under a molecular symmetry operation.

Consider now ψ to be a member of a degenerate set of electronic states $\{\psi_1 \ldots \ldots \psi_j\}$. Then

$$\hat{A}\begin{pmatrix} \psi_1 \\ \vdots \\ \psi_j \end{pmatrix} = \begin{pmatrix} a_{11} & \cdots & a_{ij} \\ & & \\ \vdots & & \vdots \\ & & \\ a_{j1} & \cdots & a_{jj} \end{pmatrix} \begin{pmatrix} \psi_1 \\ \vdots \\ \psi_j \end{pmatrix} \tag{7}$$

Of course we could have used as our 'basis set' another set $\{\chi_1 \ldots \ldots \chi_j\}$ related to $\{\psi_1 \ldots \ldots \psi_j\}$ by a unitary transformation. Then

$$\hat{A}\begin{pmatrix} \chi_1 \\ \vdots \\ \chi_j \end{pmatrix} = \begin{pmatrix} b_{11} & \cdots & b_{1j} \\ & & \\ \vdots & & \vdots \\ & & \\ b_{j1} & \cdots & b_{jj} \end{pmatrix} \begin{pmatrix} \chi_1 \\ \vdots \\ \chi_j \end{pmatrix} \tag{8}$$

However, since the two sets are related to each other by a unitary transformation the traces of the two transformation matrices will be the same. Hence

$$\sum_{i=1}^{i} a_{ii} = \sum_{i=1}^{j} b_{ii},$$ and we can use the value of the trace to

characterize the set $\{\psi_1 \ldots \ldots \psi_j\}$ under a certain symmetry operation.

For any molecule there are a certain number of symmetry operations that commute with $\mathcal{H}$, and therefore leave the molecule 'unaltered'. These are those operations which rotate or reflect the molecule into itself. Of all these operations (including leaving the molecule alone) we have that any two of them are

equivalent to a third.   Such a set of symmetry operations
constitutes a <u>group</u>, and <u>group theory</u> can tell us how the
permitted molecular states must behave.   We can therefore tell
what kind of states can occur in a molecule of a given shape.

In an octahedral molecule we have 24 symmetry operations, viz.

$$\hat{E} \qquad 8\hat{C}_3 \qquad 3\hat{C}_2 \qquad 6\hat{C}_4 \qquad 6\hat{C}_2$$

Here $\hat{E}$ stands for 'doing nothing' (the identity operation), $\hat{C}_4$,
$\hat{C}_3$ and $\hat{C}_2$ are rotations by $\frac{\pi}{2}$, $\frac{2\pi}{3}$ and π respectively.   Provided
an inversion centre is present in the molecule, the inversion
operator i is introduced, and the number of symmetry operations
are doubled.

Group theory now tells us that there are only five kinds of
states which are permissible in this molecular symmetry, and that
their traces under the symmetry operations are as follows:-

	E	$8C_3$	$3C_2$	$6C_4$	$6C_2'$
A_1	1	1	1	1	1
A_2	1	1	1	-1	-1
E	2	-1	2	0	0
T_1	3	0	-1	1	-1
T_2	3	0	-1	-1	1

In cases of the octahedron having inversion symmetry $\hat{\imath}$ can
distinguish between whether a state is even (g), that is goes into
itself or odd (u), that is goes into minus itself.

As long as no spin-coordinates occur in the Hamiltonian $\hat{S}^2$,
where S is the total spin-angular momentum, will always commute
with $\mathcal{H}$.   The molecular states can in that case be characterized
by their orbital transformation properties and the values of 2S+1
and S_z.

Consider now a Hamiltonian containing also a spin-orbit
coupling term

$$\mathcal{H}_{so} = \sum_i \boldsymbol{\xi}_i(r) l_i \cdot s_i$$

Since $\mathcal{H}_{so}$ is a scalar product of l and s it is invariant if the
same symmetry operations are applied in both spin-space and
electronic position space simultaneously.   Hence $[\hat{A}\mathcal{H}_{so}] = 0$.

For rotating by an angle α around the z axis we have with $\psi = e^{im_s\phi} f(\theta,r)$, where $|m_s| = \frac{1}{2}$, that the transformation matrix is

$$
\begin{pmatrix}
e^{i\frac{1}{2}\alpha} & 0 \\
0 & e^{-i\frac{1}{2}\alpha}
\end{pmatrix}
\tag{9}
$$

having a trace, Tr. $= 2 \cos \frac{1}{2}\alpha$. For $\alpha = 0$, Tr. $= 2$ but for $\alpha = 2\pi$, Tr. $= -2$. The functions having half integral values of angular momentum are therefore in general 'double-valued'. However, for $\alpha = \pi$ the functions are single-valued since $\alpha = \pi$ and $\alpha = 3\pi$ both give Tr. $= 0$. To avoid this double-valuedness of the half angular momentum functions we take the period of space rotation to be 4π instead of 2π. Therefore, only on rotation of 4π is the system to go into itself.

The number of symmetry operations in the so-called <u>double group</u> is therefore twice as many as in the single group. Calling the rotation by 2π for R we get in the octahedral group

	E	R	$8C_3$	$8C_3R$	$6C_2$	$6C_4$	$6C_4R$	$12C_2$	Be the notation
A_1	1	1	1	1	1	1	1	1	Γ_1
A_2	1	1	1	1	1	-1	-1	-1	Γ_2
E	2	2	-1	-1	2	0	0	0	Γ_3
T_1	3	3	0	0	-1	1	1	-1	Γ_4
T_2	3	3	0	0	-1	-1	-1	1	Γ_5
$E_{\frac{1}{2}}$	2	-2	1	-1	0	2	-2	0	Γ_6
$E_{5/2}$	2	-2	1	-1	0	-2	2	0	Γ_7
U	4	-4	-1	1	0	0	0	0	Γ_8

The aforementioned set of functions with $|m_s| = \frac{1}{2}$ is seen to transform like $E_{\frac{1}{2}}$. The transformation properties of a ^2E electronic state is found by multiplying the transformation properties of an E state with that of an $E_{\frac{1}{2}}$ state. Direct use of the above table yields a U type function.

A point to note is that the double-valued representations are all of even dimension. Hence any system containing an odd

number of electrons will have levels which are at least two-fold
degenerate.   This holds true as long as no magnetic field is
present (<u>Kramers degeneracy</u>).

The symmetry designations represent what are frequently
called 'good quantum numbers'.   Their importance lies in the
fact that a molecular state can be characterized exactly, even if
one can only perform an approximate calculation.   Only provided
two states have exactly the same characterizations can they
furthermore interact under the Hamiltonian.   It is this last
feature which enables us to draw the well known correlation
diagrams in molecular orbital theory, and especially in crystal
field theory.

STATE ENERGIES

The electronic energies of the molecular states for a single
molecule (tight binding) are evaluated using the electronic
Schrödinger equation at the equilibrium point of the ground state.

$$W = \frac{\int \Psi^* \mathcal{H} \Psi dr}{\int \Psi^* \Psi dr} \tag{1}$$

Ψ is approximated by a (sum of) Slater determinant(s) made up of
single electronic functions, the molecular orbitals (MOs).   The
molecular Hamiltonian is of the form

$$\mathcal{H} = \sum_{i=1}^{N} \hat{h}(i) + \sum_{i<j} g(i,j) \tag{2}$$

where $\hat{h}(i)$ is the kinetic energy operator plus the nuclear
attraction terms for the i'th electron, and where $g(i,j)$ represents the Coulomb repulsion between electrons i and j.

The Hartree-Fock method minimizes the total energy W of the
ground state (or an excited state) with respect to a variation of
any occupied MO.   The only condition is that the MOs shall form
an orthonormal set,

$$\int \psi_m^*(1) \psi_n(1)\ dr_1 = \delta_{m,n} \tag{3}$$

Hence

$$\delta W = \int \delta \Psi^* \mathcal{H} \Psi dr + \int \Psi^* \mathcal{H} \delta \Psi dr = 0 \tag{4}$$

We define the Coulomb, $\hat{J}$, and exchange, $\hat{K}$, operators as

$$J_k \psi_t(1) = \int \psi_k^*(2) \psi_k(2)\ g(1,2) \psi_t(1) dr_2 \tag{5}$$

$$\hat{K}_k \psi_t(1) = \int \psi_k^*(2) \psi_k(1) \, g(1,2) \psi_t(2) \, dr_2 \tag{6}$$

The conditions (3) and (4) then lead to the following equation for calculating the orbitals, provided the total wavefunction is a closed shell:

$$\hat{F}(1) \psi_i(1) = w_i \psi_i(1) \tag{7}$$

with

$$\hat{F}(1) = \hat{h}(1) + \sum_{k=1}^{N} (2\hat{J}_k - \hat{K}_k) \tag{8}$$

The orbital energy w_i is therefore given by

$$w_i = \bar{h}_{ii} + \sum_{k=1}^{N} (2J_{ki} - K_{ki}) \tag{9}$$

Attention is drawn to the fact that for a system having open shells, the concept of orbital energy becomes rather complicated.

For a closed-shell system we can easily show that $-w_i$ is equal to the ionization potential for removing an electron from the orbital ψ_i. This assumes that the other orbitals remain unaltered; the electrons must not relax. First demonstrated by Koopmans, this consequence of the Hartree-Fock equations (Koopmans theorem) plays an important role in photoelectron spectroscopy.

In so-called ab initio calculations all electrons in the system are considered. In the semi-empirical theories the electrons which are not considered to take part in the 'bonding' are lumped together as part of the atomic 'cores' and we define a Hamiltonian for the i valence electrons

$$\mathcal{H}_{val} = \sum_i \hat{h}_{core}^{(i)} + \sum_{i<j} g(i,j) \tag{10}$$

now minimizing W from

$$\mathcal{H}_{val} \Psi = W \Psi \tag{11}$$

In order to get good results it is necessary to keep the valence orbitals orthogonal to the core orbitals.

The valence orbitals in a metal complex are usually written

$$\psi = N(\chi_M + \lambda \sum_{i=1}^{L} c_i \chi_i) \tag{12}$$

where χ stands for an atomic orbital, and λ and c_i are variational parameters. N is the normalizing constant. The ligand combination of atomic orbitals

$$\sum_{i=1}^{L} c_i \chi_i$$

is usually completely determined by symmetry.

Consider, for example, the e_g orbital's in O_h. With the four   ligand orbitals centred as shown in Figure 2, the 'analytical continuation' of $d_{x^2-y^2}$ is easily seen to be

$$d_{x^2-y^2} + \sigma_1 - \sigma_2 + \sigma_3 - \sigma_4 \; .$$

Neglecting overlap between the ligand functions

$$\psi_{x^2-y^2} = N(d_{x^2-y^2} + \lambda \tfrac{1}{2}(\sigma_1 - \sigma_2 + \sigma_3 - \sigma_4)) \tag{13}$$

For d_{xy} we get in the same way (Figure 3)

$$\psi_{xy} = M(d_{xy} + \sigma \tfrac{1}{2}(\pi_1 + \pi_2 + \pi_3 + \pi_4)) \tag{14}$$

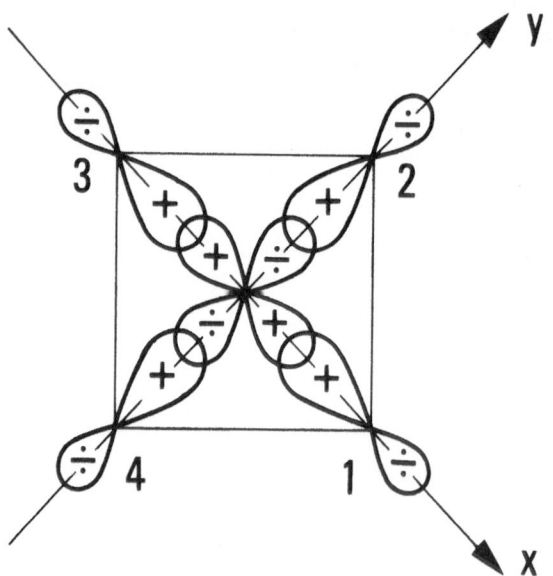

Figure 2. $d_{x^2-y^2}$ and $\sum c_i \chi_i$ in O_h.

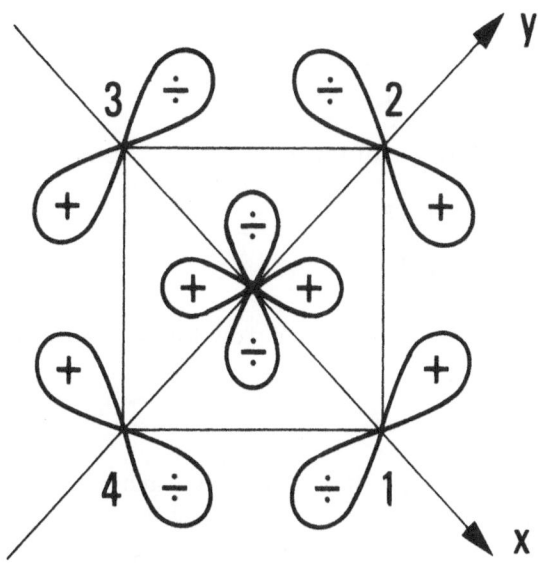

Figure 3. d_{xy} and $\sum c_i \chi_i$.

    The remaining LCAO can be generated by symmetry operations,
e.g. by using the operator $\hat{C}_3$ in the molecular point group.   The
Hartree-Fock type orbitals contain therefore in this approximation
only the variational parameters λ and δ.

    To gain a little insight, let us look at the system (core)
$(\psi_\gamma)^2$.   The molecular orbital we want to calculate is

$$\psi_\gamma = \alpha \chi_M + \beta \chi_L \tag{15}$$

with

$$\chi_L = \sum c_{iL} \chi_i$$

Using the Hartree-Fock operator $\hat{F}$ we get the secular equation

$$\begin{vmatrix} F_{MM} - w & F_{ML} - wS_{ML} \\ F_{ML} - wS_{ML} & F_{LL} - w \end{vmatrix} = 0 \tag{16}$$

with

$$F_{MM} = \int \chi_M^* \hat{F} \chi_M \, dr \tag{17}$$

and etc.

To good approximation we have for the bonding root

$$w = F_{LL} - \frac{(F_{ML}-F_{MM}S_{ML})^2}{F_{MM}-F_{LL}} \tag{18}$$

and for the antibonding solution

$$w^* = F_{MM} + \frac{(F_{ML}-F_{LL}S_{ML})^2}{F_{MM}-F_{LL}} \tag{19}$$

Provided, as is normally the case for transition metal complexes, that

$$F_{MM} - F_{LL} \gg |F_{ML} - F_{LL} S_{ML}| \tag{20}$$

the bonding root corresponds to the wavefunction

$$\psi_\gamma = \chi_L \tag{21}$$

and the antibonding solution to (16) to

$$\psi_\gamma^* = \frac{1}{\sqrt{1-S_{ML}^2}} \; (\chi_M - S_{ML}\chi_L) \tag{22}$$

We can therefore write

$$\hat{F}\chi_L = w\chi_L \tag{23}$$

and

$$\hat{F}(\chi_M - S_{ML}\chi_L) = w^*(\chi_M - S_{ML}\chi_L) \tag{24}$$

Combining (23) and (24) we have

$$\hat{F}\chi_M + (w^* - w) S_{ML}\chi_L = w^*\chi_M \tag{25}$$

Take $w^* - w = w_{ML}^o$ and define the pseudo-potential $\hat{U}$:

$$U\chi_M(1) = w_{ML}^o \int \chi_M(2)\chi_L(2)dr_2 \chi_L(1) \tag{26}$$

and we can rewrite (25) as

$$(\hat{F} + \hat{U})\chi_M = w^*\chi_M \tag{27}$$

The definition of the pseudo-potential $\hat{U}$ runs parallel to the definition of the exchange operator. $\hat{U}$ plays a role similar to the 'crystal field potential' $\mathcal{V}$ included in $\hat{F}$ as part of the term $\hat{h}_{core}$. We can then write for the orbital energy difference Δ

$$\Delta = <e_g^*|\hat{F}+\hat{U}|e_g^*> - <t_{2g}^*|\hat{F}+\hat{U}|t_{2g}^*> \tag{28}$$

$$\simeq <e_g^*|\mathcal{V}|e_g^*> - <t_{2g}^*|\mathcal{V}|t_{2g}^*> + w_{ML}^o(S_{e_gL}^2 - S_{t_{2g}L}^2)$$

With $S_{e_gL} > S_{t_{2g}L}$ which will usually be the case in octahedral complexes, we see that Δ is expected to be positive.

The crystal field theory concentrates upon the behaviour of the electrons in the antibonding orbitals $\psi_{e_g}^*$ and $\psi_{t_{2g}}^*$. We can in accord with (28) define both the one-electron orbital difference

$$10 \ Dq = w^*(e_g) - w^*(t_{2g}) \tag{29}$$

and a zero of energy. This last is taken as the one-electronic energy of the configuration

$$W\left[(t_{2g})^6(e_g)^4\right] = 0$$

or

$$6w^*(t_{2g}) + 4w^*(e_g) = 0 \tag{30}$$

Hence

$$w^*(e_g) = 6 \ Dq \tag{31}$$

$$w^*(t_{2g}) = -4 \ Dq \tag{32}$$

We can therefore reinterpret $w^*(e_g)$ and $w^*(t_{2g})$ as being equal to the core integrals $\bar{h}_{e_g}^*$ and $\bar{h}_{t_{2g}}^*$.

Most of the electronic repulsion integrals coming from the $g(i,j)$ term can be expressed as Coulomb integrals, J, and exchange integrals, K. Fifteen different J-integrals, ten different K-integrals and nine other two-electron integrals can be met with in a $(\psi_{xz}, \psi_{yz}, \psi_{xy}, \psi_{x^2-y^2})$ set of molecular orbitals. The symmetry of the molecule will, however, impose certain restrictions on the independence of these 34 integrals. In octahedral (O_h) symmetry we encounter for instance only 10 independent two-electron integrals.

The configuration $|\psi_{xz}^{\alpha} \psi_{yz}^{\alpha} \psi_{xy}^{\alpha}|$ gives in O_h symmetry rise to a $^4A_{2g}$ state, and a component of an excited $^4T_{2g}$ state is given by $|\psi_{xz}^{\alpha} \psi_{yz}^{\alpha} \psi_{x^2-y^2}^{\alpha}|$. We find

$$W(^4T_{2g}) - W(^4A_{2g}) = \Delta + 2J_{xz,x^2-y^2} - 2J_{yz,xz} - 2K_{xz,x^2-y^2} + 2K_{yz,xz} \qquad (33)$$

where Δ is defined in equation (28).   This is as far as the
problem can be reduced in octahedral symmetry.   Using a symmetry
operation which is <u>not</u> consistent with an O_h symmetry we can find

$$2J_{xz,x^2-y^2} - 2J_{yz,xz} + 2K_{yz,xz} - 2K_{xz,x^2-y^2} = 0$$

This result can of course also be obtained by going to spherical
symmetry.   Therefore, only by moving outside octahedral symmetry
is it possible to identify a measured energy difference with Δ.
We do not know how great an error we perpetrate by such an
approximation.

The interpretation and definition of the core integrals, so
simple and appealing in crystal field theory, are indeed beset
with dangers in molecular orbital theory.   Let us write for an
antibonding and bonding orbital

$$\psi_a = N_a \left(\chi_d - \lambda_a \sum_{i=1}^{L} c_{iL} \chi_i \right) \qquad (34)$$

$$\psi_b = N_b \left(\sum_{i=1}^{L} c_{iL} \chi_i + \lambda_b \chi_d \right) \qquad (35)$$

Because of orthogonality we must have

$$S_{ML} - \lambda_a + \lambda_b - \lambda_a \lambda_b S_{ML} = 0 \qquad (36)$$

where S_{ML} is the M-L overlap integral.   This equation gives us
a relation between the coefficients in the bonding and antibonding
orbitals.   These coefficients must be evaluated for a specific
electronic configuration in some suitable Hartree-Fock scheme.
Therefore, the traditional core orbitals and the antibonding 'd-
orbitals' will exhibit dependence upon the electronic state under
consideration, and each configuration will possess a different
set of h_{aa}^{core} values.   As realized by Watson and Freeman this
makes an average over all possible configurations prerequisite
to the extraction of one value of a splitting parameter.

The quantity 10 Dq is therefore a parameter <u>specific</u> to
crystal field theory as contrasted to ligand field theory.   If
we try to expand crystal field theory to include the ligand
electronic motions, we cannot retain the parameter 10 Dq and
bestow a well defined meaning on it, and all attempts to divide
an experimental number up into various 'effects' are of course
only <u>ad hoc</u>.

The orbital level diagram used in conventional crystal/ligand field diagrams pictures the $\bar{h}_{core}$ quantities (-4 Dq, 6 Dq). It is important to realize that the w_i's given by photoionization experiments using Koopmans' theorem are related to $\bar{h}_i$ via (8), and that therefore the crystal field theory level order cannot be assumed to follow the S.C.F. w_i order.

One further point. Comparing those measured energy differences in a complex which are dependent upon electronic repulsion terms with the values one can extrapolate from atomic theory, a certain reduction of the 'molecular value' over that of the 'atomic value' is often observed. For instance, the energy difference between the $^2E_g(t_{2g})^3$ and $^4A_{2g}(t_{2g})^3$ states in Cr^{+3} complexes is found to be $3K_{xz,yz}$. With the wavefunctions for ψ_{xz} and ψ_{yz} given by (22) we get correct to S_{ML}^2

$$K_{xz,yz} \simeq \frac{1}{1-2S_{ML}^2} \iint d_{xz}(1)d_{yz}(1)\frac{1}{r_{1\,2}} d_{xz}(2)d_{yz}(2)dr_1dr_2$$

$$- 4S_{ML} \iint d_{xz}(1)d_{yz}(1)\frac{1}{r_{1\,2}} d_{yz}(1) \chi_L(2)\, dr_1\, dr_2 \qquad (37)$$

In the same approximation

$$\chi_L \simeq S_{ML}\, d_{xz} \qquad (38)$$

Hence

$$K_{xz,yz}^{molecular} \simeq (1-2S_{ML}^2)\, K_{xz,yz}^{atomic} \qquad (39)$$

THE SCF-Xα METHOD

The presence of the exchange operator in the Hartree-Fock equations (8) makes the number of integrals which have to be treated large and the numerical evaluation difficult. It would therefore be nice if it were possible to replace the $\hat{K}$ operator with another operator which would incorporate the exchange effect but would be easier to handle.

Let us assume that an exchange term may be found as the potential at the point $\underline{x}$, arising from a spherical distribution of uniform charge density, holding one electron. The distribution is centred at x and has an electronic density equal to $\rho(x)$. The exchange potential - $U(x)$ is then

$$-U(x) = 4\pi \int_o^R \frac{1}{r}\, \rho(x)\, r^2 dr = 2\pi\rho(x)R^2 \qquad (40)$$

R being the radius of the spherical distribution. Since a sphere of radius R contains one electron we have also

$$\frac{4\pi}{3} R^3 \rho(x) = 1 \tag{41}$$

Eliminating R from (40) and (41) leads to

$$U(x) = - \sqrt[3]{\frac{9\pi}{2}\rho(x)} \tag{42}$$

Introducing a constant α we can use a more flexible exchange potential

$$U_{X\alpha}(x) = - \alpha \left[\rho(x)\right]^{\frac{1}{3}} \tag{43}$$

In applying this method to a molecule one performs a geometrical partitioning of the space of the molecule into three types of region:

I.   Atomic: the regions within non-overlapping spheres centred on the constituent atoms

II.   Interatomic: the region between the 'inner' atomic spheres and an 'outer' sphere surrounding the entire molecule

III.   Extramolecular: the region exterior to the outer sphere.

With this space partitioning one constructs a secular determinant over the various components of ψ_i. The energy occurs as a parameter, and the determinant must therefore be calculated over a range of energies bracketing each orbital eigenvalue. The zeros are then found by interpolation.

The Hartree-Fock equations are solved in a self-consistent manner under the assumption that $\rho(x)$ is spherically symmetric within each of the spheres of region I and also spherically symmetric in region III. In region II it is assumed that $\rho(x)$ is a constant. Different values for the parameter α may be chosen for different spheres and regions. The method is therefore of a semi-empirical nature.

The assumption of a 'statistical' exchange operator is crucial for this formalism. It seems however, to work very well. The $X\alpha$ method as a one-electron model should not be expected to give results that depend upon correlation energies.

The one-electron energies can be considered to be derivatives of the total energy with respect to the occupation numbers of the orbitals. The difference between the ground state energy and an excited state may, to very good approximation, be described by using the so-called <u>transition state concept</u>. In this scheme the excitation of an electron from orbital ψ_i to orbital ψ_t is described by taking <u>half an electron</u> from orbital ψ_i and exciting it to orbital ψ_t. The new configuration is then solved self-consistently. The excitation energy is given by the difference of the new one electron energies of ψ_i and ψ_t.

The excitation energy obtained in this way will, however, be a weighted average of all the transitions to states contained in the direct product $\Gamma(\psi_i) \times \Gamma(\psi_t)$. In particular, having a spin-singlet as ground state the energies of the excited spin-singlets and triplets will be averaged out.

SPIN-ORBIT COUPLING

A more complete treatment of the electronic motions must, as we have anticipated in the introduction of the double group notation, introduce a term into the molecular Hamiltonian which couples the motion of the electrons to the spin angular momentum. The spin-orbit coupling is a relativistic phenomenum and, restricting ourselves to the interaction between the spin and orbital motions of the same electron, it may be written

$$\mathcal{H}^{(1)} = \sum_j \frac{1}{2m^2c^2} \; (\text{grad}\, \mathcal{V}_j \times \vec{p}_j) \cdot \vec{s}_j \qquad (1)$$

Here, $\mathcal{V}_j$ is the potential electron experience, m is the mass of the electron, c the velocity of light, $\vec{p}_j$ the linear momentum, and $\vec{s}_j$ the spin momentum operator of the electron j. The summation runs over all the electrons in the molecule.

Let us investigate the properties of $\mathcal{H}^{(1)}$, and estimate the spin-orbit coupling energies. Using perturbation theory, we must evaluate matrix elements of the form

$$H_{kl} = \frac{1}{2m^2c^2} \int\int \Psi_k^* \sum_j (\text{grad}\, \mathcal{V}_j \times p_j) \cdot s_j \Psi_l \, dr d\gamma_{\text{spin}} \qquad (2)$$

Since the wavefunctions vanish at the integration limits, we can use partial integration and obtain

$$H_{kl} = \frac{-1}{2m^2c^2} \int\int \sum_j \mathcal{V}_j (\text{grad}_j \Psi_k^* \times \vec{p}_j) \vec{s}_j \Psi_l \, dr d\gamma_{\text{spin}} \qquad (3)$$

Substituting for the linear momentum $p_x = -i\hbar\frac{\partial}{\partial x}$, etc., we get for the x component of H_{kl}

$$H_{kl}^{x} = \frac{i\hbar}{2m^2c^2} \iint \sum_j \mathcal{V}_j \left(\frac{\partial \Psi_k^*}{\partial y_j} \frac{\partial}{\partial z_j} - \frac{\partial \Psi_k^*}{\partial z_j} \frac{\partial}{\partial y_j} \right) \hat{s}_{xj} \Psi_l \, dr d\tau_{spin} \qquad (4)$$

Clearly, the 'orbital differential operator' inside the parenthesis will transform like R_x, that is, like the x component of the rotation of the molecule as a whole. Since $\mathcal{V}_j$ transforms like a totally symmetric representation in the point group of the molecule, it is easy to see by symmetry arguments which molecular states will show first order spin-orbit coupling: a term whose orbital symmetry is Γ, say, will exhibit spin-orbit coupling only if the direct symmetric product representation $\Gamma \times \Gamma(R) \times \Gamma$ contains the totally symmetric representation. Further, two states having the orbital representations of Γ_i and Γ_j may be coupled together via the spin-orbit operator only provided $\Gamma_i \times \Gamma_j$ contains the representation $\Gamma(R)$. We remark in passing that since the spin-orbit Hamiltonian must transform like a totally symmetric representation in the point group of the molecule, it follows that s likewise transforms like $\Gamma(R)$.

We can then write for $\mathcal{H}^{(1)}$

$$\mathcal{H}^{(1)} = \mathcal{V}(\hat{R}_x \hat{S}_x + \hat{R}_y \hat{S}_y + \hat{R}_z \hat{S}_z) \qquad (5)$$

or alternatively

$$\mathcal{H}^{(1)} = \mathcal{V}(\tfrac{1}{2}\hat{R}_+ \hat{S}_- + \tfrac{1}{2}\hat{R}_- \hat{S}_+ + \hat{R}_z \hat{S}_z) \qquad (6)$$

where $\hat{R}_+ = \hat{R}_x + i\hat{R}_y$, $\hat{R}_- = \hat{R}_x - i\hat{R}_y$, $\hat{S}_+ = \hat{S}_x + i\hat{S}_y$, and $\hat{S}_- = \hat{S}_x - i\hat{S}_y$. It is seen that the spin-orbit coupling operator may couple states together which differ in the quantum number M_S by one or zero. Of course the double group classification will yield good quantum numbers.

The spin-orbit coupling term assumes importance in molecular spectroscopy by virtue of its ability to 'mix' the spin states which differ by $S = 1$. This feature breaks down the validity of the spin quantum number S. In the so-called charge-transfer states of inorganic complexes it may cause a measurable splitting of the spin-multiplets. It may also mix excited states into the ground state. This shows up in the magnetic features of the molecule.

Consider the case where $\mathcal{V}_j$ is a central field, depending only upon the distance r from a centre. Then $\mathcal{V}_j = \mathcal{V}_j(r)$ and

$$\text{grad } \mathcal{V}_j = (\underline{i}\frac{x}{r} + \underline{j}\frac{y}{r} + \underline{k}\frac{y}{r})\frac{d\mathcal{V}}{dr} \qquad (7)$$

The spin-orbit operator (1) is in that case

$$\mathcal{H}^{(1)} = \sum_j \frac{1}{2m^2c^2} \left\{ \frac{1}{r_j} \frac{dV_j}{dr_j} \right\} \vec{l}_j \cdot \vec{s}_j \tag{8}$$

where $\vec{l}$ is the angular moment of electron j. Measuring $\vec{l}$ and s in units of $\hbar$, we can write for a single electron

$$\mathcal{H}^{(1)} = \zeta(r) \vec{l} \cdot \vec{s}$$

with

$$\zeta(r) = \frac{\hbar^2}{2m^2c^2} \frac{1}{r} \frac{dV}{dr} \tag{9}$$

Assuming a Coulomb potential $V(r) = -\dfrac{Ze^2}{r}$ and by use of a hydrogenic radial wavefunction we calculate

$$\zeta_{n,1} = \int \psi_{n,1,m} \zeta(r) \psi_{n,1,m} dr = \frac{e^2 \hbar^2}{2m^2 c^2 a_0^3} \frac{Z^4}{n^3 1(1+\frac{1}{2})(1+1)} \tag{10}$$

The ratio of $\zeta_{n,1}$ to the electronic energy $E_n = \dfrac{Z^2 e^2}{2n^2 a_0}$ is seen to be proportional to $Z^2 \alpha^2$, where $\alpha = \dfrac{e^2}{c\hbar}$ is the so-called fine structure constant, approximately equal to 1/137. For small values of the atomic number Z the spin-orbit coupling term is seen to be small compared with the electronic energy of the system. In these cases we can, therefore, treat it as a perturbation. However, the heavier the atom, the more important the spin-orbit coupling term in the Hamiltonian. In general we can write

$$\mathcal{H}^{(1)} = \sum_j \frac{i\hbar}{2m^2c^2} \left(\frac{\partial V}{\partial y_j} \frac{\partial}{\partial x_j} - \frac{\partial V}{\partial z_j} \frac{\partial}{\partial y_j} \right) \hat{s}_{x_j}$$

$$+ \left(\frac{\partial V_j}{\partial z_j} \frac{\partial}{\partial x_j} - \frac{\partial V_j}{\partial x_j} \frac{\partial}{\partial z_j} \right) \hat{s}_{y_j} \tag{11}$$

$$+ \left(\frac{\partial V_j}{\partial x_j} \frac{\partial j}{\partial y_j} - \frac{\partial V_j}{\partial y_j} \frac{\partial}{\partial z_j} \right) \hat{s}_{z_j}$$

With N nuclei in the molecule we may approximate V as a sum of N spherical potentials, located on the N centres

$$V = V_1(r_1) + V_2(r_2) + \ldots V_N(r_N) \tag{12}$$

Hence

$$\mathcal{H}^{(1)} = \sum_N \sum_j \frac{1}{2m^2c^2} \left\{ \frac{1}{r_{Nj}} \frac{\partial V_N}{\partial r_{Nj}} \right\} \hat{l}_{jN} \hat{s}_{jN} \tag{13}$$

Evaluating the spin-orbit coupling energy with the operator (13) in a LCAO-MO set, one may to good approximation consider only the one-electron terms. The total spin-orbit coupling in a molecule is therefore, with these approximations, given as a sum of atomic spin-orbit coupling parameters multiplied by the squares of the appropriate coefficients in the LCAO-MO.

It is however clear that the approximation (13) is quite drastic, and completely does away with the charge distributions 'between' the atoms. Nevertheless those are the ones which are characteristic for the molecule! That V changes drastically close to the nuclei is well known, and is of course the basis for the above approximation. However, grad V may also change significantly in between the nuclei, and one should therefore not use the above approximations to more than is necessary to give one an order of magnitude for the molecular spin-orbit splitting. To build theories upon deviations from expected 'atomic' behaviour is a thing to avoid, and indeed the extraction of 'chemical knowledge' from variations of spectroscopic parameters is very deceptive and dangerous.

Consider now an octahedral or tetrahedral complex. The orbital angular momentum operator transforms under T_{1g} of the group O_h and under T_1 in T_d. It follows from our previous remarks that the multiplet structure is expected to be pronounced only for T_1 or T_2 states.

The three components of a T_1 or T_2 state can be characterized as Φ_1, Φ_0 and Φ_{-1} in direct analogy with an atomic P state.

Defining the total angular momentum $\vec{J} = \vec{L} + \vec{S}$ we have

$$\vec{L} \cdot \vec{S} = \tfrac{1}{2}(\hat{J}^2 - \hat{L}^2 - \hat{S}^2). \tag{14}$$

The energies in the spin multiplet $^{2S+1}T_1$ or $^{2S+1}T_2$ may then be written

$$W_{J,S} = \tfrac{1}{2}\lambda\left[J(J+1) - 1.2 - S(S+1)\right] \tag{15}$$

With say, $S = \frac{3}{2}$, J can be $\frac{5}{2}$, $\frac{3}{2}$, or $\frac{1}{2}$. Hence a ^4T is split into three levels: one six-fold degenerate $W_{5/2} = \frac{3}{2}\lambda$, one four-fold $W_{3/2} = -\lambda$ and one two-fold degenerate $W_{1/2} = -\frac{5}{2}\lambda$. The spin-orbit components of the ^4T should therefore occur with energy separations in the ratio 5:3.

Suppose that we have diagonalized the molecular orbitals under the spin-orbit operator (1) or, in other words, have found a set of molecular orbitals which transform like the representations of the double group. The total electronic wavefunction is a determinant and since the spin-orbit coupling term is a one-electron operator, and we get

$$W_{J,S} = \sum_{t=1}^{j} H_{tt} \tag{16}$$

where $H_{tt} = \int \psi_t^* \mathcal{H}^{(1)} \psi_t \, dr$ and the summation over t runs through all the occupied double group molecular orbitals ψ_t.

The summation in (38) over a filled set of double group orbitals, with parentage in an irreducible representation of the simple group, yields zero. This follows because the 'centre of gravity' cannot be moved by the spin-orbit perturbation. Hence only 'open shells' will contribute.

In an inorganic complex the double group orbitals will be of the type

$$\psi = c_M \chi_M(\text{Metal}) + c_L \sum_N a_N \chi_N(\text{Ligands}) \tag{17}$$

The low-lying 'crystal field states' will only have partially filled orbitals of the type $c_M \sim 1$, $c_L \sim 0$. According to (13) the spin-orbit splittings of the 'crystal field states' will then be dominated by $\zeta(\text{Metal})$. In the higher-lying charge-transfer states an electron has been moved from an orbital primarily of ligand character $(c_L \sim 1)$ to an orbital primarily of metal character $(c_M \sim 1)$. Hence a hole has been created in what was before a filled shell, and a contribution to the spin-orbit splitting of the ensuing state arises from the ligand subshells. If one (or more) of the ligands are heavy atoms or ions with a high value of Z, the contribution to the spin-orbit splitting from that particular ligand will be dominant. In contrast to the 'crystal field' bands the charge transfer bands may therefore show large spin-orbit coupling splittings provided $\zeta(\text{Ligand}) \gg \zeta(\text{Metal})$.

A CHEMISTS' GUIDE TO THE BAND THEORY OF SOLIDS

J.M. Thomas

Edward Davies Chemical Laboratories, University
College of Wales, Aberystwyth, SY23 1NE, U.K.

1. INTRODUCTION

This article is intended to cover some rudimentary aspects
of the band theory of solids in such a way as to outline how a
range of interesting phenomena relating to simple inorganic solids
may be interpreted and unified by means of concepts which have
long proved successful, in the hands of physicists and metallurgists, in explaining a wide range of physical properties of
crystalline solids.

We shall first be preoccouped with the grammar of the
subject – indeed, at times, we shall dwell on the basic glossary
and vocabulary – in order, later, to be in a position to appreciate
something of the romance of band-theory literature. Though it
appears to be a physical rather than a chemical method of
proceeding, we shall find it convenient to classify solids
according to their electrical properties. The magnitude of the
electrical conductivity, σ, and, in particular, the nature of the
temperature dependence, $\sigma(T)$, turns out to be a crucially
important experimental criterion in distinguishing between the
freely conducting metals on the one hand (e.g. copper has a σ of
ca. 6×10^5 ohm^{-1} cm^{-1} at 300 K) from the insulating solids on
the other copper phthalo-cyanines with a σ under comparable
conditions of less than 10^{-13} ohm^{-1} cm^{-1}. Semiconductors, such
as silicon or Cu_2O apart from possessing intermediate values of
conductivity also obey a temperature dependence given approximate-
ly by $\sigma = \sigma_o \exp(-E/RT)$ where σ_o and E are constants (see later).
To fix our ideas it is instructive to note that the chain compound
$(SN)n$ has a room temperature conductivity of ca. 10^3 whilst n-type
Si is close to 10^1 and Cu_2O to 10^{-6} (all in units of ohm^{-1}cm^{-1}).

P. Day (ed.), Electronic States of Inorganic Compounds. 27–58. All Rights Reserved.
Copyright © 1975 by D. Reidel Publishing Company, Dordrecht-Holland.

The conductivity of $(SN)_n$ decreases with increasing temperature, in contrast to the behaviour of the other two solids, and so is, by definition, a one-dimensional metal.

2. EARLY THEORIES

2.1 The free-electron approximation (classical)

The lineage of modern band theory may be traced back to Drude and Lorentz and their assumption that the outer, valence electrons of the constituent atoms in a metallic solid are totally free to move about the entire bulk, much like the molecules of a perfect gas in a container. It is easy to dismiss such a picture as hopelessly simplistic, but there is, however, considerable merit in noting some of the derived relationship of this early particulate approach, since the equations that then emerged still form the basis of our descriptions of electronic behaviour even in quantal treatments of solids. The conductivity, δ, is related to the electron mobility μ and the electronic charge q, by

$$\delta = n\mu q \tag{1}$$

and we may also write

$$\delta = nq \, (q/_m) \, \tau \tag{2}$$

since the product of the mobile charge density, the ability of the field to accelerate a charged particle of mass $\underline{m}$, and the relaxation time τ (which is akin to the mean time between collisions for the conduction electrons) must obviously equal the conductivity. Likewise

$$\mu = q\tau/_m \tag{3}$$

and the ratio of the thermal conductivity, K, to the electrical conductivity is given by

$$K/\delta = \frac{\pi^2}{3} \left(\frac{k}{q}\right)^2 T \tag{4}$$

i.e. a constant at a fixed temperature, which offered an interpretation of the so-called Wiedermann-Franz law noted empirically since 1853. Other important relationships were

$$\nu_p^2 = nq^2/m\varepsilon_0 \text{ and } \delta = \pi\tau \nu_p^2 \tag{5}$$

where ν_p is the plasma frequency and ε_0 is the permittivity. The observed bulk plasma frequencies for a range of simple solids have been found, from electron-loss measurements to be (in eV):

for Na$_2$ 5.7, Mg 10.5, Al 15.0, Si 16.9 and Sb 15.3. The values calculated from equation (5) are, respectively, 5.9, 10.9, 15.7, 16.6 and 15.1, showing that the free-electron picture, for these solids at least, is close to the truth. (It is possible to extract δ indirectly from the optical reflectivity of the solid, for which an expression was also deduced.)

All these derivations seemed to fit in neatly with observed fact – but there was one fatal flaw. The predicted specific heat per mole of valence electron of a metallic solid ought clearly to be 3/2 R (from the kinetic theory): experimentally, however, it was found to be some two orders of magnitude smaller than this. Apart from this obvious inadequacy, the Drude approach could not rationally explain why some solids were insulators whilst others were metals.

2.2 Free-electron (quantal)

In Sommerfeld's variant of the free-electron model it is recognized at the outset that the solution to the Schroedinger equation under such circumstances is a travelling wave

$$\psi = \exp(i \vec{k} \vec{r}) \tag{6}$$

with all values of $\vec{k}$, the wave vector, allowed ($k = 2\pi/\lambda$, where λ is the wavelength of the electron). If boundary conditions, deducible from the fact that the metal is a cube, say, of side L), are imposed, the periodicity requirement is:

$$\psi(x + L, y, z) = \psi(x, y, z) \tag{7}$$

with analogous equations for y and z. Acceptable solutions are of the form:

$$\psi \sim \exp\left[i \frac{2}{L} (n_x x + n_y y + n_z z) \right] \tag{8}$$

where the n's are positive or negative integers. 'Allowed' values of the wavevector k are, therefore, given by

$$\vec{k} = \frac{2\pi}{L} \vec{n} \tag{9}$$

where

$$n^2 = n_x^2 + n_y^2 + n_z^2.$$

Reverting to the totally free, unconstrained electron, its energy E is related to its momentum $\underline{p}$ by

$$E = p^2/2m \tag{10}$$

and the momentum is related to wave number by:

$$\vec{p} = h\vec{k}$$

so that:

$$E_k = (\hbar^2/2m)k^2 \tag{11}$$

and the parabolic dependence of E on k is evident. Now in the quantum-mechanical variant of the free electron (in a metal), equation (11) holds also, but the allowed values of k are given by equation (9). (There are, naturally, two possible values of spin for each allowed state.)

The real problem in a typical crystalline solid is to find solutions for a periodically varying potential, V (since we may no longer set V = 0, i.e. ignore the potential arising from the ion cores). Moreover we need fully to exploit the consequences of the periodicity of the lattice. The problem was first solved by Bloch in 1928 who found that solutions (in one dimension)were of the form:

$$\psi_k = u_k(x) \exp (\underline{i}kx). \tag{12}$$

The plane wave is modulated by the function of $u_k(x)$ which has the periodicity of the lattice, $u_k(x)$ being a periodic function with the same periodicity as the potential V of the lattice. Bloch functions are eigenfunctions which have the property:

$$\psi(x \pm a) = \psi(x) \exp(\pm \underline{i}ka) \tag{13}$$

since $u_k(x + a) = u_k(x).$ (14)

If we suppose that there are N lattice sites, and we require as a boundary condition that $\psi(x + Na) = \psi(x)$ then exp (ikNa) = 1, which will hold good if, in turn,

$$k = 2\pi n/Na. \tag{15}$$

The allowed values of k are, therefore, given by equation (15) with n = 0, $\pm$ 1, $\pm$ 2 . . .

At values of k equal to $\pm$ π/a, $\pm$ 2π/a, $\pm$ 3π/a . . . $\pm$ nπ/a discontinuities in the energy appear: there are thus allowed and forbidden ranges of the energy (Figure 1).

2.3 Nearly-free electrons

If we examine more closely the behaviour of free electrons and nearly-free electrons (without defining the latter term too

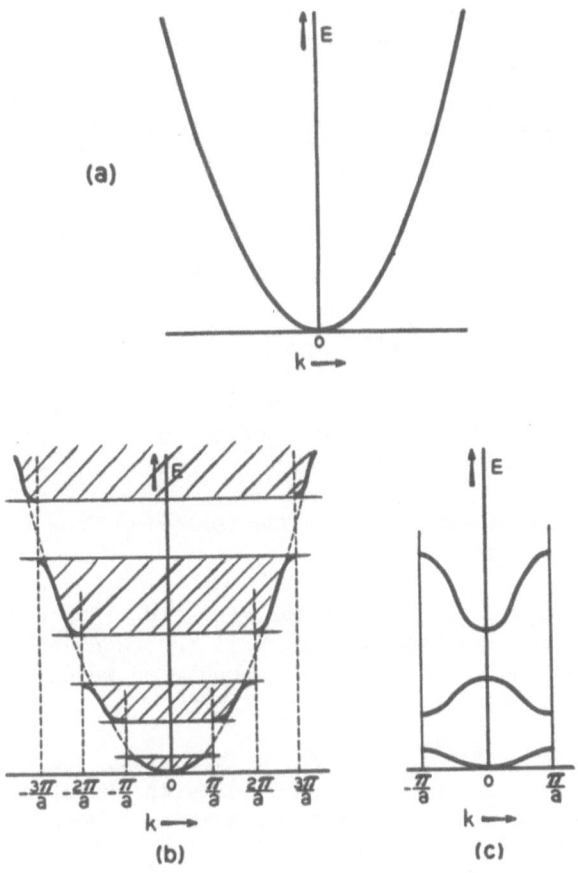

Figure 1. E vs. k (one dimension) for (a) free electrons, (b) electrons moving in a periodic potential, (c) E vs. reduced wave number. (Taken from 'Semiconductors' edited by N.B. Hannay, Reinhold, 1960.)

carefully at this stage) we can readily understand why the discontinuities arise in the E/k plots. The critical value of the electron wavelength λ ($k = 2\pi/\lambda$) for which the discontinuity in allowed energy states appears is just that for which one expects Bragg reflection (i.e. an electron impinging on the crystal lattice with this particular wavelength will be totally reflected, and cannot, therefore, penetrate into the crystal. At $k < \pi/a$, the electron behaves as a travelling wave. The Bragg condition $n\lambda = 2a \sin\theta$ reduces to

$$k = n\pi/a \tag{16}$$

since θ, the angle between the incident wave and crystal plane is
90°.   Clearly when equation (16) is obeyed reflection of the
waves takes place, and there is constructive interference of
scattered radiation from successive atom planes.   As λ approaches
2a (i.e. k→ π/a) the electron can no longer be represented as a
travelling wave.   Because of reflection it must be regarded as
a standing wave made up equally of travelling waves with opposite
directions.   These waves ($\exp(i\vec{k}\vec{x})$) and $\exp(-i\vec{k}\vec{x})$) give two
solutions to the Schroedinger equation at k = π/a namely

$$u_k \left[\exp(i\vec{k}\vec{x}) + \exp(-i\vec{k}\vec{x}) \right]$$

and

$$u_k \left[\exp(i\vec{k}\vec{x}) - \exp(-i\vec{k}\vec{x}) \right].$$

These solutions correspond to the two allowed energies shown in
Figure 1(b) at k = π/a.   Similar conditions obtain at k =±2π/a
...± nπ/a.

Summarizing, we have learned the following from the simple
one-dimensional treatment:

(i) For most values of the k-vector, the electrons behave as if
they were more or less free and are well represented by plane
waves.   But at the so-called zone* boundaries we need to invoke
the picture of a standing wave.

(ii) At values of k far from ± π/a, ± 2π/a etc., the energy may
be quite closely given by equation (11), with the electron mass
being equal to the actual electron nest mass.   Near ± nπ/a,
however, equation (11) requires modification to take account of
the momentum transfer between lattice and electrons.   The form
of equation (11) is retained but we replace the ordinary mass
m by the effective mass m*.   (By analogy with Newtonian laws
of classical motion, m* may be defined:

$$m^* = \hbar^2 / \left(\frac{d^2 E}{dk^2} \right) \tag{17}$$

It is instructive to note, in passing, that there is a kinship
between the concept of effective mass as used here and that of
thermodynamic activity as used elsewhere to represent effective
concentration.)   Equation (11) therefore becomes:

$$E_k = \frac{h^2 k^2}{2m^*} \tag{18}$$

------

* In k-space the first Brillouin zone (BZ) is the region between
$-\pi/a$ → 0 and 0 → $+\pi/a$.   The second BZ spans the region $-2\pi/a$ →
$-\pi/a$ and $+\pi/a$ → $2\pi/a$.

and the analogue of equation (3) is is

$$\mu = q\tau/m^*. \tag{19}$$

Note that m^* is related to the curvature of the E/k plot, and that m^* values may range from $-\infty$ to $+\infty$. Negative values of effective mass may appear strange, but negative values of m^* for negative electrons may be regarded as equivalent to positive masses for positive holes, which are created when an electron leaves a vacancy in an occupied energy level.

(iii) The size of the discontinuities (gaps of forbidden energy) obviously depends on the magnitude of the variation in the periodic potential; and the number of states in the bands is proportional to N.

(iv) The periodicity of a crystalline solid allows us to express energies in terms of reduced wavevectors lying in the range $-\pi/a \leqslant k \leqslant +\pi/a$ rather than to consider, each time the entire regions of k from $0 \rightarrow +\infty$. (This arises because the replacement of k by $(k + 2\pi n/a)$ in ψ_k preserves the Bloch function.) We may thus describe the energy of a state by specifying a k value between $-\pi/a$ and $+\pi/a$ together with a band number to describe the band to which the state belongs. This becomes clear in Figure 1(c) where the first three bands of energies are shown.

The principles extracted from a consideration of one-dimensional systems may be readily extended to embrace two- and three-dimensional situations. E/k plots in two dimensions, depending upon the symmetry of the solid, will in general be different along the k_x and k_y directions; and Figure 2 shows possible energy bands for a hypothetical two-dimensional square lattice. Across the top of this figure are given the first two bands along symmetry lines as conventionally represented. Across the bottom is an outline of the bands throughout the BZ, energy being plotted vertically. The bands are here shown to overlap (the second being lower at X than the first at W). For some solids there is degeneracy at certain points in the zone, i.e. the first and second bands may meet at T on certain axes of symmetry, or at the zone boundaries.

In three dimensions we may have surfaces of constant energy, as symbolized in Figure 3, which focusses attention on face-centred cubic metals such as copper, silver and gold.

It is evident, therefore, that in representing energy levels in solids extensive use is made of momentum (reciprocal- or k-) space rather than the real-space representations which theoretical chemists frequently employ for the description of isolated molecules. There are obvious advantages in doing so: optical

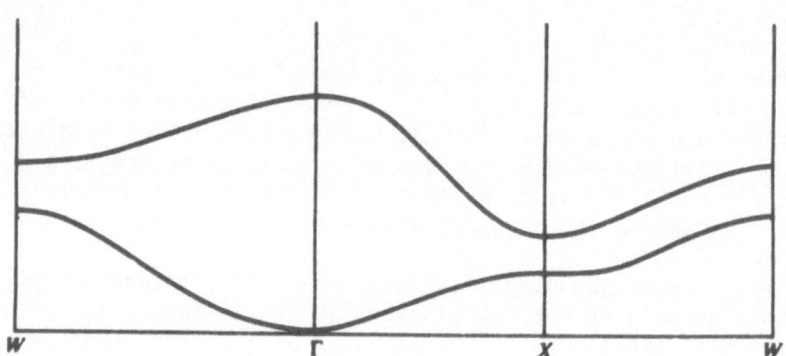

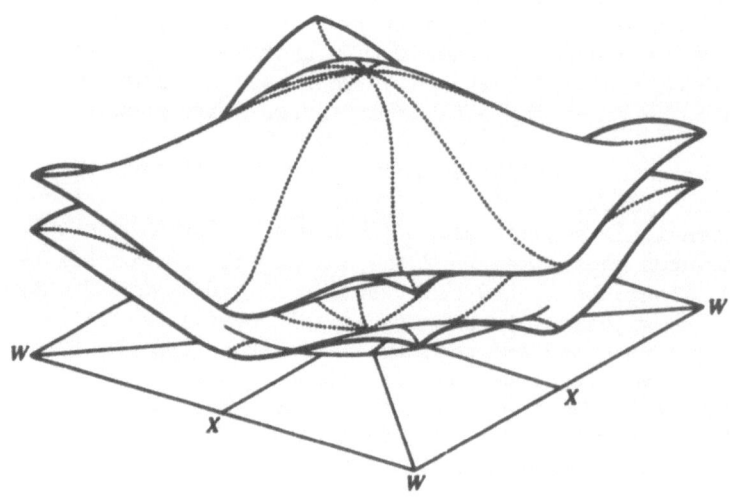

Figure 2.   Possible energy bands for a two-dimensional square
lattice.

and spectroscopic properties are concisely illustrated by or
deducible from such E/k plots, the various symmetry-allowed
optical transitions being read off directly.   Latterly, however,
ever since it has become possible to probe directly by various
techniques (described elsewhere in this book) the density-of-
states (DOS), or a property directly related to DOS, it has

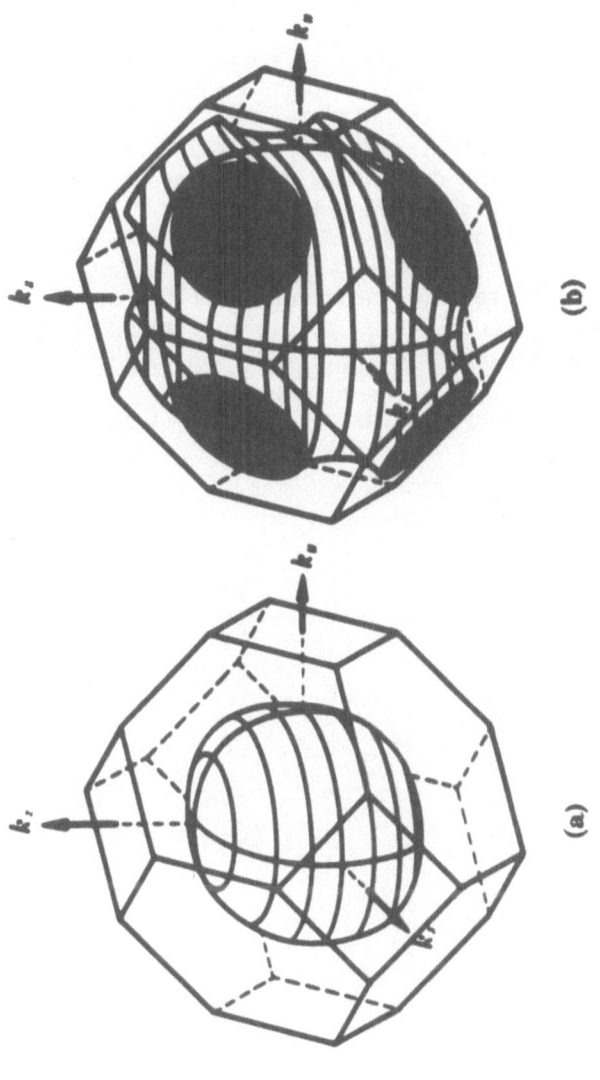

Figure 3. The first Brillouin zone of face-centered cubic structure with surfaces of constant energy of electrons shown for (a) nearly free electrons near bottom of zone and (b) electrons at zone boundary. (Taken from N.F. Mott and H. Jones, 'Theory of the Properties of Metals and Alloys', Dover, 1964.)

become equally popular to convert the E/k plots to their DOS
analogues.   (Indeed calculation, of the kind discussed briefly
later, is nowadays conducted in such a way as to arrive directly
at DOS curves.)   A typical example of the relationship between
E/k and DOS formulation is shown in Figure 4, taken from the
work of Murray and Williams,[1] on MoTe$_2$.

Even when DOS curves are not directly calculated, some
useful qualitative statements may be readily made on the basis of
the available E/k plots.   Thus where an energy band is flat, the
DOS is high.   Moreover since mobility is related to effective
mass (see equations (16) to (18)), we see that when we have a
narrow band m* is expected to be large and there will therefore
be low mobility of the charged carriers.   Conversely we may
expect high mobilities to be associated with wide bands.

In considering the occupancy of available states it is

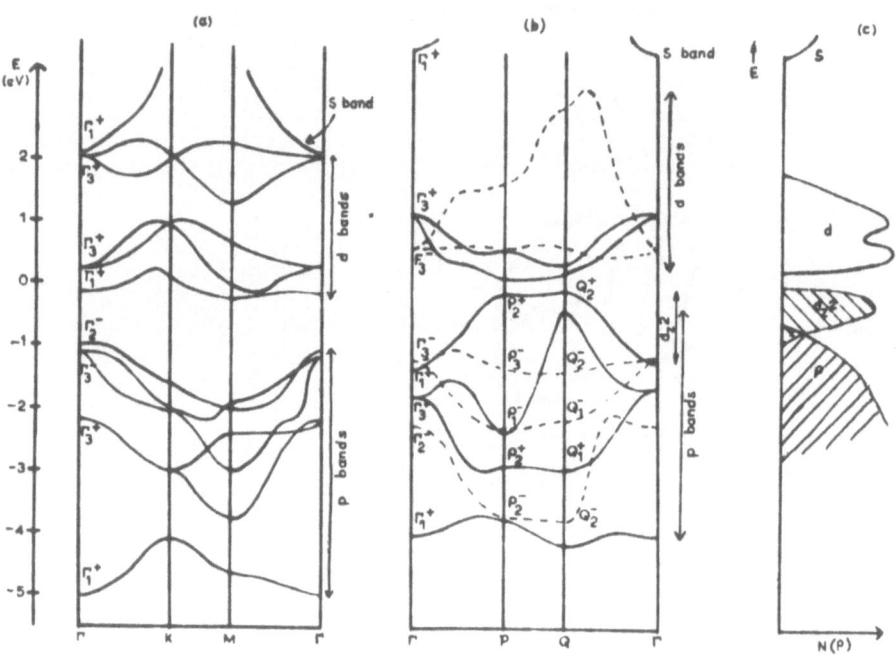

Figure 4.   Band structures calculated using the tightbinding
method for (a) a hypothetical undistorted form of octahedral
MoTe$_2$ and (b) α-MoTe$_2$.   N.B. The zero on the energy scale
corresponds to the Fermi level position.   Also shown in (c) is
the model for the density of states in α-MoTe$_2$ as given by Hughes
and Liang: the energy gap between the d bands and the gap between
the d and s conduction bands are clearly visible.

necessary to invoke the Fermi-Dirac distribution function.
Filling of all states must obviously proceed according to the
Pauli exclusion principle. At zero Kelvin, the cutoff between
filled and empty states, and the highest filled state occurs at
the so-called Fermi energy,[*] E_F. Fermi-Dirac statistics tell us
that the probability $f(E)$ that a state of energy E is occupied is
given by

$$f(E) = \frac{1}{\left\{ \exp\left(\frac{E-E_F}{kT}\right) + 1 \right\}} \tag{20}$$

At zero Kelvin $f(E)$ is unity at all energies up to E_F, where it
drops to zero. The rectangular section representing $f(E)$ versus
E changes to a distorted shape with a progressively larger tail
as temperature increases. It follows that, for finite temperature, $f(E) = \frac{1}{2}$ at $E = E_F$, and that the number of electrons which
lie above E_F, proportionately very small, will increase with
increasing temperature. These are the only electrons which can
contribute to the heat capacity and we now appreciate the origins
of the failure of the simple Drude theory, mentioned earlier.

The amount of energy required to remove an electron from the
Fermi level to a point at nest just outside the metal (in a
vacuum) is termed the work function, ϕ. It can be useful to
think of ϕ as signifying both the ionisation potential, when
discussing emission, and the electron affinity, when dealing with
electron capture. For metals the magnitude of the work function
falls within the range 3 to 6 eV and may vary from one crystallographic face to another. But the most distinguishing feature of
a metal is that it has a band (or overlapping bands) which are
incompletely filled. This is the characteristic which explains
why metals are good electrical conductors. Upon application of
a field electrons may readily gain energy and ascend to higher
unoccupied states. The fact that, for a metal, $\underline{\phi}$ decreases with
increasing temperature follows naturally since the mobile electrons will suffer enhanced scattering owing to the increased
vibrations of the lattice at elevated temperatures.

When a fully occupied band is separated in energy space from
another, empty one, electrical conductivity is low, and this is
the situation which exists with insulators and intrinsic semiconductors, which differ, qualitatively, in that the forbidden
gap is larger for insulators than for semiconductors. Clearly
the number of carriers n (equation (1)) will increase, and hence
the conductivity will mix experimentally as temperature is

---

[*] It may be readily shown that the Fermi energy, E_F, is synony-
mous with the electrochemical potential (the partial molar free
energy) of the electrons.

increased (or photon energy provided) to promote electrons from
a full to an otherwise empty band (see later - equation (29)).

3. THEORETICAL APPROACHES TO THE EVALUATION OF ELECTRONIC BAND
    STRUCTURES

    Basically the problems involved here, as with free molecules,
centre around the various ways of solving the Schroedinger
equation so as to yield acceptable one-electron solutions for a
many-body situation.   Fundamentally one is faced with an
appropriate choice of potential and of coping with exchange interaction and electron correlation.   Many of the detailed problems
are computational, and many approximations and assumptions
necessarily have to be made.

    One proceeds with one-electron approaches which, fortunately,
appear to be well justified for solids in general except when
dealing with special phenomena such as superconductivity and,
in principle, the strategy entails making a suitable choice of
Hamiltonian (based on the known crystal structure) and then
proceeding, by working within a self-consistent loop involving:
charge density ――――→ periodic Hamiltonian

        wave functions and energies

In almost all band-structure calculations, the core electrons are
regarded as outside the area of interest, the assumption being
that only the outer electrons are responsible for the solid-state
properties.

    Two main approaches to the question of band-structure
calculation have been adopted.   In the tight-binding (TB)
approximation it is argued that the isolated, uncoupled atoms
provide a good model for the solid itself, a situation likely to
be true when the potential in the neighbourhood of some or all of
the ions is very strong compared with that in the interstitial
areas between them.   Under these circumstances the eigenfunctions
of the isolated atom are suitable basis states for the description
of the band structure.   (This procedure is the solid state
scientist's analogue of the chemist's LCAO and related methods
in M.O. theory.)   In the nearly-free electron (NFE) approximation the natural basis functions are taken as plane waves which
are weakly mixed by the potential.   This approximation is valid
when the individual atomic potentials in a crystal overlap so
much that the net potential seen by a valence electron has little
atomic character.

    Theoretical analysis of some solids falls naturally into
one or other of these two approximations.   Na, Al (and the other

elements mentioned in section 2.2 above) are well described by
the NFE formulation, whereas alkali halides are quite well dealt
with in the framework of the TB approximation.   Some solids,
however — and transition metals rank amongst these — require a
composite approach since some of the electron states are tightly
bound, whilst others, nominally associated with the same atom,
take on an itinerant character.   Full details of the calculations
are given in several textbooks (see Aetmann,[2] Ziman,[3] Seitz[4] and
Harrison[5]), but some skeletal aspects are considered here.

The cellular method of Wigner and Seitz (1934),[6] who
considered alkali metals, focusses attention on the lowest lying
state in the band, i.e. the k = 0 state.   The wavefunction for
this state is simply the Bloch function $u_0(r)$ which has the full
symmetry of the lattice.   The crystal is divided into cells,
such that the cell associated with each atom contains all points
closer to that atom than to other atoms.   It then follows that,
in simple structures, the normal component of the gradient $u_0(r)$
vanishes at all atomic-cell boundaries.   Wigner and Seitz took
the free-ion potential as the valid one for the solid, it being
taken to be spherically symmetric within each cell.   To treat
other states (besides k = 0) in the band, difficulties arise
because of the need to match the wavefunctions at the boundaries.

In the plane-wave method one takes as the crystal potential
a superposition of free atom potentials, each of which should
include a free-electron exchange potential based upon that free-
atom density.   Herring[7] improved the rather laborious simple-
plane-wave method by introducing the orthogonalized-plane-wave
method in which plane waves that have been made orthogonal to the
core states are used.   In the resulting set of simultaneous
equations one has to solve, owing to favourable matrix-element
coupling, the number of effective equations is considerably
reduced.

Earlier (in 1937) Slater had proposed another type of function for expansion of the wavefunctions, i.e. augmented plane
waves (APW).   In this method, one first approximates the
potential that will be used in the calculation.   Near each
nucleus we expect the potential to be rather spherical, and at
positions in between the nuclei it is expected to be relatively
flat.   We can therefore construct a sphere around each nucleus
making the radii of the spheres sufficiently small that they do
not overlap each other.   One also assumes that the potential
within the sphere is atomic-like, precisely spherically
symmetric — and that, between the spheres, it is constant.*

---

* Note that, by prescribing this form of potential one is
prevented from doing an accurate self-consistent calculation.

(One now sees why – in North America at least – the term muffin-tin potential is used.)   It is evident that, in the APW method one seeks to match travelling plane waves in the interstitial region to sets of spherical harmonics and radial functions satisfying the Schroedinger equation inside each atomic sphere. In another, related variant – the Kohn, Korringa, Rostoker (KKR) method – each ion sphere is treated as a source of scattering for the plane waves in the interstitial space.

Perhaps the most currently popular approach is that termed the pseudopotential method, first outlined by Hellmann and resurrected by Cohen, Heine, Phillips and others.   The ethos of this method is that, in practice, whereas the true crystal potential does not satisfy the criteria for the applicability of the NFE approximation, there are much weaker equivalent potentials (or pseudopotentials) which do.   These pseudopotentials produce the same band structure for the valence and conduction electrons – though not necessarily the same wavefunctions.   The difference is that the true potential must of necessity be strong enough to bind core states (of lower energy) but that the pseudopotential need not.   In this technique one effectively cancels out the strong 'inner' potential.   It has experienced great success with sp metals, and, recently in the hands of Harrison and others, even with some transition metals.

Notwithstanding the various idiosyncracies of each of the above-mentioned methods of attack, it is important to bear in mind that:

(a)   the width of a given band is largely independent of the number of atoms (or unit cells) that make up the solid (once the number itself becomes large);

(b)   the DOS within the band is governed by the symmetry and separation distances of the crystallographic structure;

(c)   the magnitude of the energy gap is determined by the strength of the crystal potential, and

(d)   the location of the energy discontinuities in k space is determined by the translational symmetry of the lattice.

4.   BASIC PRINCIPLES OF BAND-STRUCTURE PROPERTIES OF SEMI-
      CONDUCTORS

Having earlier outlined the difference in electronic structure between metals and non-metallic solids, we now enquire more deeply into certain thermodynamic aspects of semiconductors. We recall that the conductivity (cf. equation (1)) is given by:

$$\mathbf{\sigma} = q(n_e\mu_e + n_h\mu_h) \tag{21}$$

where the subscripts e and h refer to electrons and positive holes
respectively.   We shall see that the number of changed carriers
(n_e and n_h) depend steeply upon temperature.   No conduction is
expected at absolute zero when there is a finite separation E_g
(see Figure 5) between the top of the valence band and the bottom
of the conduction band.   It follows that intrinsic behaviour
for all solids possessing E_g greater than a few tenths eV is to
be expected only with exceptionally pure materials.   Thus in
E_g (ca. 0.76 eV) n_e, the number of electrons promoted to
the conduction band at room temperature is ca.10^{13} cm^{-3} (cf.
total number of Ge atoms which is ca. 5 x 10^{22} cm^{-3}).   Hence a
Group V impurity, such as As or N (which could release an electron
easily) of only in 10^8 would yield 5 x 10^{14} <u>extrinsic</u> electrons,
and thereby mask the intrinsic electronic properties.

4.1 Location of Fermi energy in an intrinsic semiconductor

     The precise location of the Fermi energy E_F of a semi-
conductor is a matter of great practical importance since it
decides the direction of electronic flow when a metal is placed
in contact with it and the magnitude of the potential barrier
thereby formed.   It is often calmly asserted that E_F falls in

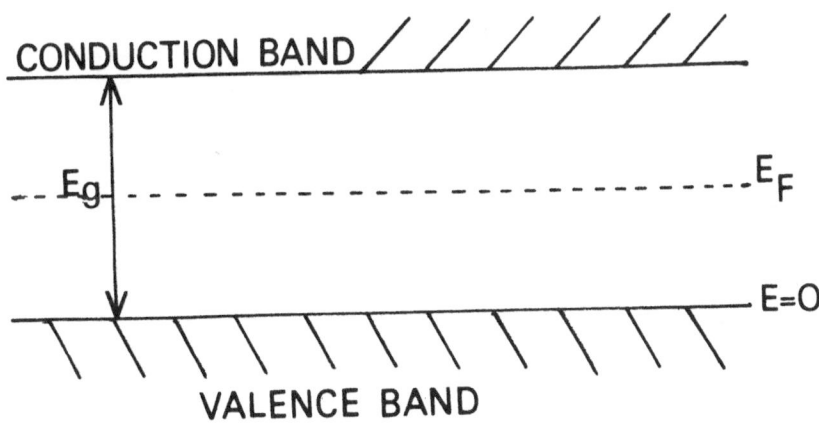

Figure 5.   Energy bands in a semiconductor, showing the gap, Eg,
between the valence and the conduction band.   At zero Kelvin the
valence band is full and the conduction band is empty, so that
the materialsbehaves as an insulator.

the centre of the forbidden energy gap. We shall now briefly demonstrate that this is indeed so, under certain conditions.

Recalling equation (19) we say for simplicity, that the number n_e of electrons in the conduction band is approximately equal to the product of the Fermi-Dirac distribution function, $f(E)$, and a number Nc of available states at the band edge, i.e.

$$n_e = N_c/(H \exp(E_g-E_F)/kT) \sim N_c \exp [E_F-E_g]/kT \tag{22}$$

if $E_g-E_F \gg kT$, and where (see Figure 5) the energy zero is taken to be the top of the valence band. Likewise,

$$n_h = N_V(1-f(E = o)) \tag{23}$$

where N_V is the number of available states at the top of the valence band and $f(E = o)$ denotes the probability of occupancy at the top of the valence band.

$$n_h = N_V/(1 + \exp E_F/kT) \tag{24}$$

and, if we may now take $E_F \gg kT$,

$$n_h \sim N_V \exp(-E_F/kT) \tag{25}$$

Since for an intrinsic semiconductor, $n_e = n_h$, we have from equation (22) and (25)

$$N_c \exp[(E_F-E_g)/kT] = N_V \exp(-E_F/kT) \tag{26}$$

so that

$$E_F = (E_g/2) + (kT/2)\ln(N_V/N_c) \tag{27}$$

or, if $N_V = N_e$,

$$E_F = E_g/2.$$

A more precise derivation would have shown that:

$$N_c = 2(2\pi m_e^*)kT/h^2)^{3/2} \quad \text{and}$$

$$N_V = 2((2\pi m_h^*)kT/h^2)^{3/2}$$

and which would also yield

$$E_F = \frac{E_g}{2} + \frac{3}{4} kT\ln \frac{m_h^*}{m_e^*} \tag{28}$$

This last equation shows us that the Fermi energy is halfway up the band gap in an intrinsic semiconductor only when $m_h^* = m_e^*$. In some solids this equivalence of effective masses is not met (e.g. in InSb $m_h^*/m_e^* \sim 20$) and E_F consequently varies markedly with temperature as well as being shifted well towards the bottom of the conduction band at room temperature.

Note also that the product $n_e \cdot n_h$ (from equations (22) and (24)) is

$$n_e \cdot n_h = (N_v N_c) \exp\left\{-E_g/kT\right\}. \tag{29}$$

But since, for an intrinsic semiconductor,

$$n_e = n_h = n_i \text{ (say), then}$$

$$n_i = (N_v \cdot N_c)^{\frac{1}{2}} \exp(-E_g/2kT) \tag{30}$$

which explains why conductivity (see equation (21)) varies exponentially with temperature.

4.2 The band gap viewed as an enthalpy charge

We now demonstrate how the result of equation (30) falls in with ordinary chemical thermodynamic formulations, provided we regard the electron and the positive hole each as a chemical entity generated according to an equilibrium crudely represented by:

Crystal (e.g. silicon or germanium) $\rightleftharpoons n_e + n_h$

so that a thermodynamic equilibrium constant K may be written (assuming unit thermodynamic activity coefficients) as

$$K = [n_e] \cdot [n_h]. \tag{31}$$

But the standard Gibbs free energy charge ΔG^O yields:

$$[n_e] \cdot [n_h] = \exp\left(\frac{-\Delta G^O}{RT}\right) = \exp\left(\frac{-\Delta H^O}{RT} + \frac{\Delta S^O}{R}\right) \tag{32}$$

where ΔH^O and ΔS^O are the standard enthalpy and entropy change pen mole for the reaction (which generates a pair of changed carriers). We may rewrite equation (32) with respect to individual, rather than a mole of charge carriers so that

$$[n_e][n_h] = \exp\left(\frac{-\Delta h^O}{kT}\right) \cdot \exp\left(\frac{\Delta s^O}{kT}\right) \tag{33}$$

where Δh^O and Δs^O now refer to the respective thermodynmmic changes per charged species. As Boltzmann's equation gives

$$\Delta s^{o} = k \ln W \tag{34}$$

where W refers to the number of ways in which the hole and the electron can be accommodated, we also have:

$$\Delta s^{o} = \Delta s_{e} + \Delta s_{h} = k \ln N_{c} + k \ln N_{v} = k \ln(N_{v}N_{c}) \tag{35}$$

so that:

$$[n_{e}] \cdot [n_{h}] = (N_{c} \cdot N_{v}) \exp\left(\frac{-\Delta h^{o}}{kT}\right) \tag{36}$$

from which with equation (29) it is clearly seen that $E_{g} = \Delta h^{o}$. In other words, the band gap is simply the enthalpy change per electron (or hole) created.

4.3 Localized levels within the band gap

Consider a Group V element substitutionally accommodated within a diamond-like lattice (of Si or Ge for example). The impurity atom forms four covalent bonds with its neighbours, and therefore leaves one electron in excess which experiences an electrostatic attraction to the impurity because of the extra positive charge of the parent ion. This situation resembles the hydrogen atom, but one in which the potential due to the positively changed nucleus is modified by the presence of the surrounding Si or Ge lattice. The atoms of the host are electro-statically polarised by the excess positive charge of the ionized guest (P , As etc.) and their effect on the electrostatic potential can be crudely approximated by the introduction of a dielectric constant, ε .

Pursuing the analogy with the H atom and, in particular the simple Bohr treatment of the latter, we have, for an electron of effective mass m^{*} in an orbit of principal quantum number n, an energy E given by:

$$E = \frac{-m^{*}}{m} \frac{R}{\varepsilon^{2}n^{2}} \tag{37}$$

where R is now the Rydberg constant, and m is the nest mass of the electron. Also for the radius, r, of the electron orbit:

$$r = (m/m^{*})\varepsilon r_{o} \tag{38}$$

where r_{o} is the Bohr radius. Knowing that m^{*}/m is ca. 0.2 (based on experimental observation) and $r_{o} = 0.53\text{\AA}$, we have E being typically 0.01 eV for the lowest (n=1) state. (The experimental value for a Group V impurity in Ge is 0.012 eV.) It is clearly legitimate to regard E as the energy ultimately required (n= ∞) to free electrons from donor centres into the conduction band.

It is interesting to reflect that a donor level situated only
0.01 eV below the bottom of the conduction band implies that a
temperature of 120 Kelvin would suffice to 'ionise' (into the
conduction band) an element such as phosphorus when it is
substitutionally embedded in a host lattice such as Si or Ge.
It is also to be noted how, in relative terms, enormous these
donor atoms are.   With ε = 16 (for Ge) we have, from equation
(38), r ~ 40 Å.

The energy levels of Group III substitutional impurities
Si or Ge may also be estimated theoretically by taking the
interaction of the hole which is bound to the negatively charged
impurity (B⁻.h or Ga⁻.h) as an 'inside-out' hydrogen atom consisting of a negatively charged 'nucleus' with a positive hole 'h'
in orbit around it.   Typical values (experimentally determined)
of localised levels in the band gap are shown in Figure 6.

It is important to recognise that, as well as impurities,
structural imperfections such as point defects and dislocations
give rise to energy levels within the band gap of both semiconductors and insulators.   We shall return to this fact later.

4.4 Excitons in elemental semiconductors

To understand how these exciton states arise it is first
prudent to consider briefly light absorption processes in semiconductors, since there is here operative a selection rule
connected with the conservation of linear momentum.

A photon of energy hν carries momentum hν/c which is
negligible compared with the momentum of a particle of non-zero
rest mass (such as an electron) of the same energy.   Therefore
the momentum of an electron excited into the conduction band must
be the same as it was in the valence band before the absorption
of the photon.   Quantum mechanical analysis shows that it is the
crystal momentum hk which is conserved, giving a selection rule
Δk = 0.

If the maximum of the valence band and the minimum of the
conduction band both occur at k = 0 no difficulty arises.   This
applies to the so-called direct semiconductors of which InSb and
GaAs are examples, and the band-to-band absorption determines the
magnitude of the band gap E_g.   If, however, as in Ge, Si and GaP
(see Figure 7) the conduction band has only a subsidiary minimum
at k = 0 (the situation with indirect-semiconductors) the deeper
minimum occurring at a finite value of k, then vertical transitions
in the vicinity of k = 0 do not determine the minimum value of E_g.

Transitions such as that so-called indirect transition marked
Δk ≠ 0 are allowed (though much weaker) provided the lattice can

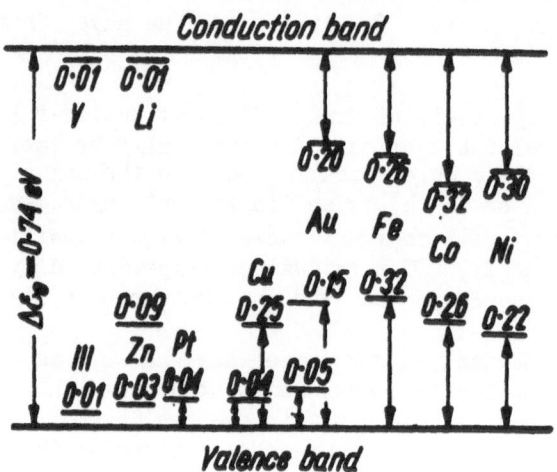

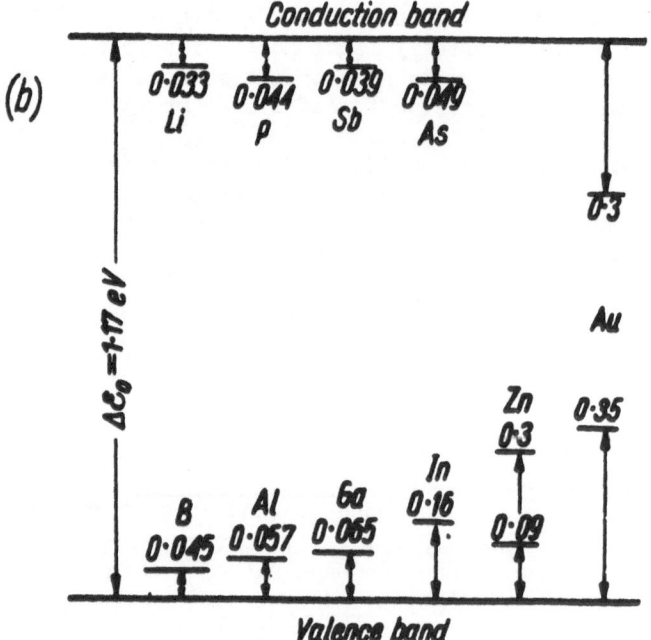

Figure 6.   Position of impurity levels in energy gap: (a) of
germanium;   (b) of silicon.   (Taken from Ambroziak.)

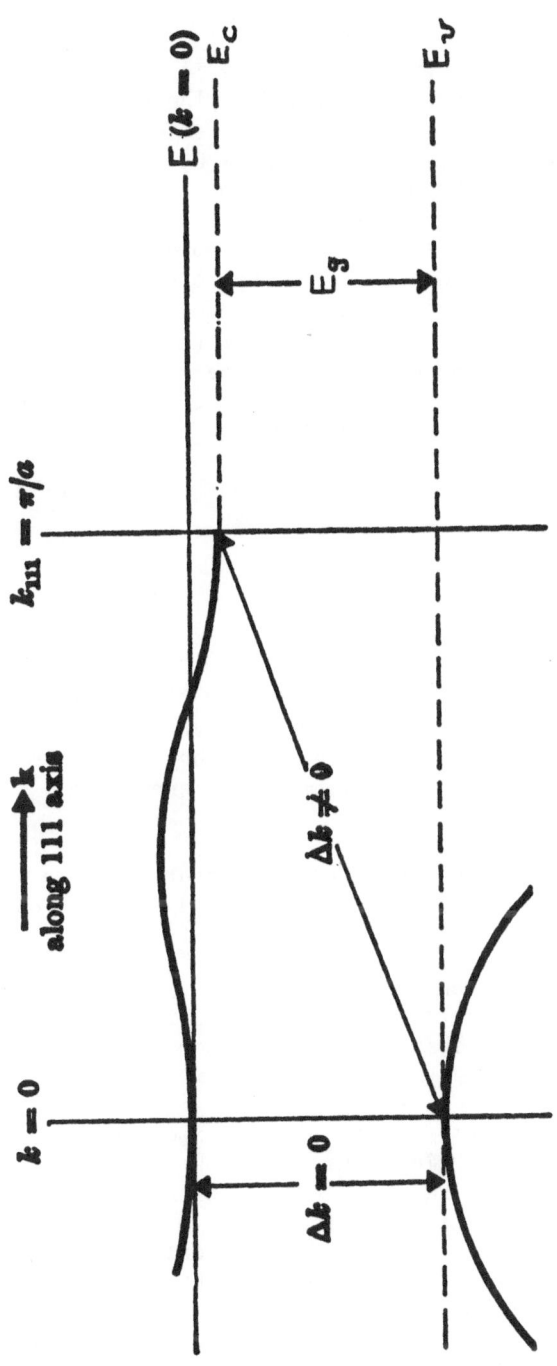

Figure 7. Shape of the band edges against crystal momentum k for germanium. The momentum of a photon is negligible, so that there can be no net transfer of momentum on absorption. Either Δk = 0 for the electron, or the difference in momentum when Δk ≠ 0 must be taken up by the creation or destruction of a phonon. For germanium Eg = 0.75 eV at 0°K, but E (k = 0)−Ec ~ 0.14 eV, so that the Δk ≠ 0 transitions give a fine structure on the absorption edge. (Taken from Bleaney and Bleaney.)

supply or take up the momentum required to make the total momentum of lattice plus electron unchanged.   This involves the creation or destruction of a <u>phonon</u>.

When an electron is excited from the valence to the conduction band by a direct transition, a hole is created in the valence band, the momentum of which must be equal and opposite to that of the electron in the conduction band (so that $\Delta k = 0$).   The electron and hole move apart in opposite directions.   In the vicinity of $k = 0$, however, they will separate rather slowly and their mutual coulomb attraction begins to play a role;   finally at $k = 0$ itself, the electron and hole will stay together.   The e.h bound pair is known as an exciton, and it resembles the donor impurity atom and negative charge discussed above (and therefore the hydrogen atom).   Clearly the electron-hole pair may now move in discrete (Bohr) orbits about the mutual centre of mass giving rise to an exciton series which may be written:

$$E_n = E_\infty - K\frac{R}{n^2} \qquad\qquad (39)$$

with $K = \dfrac{mr}{m} \cdot \dfrac{1}{\varepsilon^2}$

mr being the reduced mass: $\dfrac{1}{m_r} = \dfrac{1}{m_e^*} + \dfrac{1}{m_h^*}$ $\qquad\qquad (40)$

and R is again the Rydberg constant.   When $E_n = 0$ (in equation (39) the electron and hole are separated to such large distances that their mutual attraction is negligible (i.e. they are ionized, and it corresponds to arrival at the bottom of the conduction band (see Figure 8).

The exciton spectrum is identified from its hydrogen-like nature in semiconductors such as Cu_2O (see Figure 9), ZnS, Ge and Si.

Excitons of this kind are called Mott-Wannier excitons. They are of large radius and are essentially delocalized: the e.h pair separation is large compared with the interatomic distance.

There are other, more tightly bound and localized excitons, first conceived by Frenkel and Peierls in their discussion of solids such as xenon and alkali halides.   They defined the exciton as a quantum of electronic excitation energy travelling in the periodic structure of a crystal.   Frenkel excitons have radii which, in alkali halides are comparable to the interatomic distances in such solids, and their binding energy may amount to several eV.   Davydov excitons which occur in molecular crystals are more akin to Frenkel rather than to Mott-Wannier excitons and they have crucial roles in governing the electro-optical properties

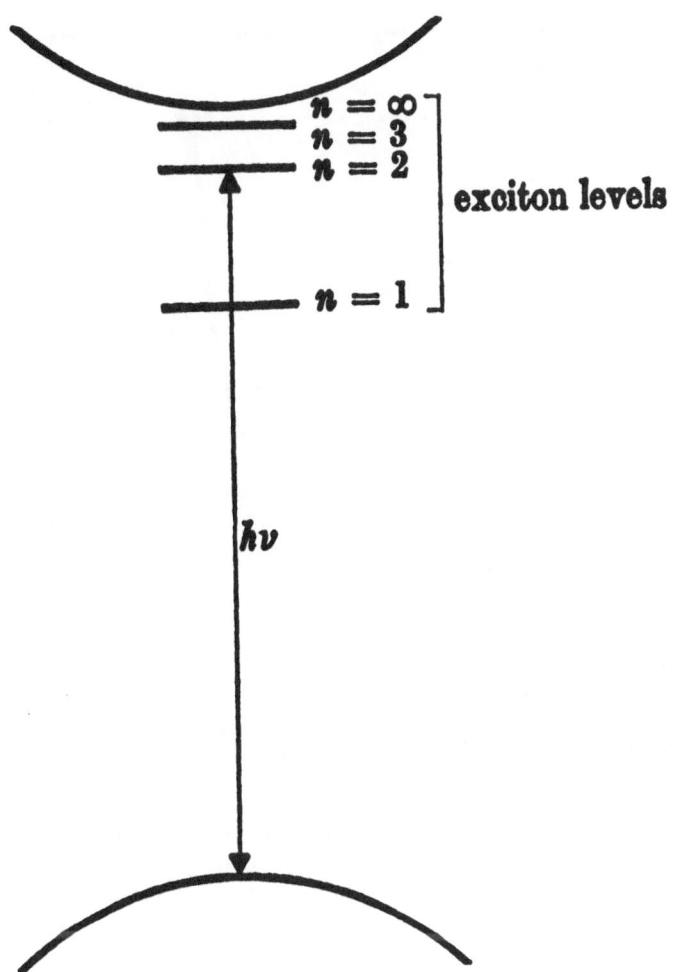

Figure 8.   Exciton levels lying just below the conduction band
in the vicinity of k = 0.   The quantum of energy shown is that
required to excite an electron from the valence band to the n = 2
level.

of certain molecular crystals.

5.   SELECTED APPLICATIONS OF BAND THEORY

    We are now in a position first to appreciate how well band
theory can account for a variety of experimentally observed
properties; and, second, to outline how well the experimentally
determined properties of energy bands - features such as DOS,

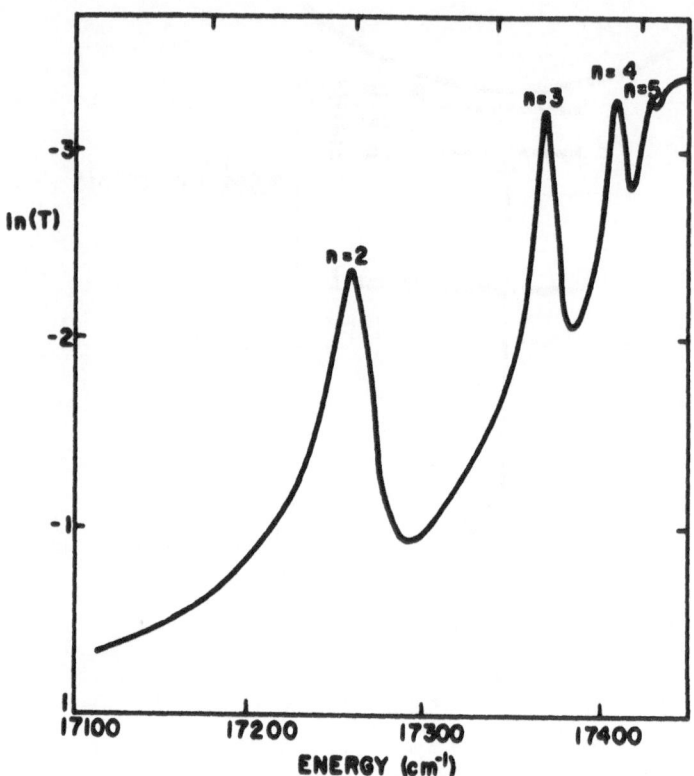

Figure 9.   The absorption spectrum of Cu_2O at $77^\circ K$ showing the
exciton lines corresponding to several values of the quantum
number n.   (Reprinted from P.W. Baumeister, Phys. Rev. 121, 359
(1961).)

band widths, gaps, Fermi energies etc. – tally with the theoretically computed properties.   Since other chapters in this volume
deal with the direct application of XPS, UPS and SXS to band-
structure determination (see contributions by Wertheim, Orchard
and Urch respectively), we shall concentrate on the first category
of applications, no attempt being made to be complete or comprehensive.

Familiarity with the language of band theory enables us to
understand the various inter-relationships between ostensibly
different properties.   We now know – see section 2.1 – how
optical reflectivity and electrical conductivity are directly
related;   how photon absorption and electrical conductivity are
linked (this is the phenomenon of photoconductivity);   and how
the passage of an electric current, or irradiation with electrons,

leads to photon emission (i.e. electroluminescence or cathodo-
luminescence respectively).   Conversely, we can rightaway grasp
how, qualitatively it comes to be that photon irradiation leads
to the separation of charge and the generation of electrical
power[9] (as in a solar battery), or surges of current as a result
of photo-detrapping when charged carriers are excited from their
localized traps to the transport bands.[10]

5.1   One- and two-dimensional metals

Considerable interest has been aroused recently by the
experimental demonstration that at least two distinct types of
solids - $(SN)_n$ and $K_2Pt(CN)_4 \cdot Br_{0.30} \cdot 3H_2O$ - display exceptionally
high electrical conductivity which appears to be highly aniso-
tropic and restricted along the chain directions.   Whilst some
doubt still exists concerning the full band-structure interpretation of $(SN)_n$ (see ref. 11), the reasons why the platinum
complex freely conducts stand   out more clearly.   The Pt-Pt
distance in the chain is ca. 2.89 Å so that the $5d_{z^2}$ atomic
orbitals on each Pt strongly overlap and form a band, which is
completely filled in the stoichiometric compound $K_2Pt(CN)_4$, since
each Pt atom contributes two electrons thereby filling each energy
state in the band.   Upon slight oxidation, some Pt(IV) are added
to the Pt(II) so that, in essence, a few electrons are removed
from the previously full band.   Metallic conductivity ensues[12]
(see Figure 10).   Remarkable one-dimensional quasi-metallic
conductivity has also been reported,[13] but not yet fully rationalized, in deformed single crystals of CdS.

Amongst the two-dimensional facile conductors a good
proportion belong to the intensively studied transition metal
chalcogenides (Figure 11), most of which occur in two or more
polytypes (Yoffe[14]).   It will be recalled (see Figure 4) that
the band structures of these solids have been calculated, and much
insight into their electronic behaviour has thereby been gained.
Thus it transpires that some (depending principally on the d-
electron configuration of the metal atom) are metals (e.g. $NbSe_2$
which has a partially filled d_{z^2}-band), others are semiconductors
(e.g. MoS_2 which has a full d_{z^2}-band).   Other chalcogenides
display even nicer refinements, and the most exquisite example is
$MoTe_2$ (Figure 4) where a change of polytype is predicted to
change the solid from a semiconductor (α-form) to a metal (β-
form).   Note that the difference between α and β forms amounts
only to a change from octahedral to trigonal biprismatic coordination of the chalcogen around the transition metal.   In a
beautiful UPS study, Williams and Shepherd[15] have demonstrated
that, as expected theoretically (see ref. 1), there is a finite
density-of-states at the Fermi energy in the β (metallic) form,
but not in the α(semiconducting) analogue (Figure 12).

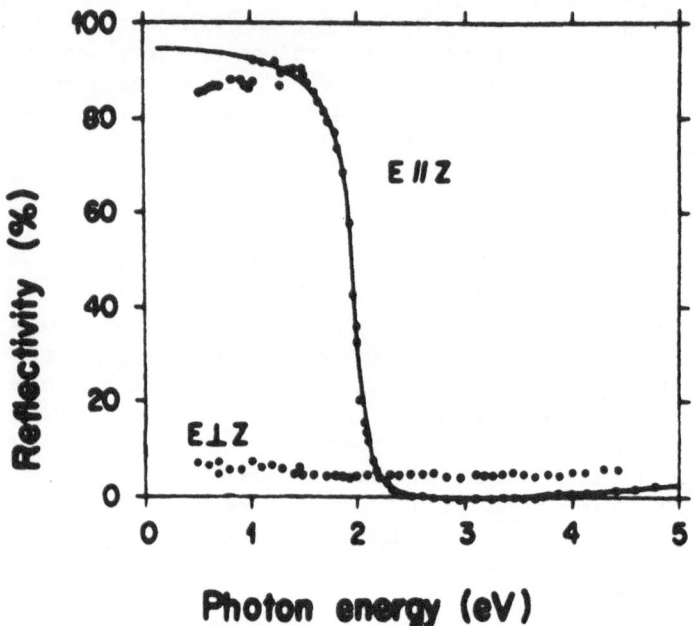

Figure 10.   Reflection spectra of $K_2Pt(CN)_4Br_{0.3}.(H_2O)_n$ measured
at room temperature with the light polarised parallel $(E \parallel Z)$ and
perpendicular $(E \perp Z)$ to the tetragonal optical crystal axis.
Open circles represent experimental data.   The solid curve is a
reflection spectrum calculated from Drude's free-electron theory.
(From Ref. 12.)

5.2 Three-dimensional metals and alloys

It was discovered by Hume-Rothery that the stability and
phase-relationships which existed in numerous alloys formed from
Cu or Ag could be simply (and at first empirically) rationalised
in terms of electron/atom ratios, this ratio being a kind of
universal parameter for the description of the properties of
alloys (see Table 1).   It seems as if in the alloy systems an
intermediate-phase crystal structure is determined by the estab-
lishment of a certain electron/atom ratio.   At first this
constituted something of a mystery, until Hume-Rothery himself
(and, later, others - see Altmann[2] for fuller details) realized
that there is a close connection between the electron concentration
at which a new phase appears and the electron concentration at
which the Fermi surface makes contact with the Brillouin zone
boundary.   In other words the general thesis is that it is
expensive energetically to add further electrons once the filled
states contact the zone boundary.   Additional electrons can be

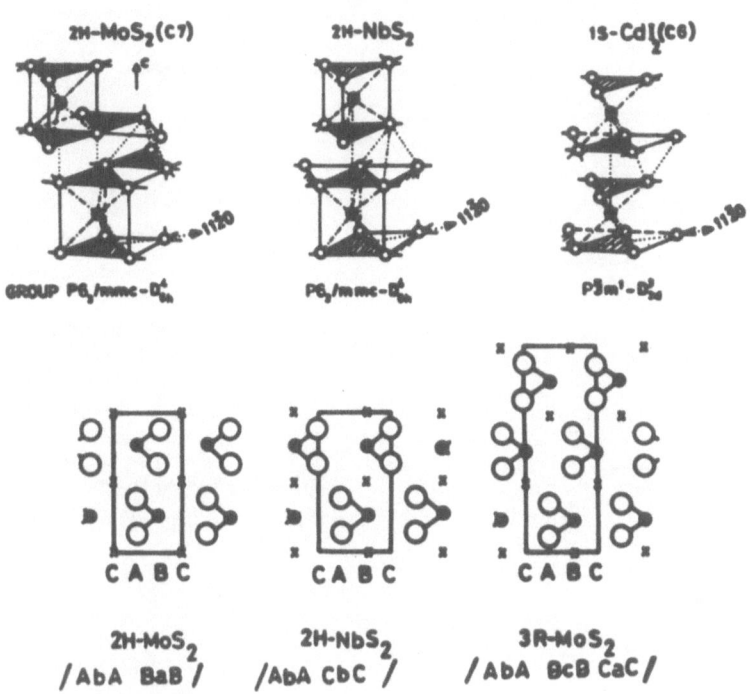

Figure 11. The structures of $2H-MoS_2$, $2H-NbS_2$, and CdI_2 (octahedral of ZrS_2) layer polytypes; also given are $11\bar{2}0$ sections through various atacking polytypes in $2H-MoS_2$, /AbA BaB/, $2H-NbS_2$, /AbA CbC/ and $3R-MoS_2$,/AbA BcB CaC/: ● = Metal, ○ = nonmetal. (From Ref. 14.)

Table 1.   Electron/atom ratios of electron compounds. (The figures give the minimum ratios for the denoted phases.)

	α-Phase (f.c.c)	β-Phase (b.c.e)	γ-Phase (complex)	ε-Phase (h.c.p)
Alloy	Boundary	Boundary	Boundaries	Boundaries
Cu–Zn	1.38	1.48	1.58–1.66	1.78–1.87
Cu–Al	1.41	1.48	1.63–1.77	
Cu–Si	1.42	1.49		
Ag–Zn	1.38		1.58–1.63	1.73–1.75

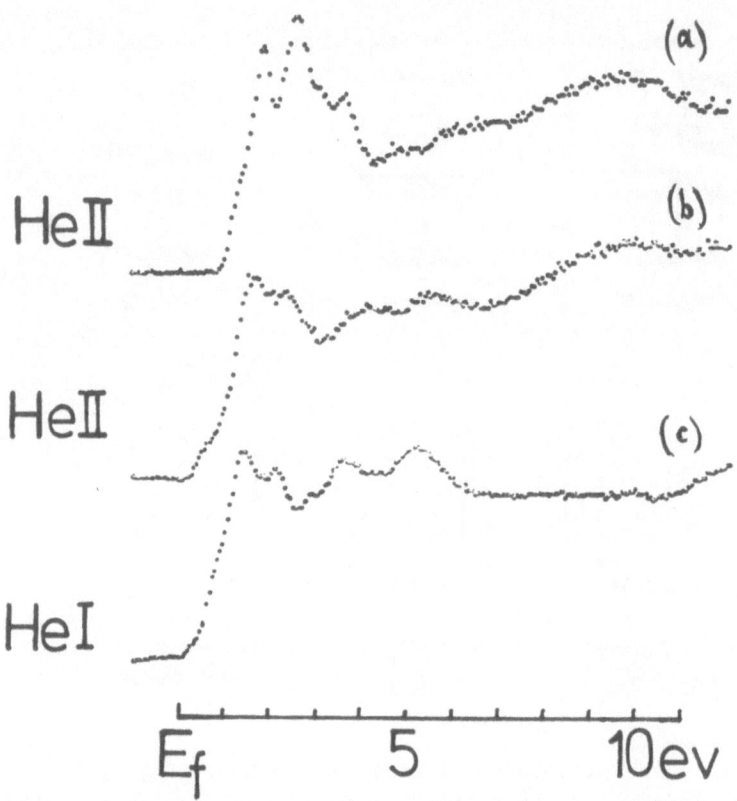

Figure 12.   UPS spectra in the valence band region of (a) α- and (b) and (c) β-MoTe$_2$.   Note differences in DOS at E$_F$.   (From Ref. 15.)

accommodated only in states above the energy gap characterising the boundary or in the states near the corners of the first zone. Since the number of states near the corners falls off markedly as a function of energy, it is often energetically favourable for the crystal structure to change, the new structure being one which contains a larger Fermi surface.

There are obviously other crucial and chemically relevant properties of metals which band theory enables us to understand. The phenomenon of filled-emission, for example, is itself of much value as an exploratory tool in surface chemistry and catalysis; and it is nowadays not without relevance in new developments in organic mass spectrometry.   Equally, the nature of the contact between a metal and an insulator, discussed many years ago by Mott and Gurney,[16] holds the key to our understanding of carrier

injection (from metallic electrodes) into insulating solids.
Figure 13 shows how, as a result of the application of quite small
fields, if $\phi - \chi$ is small, electron injection into the insulator is
facile and the extrinsic conductivity increases enormously and
generally in a non-Ohmic fashion.   This is the basis of electrical
methods of probing trapping centres in inorganic and organic
crystals.

5.3 Bulk semiconductors

One of the most basic properties of a semiconductor is the
magnitude of its band gap, E_g.   From what has been said in
section 4.4 it should be clear how easy it is to extract the value
of E_g for a particular semiconductor by simply monitoring the
absorption coefficient as a function of light frequency.   Band

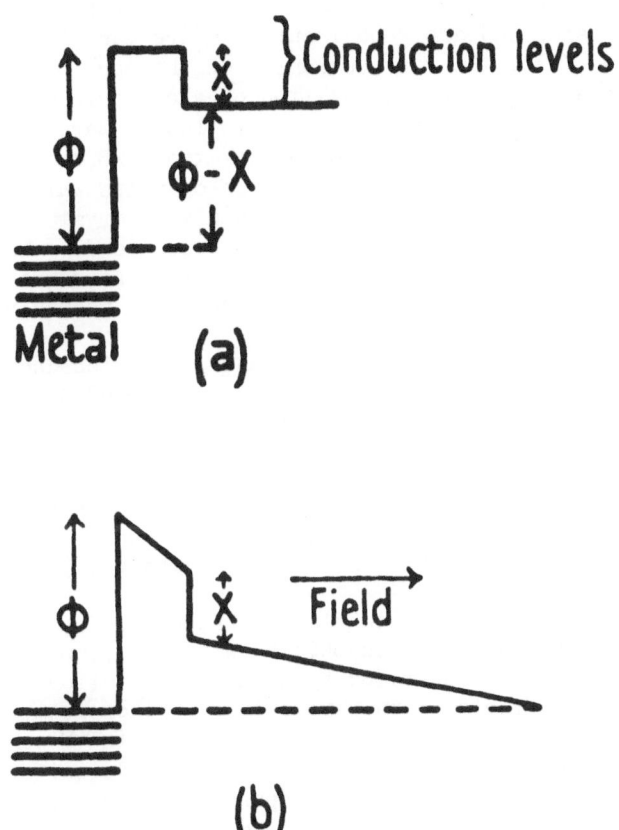

Figure 13.   Insulator in contact with a metal. ϕ is the work
function of the metal, χ the electron affinity of the insulator
(a) in the absence of a field; (b) with a field. (From Ref. 16.)

gaps ranging from ca. 0.1 eV to 2.3 eV have been routinely
determined in this fashion.   With rather large band-gap materials
it is feasible to measure directly the drift mobility by simply
timing the appearance of a charge-surge generated at the other end
of a crystal by a short-duration pulse.   The mobility yields, in
turn (equation (19)) the effective mass, and this via equations
such as (17) and (18) reflects the band-structure of the material
under investigation.

Much has been written elsewhere[5,9,17] on the topic of metal-
semiconductor functions, semiconductor-semiconductor devices of
various kinds, and the whole range of problems associated with
surface-states in elemental and compound semiconductors.   They
will not be discussed here.

5.4 Large band-gap insulating solids

The most well known inorganic solids which possess broad
electronic bands and are good insulators are the alkali halides.
These solids, from the very beginning of the application of
quantum mechanics to the computation of electronic properties of
solids have received considerable attention.   Care must sometimes
be taken to distinguish between one-electron orbital energies of
the crystal in its ground state and the energies which become
involved when an extra electron is added to a crystal (to the
conduction band) or removed from the valence band.   There is a
distinction between the bare band and the excess carrier band;
but to a first approximation they are regarded as equivalent.

As with small or intermediate band-gap semiconductors, many
localised energy levels may exist in the 'forbidden' region.
Seitz, early on, considered the various levels associated with
the distinct spectroscopic states of the thallous ion, Te^+,
present as a deliberate and spectroscopically significant substitutional impurity in KCl.   A recent picture, by Ferd Williams,[18]
shows the relative positions of the various energy levels with
respect to the broad bands also present (Figure 14).

Apart from the localised states that arise from impurities
there are others which are associated with 'intrinsic' point
defects which may be generated by simple irradiation.   These are
numerous (F, V, K, etc.) and have been comprehensively reviewed
recently.   We need note in passing that, for the F-centre, which
entails an electron trapped at an anion vacancy, quite useful
descriptions of the luminescent properties of this centre may be
derived from a simple particle-in-a-box or a point charge approach.

The unifying quality of the band-theoretical approach to
solid-state-chemistry has, it is hoped, been properly indicated.

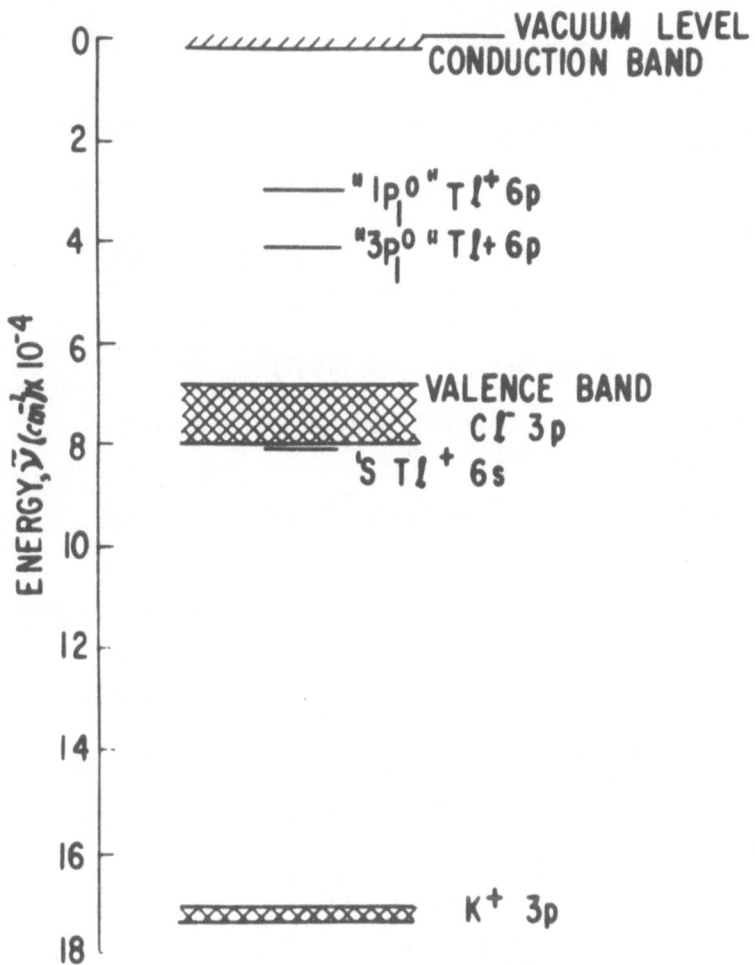

Figure 14.   Band theory model for energy level structure of KCl:Tl.   (From Ref. 18.)

I acknowledge stimulating discussions with my colleague Dr. D.E. Parry, to whom I am grateful.

REFERENCES

1.   R.B. Murray and R.H. Williams, Phil. Mag. 473 (1974).

2.   S. Altmann 'Band Theory of Metals – The Elements' (Pergamon 1970).

3.   J.M. Finman 'Principles of the Theory of Solids' (C.U.P. 1964).

4.   F. Seitz 'Modern Theory of Solids' (McGraw Hill 1940).

5.   W.A. Harrison ' Solid State Theory' (McGraw Hill 1970).

6.   E. Wigner and F. Seitz, Phys. Rev. $\underline{46}$, 509 (1934).

7.   C. Herring, Phys. Rev. $\underline{57}$, 1169 (1940).

8.   J.C. Slater, Phys. Rev. $\underline{51}$, 846 (1937).

9.   L.V. Azaroff and J.J. Brophy 'Electronic Processes in Materials' (McGraw Hill 1963).

10. G.M. Parkinson, J.M. Thomas, J.O. Williams, J. Phys.

11. D.E. Parry and J.M. Thomas, J. Phys. C: Solid State Phys. 1975, in press.

12. D. Kuse and H.R. Zeller, Phys. Rev. Lett. $\underline{27}$, 1060 (1971).

13. C. Elbaum, Phys. Rev. Lett. $\underline{32}$, 376 (1974).

14. A.D. Yoffe, Annual Review of Materials Science, $\underline{3}$, 147 (1973).

15. P.M. Williams and P. Shepherd, Phil. Mag., in press.

16. N.F. Mott and R.W. Gurney 'Electronic Processes in Ionic Crystals' (Dover 1964).

17. L. Solymar and D. Walsh 'Lectures on the Electrical Properties of Materials' (Clarendon 1970).

18. F.E. Williams, G.E. Report No. 60-RL-2588G.

INTRODUCTION TO LINEARLY POLARIZED ELECTRONIC SPECTRA OF
INORGANIC CRYSTALS

J. Ferguson

Research School of Chemistry, The Australian National
University, Canberra, Australia

I.   INTRODUCTION

1.1 Free ion terms and term energies

The energy levels of the free $3d^n$ ions are reasonably well
documented, though this is not true for the $4d^n$ and $5d^n$ ions.[1]
Each d^n configuration is characterized by a number of terms
according to the values of the total spin and orbital quantum
numbers (S and L), $^{(2S + 1)}L$, in the Russell-Saunders coupling
scheme.   This scheme is adequate for treating most features of
$3d^n$ ions.   The degeneracy of the terms is removed by repulsions
between the d electrons and the resultant energy separations are
described very well by the parameters F_2, F_4 and $\alpha L(L + 1)$, for
the $3d^n$ ions.[2]

Each (spinning) electron has an associated magnetic moment
and because of the orbital motion about the nucleus there is a
consequent magnetic field which interacts with the magnetic
moment of the electron.   This leads to a coupling of the spin
and orbital angular momenta to form the resultant total angular
momentum, denoted by J.   One observable effect is the spin-orbit
splittings of terms, e.g., $^3P_2$, $^3P_1$ and $^3P_0$.   Another important
consequence is mixing between states of the same J but different
L and S (off-diagonal spin-orbit interaction).

1.2 Ions in crystals - energy level schemes

The degeneracy of the d orbitals is removed by interaction
with the ligands in a condensed phase.   This produces a splitting

P. Day (ed.), Electronic States of Inorganic Compounds. 59−93. All Rights Reserved.
Copyright © 1975 by D. Reidel Publishing Company, Dordrecht-Holland.

of many terms into a number of ligand field states, dependent on
the symmetry of the ligand field.   The energies of these states
can be described by ligand field parameters (Dq plus others if
the symmetry is less than cubic) together with those of the free
ion theory.   The ligands also have their own set of electrons
and overlap between the orbitals of the ligands and those of the
metal ion (covalency) can lead to significant and sometimes very
large departures from the energy level spacings predicted from
simple electrostatic crystal field theory.

One reason for this is that the ligand field parameters are
state dependent, because of the involvement of the ligand orbitals
in a molecular orbital description, so that the basic simplicity
of the crystal field scheme becomes lost for covalent ligands.

Description of the d electron states can be made using
either the weak field or the strong field theory.   The choice is
dictated by convenience and the nature of the problem under study.
In order to compare the energies of all of the states arising from
the terms, using parameters which can be simply related to those
of the free ion, then a weak field description is easier to use,
because it is simpler to incorporate the Trees correction into
the weak field matrices of the ligand field.[2]   For consideration
of potential energy surfaces, magnetic interactions etc., the
strong field description is more meaningful and it is preferred.
Provided the·same terms are used in the Hamiltonian for the
system, the two descriptions are, of course, equivalent.

1.3   Propagation of light in crystals

The condensed phase which provides the most spectral
information is a single crystal in which there is a periodic
array of the atoms contained in the simplest repeating unit, the
primitive unit cell.   The propagation of electromagnetic
radiation in a crystal is dependent on the symmetry properties of
the crystal because of the (transverse) electric field of the
radiation which interacts with the electrons in the crystal,
associated, of course, with the atomic array.   A measure of this
interaction is the refractive index.   If the crystal has cubic
symmetry, the refractive index is independent of the direction of
propagation and the material is optically isotropic.   All other
crystal systems are either uniaxial or biaxial.

Uniaxial crystals belong to the trigonal, tetragonal and
hexagonal systems, so they have a three-fold, four-fold or six-
fold symmetry axis respectively.   These crystals, in contrast to
those of the cubic system, have two refractive indices.   Light
incident in an arbitrary direction will be resolved into two
linearly polarized beams, each having a different refractive index

(double refraction).   One is termed the ordinary ray refractive index and the other, the extraordinary ray refractive index. There is, however, one direction of propagation for which the two rays have the same velocity (and refractive index).   This is the optic axis and it corresponds to the axis of highest symmetry. For this direction of propagation, the symmetry of the atomic array in the plane normal to this requires that the interaction between the light and the electric charges is independent of the electric vector direction.

The remaining crystal systems fall into the biaxial class. The optical properties of these crystals are best considered by referring to a three-dimensional figure known as the optical indicatrix.[3]   This is an ellipsoid in which the three principal dimensions have lengths proportional to minor (α), intermediate (β) and major (γ) principal refractive indices.

There are two directions for which there is no double refraction and the single refractive index is equal to the inter-mediate principal refractive index.   However, these optic axes do not coincide with a crystallographic axis so that their orientation, relative to the crystal axes will vary with the wavelength of the light (dispersion).   They provide no useful purpose for general spectroscopic investigation but can be useful for microscopic identification of crystal sections.   The principal directions of the optical indicatrix are of more use for spectroscopic studies.   These directions coincide with the three axes of the orthorhombic crystal system and are therefore wavelength independent.   For monoclinic crystals, one axis of the optical indicatrix coincides with one crystal axis (the two fold axis).   The extinction directions for the section normal to this propagation direction corresponding to the electric vector directions of the two polarized beams are, however, wavelength dependent and show dispersion.   Light, incident on a section which contains the symmetry axis, will be resolved into two beams and the direction of polarization of one beam will coincide with the symmetry axis for all wavelengths.   The low symmetry of the triclinic system requires no coincidence between the crystal axes and the axes of the optical indicatrix and this system is least useful for spectroscopic studies.

The interaction between the electromagnetic radiation and the crystal can also lead to the absorption of light and a change of the energy state of the crystal.   For the present purpose it is enough to neglect the interaction between the N ions in the crystal so that the interaction with the crystal is simply that between the light and an oriented atomic system which is N-fold degenerate, i.e., the oriented gas model.   The interaction, leading to electronic transitions, is treated in many texts dealing with quantum mechanics.   For our purpose, we are

interested in two mechanisms for the absorption of light. The
first is the electric dipole mechanism and the second is the
magnetic dipole mechanism.

The probability of an electronic transition can be discussed
in terms of the transition moment matrix element

$$\langle \psi_b | \underline{M} | \psi_a \rangle,$$

where $\underline{M}$ is either the magnetic dipole operator or the electronic
dipole operator.

For the electric dipole mechanism $\underline{M}$ is $e\underline{r}$ and the probability
of absorption of light is simply related to the scalar product of
this vector and the electric vector of the light. It follows
then that the intensity of absorption is proportional to the
square of the cosine of the angle between the transition moment
vector and the electric vector of the light wave.

For the magnetic dipole mechanism the operator is $(e/2mc)\underline{l}$
and the intensity of absorption is proportional to the square of
the cosine of angle between the transition magnetic moment and
the magnetic vector of the light wave.

We note that the electric dipole moment operator has odd
parity while the magnetic dipole moment operator has even parity.

II. ELECTRONIC TRANSITIONS IN CRYSTALS

2.1 Single ion mechanism – selection rules

Group theory can now be used to determine the selection rules
for the absorption (and emission) of light for ions (or molecules)
in crystals, assuming the oriented gas approximation. For this
purpose the crystallographic point group symmetry of the site of
the transition metal ion must be known. In many cases this will
be the same as the symmetry of the unit cell, but in others it
will be a sub-group.

It is a simple matter to construct tables of selection rules
for which dipole transitions are allowed by symmetry and some of
the most frequently used are included in Table 1. We note that
this Table also includes the representations of the double groups,
needed for consideration of spin-orbit effects for ions having an
odd number of electrons.

As an example of the use of Table 1, we consider the case
of the rutile structure $(TiO_2$ or $MF_2)$ which has the space group

TABLE 1. Electric and magnetic dipole selection rules

O_h	r	L
A_{1g}	T_{1u}	T_{1g}
A_{2g}	T_{2u}	T_{2g}
E_g	$T_{1u}T_{2u}$	T_{1g}
T_{1g}	$A_{1u}E_uT_{1u}T_{2u}$	$A_{1g}E_gT_{1g}T_{2g}$
T_{2g}	$A_{2u}E_uT_{1u}T_{2u}$	$A_{2g}E_gT_{1g}T_{2g}$
Γ_{6g}	$\Gamma_{6u}\Gamma_{8u}$	$\Gamma_{6g}\Gamma_{8g}$
Γ_{7g}	$\Gamma_{7u}\Gamma_{8u}$	$\Gamma_{7g}\Gamma_{8g}$
Γ_{8g}	$\Gamma_{6u}\Gamma_{7u}\Gamma_{8u}$	$\Gamma_{6g}\Gamma_{7g}\Gamma_{8g}$

T_d	r	L
A_1	T_2	T_1
A_2	T_1	T_2
E	T_1T_2	T_1T_2
T_1	$A_2ET_1T_2$	$A_1ET_1T_2$
T_2	$A_1ET_1T_2$	$A_2ET_1T_2$
Γ_6	$\Gamma_7\Gamma_8$	$\Gamma_6\Gamma_8$
Γ_7	$\Gamma_6\Gamma_8$	$\Gamma_7\Gamma_8$
Γ_8	$\Gamma_6\Gamma_7\Gamma_8$	$\Gamma_6\Gamma_7\Gamma_8$

(continued)

D_{4h}	z	x,y	L_z	L_x, L_y
A_{1g}	A_{2u}	R_u	A_{2g}	E_g
A_{2g}	A_{1u}	E_u	A_{1g}	E_g
B_{1g}	B_{2u}	E_u	B_{2g}	E_g
B_{2g}	B_{1u}	E_u	B_{1g}	E_g
E_g	E_u	$A_{1u}A_{2u}B_{1u}B_{2u}$	E_g	$A_{1g}A_{2g}B_{1g}B_{2g}$
Γ_{6g}	Γ_{6u}	$\Gamma_{6u}\Gamma_{7u}$	Γ_{6g}	$\Gamma_{6g}\Gamma_{7g}$
Γ_{7g}	Γ_{7u}	$\Gamma_{6u}\Gamma_{7u}$	Γ_{7g}	$\Gamma_{6g}\Gamma_{7g}$

D_{3d}	z	x,y	L_z	L_x, L_y
A_{1g}	A_{2u}	E_u	A_{2g}	E_g
A_{2g}	A_{1u}	E_u	A_{1g}	E_g
E_g	E_u	$A_{1u}A_{2u}E_u$	E_g	$A_{1g}A_{2g}E_g$
Γ_{4g}	Γ_{5u}	Γ_{6u}	Γ_{5g}	Γ_{6g}
Γ_{5g}	Γ_{4u}	Γ_{6u}	Γ_{4g}	Γ_{6g}
Γ_{6g}	Γ_{6u}	$\Gamma_{4u}\Gamma_{5u}\Gamma_{6u}$	Γ_{6g}	$\Gamma_{4g}\Gamma_{5g}\Gamma_{6g}$

C_{2h}	z	x,y	L_z	L_x, L_y
A_g	A_u	B_u	A_g	B_g
B_g	B_u	A_u	B_g	A_g
Γ_{3g}	Γ_{3u}	Γ_{4u}	Γ_{3g}	Γ_{4g}
Γ_{4g}	Γ_{4u}	Γ_{3u}	Γ_{4g}	Γ_{3g}

(continued)

C_{4v}	z	x,y	L_z	L_x, L_y
A_1	A_1	E	A_2	E
A_2	A_2	E	A_1	E
B_1	B_1	E	B_2	E
B_2	B_2	E	B_1	E
E	E	$A_1 A_2 B_1 B_2$	E	$A_1 A_2 B_1 B_2$
Γ_6	Γ_6	$\Gamma_6 \Gamma_7$	Γ_6	$\Gamma_6 \Gamma_7$
Γ_7	Γ_7	$\Gamma_6 \Gamma_7$	Γ_7	$\Gamma_6 \Gamma_7$

D_{2d}	z	x,y	L_z	$L_{x,y}$
A_1	B_2	E	A_2	E
A_2	B_1	E	A_1	E
B_1	A_2	E	B_2	E
B_2	A_1	E	B_1	E
E	E	$A_1 A_2 B_1 B_2$	E	$A_1 A_2 B_1 B_2$
Γ_6	Γ_7	$\Gamma_6 \Gamma_7$	Γ_6	$\Gamma_6 \Gamma_7$
Γ_7	Γ_6	$\Gamma_6 \Gamma_7$	Γ_7	$\Gamma_6 \Gamma_7$

D_{2h}	z	x	y	L_z	L_x	L_y
A_{1g}	A_{2u}	B_{2u}	B_{1u}	A_{2g}	B_{2g}	B_{1g}
A_{2g}	A_{1u}	B_{1u}	B_{2u}	A_{1g}	B_{1g}	B_{2g}
B_{1g}	B_{2u}	A_{2u}	A_{1u}	B_{1g}	A_{2g}	A_{1g}
B_{2g}	B_{1u}	A_{1u}	A_{2u}	B_{1g}	A_{1g}	A_{2g}
Γ_{5g}	Γ_{5u}	Γ_{5u}	Γ_{5u}	Γ_{5g}	Γ_{5g}	Γ_{5g}

D_{4h}^{14}. There are two formula units per cell and the site group of M is D_{2h} (see Figure 1). A four-fold screw axis interchanges the corner and centre ions and thus interchanges the x and y D_{2h} coordinate axes. The site group of the metal ions is therefore a sub-group of the unit cell group, which is tetragonal so the crystal is uniaxial.

By using Table 1 we see that electronic transitions between g states are forbidden for the electric dipole mechanism. They are however allowed for the magnetic dipole mechanism, e.g., A_{1g} can combine with A_{2g} (H // L_z), B_{2g} (H // L_x) and B_{1g} (H // B_1L_y). By convention, the electric vector direction is used to characterize the incident light, so that H // L_z corresponds to the electric vector perpendicular to the z axis (in this case the crystal c axis) and is denoted by σ. H // L_x and H // L_y correspond to the electric vector parallel to c and are denoted by π. These conventions, commonly used for a uniaxial crystal, are shown in Figure 2. A selection rule matrix for magnetic dipole absorption is then obtained and this is given in Table 2. We note that identical selection rules apply to σ and π spectra, so that these two spectra will be identical.

TABLE 2. Magnetic dipole selection rules for the MF_2 (rutile) structure

	A_{1g}	A_{2g}	B_{1g}	B_{2g}
A_{1g}	–	σ	σ,π	σ,π
A_{2g}	σ	–	,	,
B_{1g}	σ,π	σ,π	–	σ
B_{2g}	σ,π	σ,π	σ	–

Although group theory is extremely useful in deriving the selection rules, it tells us only whether the transition moment is non-zero and nothing about its magnitude. Fortunately, the theoretical calculation of the magnetic dipole transition moment intensity is straightforward and reasonably exact. For spin-allowed transitions the allowed magnetic dipole oscillator strengths can be of the order 10^{-5} to 10^{-6}. [4,5]

If the site group of the metal ion in the crystal contains a centre of inversion then pure electronic transitions between the d orbitals are forbidden for the electric dipole mechanism.

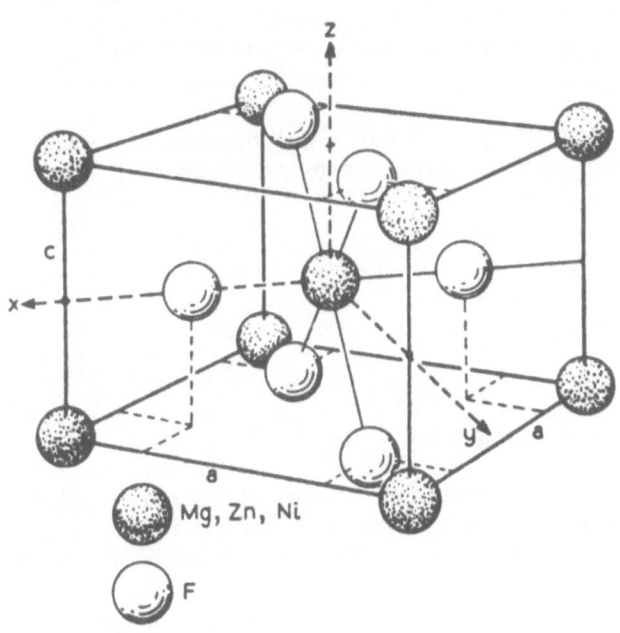

Figure 1.   Rutile structure of MF_2.

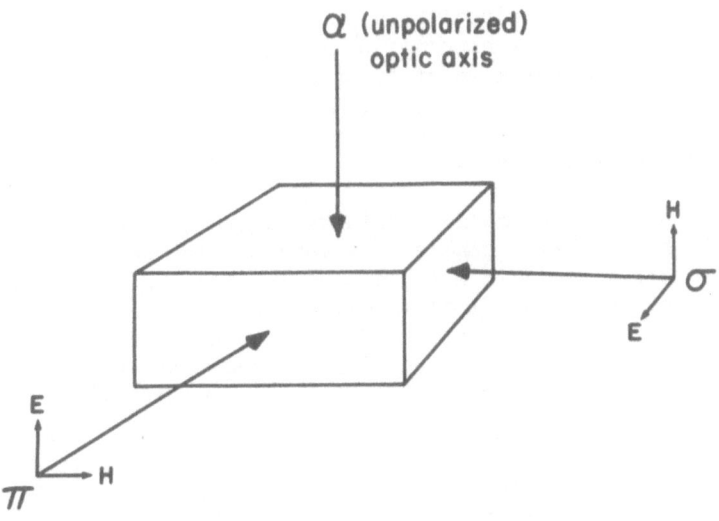

Figure 2.   The three methods of measuring the absorption spectrum of a uniaxial crystal, denoted by α, δ and π.

However, electric dipole transitions are observed in such cases
and they occur through the participating of ungerade, or odd,
vibrations of the chromophoric system.   The theory for this
mechanism was developed by Herzberg and Teller[6,7] in their
consideration of forbidden electronic transitions in molecules.

The effect of the nuclear displacement is usually treated
by first order perturbation theory in terms of a mixing of other
electronic functions,

$$\Psi_{be}(Q) = \psi^o_{be} + \sum_i \lambda_i(Q)\,\psi^o_{ie}$$

$$\lambda_i = \langle \psi^o_{be} \mid H'(Q) \mid \psi^o_{ie}\rangle / (E^o_i - E^o_b)$$

$$H'(Q) = \left(\frac{\partial H}{\partial Q}\right)_o Q$$

Q represents the nuclear coordinate, the superscript o indicates
the electronic wavefunction in the absence of any vibration and
$H'(Q)$ is the perturbation.   The transition moment is then

$$\langle \Psi_{be} \mid \underline{M} \mid \psi^o_{ae}\rangle = \langle \psi^o_{be} \mid \underline{M} \mid \psi^o_{ae}\rangle + \sum_i \lambda_i(Q)\langle \psi^o_{ie} \mid \underline{M} \mid \psi^o_{ae}\rangle \qquad (1)$$

A similar expression follows from the consideration of the effect
of the u vibration in the ground state.   In this case the
perturbation is temperature dependent and is absent at very low
temperatures.   This intensity is termed the 'hot band' contribution and a decrease of absorption intensity of an absorption band
on lowering the temperature of the crystal is an indication that
a vibronic electric dipole mechanism is involved.   In passing we
note that the absorption intensity of an allowed electronic band
is temperature independent.

It sometimes occurs that, although a transition metal ion is
at a site which lacks a centre of inversion, the departure from a
centro-symmetric field is not great.   In this case the first term
of (1) is not zero, but it still may be quite small and of the
same order as the second term.   The electronic transition will
then have features which can be considered as either allowed or
forbidden through the involvement of both terms of (1) in
comparable amounts.

We now consider the application of group theory to the second
term in (1).   If we label the representations of ψ^o_{ae}, ψ^o_{be} and
ψ^o_{ie} by Γ_a, Γ_b and Γ_i respectively and the representations of M
and $H'(Q)$ by Γ_M and Γ_Q respectively, then the requirement for the
transition moment matrix element to be non-zero is that $\Gamma_a \times \Gamma_M \times \Gamma_i$
contains the totally symmetric representation, while the require-

ment for λ_i to be non-zero is that $\Gamma_i \times \Gamma_Q \times \Gamma_b$ also contains the totally symmetric representation. It is not usually difficult to find more than one combination of $H'(Q)$ and $\underline{\underline{M}}$ to satisfy these conditions for a given ψ^o_{ae} and ψ^o_{be}. The problem is that group theory gives no indication of the magnitude of a non-zero integral, which must come from a theoretical calculation.

Vibronically induced electronic transitions are common and generally the oscillator strengths fall in the range 10^{-4} to 10^{-6} and thus overlap with the values observed for allowed magnetic dipole transitions. There is then a need to distinguish between the two mechanisms experimentally. This is most readily done by a consideration of their dichroic properties in uniaxial and bi-axial crystals, other methods being used for cubic crystals.

For uniaxial crystals, it is necessary to measure the absorption spectrum in each of the three possible ways, shown in Figure 1. If we denote the two absorption intensities for electric dipole absorption by $e_{//}$ and $e_\perp$ (parallel and perpendicular to the optic axis, respectively and the corresponding quantities for a magnetic dipole transition by m and m , then the three possible spectra are

$$\alpha = e_\perp + m_\perp$$

$$\pi = e_{//} + m_\perp$$

$$\sigma = e_\perp + m_{//}$$

There are two limiting cases: (a) $e_{//} + e_\perp = 0$ and (b) $m_{//} = m_\perp = 0$. For the first $\alpha = \pi \neq \sigma$ (magnetic dipole allowed) and for the second $\alpha = \sigma \neq \pi$ (electric dipole allowed).

For biaxial crystals it is necessary to determine two polarized spectra for light incident on each of three orthogonal crystal faces, giving a total of six spectra. If x, y and z denote an orthogonal axis system, the six spectra are (where $\underline{\underline{k}}$ is the propagation direction of the light)

$$\underline{\underline{k}} // x \begin{cases} \underline{\underline{E}} // y \quad \underline{\underline{H}} // z \quad A_{xy} = e_y + m_z \\ \underline{\underline{E}} // z \quad \underline{\underline{H}} // y \quad A_{xz} = e_z + m_y \end{cases}$$

$$\underline{\underline{k}} // y \begin{cases} \underline{\underline{E}} // z \quad \underline{\underline{H}} // x \quad A_{yz} = e_z + m_x \\ \underline{\underline{E}} // x \quad \underline{\underline{H}} // z \quad A_{yx} = e_x + m_z \end{cases}$$

$$\underline{k} \parallel z \begin{cases} \underline{E} \parallel y \quad \underline{H} \parallel x \qquad A_{zy} = e_y + m_x \\ \underline{E} \parallel x \quad \underline{H} \parallel y \qquad A_{zx} = e_x + m_y \end{cases}$$

For orthorhombic crystals we naturally choose the propagation directions to coincide with the crystal axes. Examples are shown in Figure 3 and Figure 4 for BaNiF$_4$.[8] In Figure 3 it can be seen that the transition involves both electric and magnetic dipole contributions and six different spectra are obtained, while in Figure 4 the magnetic dipole contribution is zero so that only three spectra are observed.

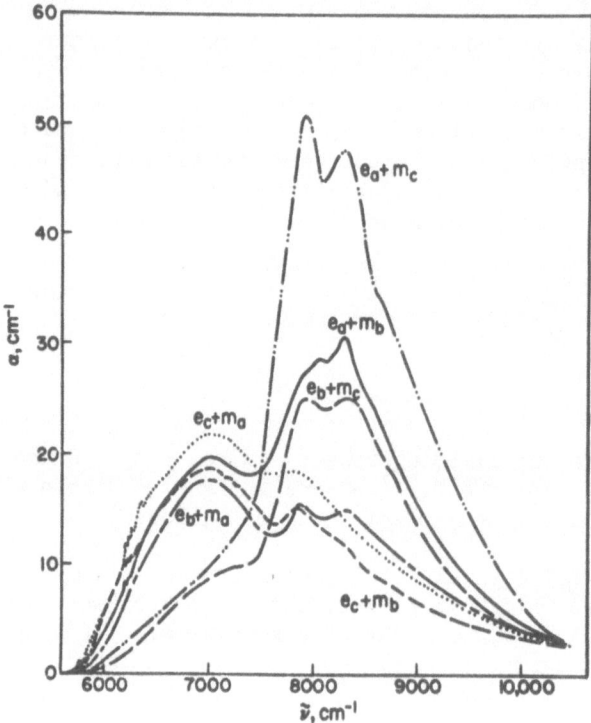

Figure 3. Absorption spectrum (10K) of BaNiF$_4$ in the near infrared region corresponding to the transition $^3A_{2g} \rightarrow {}^3T_{2g}$. The involvement of both electric dipole and magnetic dipole mechanisms leads to six different spectra for polarized light propagating down the three orthorhombic crystal axes.

For monoclinic crystals, the choice of propagation directions is only determined by symmetry for one direction, i.e. the axis of symmetry (b) and the other two directions are arbitrary.

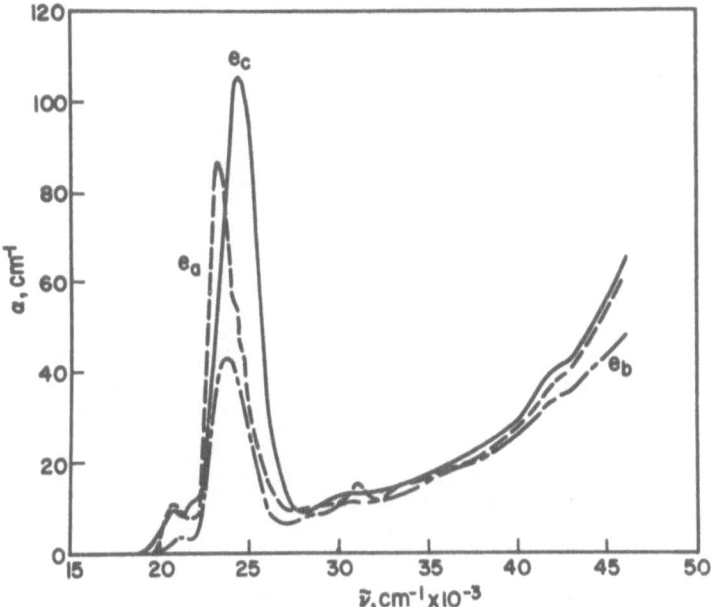

Figure 4. Absorption spectrum (10K) of $BaNiF_4$ in the spectral region corresponding to the transition $^3A_{2g} \rightarrow {}^3T_{1g}$. The absence of magnetic dipole intensity gives rise to only three spectra.

Ideally, it is best to choose the two extinction directions of the ac face and assume that there is no significant dispersion of these directions. However, this is not always convenient, perhaps because of some marked cleavage, and another choice has to be made. In either case, six spectra must be measured and compared carefully before the relative intensities of the electric and magnetic dipole mechanisms can be determined.

As an example we consider the absorption spectrum of $CoCl_2.6H_2O$ in the near infrared region[9] corresponding to the (cubic) transition $^4T_{1g}(^4F) \rightarrow {}^4T_{2g}(^4F)$, which is allowed for the magnetic dipole mechanism. The space group is C_{2h}^3 and the site group of the Co^{2+} is C_{2h}. The crystal shows marked cleaveage parallel to (001) and it is convenient to choose the ab, ac' and bc faces for spectral study. The projections of the $CoCl_2(H_2O)_4$ unit on these faces is shown in Figure 5, together with the orientation of the molecular axes. The z axis is chosen to lie along $Cl - Co - Cl$ and the x axis in the plane of the water molecules, parallel to the b crystal axis. We need to be able to transform the molecular coordinate system into the crystal axis

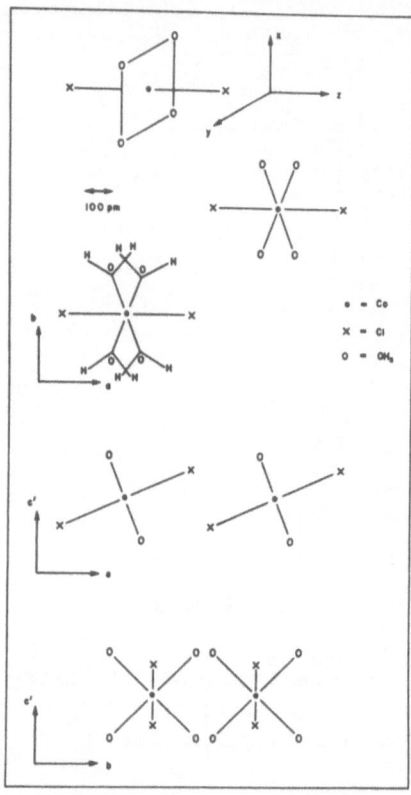

Figure 5.   Projections of the $CoCl_2.4H_2O$ molecular unit on
the ab, ac' and Bc' crystal faces.   The choice for the molecular
axes is included.

system in order to relate crystal absorption intensities to the
oriented gas model intensities.   This involves the squares of
the projections of the molecular transition moments and these are
given in Table 3.   We see then that the b and x axes are equivalent, while the y and x molecular axes project principally on to
the a and c' crystal axes, respectively.

    The absorption spectra of the three faces are given in Figure
6 and the band positions and intensities are collected in Table 4.
Uiing these date we can arrive at the following approximate ranges
for the magnetic (m_i) and electric (e_i) dipole oscillator strengths
(units of 10^{-6})

$$m_a = 0 - 2 \qquad\qquad m_b \sim m_{c'} = 3 - 5$$

$$e_a = 7\text{-}5 - 9.5 \qquad\qquad e_b \sim e_{c'} = 0.2$$

TABLE 3.   Squares of unit vector projections on the crystal axes of $CoCl_2 6H_2O$

$CoCl_2(H_2O)_4$ axis			
	a	b	c'
X	0	1	0
Y	0.158	0	0.842
Z	0.842	0	0.158

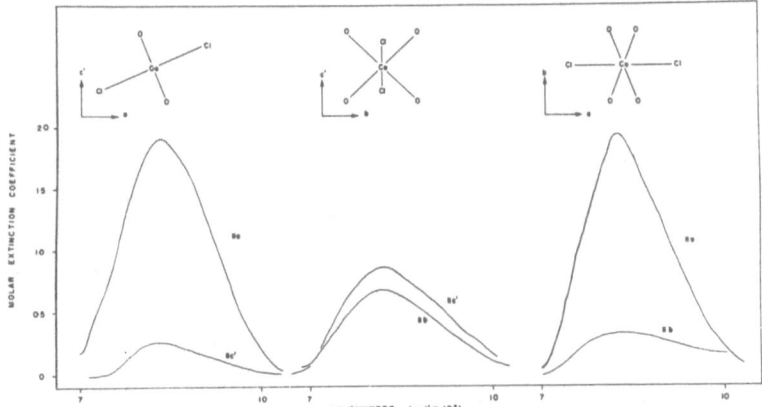

Figure 6.   Polarized absorption spectra of $CoCl_2.6H_2O$ in the near infrared region (10-40K).   The water overtone vibrations have been removed for clarity.

The errors in these figures are relatively large but they show clearly that there is a dominant bc'(xy) polarized magnetic dipole transition and a (z) polarized electric dipole (vibronic) band.   The dependences of the band intensities on temperature (Table 4) support this conclusion.   The absorption intensity then arises from contributions from the two mechanisms in quite aniso- tropic ways, the magnetic dipole contribution being polarized in

TABLE 4.   Infrared bands in $CoCl_2.6H_2O$ and $CoCl_2.6D_2O$

Polarization*	Position+	f_{10}++	f_{300}/f_{10}
ab ∥ a	8330 ± 30	12.5	2.0
ac' ∥ a	8330 ± 50	12.5	2.0
ab ∥ b	8065 ± 60	2.5	1.7
bc' ∥ b	8160 ± 60	5.5	1.1
ac' ∥ c'	8200 ± 50	2.0	1.6
bc' ∥ c'	8200 ± 60	5.0	1.1

* Crystal face and polarization of electric vector

+ Franck-Condon maxima at 10K (cm^{-1})

++Units of 10^{-6}.   Larger values ± 10%, smaller values ± 20%.

the molecular xy plane and the electric dipole part appearing strongly polarized in the z direction.   The magnetic dipole polarization is consistent with the approximately tetragonal component $^4A_{2g}$ of $^4T_{1g}$ lying lower in energy than ^4E_g, an assignment which is confirmed by a more complete spectral analysis.[9]

2.2 Vibrational structure of electronic transitions

The spectrum of an ion in a crystal differs in a very important way from the free ion spectrum and this, of course, is the band width associated with the electronic transition from one state to another.   The band width is explained by changes in vibrational quanta which can occur in the absorption or emission process.   The number and symmetry which can be involved in either process is determined by a consideration of the equilibrium (zero point) positions of the nuclei in the ground and electronically excited state.   For the present purpose we assume that the symmetry of each state is the same.   However, it is likely that the metal-ligand distances will be different so that the two nuclear potential energy surfaces are displaced along a totally symmetric coordinate.   If we assume Born-Oppenheimer separability of the electronic and vibrational wave functions then

$$\Psi = \psi_e \psi_v$$

and it is easy to show that the form of the transition moment
integral can be expressed as

$$\langle \psi^o_{be} | \underline{M} | \psi^o_{ae} \rangle \langle \psi_{by} | \psi_{av} \rangle$$

The first integral is the electronic transition moment while
the second is the vibrational overlap integral. This latter
integral determines the way in which the electronic transition
intensity is distributed over the various vibrational sub-states.
We expect therefore, that an allowed electronic transition will
be composed of a pure electronic (zero phonon) line, together with
a progression in one or more totally symmetric modes of vibration.
In the case of a forbidden (electric dipole) transition, the
origin for the totally symmetric progression will be a false-
origin which corresponds to one quantum of a perturbing ungerade
vibration. In this case there can be more than one perturbing
vibration, and the progressions will overlap.

So far we have neglected the role of spin-orbit coupling
which must be considered in order to make a quantitative analysis
of the electronic spectrum, particularly at low temperatures.
The spin-orbit coupling enters in two important ways, through the
splitting of the ligand field states and through the mixing of
states of different multiplicity. The various features mentioned
in the previous paragraphs then apply to the transitions to each
of the spin-orbit components and the overall band structure can
become too complicated for detailed analysis, unless the symmetry
of the system is high. For example, the effect of spin-orbit
coupling and a ligand field of symmetry D_{2h} completely removes
all of the degeneracy of the ^3T states of Ni^{2+} and each has nine
non-degenerate spin-orbit components.

The other important property of siin-orbit coupling is the
off-diagonal interaction which can mix states of different
multiplicity, i.e. S with $S' = S \pm 1$. It is through this interaction that single ion spin forbidden transitions can take place
because the selection rule $S = 0$ is a rigorous one. The
electronic transitions can be considered to involve those
components which have the same spin function, e.g. the transition
to ^1E_g of Ni^{2+} in an octahedral site occurs via the magnetic
dipole mechanism because of spin-orbit coupling with $^3T_{2g}$ and via
the electric dipole mechanism through spin-orbit mixing with $^3T_{1g}$
(and an ungerade vibration).

An instructive example of the difference between allowed and
forbidden electric dipole transitions in two loosely related
crystal systems is given by consideration of the absorption
spectra of K_2NiF_4 and $K_3Ni_2F_7$. Both crystals belong to the
space group D^{17}_{4h} and their structures are related to the cubic

perovskite fluoride $KNiF_3$ (see Figure 7).   The cubic components
of the crystal fields are virtually identical and the tetragonal
fields are relatively small, so that it is impossible to detect
any difference in colour between the two tetragonal crystals.
There is, however, a very significant difference between the site
symmetries of the Ni^{2+}.   In K_2NiF_4 it is D_{4h}, while in $K_3Ni_2F_7$
it is C_{4v}. Pure electronic transitions are allowed for the latter
but forbidden for the former.

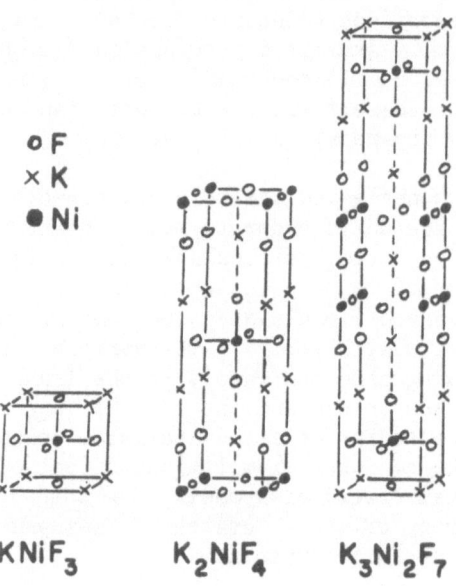

o F
× K
● Ni

KNiF$_3$ K$_2$NiF$_4$ K$_3$Ni$_2$F$_7$

Figure 7.   Unit cell structures of $KNiF_3$, K_2NiF_4 and $K_3Ni_2F_7$.

We need to know the correlations between the cubic and
tetragonal representations and these are given in Table 5, where
Γ is used to label the representations, indicating that the spin-
orbit states are under consideration.

For the cubic case the $^3T_{1g}$ state is split by spin-orbit
interaction into four states belonging to the representations
Γ_{1g}, Γ_{3g}, Γ_{4g} and Γ_{5g} while the ground state has symmetry Γ_{5g}.
The corresponding tetragonal symmetry representations for K_2NiF_4
and $K_3Ni_2F_7$ are obtained from Table 5.   As a result of the
tetragonal field and spin-orbit coupling, the $^3T_{1g}$ state splits
into seven states, two of which are double degenerate.

TABLE 5.   Correlations between cubic and tetragonal representations

O_h	D_{4h}
$\Gamma_{1g}(A_{1g})$	$\Gamma_{1g}(A_{1g})$
$\Gamma_{2g}(A_{2g})$	$\Gamma_{3g}(B_{1g})$
$\Gamma_{3g}(E_g)$	$\Gamma_{1g} + \Gamma_{3g}(A_{1g} + B_{1g})$
$\Gamma_{4g}(T_{1g})$	$\Gamma_{2g} + \Gamma_{5g}(A_{2g} + E_g)$
$\Gamma_{5g}(T_{2g})$	$\Gamma_{4g} + \Gamma_{5g}(B_{2g} + E_g)$

We consider $K_3Ni_2F_7$ first because electric dipole transitions are allowed.   If the ground state is Γ_4 then we find from Table 1 that transitions can occur to one of the spin-orbit components of $^3T_{1g}$ (Γ_4) for z (π) polarization and to two components (Γ_5, Γ_5) for x,y polarisation (δ,α).   If the ground state is Γ_5 then transitions can occur to two components (Γ_5, Γ_5) for z polarization and to the other five components (Γ_1, Γ_1, Γ_3, Γ_2, Γ_4) for x,y polarization.

The axial (α) spectrum and its MCD are shown in Figure 8. The appearance of at least four zero-phonon transitions, confirmed by their MCD, in the α spectrum indicates that the ground state of $K_3Ni_2F_7$ at low temperatures is Γ_5.   This is entirely in agreement with the effect of a relatively strong exchange interaction between nearest neighbour nickel ions.   This interaction, at temperatures much below the Neel temperature, can be considered as an effective internal magnetic field (the molecular field) which gives rise to a large Zeeman splitting of the $_5$ state.   There are then two magnetic sub-lattices which are degenerate: in one the spin moments lie along +z while they lie along −z in the other and the crystal is antiferromagnetic.

At the present stage an analysis of the vibrational structure is lacking, although it is clear that much of it involves totally symmetric phonons.   However, the intensity of the band system is not greater than the intensity of the spectrum of $KNiF_3$ by more than a factor of two or three, so that it should be possible to identify phonon structure which corresponds to the 'forbidden' component, i.e. the second term of equation (1).

The corresponding electronic transitions are forbidden in K_2NiF_4 and the absorption intensity must come from the vibronic mechanism, the second term of equation (1).   The δ and π polarized

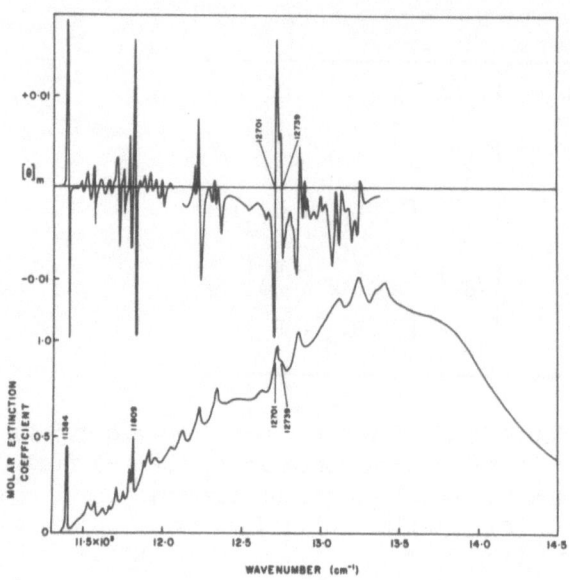

Figure 8. Absorption spectrum (lower, 10K) and MCD spectrum (upper, 8.5K) for the transition $^3A_{2g} \rightarrow {}^3T_{1g}^a$ of $K_3Ni_2F_7$. The units of $[\theta]_m$ are deg dl dm^{-1} M^{-1} G^{-1}.

spectra are shown in Figure 9, along with the MCD of the α spectrum and the absence of zero phonon lines is in agreement with this expectation. The vibrational structure is broad, in part due to the expected overlapping of many band systems associated with many false-origins.

A group theoretical analysis of the phonon modes at $\underline{\underline{k}} = 0$ shows that there are the following (optical) u vibrations: $4e_u$, $3a_{2u}$ and $1b_{2u}$. These are the tetragonal analogues of the $3t_{1u}$ and $1t_{2u}$ phonons of $KNiF_3$ at $\underline{\underline{k}} = 0$. By consideration of the symmetry requirements for the non-vanishing of the second terms in equation (1) we can easily see that e_u vibrations can induce π polarization for transitions from Γ_{5g} to Γ_{1g}, Γ_{2g}, Γ_{3g} and Γ_{4g} and δ polarization for transitions to Γ_{5g}. Similarly, a_{2u} and b_{2u} phonons will provide δ polarization for transitions to Γ_{1g}, Γ_{2g}, Γ_{3g} and Γ_{4g} and π polarization for transitions to Γ_{5g}.

Unlike $K_3Ni_2F_7$ therefore, the axial (or δ) spectrum should record transitions to all of the spin-orbit components of $^3T_{1g}$. Although a detailed analysis of the vibrational structure in Figure 9 has not been made yet, it is relatively easy to identify

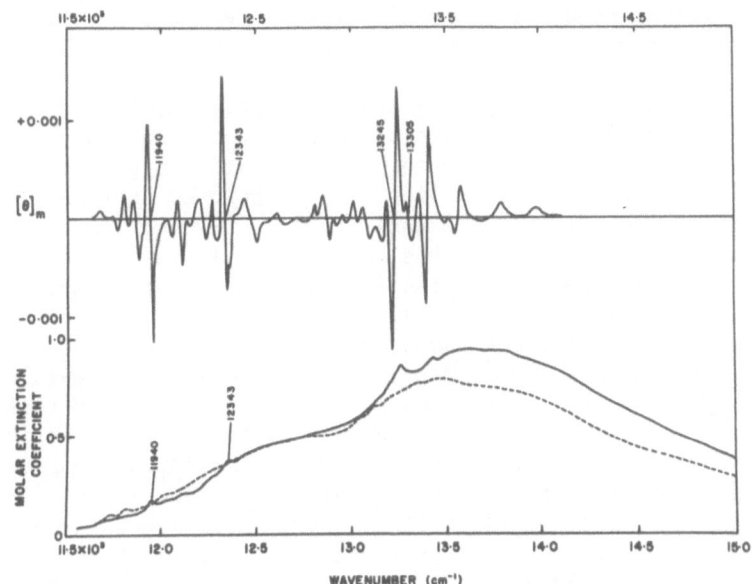

Figure 9.   Absorption spectra (---π, ___δ, 10K) and the MCD spectrum (upper, 8.1K) for the transition $^3A_{2g} \rightarrow \, ^3T_{1g}^{a}$ of K_2NiF_4. The units of $[\theta]_m$ are deg dℓ dm^{-1} M^{-1} G^{-1}.

false-origins due to e_u vibrations (π polarized) and a_{2u} (and perhaps b_{2u}) (δ polarized) in the region of the lowest energy spin-orbit component of the band system.

A quantitative description of the vibrational structure in the electronic transitions of ionic crystals is difficult because of the band structure associated with the phonon (as well as the exciton to some degree).   For a pure crystal the phonon frequency spectrum falls into alternately allowed and forbidden bands. The number of allowed bands is three times the number of atoms in the primitive cell and the number of frequencies in a band equals the number of cells in the crystal, which is a very large number. Each normal mode is characterized by a wavevector $\underline{k}$ and the energy is a function of $\underline{k}$.   The infrared and Raman spectra, involving single quanta of these phonons, correspond to $\underline{k} \sim 0$.   However, for electronic excitations involving vibrational quanta the corresponding selection rule is $\underline{K} = \underline{k} + \underline{k}' \sim 0$, where $\underline{k}'$ is the wavenumber of the exciton.   Normally we expect that the phonon band will show dispersion, i.e. variation of the energy with $\underline{k}$, to a much greater extent than the exciton and the energy corresponding to $\underline{K} \sim 0$ will be close to combined values of the

energies of the phonon and exciton near the Brillouin zone
boundary, where the density of phonon energy states is high.
The phonon energies which appear in the electronic absorption
spectrum will therefore in general, be different from those
fundamental frequencies observed in the infrared and Raman spectra.

For the case of an impurity in a host crystal, the description can become more complicated depending on the nature of the
impurity. It is possible for the impurity atom to participate
in vibrations whose frequency lies within the band corresponding
to the normal mode of the pure host crystal or which lies in one
of the forbidden bands of the host crystal. In the latter case
only the atoms of the crystal located near the impurity are
involved and the resultant vibration is known as a localized mode.
There are also intermediate cases in which it is useful to
consider 'pseudo-localized modes'. These problems are outside
the scope of the present work and, in fact, fall into the realm
of research topics in their own right. Of more concern to the
chemist is the crystal which is made up from molecular complex
ion units which have their own internal vibrational modes. In
such cases the coupling between the molecules is generally small,
so that the dispersion of the corresponding band in the phonon
spectrum is negligible and the vibrational energies can more
easily be assigned to vibrations of the metal ion and its ligands.
The oriented gas model is then a good approximation for considerations of both electronic and vibrational excitations.

III. ABSORPTION OF LIGHT BY EXCHANGE COUPLED PAIRS

Spin forbidden transitions occur through the effect of spin-orbit coupling and their intensities can be fairly accurately
estimated from consideration of the off-diagonal elements of the
spin-orbit interaction. This applies to excitation of a single
ion however, because there is another mechanism which allows a
pair of magnetic ions to absorb energy from an electromagnetic
radiation field. In this case, one or both ions can make an
electronic transition between states of the exchange coupled pair
through off-diagonal elements of the exchange interaction. The
selection rule is $\Delta S = 0$, where S is the total spin quantum state
of the pair, coupled by Heisenberg exchange described by the
Hamiltonian $J_{ab}S_a \cdot S_b$.

This mechanism was used first by Tanabe[10] to describe the
absorption of light by pairs of Mn^{2+} in $KZnF_3$ and it has since
been studied in more detail so that the features of the mechanism
are well established. The same mechanism is responsible for the
phenomenon of spin-wave (magnon) side-bands, which are observed
in the spectra of many antiferromagnetic crystals. The mathematical description of the co-operative (exciton + magnon)

transition is complicated by the translational symmetry of the crystal and a quantitative account will not be given here. The essential details can be more easily understood from a discussion of the pair spectra.

In order to keep the account as simple as possible we limit the treatment to transitions between states which correspond to spin flips. The description can easily be enlarged to include transitions which involve orbital changes, as given by Gondaira and Tanabe.[11]

The interaction between the pair and the radiation field can be described by the following perturbation term

$$H' = \sum (\underline{P}_{a_i b_j} \cdot \underline{E}) \; (\underline{s}_{a_i} \cdot \underline{s}_{b_j})$$

This term is too general though for specific discussion of pair spectra and we need to specify the form of $\underline{P}_{a_i b_j}$. Tanabe[10] has shown that there are three terms which contribute to $\underline{P}_{a_i b_j}$

$(\underline{P}_{a_i b_j} = \underline{\pi}^1_{a_i b_j} + \underline{\pi}^2_{a_i b_j} + \underline{\pi}^3_{a_i b_j})$. These are given by

$$\pi^1_{a_i b_j} = - 4\langle a_i |\underline{P}| b_j\rangle h(a_i \leftarrow b_j)/\Delta E) a_i \leftarrow b_j)$$

$$\pi^2_{a_i b_j} = -\sum_\mu 2\langle \mu |\underline{P}| a_i\rangle\langle a_i b_j | b_j \mu)(2-n_\mu)\Delta E(\mu \leftarrow a_i)$$

$$\pi^3_{a_i b_j} = \sum_\nu 2\langle a_i |\underline{P}| \nu\rangle\langle \nu b_j | b_j a_i\rangle n_\nu /\Delta E(a_i \leftarrow \nu)$$

+ expressions for each obtained by interchanging a_i and b_j

Here p is the dipole moment operator, $n_{\mu,\nu}$ the number of electrons in the orbitals μ and ν, $\Delta E(a_i \leftarrow b_j)$ is the energy for the transfer of an electron from b_j to a_i, $h(a_i \leftarrow b_j)$ is the transfer integral and

$$\langle a_i b_j | b_j \mu\rangle = \int a_i(1) b_j(2) \frac{e^2}{r_{12}} b_j(1) \mu(2) \, d\tau_1 \, d\tau_2$$

The summations over μ and ν are carried out over all empty, singly occupied and filled orbitals except for a_i and b_j.

We now consider the application of this mechanism to the
case of a pair of like ions with the highest symmetry. The metal
ions are then located in sites of O_h symmetry, sharing a common
ligand in a linear arrangement. We take the z axis of the coordinate system along the pair axis, and the pair has D_{4h} symmetry.
If g and e denote the ground and excited electronic states of each
ion, then the interaction matrix element is

$$\langle \psi_a^g \, \psi_b^g \mid H' \mid \psi_a^e \, \psi_b^g \rangle$$

for the case in which the electronic excitation occurs in atom a.

From the form of H' we note that the operator $\underset{=a_i}{s} \cdot \underset{=b_j}{s}$

provides the selection rule $\Delta S = 0$. Next we examine the terms
which contribute to $P_{a_i b_j}$. We note that the covering symmetry

D_{4h} applies to $\pi_{a_i b_j}^1$ and also to $\pi_{a_i b_j}^{2,3}$ if μ and ν are metal ion

orbitals so that all three terms vanish. If however, μ and ν
are ligand orbitals the covering point group symmetry is C_{4v} and
the terms $\pi_{a_i b_j}^{2,3}$ do not necessarily vanish. In these cases the

off-diagonal exchange interaction allows intensity to be stolen
from metal to ligand ($\pi_{a_i b_j}^2$) and ligand to metal ($\pi_{a_i b_j}^3$) electron

transfer transitions. The selection rules can be derived using
Table 1 and we use the manganese pair as an example to illustrate
them. In order for $P_{a_i b_j}$ to be non-zero the direct product of

the characters of the representations of ψ_a^g and ψ_a^e must contain
A_1 or E for the point group C_{4v}. The first corresponds to z
polarization and the second to x,y polarization. The ground
state of Mn^{2+} is $^6A_{1g}$ so that z polarized pair transitions can
occur to $^4A_{1g}$ and ^4F_g (see Table 5 for O_h to C_{4v} correlations),
while x,y polarized pair transitions can occur to $^4T_{1g}$ and $^4T_{2g}$
states (in both cases to the ^4E components in C_{4v}).

Next we consider the case of a pair of unlike ions arranged
in a similar way. The symmetry which covers all three terms is
now C_{4v} and, most important, intensity can be stolen from the metal
to metal electron transfer transitions. It is expected that
these contributions will, in most cases, dominate the intensity
stealing mechanism and it is therefore not surprising that for the
case of the $Mn^{2+} - F - Ni^{2+}$ pair in $KZnF_3$, the intensity of the
transition associated with the excitation of the Mn^{2+} to ^4E_g, $^4A_{1g}$
is more intense ($f \sim 10^{-5}$) than the corresponding transition in the
$Mn^{2+} - F - Mn^{2+}$ pair by two orders of magnitude.

The same selection rules apply as for the like pair, so that the excitation of the Mn^{2+} to $^4A_{1g}$ or ^4E_g should have z polarization. Let us now examine the spectrum of the manganese nickel pair in the crystal $K_3Zn_2F_7$ (see Figure 7 for the crystal structure). In this structure there can be two kinds of pairs. One in which both ions are in the same layer and the other in which each ion is in a different layer. The first pair has C_{2v} symmetry while the latter has C_{4v} symmetry. We note that statistically, there should be more pairs of the intra-layer type than of the inter-layer type, by approximately 4:1. In both cases, the pair excitation involving transition of the Mn^{2+} to ^4E_g and $^4A_{1g}$ should be polarized along the pair axis, i.e. δ polarization for the intra-layer pair and π polarization for the inter-layer pair.

The energy levels of the pair are shown in Figure 10 and observed spectra are given in Figure 11 and we see that there are two major lines in polarization, corresponding to excitation of the Mn^{2+} to ^4E_g and $^4A_{1g}$ and two lines in the spectrum, which have corresponding assignments. Each of these four lines arises from the S = 3/2 level of the ground state and they each decrease in intensity on raising the temperature as the S = 5/2 and S = 7/2 levels become populated. At the same time there appear four lines which involve the S== 5/2 levels, two with δ polarization and two with π polarization.

If we examine other spectral regions, correspondingly to the electronic excitation of the manganese, we find that only one other region shows a pronounced intensity enhancement. This is the region near 330 nm in which the Mn^{2+} is excited to $^4E_g^b$. The selection rules indicate, however, that $P_{\underset{a_ib_j}{=}}$ is non-zero for

transitions involving the $^4T_{1g}$ and $^4T_{2g}$ states of the Mn^{2+}, but no absorption due to the manganese nickel pair can be observed in these regions. This is a good example of the fact that group theory only provides us with information about non-zero transition moments and not magnitudes. However, if we examine the form of $\pi^1_{\underset{a_ib_j}{}}$ given above, along with the analogous term when there is an

orbital change, given by Gondaira and Tanabe,[11] which is

$$-2 \sum_{j \neq p, p'} \left\{ \langle \phi_{p'} | \underline{\underline{P}} | a_j \rangle h(a_j \leftarrow \phi_p)/\Delta E(j \leftarrow p) + h(\phi_{p'} \leftarrow a_j) \right.$$

$$\left. \langle a_j | \underline{\underline{P}} | \phi_p \rangle /\Delta E(j \leftarrow p) \right\} \underline{\underline{S}}_{p' \leftarrow p} \cdot \underline{\underline{S}}_j$$

we can see why this is so. To begin with, for $h(a_j \leftarrow \phi_p)$ to be non-zero a_j and ϕ_p must have the same symmetry. In the first term of this expression we can see that if a_j and ϕ_p are d_{z^2}

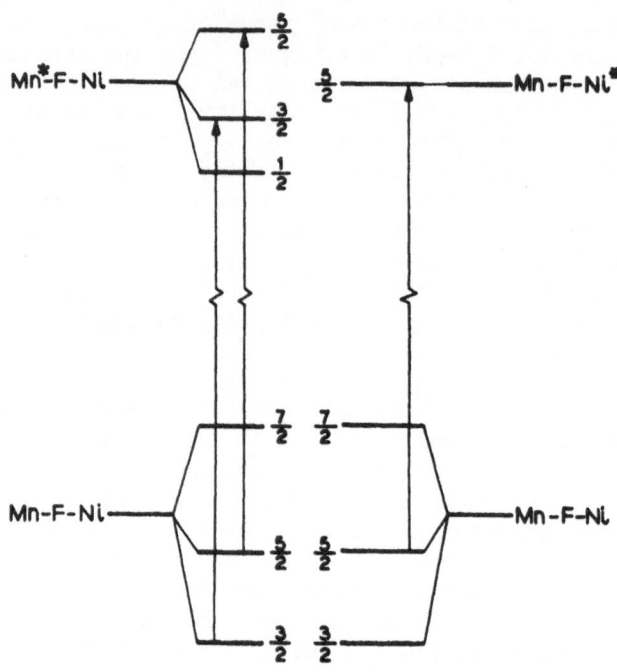

Figure 10.   Energy levels of a manganese-nickel pair.   Allowed
electronic transitions are indicated by vertical arrows.

orbitals (a_1 symmetry) than ϕ_p, must have e symmetry, i.e. the
d_{xz}, d_{yz} pair and the transition moment matrix element $\langle \phi_p{}' | \underline{p} | a_j \rangle$
will be very small because of the poor overlap.   Similar
arguments apply to other choices for a_j and p and also for the
second term in the expression.   On the other hand, for the $^4E_g^{a,b}$
and $^4A_{1g}$ states, which involve spin flips, the overlap conditions
are large so that $\pi^1_{a_i b_j}$ will have a much higher value than is the
case for the transitions to the ^4T_g states.

In addition to the electronic excitation of one ion of a pair,
it is possible to have a simultaneous electronic excitation of the
ions.[12]   In this case the interaction matrix element is

$$\langle \psi_a^g \psi_b^g | H' | \psi_a^e \psi_b^e \rangle$$

The selection rule $\Delta S = 0$ still applies and for $\underline{P}_{a_i b_j}$ to be non-

zero the direct product of the characters of the representations
of all four electronic wavefunctions must contain A_1 (z polarization) or E (xy polarization) for both like and unlike pairs.

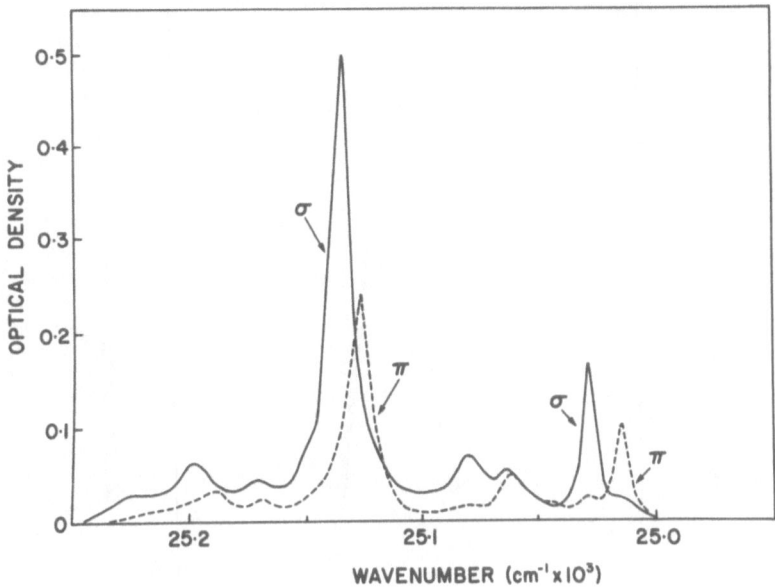

Figure 11.   Absorption spectrum of $K_3Zn_2F_7$ containing Ni and Mn for σ and π polarization for the region of excitation of the Mn^{2+} to $^4A_{1g}$ and $^4E_g^a$.

In addition for like ions the matrix element will be zero if the excited states are the same in each ion.   Here again the $\underline{\underline{P}}$ operator can be expressed by a number of $\underline{\pi}$ terms, analogous to the case of the single electronic excitation and Fujiwara[13] has given a detailed treatment for the manganese pair.   For the manganese-nickel pair we may note that the $\underline{\underline{\pi}}$ terms which involve the transfer integral will contribute strongly only for those excitations which involve spin–flips.   For this reason there are only two intense regions of absorptions in the ultraviolet involving simultaneous electronic excitation, one associated with the states $^4E_u^a\ ^1E_v$ and $^4A_1\ ^1E_v$ and the other with $^4E_u^b\ ^1E_v$.   In each case the transition should be polarized along the pair axis. These two spectral regions are shown in Figures 12 and 13 for the manganese nickel pair in $K_3Zn_2F_7$.   The temperature dependences of the σ and π lines are different indicating that the two pairs have different exchange parameters in the ground state.   Analyses of these dependences show that the intra-layer pair has an exchange energy parameter $J = 22\ cm^{-1}$, while the inter-layer pair has a value $J = 27\ cm^{-1}$.   For the cubic lattice $KZnF_3$ $J = 18\ cm^{-1}$.[12]

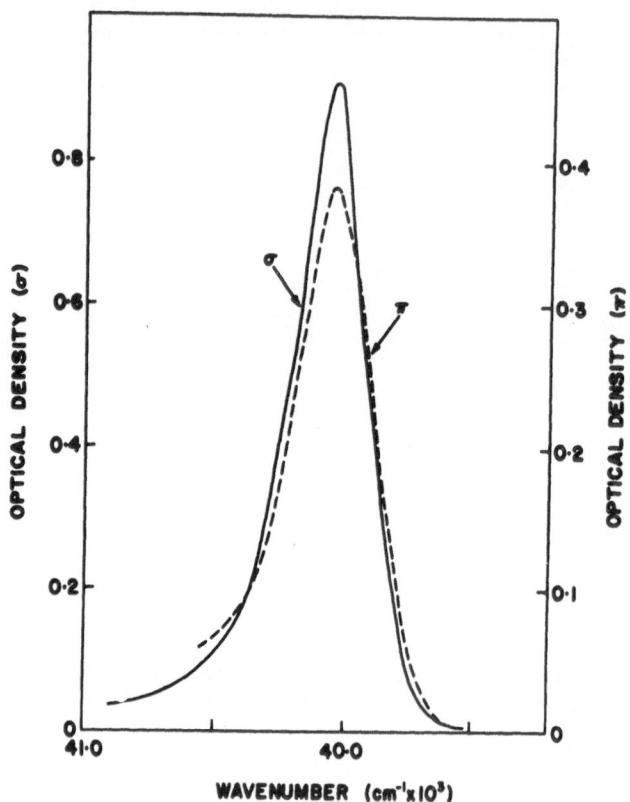

Figure 12.   Absorption spectrum of $K_3Zn_2F_7$ containing Ni and Mn for σ and π polarization in the region of the double electronic excitation of Mn^{2+} to $^4A_{1g}$, $^4E_g^a$ and Ni^{2+} to ^1E_g.

Finally, we note that the exchange induced mechanism does not replace the single ion mechanism and absorption by light can occur by either mechanism.   It is only the special situation existing for the d^5 configuration, for which both Laporte and spin prohibitions apply, which enables us to see the exchange induced mechanism so clearly because of the inherent weakness of the single ion mechanism.   The most interesting example of single ion and pair mechanisms is for the case of the transition $^3A_{2g} \rightarrow {}^1E_g$ in $KNiF_3$, K_2NiF_4 and $K_3Ni_2F_7$.   In these cases the overall absorption band has contributions arising from

(a) the magnetic dipole single ion mechanism via spin-orbit coupling through to $^3T_{2g}$

(b) the electric dipole single ion mechanism via spin-orbit

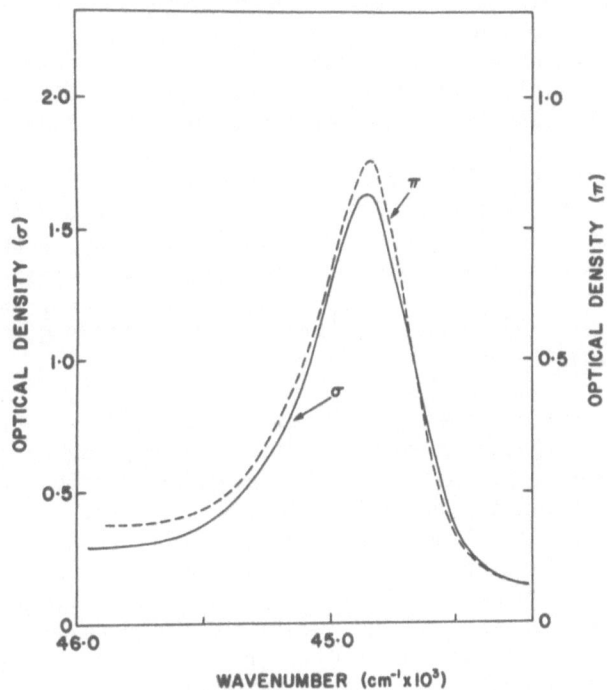

Figure 13.   Absorption spectrum of $K_3Zn_2F_7$ containing Ni and Mn for σ and π polarization in the region of the double excitation of Mn^{2+} to $^4E^b_g$ and Ni^+ to ^1E_g.

coupling through to $^3T_{1g}^a$ and

(c) the exchange induced electric dipole pair mechanism.

IV.   NOTES ON TECHNIQUES

In order to measure linear dichroism it is of course necessary to have at hand single crystals and good polarizers. The simplest type of polarizer to use is Polaroid sheet which is available for near ultraviolet, visible and near infrared regions. However, more useful is a crystal birefringent polarizer because it can cover a wider wavelength range.   Two types are commonly used: a Glan-Taylor calcite polarizer, which is a modification of the Glan-Foucault design, and a Rochon polarizer.

Glen-Taylor prisms can be obtained in matched pairs which are ideal for doublebeam absorption spectroscopy in a Cary 14 or 17. Only one beam is transmitted and if the polarizers are placed in the wall of the sample and reference compartments of the Cary 14 or 17, very good polarization behaviour and good baselines can be obtained to wavelengths as low as 215 nm for polarizers of small dimension. These polarizers can be used throughout the visible, ultraviolet and near infrared regions.

The Rochon polarizer has the disadvantage of transmitting two beams so the optics have to be arranged to remove the deviated beam at some point. This makes the polarizers unsuitable for double beam spectroscopy in most cases, but the prism is useful as a polarizer in modulation linear dichroism, especially for operation in the short wavelength ultraviolet region where MgF_2 is a suitable material.

The modulation linear dichroism technique[14] is very convenient for measuring weak dichroic spectra because of the very high sensitivity of the detecting system for measuring the difference between two linearly polarized signals. The arrangement can be either a linear polarizer, modulator and a fixed quarter wave polarizer (e.g. a Fresnel rhomb) or a linear polarizer and modulator. In the first case the signal appears with the same frequency as the modulation frequency while in the seocnd it has twice the frequency. The technique is particularly good for cases in which the two polarized spectra are broad and the dichroism is weak. As an example the polarized absorption spectrum of MnF_2 for the $^6A_{1g} \rightarrow {}^4T_{1g}^a$ band is shown in Figure 14, along with the (modulation) linear dichroism.

In order to obtain the most useful data it is necessary to cool the crystal sample to temperatures near to liquid helium temperature. Traditionally this has involved the use of large metal, glass or silica dewars with their accompanying pumping systems, and the need for careful pre-cooling of the cryostats with liquid nitrogen before transfer of the liquid helium. Also, often there is the need to measure the dependence of absorption (or luminescence) on temperature above liquid helium temperature. This requires a relatively elaborate exchange gas system to be added to the oryostat. Finally, changing of samples once the cryostat has been filled with liquid helium is usually not possible, or if it is care must be taken to ensure that no air is admitted during the change.

A much simpler technique for low (variable) temperature studies exists in the form of a gas flow tube which can be made from silica in a number of shapes and sizes. The flow tube is a double wall (evacuable) tube with a long narrow cross-section tail which is inserted into a liquid helium or nitrogen storage

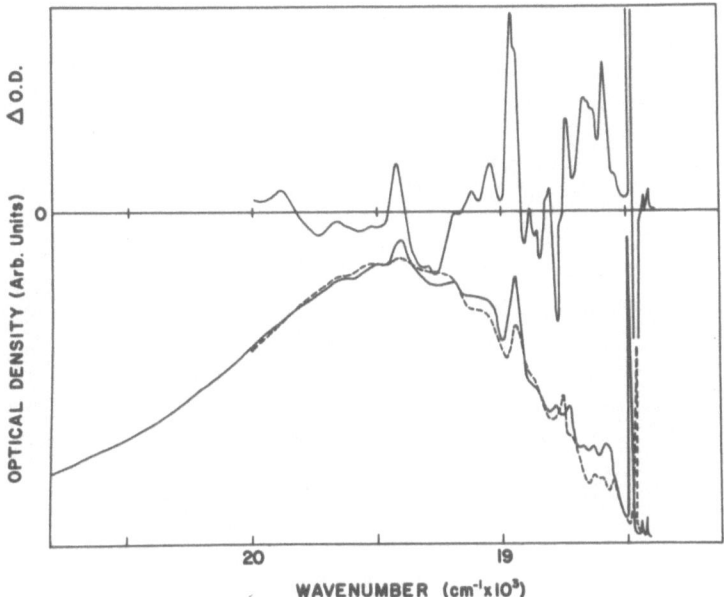

Figure 14. Absorption spectrum (lower) of MnF_2 for the transition $^6A_{1g}$ $^4T_{1g}^a$ ($---\pi$, ___ ℓ) and linear dichroism (upper) measured at 10K.

vessel (Figure 15). A small resistor (wire wound) is placed in the liquid prior to insertion of the tube so that application of a controllable amount of current enables cold boil-off gas from the liquid to flow up the tube and past a crystal sample, mounted in such a way that its absorption or emission spectrum can be measured. Using this technique it is easy to maintain the sample temperature in the range 6-300K, and as the crystal is in the flowing cold gas, the temperature responds very quickly to a change in the boil-off rate so that temperature dependent measurements are easy to carry out. The important advantage of this technique is the simplicity of sample change, which can be carried out in a few minutes. Finally, this device leads to a very efficient use of liquid helium.

It is usual that crystals, which are of interest to the inorganic chemist, are very small, perhaps having dimensions of one or two nm. It is relatively easy to measure absorption spectra of such crystals in polarized light with a double beam instrument such as the Cary 14 or 17, provided the photomultiplier tube is replaced by a specially selected tube which has very high

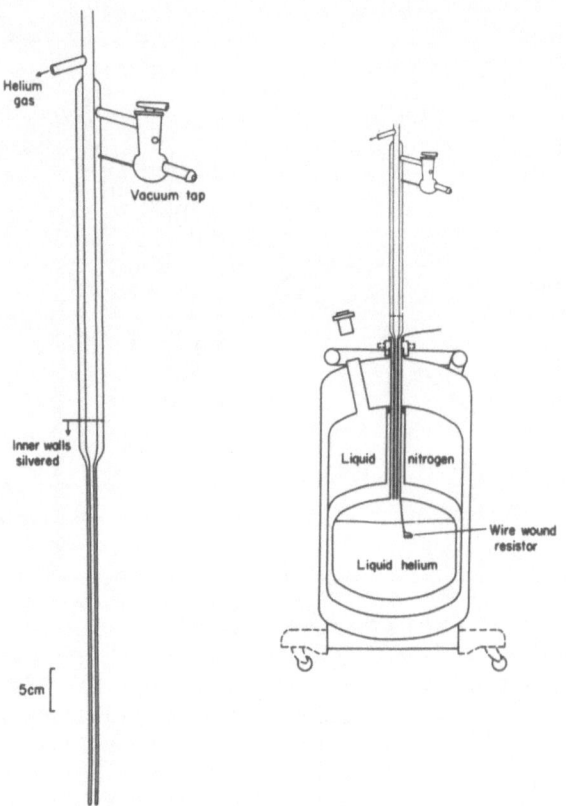

Figure 15.   Typical silica flow tube.

gain and the most suitable spectral response.   It would be
necessary to stop down the reference beam with a variable
attenuator and the principal advantage of the Cary 14 or 17 is
that it has very low stray light levels because of the double
monochromator construction.   If the crystal is mounted in a gas
flow tube then the temperature dependence of its polarized
absorption spectrum can be measured easily.

Measurements of absorption spectra of single crystals with
dimensions as small as 0.1 nm can be made by double beam spectroscopy in a Cary 14 or 17 by using condensing optics in the form of
a pair of reflecting objectives.[15]   The arrangement is shown in
Figure 16.   Objectives of power x 15 allow sufficient working
space for insertion of a flow tube so that temperature dependent
studies can be made.   For such measurements the location of the
sample in the beam is critical, particularly in the horizontal

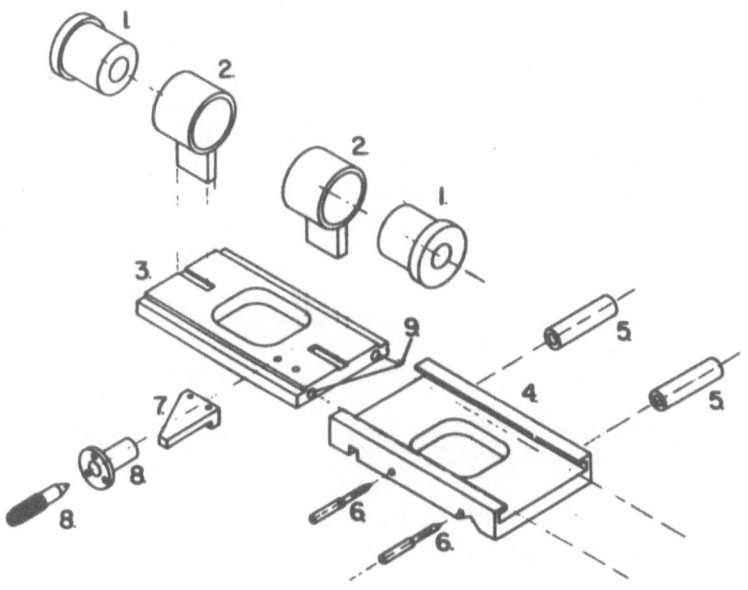

Figure 16.   Exploded diagram showing the construction of the
microscope attachment.   1 - reflecting objective;   2 - objective
mount;   3 - slide;   4 - base;   5 - spacers;   6 - locking screws;
7 - adjusting plate;   8 - adjusting screw and bush;   and
9 - slide spring holes.

plane.   This can be most easily taken care of by mounting the
storage vessel on a platform which has a translation stage.   The
objectives are achromatic and the slight convergency of the light
from the x 15 objectives is not a serious problem so that very
high polarization ratios can be measured.

In addition to studies of dichroic spectra by direct
absorption measurements, it is possible to use two other techniques
to considerable advantage.   One is the measurement of luminescence
spectra and the other is the measurement of luminescence
excitation spectroscopy.

The measurement of luminescence spectra of exchange coupled
ions is a very valuable probe for obtaining information about the
levels of the ground state of the system.   Normally these levels
are obtained from analysis of the temperature dependence of
magnetic susceptibility data.   However, the levels of the ground
state can be obtained more directly from a combination of absorp-

tion and luminescence measurements in some cases.   Crystals
containing exchange coupled Cr^{3+} ions have been studied by these
methods.[16]   If the crystals are small then the ideal photomultiplier to use is one with a very small effective photocathode
area, such as the ITT FW 130.

Luminescence excitation spectroscopy can be applied to any
system which luminesces and the method can be made into a very
powerful tool for the investigation of weakly absorbing ions such
as Mn^{2+} [17] and Fe^{3+}.   For maximum sensitivity and resolution the
light incident on the crystal should be dispersed by a double
monochromator.   The technique is particularly valuable for the
study of crystals which contain non-equivalent sets of complex
ions.[16]

REFERENCES

1.   C.E. Moore, 'Atomic Energy Levels', National Bureau of
     Standards Circular 467 (1952).

2.   J. Ferguson, 'Progress in Inorganic Chemistry', edited by
     S.J. Lippard, 12, 159 (1970).

3.   For a simple description of the propagation of light in
     biaxial crystals, see C. Bunn, 'Crystals: Their Role in
     Nature and Science', Academic Press, N.Y. (1964).

4.   J.S. Griffith, 'The Theory of Transition-Metal Ions',
     Cambridge University Press (1961).

5.   C.J. Ballhausen, 'Introduction to Ligand Field Theory',
     McGraw-Hill, New York (1961).

6.   G. Herzberg and E. Teller, Z. physik. Chem. B21, 410 (1933).

7.   A.C. Albrecht, J. Chem. Phys. 33, 156 (1960).

8.   J. Ferguson, H.J. Guggenheim and D.L. Wood, J. Chem. Phys.
     53, 1613 (1970).

9.   J. Ferguson and T.E. Wood, Inorg. Chem. in press (1975).

10. J. Ferguson, H.J. Guggenheim and Y. Tanabe, J. Phys. Soc.
     Japan, 21, 692 (1966).

11. K. Gondaira and Y. Tanabe, J. Phys. Soc. Japan, 21, 1527
     (1966).

12. J. Ferguson, H.J. Guggenheim and Y. Tanabe, Phys. Rev. 161,
     207 (1967).

13.   T. Fujiwara, J. Phys. Soc. Japan, 34, 1180 (1973).

14.   R. Gale, A.J. McCaffery and R. Shatwell, Chem. Phys. Lett.
      17, 116 (1972).

15.   J. Ferguson and W. Orr, Rev. Sci. Instrum. 44, 225 (1973).

16.   J. Ferguson and H.U. Gudel, Aust. J. Chem. 26, 505 (1973).

      J. Ferguson, H.U. Gudel and M. Puza, Aust. J. Chem. 26,
      513 (1973).

      J. Ferguson and H.U. Gudel, Chem. Phys. Lett. 17, 547
      (1972).

17.   J. Ferguson, H.U. Gudel, E.R. Krausz and H.J. Guggenheim,
      Mol. Phys. 28, 893 (1974).

VIBRATIONAL-ELECTRONIC INTERACTIONS

P.J. Stephens

Department of Chemistry, University of Southern
California, Los Angeles, U.S.A.

If nuclei were infinitely heavy and did not move, molecules
would have a rigid geometry and their electronic spectra would
consist of one sharp line per transition.   In fact, the
electronic spectrum arising from a transition generally consists
of a band of transitions, somewhere between discrete and
continuous.   This originates in the nuclear motion: as the
nuclei move the electronic state energies vary, describing the
electronic potential surfaces; the nuclear motion on these
surfaces is quantized; transitions from one particular ground
vibrational state can take place to a range of excited vibrational
states - thereby spreading out the electronic transition.

A few examples will illustrate the range of band structure
typically found in transition-metal spectroscopy.   The $^4A_2 \rightarrow {}^4T_1$
(F) $d \rightarrow d$ transition of $CoCl_4^{2-}$ (in Cs_3CoCl_5)[1] and the $^1A_1 \rightarrow {}^1T_2$
charge transfer spectrum of MnO_4^- (in $KClO_4$)[2] are shown in
Figures 1 and 2.   Quite sharp vibrational structure is seen.
In the case of systems without discrete molecular ions, structure
is usually more continuous, as in the d d spectra of Ni^{2+} in
MgO[3] (Figure 3).   In many cases, whether molecular or not,
spectra are broad but unstructured - see, for example, the
highest MnO_4^- charge-transfer band shown in Figure 2.

The purpose of examining vibrational-electronic interactions,
and their consequences for electronic spectra is twofold.   The
first is to extract the electronic state information which would
be obtained directly if nuclear motion were absent.   The second
is to elucidate the nature of electronic potential surfaces.
These goals should be kept in mind in the following broad overview
of the theoretical methods and problems encountered in treating
vibrational-electronic interactions.

P. Day (ed.), Electronic States of Inorganic Compounds. 95–112. All Rights Reserved.
Copyright © 1975 by D. Reidel Publishing Company, Dordrecht-Holland.

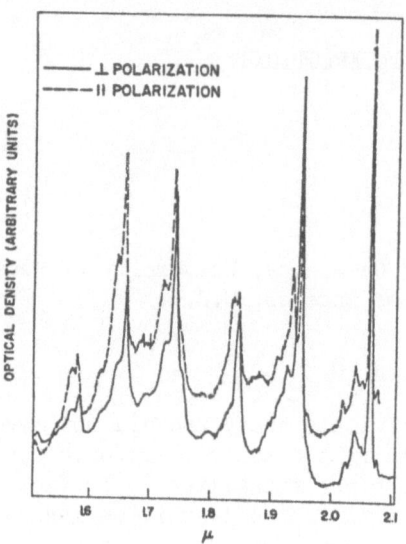

Figure 1. Polarized single-crystal spectrum of Cs_3CoCl_5 in $^4T_1(^4F)$ region at 4.2°K. Mean spectral slitwidth 5 cm⁻¹.

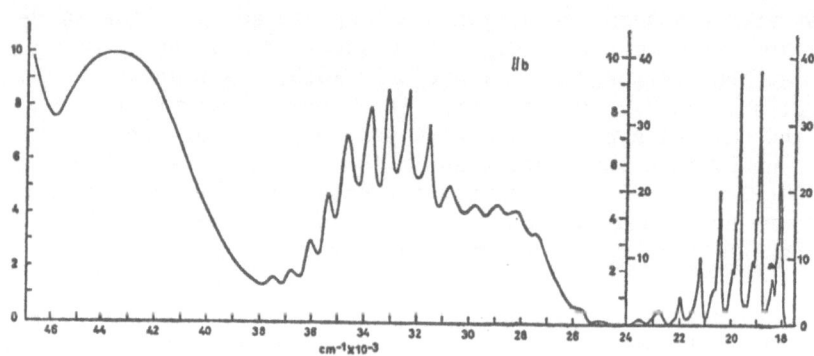

Figure 2. Absorption spectrum of $KMnO_4$ dissolved in $KClO_4$ at liquid helium temperature. The electric vector is parallel to b. Relative extinction coefficient as ordinate.

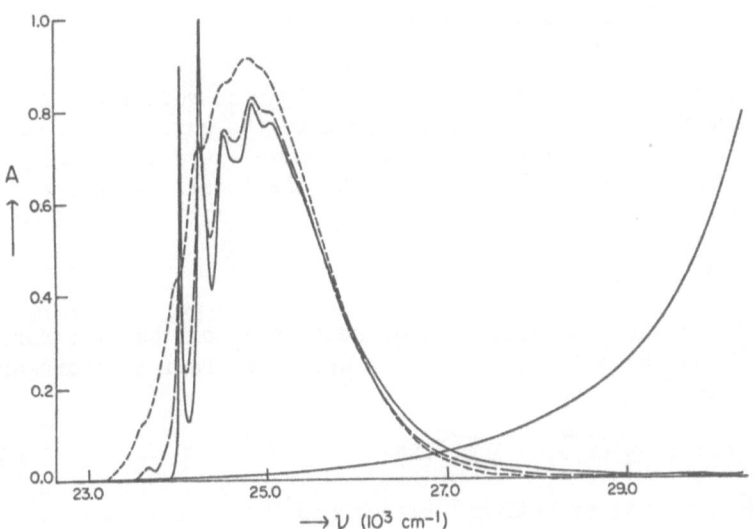

Figure 3. Absorption spectrum of $^3T_1^b$ band of MgO: Ni; solid line, 8°K; long-dashed line, 121°K; short-dashed line, 208°K. Curved baselines have been subtracted; that for the 8°K spectrum is shown.

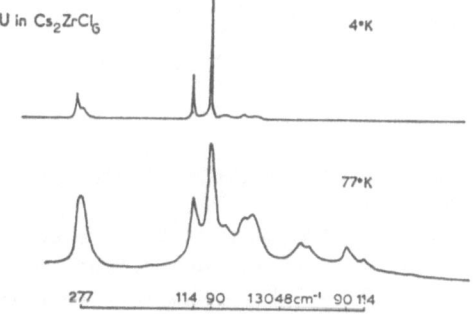

Figure 4. The vibronic spectrum of U^{4+} in Cs_2ZrCl_6 associated with the electronic level at 13,048 cm^{-1}.

1. PROBLEMS TO BE SOLVED

Let us write the Hamiltonian for our system

$$\mathcal{H}(\vec{r},\vec{R}) = \mathcal{H}_{el}(\vec{r},\vec{R}) + T_n(\vec{R}) \tag{1}$$

where $\vec{r}$ and $\vec{R}$ denote electronic and nuclear coordinates respectively, $\mathcal{H}_{el}$ is the Hamiltonian with the nuclei clamped to the geometry R - the electronic Hamiltonian - and T_n is the nuclear kinetic energy operator. The electronic Schrodinger equation is

$$\mathcal{H}_{el} \psi_K(\vec{r},\vec{R}) = W_K(\vec{R}) \psi_K(\vec{r},\vec{R}) \tag{2}$$

and produces the electronic wave-functions ψ_K of the electronic states K and the potential surfaces W_K. The full Schrodinger equation is

$$(\mathcal{H}_{el} + T_n)\Psi(\vec{r},\vec{R}) = E\Psi(\vec{r},\vec{R}) \tag{3}$$

where Ψ are the vibrational-electronic (vibronic) states. In discussing vibrational-electronic interactions we must solve equations 2 and 3.

2. THE BORN-OPPENHEIMER APPROXIMATION

The solutions to equation (3) within the Born-Oppenheimer (BO) approximation are written

$$\Psi_k = \psi_K \chi_{Kk} \tag{4}$$

where the vibrational function χ_{Kk} is obtained by solving

$$(W_K + T_n) \chi_{Kk} = E_k \chi_{Kk} \tag{5}$$

Equations (4) and (5) say that the nuclei vibrate on the potential surface of one electronic state (K). The BO approximation is a good one if state K is well separated from other electronic states and, if it is degenerate, if its degeneracy is maintained as the nuclei move.[4] If the latter is not true we have a Jahn-Teller effect - to be discussed later.

3. ALLOWED ELECTRONIC TRANSITIONS

Consider now an electronic transition between two electronic states for which the BO approximation works. The transition dipole moment for a single vibronic transition Gg→Ee is

$$\langle Gg | \vec{\mu} | Ee \rangle = \langle g | \langle G | \vec{\mu} | E \rangle | e \rangle \tag{6}$$

where $\vec{\mu}$ is the electronic electric dipole moment operator and

$$\langle G | \vec{\mu} | E \rangle \equiv \int \psi_G(\vec{r}, \vec{R}) \vec{\mu}(\vec{r}) \psi_E(\vec{r}, \vec{R}) d\vec{r} \tag{7}$$

is the electronic transition moment – a function of $\vec{R}$.
Assuming the nuclei vibrate with small amplitude $\langle G | \vec{\mu} | E \rangle$ can
be expanded in a Taylor series about the ground state equilibrium
geometry, say $\vec{R}_o$

$$\langle G | \vec{\mu} | E \rangle = \langle G | \vec{\mu} | E \rangle^o$$

$$+ \sum_i \left[\partial \langle G | \vec{\mu} | E \rangle / \partial q_i \right]_o q_i + \cdots \tag{8}$$

where q_i are suitable nuclear displacement coordinates.   In the
Franck-Condon (FC) approximation only the first term of
equation (8) is retained, when

$$\langle Gg | \vec{\mu} | Ee \rangle = \langle G | \vec{\mu} | E \rangle^o \langle g | e \rangle \tag{9}$$

$$|\langle Gg | \vec{\mu} | Ee \rangle|^2 = |\langle G | \vec{\mu} | E \rangle^o|^2 |\langle g | e \rangle|^2$$

The transition probability is then determined by (i) the $\vec{R}_o$
electronic transition moment $\langle G | \vec{\mu} | E \rangle^o$ and (ii) the overlap
integral between the ground and excited vibrational states $\langle g | e \rangle$.
The relative intensities of different vibrational transitions
are proportional to $|\langle g | e \rangle|^2$ – called the Franck-Condon factor.

The total intensity of the band is proportional to

$$\sum_{g,e} \alpha_g |\langle Gg | \vec{\mu} | Ee \rangle|^2 \tag{10}$$

where α_g is the fraction in ground vibrational state g ($\sum_g \alpha_g = 1$).
Within the FC approximation equation (10) becomes

$$\sum_g \alpha_g |\langle G | \vec{\mu} | E \rangle^o|^2 \sum_e |\langle g | e \rangle|^2 \tag{11}$$

$$= |\langle G | \vec{\mu} | E \rangle^o|^2$$

and the total band intensity is independent of ground or excited
state vibrational functions – and hence potential surfaces – and
depends only on the $\vec{R}_o$ transition moment.

4. EXCITED STATE STRUCTURE DETERMINATION

To predict the vibrational structure of an allowed electronic transition one must calculate the Franck-Condon factors. This in turn requires the ground and excited potential surfaces and vibrational functions. Conversely, these can be derived by analysis of an observed spectrum. The complexity of this process is in proportion to the extent of the vibrational function calculations. For illustration, we refer to a simple example, namely the $^1A_1 \rightarrow {}^1T_2(1)$ MnO_4^- transition, analysed by Ballhausen.[5] As shown in Figure 2, the main vibrational structure in this band is a progression in the 768 cm^{-1} a_1 vibrational mode, the symmetric stretching mode of the tetrahedron. If only this one coordinate is considered and it is assumed that its frequency (force constant) is the same in the ground and excited states, the Franck-Condon factors relate simply to the change in Mn-O bond length r. Ballhausen showed that the observed intensity pattern was explained quite well with $\Delta r = + 0.1A$ – a change from 1.6 to 1.7 A.

5. FORBIDDEN ELECTRONIC TRANSITIONS

If the transition moment $\langle G | \mu | E \rangle^0$ is zero – as, for example, for a d→d transition in O_h symmetry – equation (9) leads to the conclusion that all transitions are forbidden. In practice, this is not the case, owing to the inexactness of the FC approximation; that is, the second term of equation (8) is not exactly zero and leads to non-zero transition moments. Suppose, for simplicity, that $\langle G | \vec{\mu} | E \rangle$ is a function of only one vibrational coordinate q

$$\langle G | \vec{\mu} | E \rangle = \left[\partial \langle G | \vec{\mu} | E \rangle / \partial q \right]_0 q \qquad (12)$$

$$\equiv \langle G | \vec{\mu} | E \rangle' \, q$$

then,

$$| \langle Gg | \vec{\mu} | Ee \rangle |^2 = | \langle G | \vec{\mu} | E \rangle' |^2 \qquad (13)$$

$$\times | \langle g | q | e \rangle |^2$$

and the relative intensities of the vibrational components g→e are proportional to $| \langle g | q | e \rangle |^2$. The vibrational selection rules will thus be different from those for an allowed transition. For example, in the simplest case where ground and excited potential surfaces are identical and harmonic

$$W_G = W_G^o + \frac{1}{2} \sum_i k_i^G g_i^2 \tag{14}$$

$$W_E = W_E^o + \frac{1}{2} \sum_i k_i^G g_i^2$$

we have

$$\chi_g = \prod_i H_{n_i}^g (q_i) \tag{15}$$

$$\chi_e = \prod_i H_{n_i}^e (g_i)$$

where H_{n_i} are one-dimensional harmonic oscillator functions, and

$$\langle g | q_j | e \rangle = \left\{ \prod_{i \neq j} \langle H_{n_i}^g | H_{n_i}^e \rangle \right\} \langle H_{n_j}^g | q_j | H_{n_j}^e \rangle \tag{16}$$

$$= \left\{ \prod_{i \neq j} \delta_{n_i^g n_i^e} \right\} c \, \delta_{n_j^e, \, n_j^g \pm 1}$$

Thus, with one 'allowing vibration' q_j, from the lowest vibrational level g ($n_i^g = 0$) transitions are allowed to only one excited vibrational level, that with $n_j^e = 1$, $n_i^e = 0$ ($i \neq j$). Hot bands with $\Delta n_j = -1$ are also allowed. This contrasts with the selection rule for an allowed transition $\Delta n_i = 0$ (all i). In one case the $0 \rightarrow 0$ transition is forbidden and $0 \rightarrow 1$ (j) is allowed; in the other case $0 \rightarrow 0$ is allowed and all others forbidden.

In general, when W_G and W_E are not parallel, the calculation of vibrational structure is much more complex. However, it does remain true that for a vibration-induced transition the $0 \rightarrow 0$ (origin) transition is forbidden.

Simple examples of vibration-induced transitions are provided by UCl_6^{2-} in $CsZrCl_6$[6] (Figure 4). Since for f → f transitions the potential surface is essentially unchanged on excitation just $\Delta n = \pm 1$ transitions are observed in the three UCl_6^{2-} u vibrations: 90 (t_{2u}), 114 (t_{1u}) and 277 (t_{1u}) cm^{-1}.

6. THE JAHN-TELLER EFFECT[7]

The Jahn-Teller (JT) theorem states that an electronic state which is degenerate at some symmetrical nuclear geometry loses its degeneracy by nuclear motion causing the symmetry to be broken. (Kramers degeneracy, due to time-reversal symmetry, is

excluded.)   The potential surfaces of such a state are always
split apart, therefore, and the symmetrical configuration is not
an energy minimum.   The consequences of the JT effect are
extensive.   The BO approximation breaks down and it is no longer
possible to say that nuclear motion takes place on one potential
surface.   The vibronic states, and transitions involving them,
become more complex and must be dealt with on an individual basis.

7.   T-e: POTENTIAL SURFACES

One of the simplest JT effects occurs in the case of a T
(T_1 or T_2) state in O_h symmetry and we will embark on this first.
Let us consider an O_h MX_6 molecule with an A_{1g} ground state.
Two of the displacement (normal) coordinates of the ground state
belong to the e_g representation and describe linear combinations
of tetragonal distortions of the octahedron.   Now consider an
excited state which is T_{1u} at $\vec{R}_o$ (the ground state equilibrium
geometry) and its potential surface with respect to distortions
along these tetragonal coordinates q_1 and q_2.   For small
displacements we can expand $\mathcal{H}_{el}$:

$$\mathcal{H}_{el}(q_1, g_2) = \mathcal{H}_{el}^o + \left(\frac{\partial \mathcal{H}_{el}}{\partial q_1}\right)_o q_1 + \left(\frac{\partial \mathcal{H}_{el}}{g_2}\right)_o q_2 \qquad (17)$$

$$= \mathcal{H}_{el}^o + \delta \mathcal{H}_{el}$$

and treat $\delta \mathcal{H}_{el}$ by perturbation theory.   This leads to the zeroth-
order equations

$$\psi_i(q_1 g_2) = \sum_{\alpha=1}^{3} c_{i\alpha} \psi_{T_{1u}}^o \qquad (i = 1\text{–}3) \qquad (18)$$

$$\begin{bmatrix} W_{11}-W_i & W_{12} & W_{13} \\ W_{21} & W_{22}-W_i & W_{23} \\ W_{31} & W_{32} & W_{33}-W_i \end{bmatrix} \begin{bmatrix} c_{i1} \\ c_{i2} \\ c_{i3} \end{bmatrix} = 0$$

where $\psi_{T_{1u}}^o$ are the $\vec{R}_o$ T_{1u} electronic functions and

$$W_{\alpha\beta} = \langle T_{1u}\alpha | \mathcal{H}_{el} | T_{1u}\beta \rangle \qquad (19)$$

$$= W_{T_{1u}}^o + \langle T_{1u}\alpha | \delta \mathcal{H}_{el} | T_{1u}\beta \rangle$$

The matrix $\langle T_{1u}\alpha | \mathcal{E}\mathcal{H}_{el} | T_{1u}\beta \rangle$ can be simplified by group theory. Since $\mathcal{H}_{el}$ must be A_{1g} it follows that $(\partial \mathcal{H}_{el}/\partial q_i)_o$ transforms as does q_i. Defining q_1 and q_2 to transform like d_{z^2} and $d_{x^2-y^2}$ functions and $\psi^o T_{1u\alpha}$, $\alpha = x, y, z$ to transform like x, y and z, the non-zero matrix elements are

$$\langle T_{1u}{}^x | \mathcal{E}\mathcal{H}_{el} | T_{1u}{}^x \rangle = (-\tfrac{1}{2}q_1 + \tfrac{3}{2} q_2)l$$

$$\langle T_{1u}{}^y | \mathcal{E}\mathcal{H}_{el} | T_{1u}{}^y \rangle = (-\tfrac{1}{2}q_1 - \tfrac{3}{2} q_2)l \qquad (20)$$

$$\langle T_{1u}{}^z | \mathcal{E}\mathcal{H}_{el} | T_{1u}{}^z \rangle = q_1 l$$

where l is a numerical parameter. Thus the functions $\psi^o T_{1u\alpha}$ diagonalize $\mathcal{H}_{el}$ and equation 20 describes the first order form of their potential surfaces. These are split: that is, they exhibit a Jahn-Teller effect. For example, consider motion along q_1 ($q_1 \neq 0$; $q_2 = 0$) describing the tetragonal distortion along the z axis. $T_{1u}x$ and $T_{1u}y$ remain degenerate, but split from $T_{1u}z$ as expected for a tetragonal distortion along the z axis. At some general displacement (q_1, $q_2 \neq 0$) x, y and z states are all non-degenerate.

The full form of the T_{1u} potential surfaces in q_1, q_2 space can be obtained by extending equation (20) to higher order. We here make the simplifying assumption that the quadratic terms in q are also diagonal in   and, further, independent of   and of the form $\tfrac{1}{2}k(q_1{}^2 + q_2{}^2)$ where k is the ground state e_g force constant. That is, we assume a quadratic potential of the same form as for the ground state. Then the potential surfaces are described by

$$W_x = W^o_{T_{1u}} + \frac{1}{2}(-q_1 + \sqrt{3}q_2) + \frac{k}{2}(q_1^2 + q_2^2)$$

$$W_y = W^o_{T_{1u}} + \frac{1}{2}(-q_1 - \sqrt{3}q_2) + \frac{k}{2}(q_1^2 + q_2^2) \qquad (21)$$

$$W_z = W^o_{T_{1u}} + lq_1 + \frac{k}{2}(q_1^2 + q_2^2)$$

which can be rewritten

$$W_x = W^o_{T_{1u}} - \frac{1}{2k}^2 + \frac{k}{2}(q^2_{1x} + q^2_{2x})$$

$$W_y = W^o_{T_{1u}} - \frac{1}{2k}^2 + \frac{k}{2}(q^2_{1y} + q^2_{2y})$$ (22)

$$W_z = W^o_{T_{1u}} - \frac{1}{2k}^2 + \frac{k}{2}(q^2_{1z} + q^2_{2z})$$

where

$$q_{1x} = q_1 - \frac{1}{2k} \qquad q_{2x} = q_2 + \frac{3l}{2k}$$

$$q_{1y} = q_1 - 1/2k \qquad q_{2y} = q_2 - 3l/2k$$ (23)

$$q_{1z} = q_1 + 1/k \qquad q_{2z} = q_2$$

These equations describe 3 equivalent displaced parabaloids, as shown in Figure 5, whose minima correspond to tetragonal distortions along x, y and z axes.

8. T-e: VIBRONIC STATES

We now look for vibronic wavefunctions describing nuclear motion on the three intersecting potential surfaces of the form

$$\Psi = \sum_{\alpha=1}^{3} \psi^o_{T_{1u\alpha}} \chi_\alpha$$ (24)

It is intuitively reasonable to generalize the BO approximation (equation (4)) in a JT system to

$$\Psi_k = \sum_k \psi_{K_k}(\vec{r},\vec{R})\chi_{K_k k}(\vec{R})$$ (25)

where ψ_{K_k} are the electronic wavefunctions of the split states and $\chi_{K_k k}$ are generalized vibrational functions. Equation (24) is the 'crude' approximation to equation (25) obtained by putting $\psi_{K_k}(\vec{r},\vec{R}) = \psi_{K_k}(\vec{r},\vec{R}_o)$. Substituting Ψ in equation (3) gives

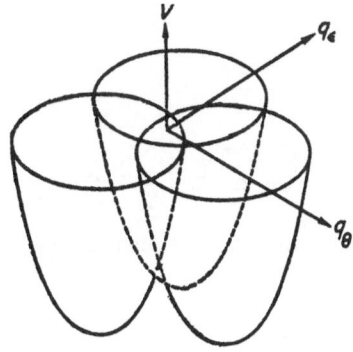

Figure 5.   T-e Jahn-Teller potential surfaces. q_θ , q_ϵ = q_1, q_2.

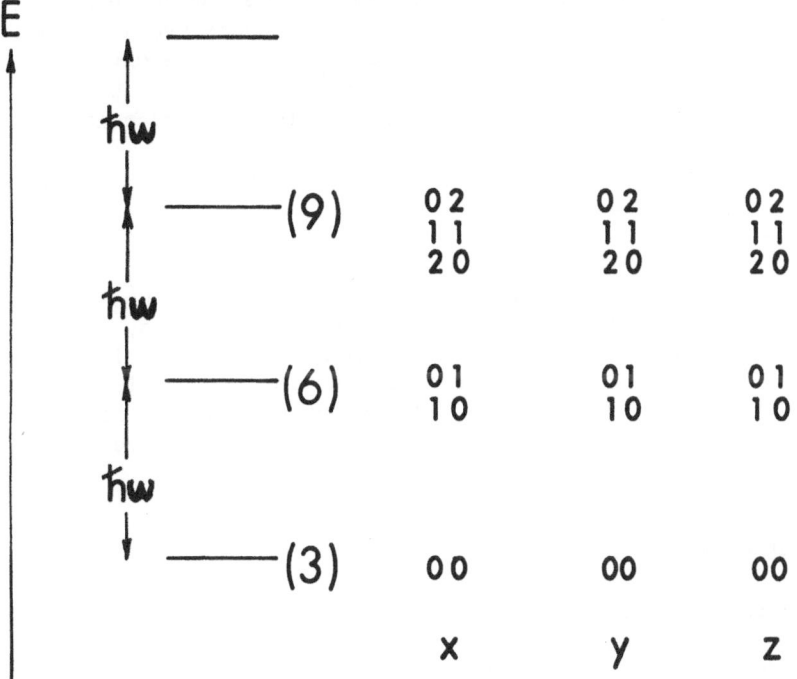

Figure 6.   T-e lowest vibronic levels. $\hbar\omega$ is the e_g mode energy.
Numbers in parentheses are level degeneracies.   The pairs of
numbers are 2-D harmonic oscillator quantum numbers.

$$\sum_{\alpha} \left[(\mathcal{H}_{el} \psi^{o}_{T_{1u\alpha}}) \chi_{\alpha} + \psi^{o}_{T_{1u\alpha}} (T_n \chi_{\alpha}) \right] \tag{26}$$

$$= \sum_{\alpha} E \psi^{o}_{T_{1u\alpha}} \chi_{\alpha}$$

which leads to

$$\sum_{\alpha} \left[\langle T_{1u\beta} | \mathcal{H}_{el} | T_{1u\alpha} \rangle + \delta_{\alpha\beta} T_n \right] \chi_{\alpha} = \sum_{\alpha} E \chi_{\alpha} \delta_{\alpha\beta} \tag{27}$$

Since $\langle T_{1u\beta} | \mathcal{H}_{el} | T_{1u\alpha} \rangle$ is diagonal, equation (27) further decouples to

$$\Psi_{\alpha} = \psi^{o}_{T_{1u\alpha}} \chi_{\alpha} \tag{28}$$

$$\langle T_{1u\alpha} | \mathcal{H}_{el} | T_{1u\alpha} \rangle + T_n \right] \chi_{\alpha} = E_{\alpha} \chi_{\alpha}$$

showing that the vibronic states are just those for nuclear motion on one potential surface at a time. These are diagrammed in Figure 6. The lowest vibronic state is 3-fold degenerate, comprising the n=0 levels of each of the three wells. Higher vibronic levels are spaced by multiples of the e_g fundamental frequency.

9.   T-t$_2$

A T_1 or T_2 electronic state of an O_h system is also split by motion along the trigonal, t_{2g} displacement coordinates, in a way very much more complex than the T-e system. Figure 7 illustrates computed vibronic state energies as a function of JT interaction.[8] Harmonic intervals are here not retained under the JT effect. This is the general rule; the T-e system discussed above is an exception in this respect.

10. BANDSHAPES

The presence of a JT effect modifies the vibronic states, and hence the bandshapes of electronic transitions thereto or -from. Calculation of bandshapes have been carried out for various specific model systems. Figure 8 illustrates one example, involving a T-t$_2$ system. The JT effect produces structure in the overall contour that is dependent on the magnitude of JT effect, and also on temperature. Similar results have been obtained for the E-e system.

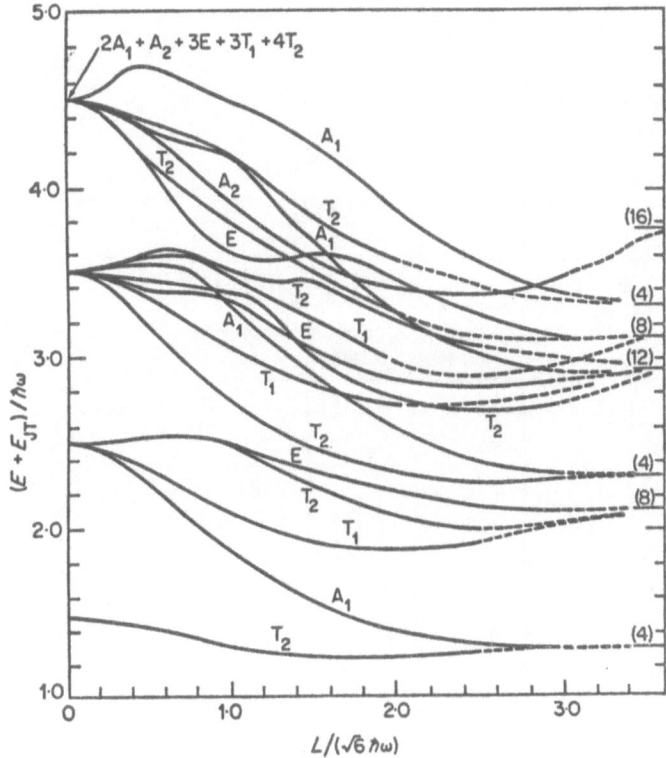

Figure 7. Vibronic energy levels as function of the coupling coefficient for an electronic triplet T_2 interacting with t_2 modes. L is a measure of the magnitude of Jahn-Teller effect.

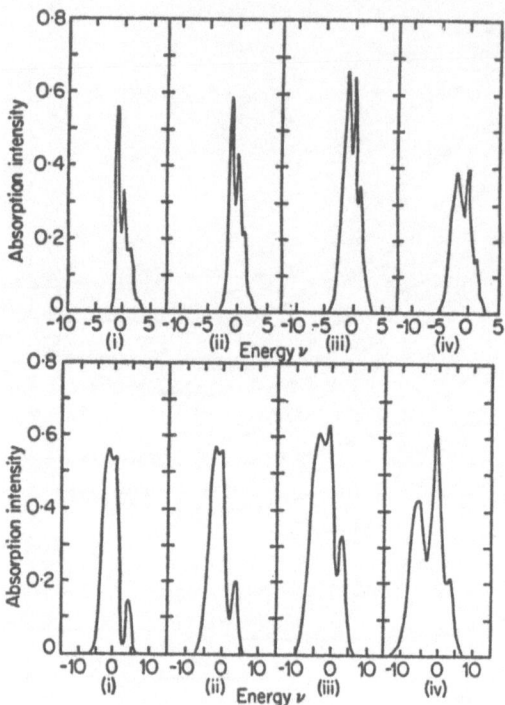

Figure 8.   Calculated transition probabilities (per unit energy)
vs. light frequency   for absorption by a T t_2 system.   All
energies are in units of the vibrational energy $\left(\hbar\omega_{t_2} = 1 \right)$.
The parameters of the figures are, with the drawings listed from
left to right:

First row: Absorption, L = $-\sqrt{6}$, α(the broadening parameter) =
0.5 (i) kT = 0, (ii) kT = 0.5, (iii) kT = 1, (iv) kT = 4

Second row: Absorption, L = 6, α = 1 (i) kT = 0, (ii) kT = 0.5,
(iii) kT = 1, (iv) kT = 4.

In principle, when the excited state of a transition suffers
a JT effect the vibrational structure again allows the excited
state potential surfaces – now multiple – to be obtained.
Owing to the complexity of dynamic JT calculations, however, few
examples of this yet exist.

11. THE HAM EFFECT

One of the important ways of studying the JT effect in an
electronic state experimentally is to investigate the consequences
of perturbations on the lowest vibronic level.   Ham showed in
1965[9] that in a T-e system the splittings due to certain types of
perturbation are quenched (reduced) increasingly as the JT effect
increases, and this type of phenomenon is now often called the
Ham effect.   We shall here briefly summarize Ham's discussion
for the T-e system.

The lowest vibronic level has wavefunctions

$$\psi_x \chi_x$$

$$\psi_y \chi_y \tag{29}$$

$$\psi_z \chi_z$$

where ψ_x, ψ_y and ψ_z are the $\psi^0_{T_{1u}\alpha}$ functions and χ_x, χ_y and χ_z

are $\chi_{\alpha 00}$.   Consider a purely electronic operator O representing

an applied perturbation.   The matrix to be diagonalized to find

the perturbed energies is

$$\langle \psi_\alpha \chi_\alpha |0|\psi_{\alpha'} \chi_{\alpha'} \rangle = \langle \psi_\alpha |0|\psi_{\alpha'} \rangle \langle \chi_\alpha |\chi_{\alpha'} \rangle \tag{30}$$

When there is no JT effect, χ_α is independent of α and

$$\langle \chi_\alpha |\chi_{\alpha'} \rangle = 1 \tag{31}$$

$$\langle \psi_\alpha \chi_\alpha |0|\psi_{\alpha'} \chi_{\alpha'} \rangle = \langle \psi_\alpha |0|\psi_{\alpha'} \rangle$$

The perturbed energies are then unaffected by the vibrational
motion.   However, in the case of a JT effect $\langle \chi_\alpha |\chi_{\alpha'} \rangle \neq 1$ when
$\alpha \neq \alpha'$ since this overlap is between spatially displaced
vibrational functions in different electronic wells.   Writing

$$\langle \chi_\alpha | \chi_{\alpha'} \rangle = 1 \quad (\alpha = \alpha')$$
$$= S \, (<1) \quad (\alpha \neq \alpha') \tag{32}$$

we have

$$\langle \psi_\alpha \chi_\alpha | 0 | \psi_{\alpha'} \chi_{\alpha'} \rangle = \langle \psi_\alpha | 0 | \psi_\alpha \rangle \quad (\alpha = \alpha') \tag{33}$$
$$= \langle \psi_\alpha | 0 | \psi_{\alpha'} \rangle S \quad (\alpha \neq \alpha')$$

Now if $\langle \psi_\alpha | 0 | \psi_{\alpha'} \rangle$ is entirely off-diagonal, the whole matrix of
0 is uniformly reduced by the factor S - and hence the splitting
caused by 0. S is in fact given by

$$S = \exp \left[-3E_{JT}/2\hbar w \right] \tag{34}$$

where E_{JT} is the depth of the JT well (relative to q = o) and
$\hbar w$ is the e_g energy.

Examples of operators that suffer quenching in this case
are $\vec{L}$, the orbital Zeeman interaction $\beta \vec{L} \cdot \vec{H}$ and, when the T
state is spin-degenerate, spin-orbit coupling $\lambda \vec{L} \cdot \vec{S}$.   One
thus expects Zeeman effects and spin-orbit splittings to be
sensitive to the JT effect in a T state.   Much experimental
work recently on the JT effect has focussed on this type of
phenomenon.   An example is the reduced spin-orbit splitting in
the $^4T_2$ state of Co^{2+}/MgO[10] (Figure 9).

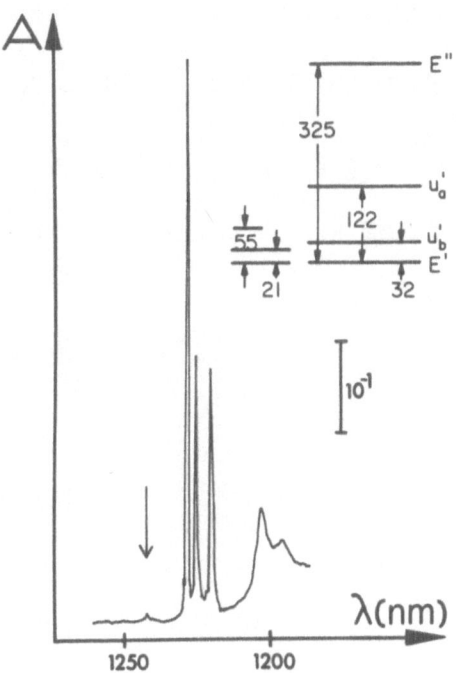

Figure 9.   The 0-0 $^4T_1 \rightarrow {}^4T_2$ transitions of Co^{2+}/MgO.   The experimental separations of the $^4T_2$ spin-orbit components are shown and compared with the predictions of ligand field theory.

REFERENCES

1.   J.P. Jesson, J. Chem. Phys. 48, 161 (1968).

2.   S.L. Holt and C.J. Ballhausen, Theor. Chim. Acta 7, 313 (1967).

3.   B.D. Bird, G.A. Osborne and P.J. Stephens, Phys. Rev. B 5, 1800 (1972).

4.   A.D. McLachlan, Mol. Phys. 4, 417 (1961).

5.   C.J. Ballhausen, Theoret. Chim. Acta 1, 285 (1963).

6.   D. Johnston, R.A. Satten and E. Wong, in 'Optical Properties of Ions in Crystals', Ed. Crosswhite and Moos, Interscience, 1967, p. 432.

7.   M.D. Sturge, Solid State Physics, 20, 91 (1967);
     R. Englman, 'The Jahn-Teller Effect in Molecules and
     Crystals', Wiley, 1972.

8.   M. Caner and R. Englman, J. Chem. Phys. 44, 4054 (1966);
     R. Englman, M. Caner and S. Toaff, J. Phys. Soc. Japan
     29, 306 (1970).

9.   F.S. Ham, Phys. Rev. 138 A, 1727 (1965).

10. J.C. Cheng, A. Mann and P.J. Stephens, Chem. Phys., in
     press.

INTERCONFIGURATIONAL AND CHARGE TRANSFER TRANSITIONS

Donald S. McClure

Department of Chemistry, Princeton University,
Princeton, New Jersey, U.S.A.

I. INTRODUCTION

In charge transfer transitions, an electron is removed from
the donor atom and goes to an acceptor. The wavefunctions of
the hole left on the donor and the electron on the acceptor may
overlap, but in many cases the overlap is small enough so that
it is a good approximation to say that an electron has actually
been transferred from one atom to another. This circumstance
leads to the possibility of relating ionization potentials and
electron affinities of atoms to processes occurring in the solid
or liquid state. The transition energy can be written:

$$\Delta E = E_A - E_D = |B_A + \delta E_A - (B_D + \delta E_D)| - e^2/r_{AD} \qquad (1)$$

where subscripts A, D refer to acceptor, donor; B are electron
binding energies and δE_a, δE_D are the crystal field or other
solid or liquid state effects on the final or initial state.
The last term is the energy of interaction between hole and
electron. It is not necessarily small at the interatomic
distances which are found in systems undergoing charge transfer
transitions. These distances could be 3–6 Bohr radii corresponding to 4.5 to 2.25 ev. The quantum mechanical expression for
this term would contain the coulomb and exchange integrals over
the appropriate molecular orbitals of the system.

The justification of equation (1) can be sought empirically
by studying a homologous series in which the only large changes
occur in B_A or B_D and can also be made in terms of molecular
orbital theory.

P. Day (ed.), Electronic States of Inorganic Compounds. 113–139. *All Rights Reserved.*
Copyright © 1975 by D. Reidel Publishing Company, Dordrecht-Holland.

The lowest energy interconfigurational transitions in
transition metal ions could be 3d → 4s and 3d → 4p and similarly
for the higher transition groups.   In the rare earths and
actinides the 4f → 5d (5f → 6d) transitions are the lowest in
energy.   Although these are well defined transitions in the
free atoms, one might wonder if 4s in a transition metal ion
would still be recognizable in the solid state.   The 5d orbital
in the rare earths is far more diffuse than the 4f-orbital, as
shown in Figure 1, so there could also be a question about its
existence as an atomic-like orbital in solids.   The experimental
data show, however, that both of these orbitals have distinct
atom-like characteristics.   It is found that the 4f → 5d and
3d → 4s transitions in a series of rare earth or transition metal
ions vary with atomic number in the way predicted by equation (1)

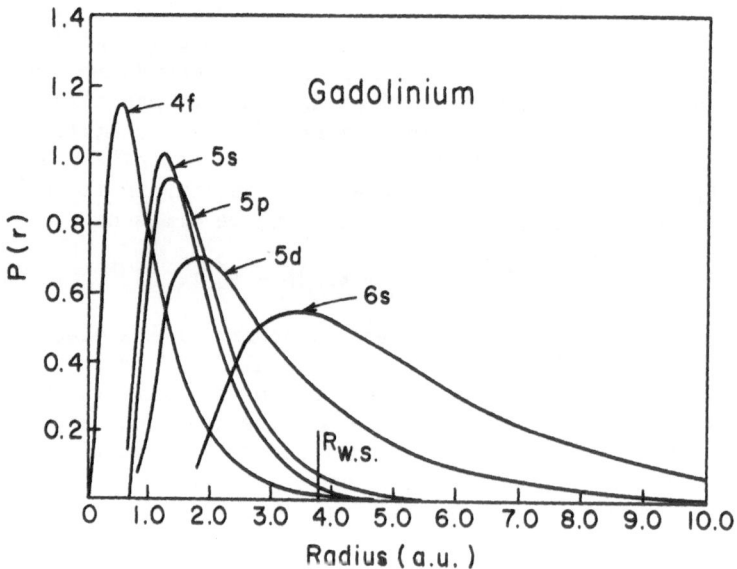

Figure 1.   The radial probability distribution $P(r) = \psi^2(r)$.
$4 \pi r^2$ of electrons in the outer parts of the 4f, 5s, 5p, 5d and
6s orbitals of atomic Gd.   The radius of the Wigner-Seitz sphere
(about one half of the nearest neighbour distance in the metal)
is marked as R_{ws}.   The radial distance is given in atomic units.
From A.S. Freeman and R.E. Watson, Phys. Rev. 127, 2058 (1962).

if B_A-B_D is interpreted to mean the energy of the interconfigurational transition. In this case, the same atom is
acceptor and donor, and B_D, δE_D apply to the ground state while
B_A, δE_A apply to the excited state.

The charge transfer transitions to be discussed in this
article involve a transition metal or rare earth ion and their
ligand groups. A spectroscopic transition in which the metal
receives an electron from the ligands will be called an acceptor
charge transfer, and that in which it loses an electron will be
called a donor transition. Interconfigurational transitions
resemble the latter in the sense that the d- or f- orbitals of
the metal donate an electron to a higher energy orbital which
extends over the ligands much more than does the initial orbital
in the spectroscopic process.

While the simple theory based on equation (1) is usually
adequate for identification of the type of transition, it does
not tell anything about the nature of the wavefunctions of the
excited state. In fact, equation (1) is useful both for ion
clusters such as NiF_6^{4-}, for crystals such as NiF_2 and for
impurities in crystals, such as $KMgF_3$:Ni. The wavefunctions
of the states of all these systems are different, even when the
elementary atomic processes are the same. We must find the
more subtle differences between these systems in the molecular
orbital or other detailed treatments of these systems. The
main emphasis of this article, however, will be to see how well
equation (1) provides a zero-order basis for understanding charge
transfer and interconfigurational transitions.

To help keep the different types of transitions in mind,
Figure 2 shows an energy level diagram for an impurity in a solid
with d → d, d → s, ligand → d, band → band and photoelectron
transitions marked on it.

II. RARE EARTHS (LANTHANIDES) AND ACTINIDES

Table I shows the ground configuration for the rare earths
in their first four stages of ionization, and their ionization
potentials. The 4f → 5d and 4f → 6s energies are shown for the
di- and trivalent ions. Most of these data have been organized
and discussed quite recently by Sugar and Reader.[1-4] Acceptor-
type transitions in the trivalent state should have the same
variation with atomic number as does the third ionization
potential. The 4f → 5d transitions in the ions in solids should
be close to those of the free ions. Figure 3 is a plot of the
4f, 5d and 6s binding energies vs. Z. For the 4f values, note
the repetition in the second half of the series of all the features
seen in the first half.

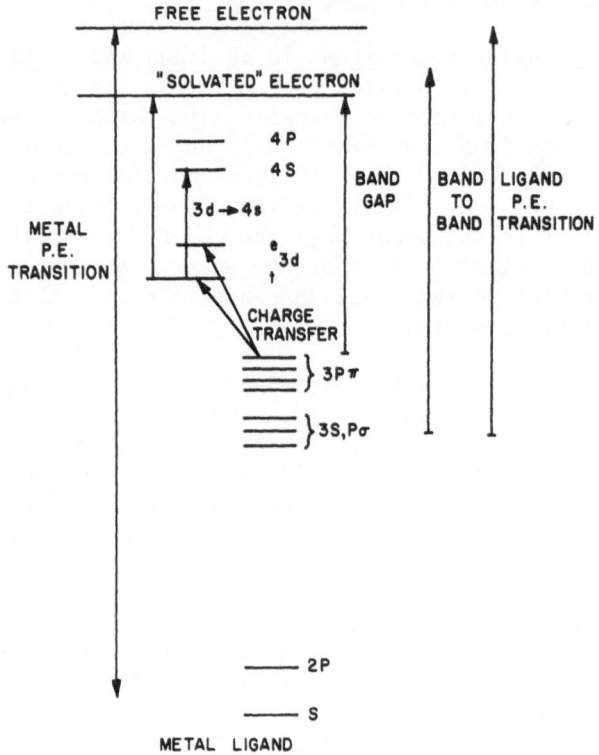

Figure 2. A general one-electron energy level diagram showing
metal and ligand orbitals, a conduction band or solvated electron
region and the free electron energy. Crystal field transitions
are 3d t → e. Some ligand → metal charge transfer transitions
into the two 3d levels are shown. Other transitions shown are
3d → 4s, 3d → conduction band, valence → conduction band and two
kinds of photoelectron transitions.

The 4f → 5d transitions of the entire series of divalent
rare earths in CaF$_2$ shown in Figure 4 were described and analysed
by McClure and Kiss.[5] In the absence of experimental data, the
Slater-Condon theory of the f^n and f^{n-1}d configurations was
applied to calculate the transition energy. The comparison
between observed and calculated values was quite good. The
comparison with the experimental free ion data of Table I is
shown in Figure 5 and serves to confirm the 4f → 5d assignment.

Table I. Ionization energies and binding energies of electrons in rare earth ions. For the neutral and +1 species and ground configuration is given. For the +2 and +3 species the energy of the lowest term of a configuration below the ionization energy is given. (For 4f columns this is I_2 or I_3 with the exception of Gd++, La++ and Lu++ entries.)

	Neutral				+1				+2			+3		
	f	d	s	I_2	f	d	s	I_2	4f	5d	6s	4f	5d	6s
La	0	2	1	44.98	1	0	1	89.2		154.6	141.1	296.5	246.8	209.9
Ce	1	1	1	44.09	2	0	1	87.5	162.9	159.6	143.6	314.4	253.2	214.1
Pr	3	0	2	43.73	3	0	1	85.1	174.4	161.6	146.1	325.9	255.8	217.0
Nd	4	0	2	44.27	4	0	1	86.5	178.6	162.6	148.3	331.4	258.1	220.4
Pm	5	0	2	44.80	5	0	1	87.9	180.0	163.6	150.6	333.7	260.0	223.5
Sm	6	0	2	45.42	6	0	1	89.3	189.0	164.5	152.8	344.0	262.2	226.8
Eu	7	0	2	45.81	7	0	1	90.7	199.2	165.3	154.9	355.0	263.8	229.6
Gd	7	1	2	49.53	7	1	1	97.9	164.0	166.4	157.2	320.9	266.0	232.8
Tb	9	0	2	47.20	9	0	1	92.9	176.7	167.7	159.0	334.5	268.2	234.4
Dy	10	0	2	47.82	10	0	1	94.1	183.8	166.9	160.7	342.6	268.4	237.9
Ho	11	0	2	48.54	11	0	1	95.2	184.2	166.1	162.4	344.0	268.6	240.4
Er	12	0	2	49.21	12	0	1	96.2	183.4	166.4	164.1	344.0	268.6	243.0
Tm	13	0	2	49.84	13	0	1	97.2	191.0	168.1	165.7	344.3	270.0	243.0
Yb	14	0	2	50.44	14	0	1	98.2	201.9	168.5	167.2	352.8	272.6	245.3
Lu	14	1	2		14	0	1			169.0	174.7	364.5	274.1	247.7

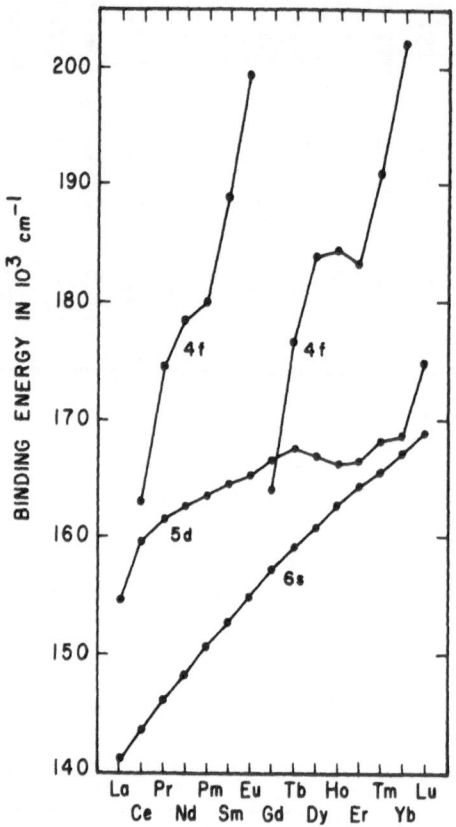

Figure 3.   Binding energies of 4f, 5d and 6s electrons in
divalent rare earth ions, measured as the difference in energy
between the lowest state of the configurations $4f^n$, $4f^{n-1}$ 5d or
$4f^{n-1}6s$ and the ionization energy.   Data from ref. 4.

Figure 6 shows the observed transition energy in CaF_2 minus the
$4f \rightarrow 5d$ free-atom transition energy plotted against atomic number.

         In order to see what these results mean, Figure 7 shows the
calculated energy of the lowest state of the f^n configuration
and of the lowest state of the $f^{n-1}d$ configuration.   This Figure
shows that the variation of the transition energy with atomic
number is due mainly to the changes in the f-electron electrostatic splitting, and that the variations in the f-d energy are
small.   The small f-d exchange energy is due to the rather small
f-d overlap.   These two kinds of wavefunctions have quite
different radial distributions, as shown in Figure 1, and this
accounts for the small overlap.

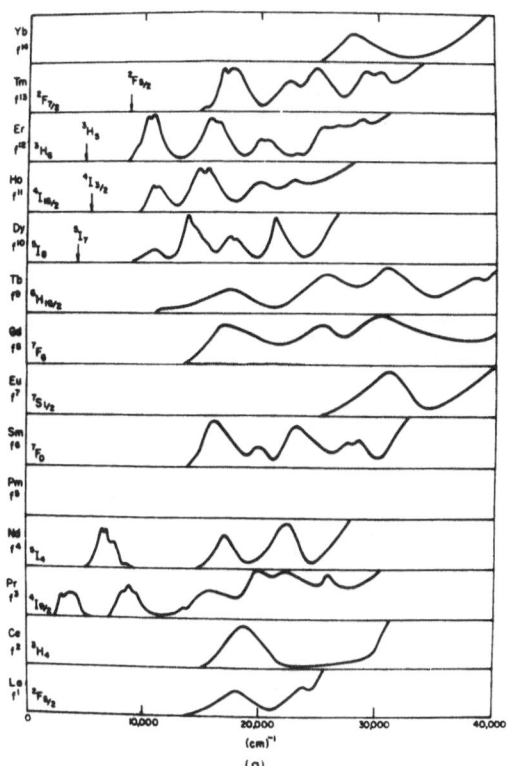

Figure 4. Low resolution spectra of the divalent rare earth ions in CaF$_2$. The broadbands are 4f → 5d transitions. Some 4f → 4f transitions are marked. The spectra shown for La, Ce, Gd and Tb are not those of the divalent ion but correspond to a trivalent ion–F-centre complex (Anderson and Sabisky, Phys. Rev. B3, 527 (1971).) Data from ref. 5.

Returning to Figure 6, the nearly linear dependence of the difference between the lowest f → d transition in solid state and in free ions is worth some comment. We must first note that the transition energy for Ce^{++} which has a 5d-derived ground state is counted negative. With this understanding the transition energy can always be written as T = E(d) − E(f) − Δ where Δ is 6 Dq, the crystal field splitting of the 5d-electron. The most likely explanation of the atomic number dependence of T (free ion) − T (solid state) shown in Figure 6 is a change in the value of Δ with atomic number. This type of change is

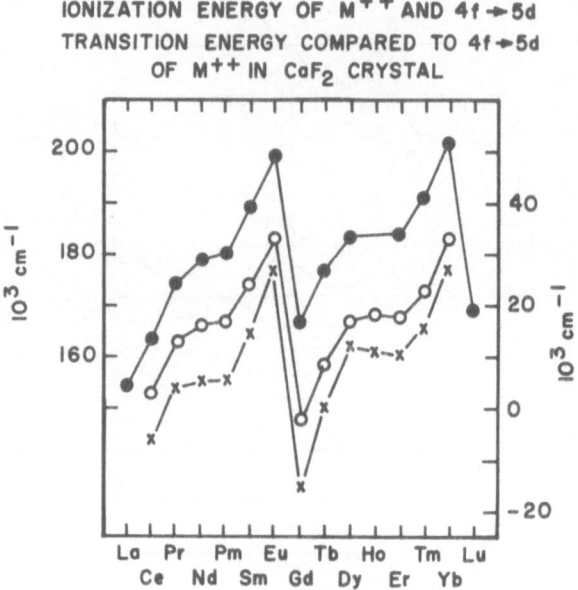

IONIZATION ENERGY OF M^{++} AND 4f → 5d
TRANSITION ENERGY COMPARED TO 4f → 5d
OF M^{++} IN CaF$_2$ CRYSTAL

Figure 5.   Comparison of ionization energy of M^{++}, 4f → 5d
transition energy in free M^{++} and 4f → 5d transition energy of
M^{++} in CaF$_2$.   Upper curve, solid dots, is ionization energy,
using lefthand side.   Open dots is free ion 4f → 5d energy;
X is 4f   5d in CaF$_2$.   The latter two use righthand scale.
Data from refs. 4 and 5.

familiar from transition-metal spectroscopy.[6]   In the case of
rare earths it is much clearer because of the absence of Jahn-
Teller effects.   Figure 6 requires 6 Dq (Ce^{++}) = 9500, and a
decrease by 375 cm^{-1} for each subsequent ion.   The decrease
over the entire series is by a factor of two, somewhat larger
than the decrease in the transition metal series, usually a factor
of 1.5.[6]

The 4f → 5d transitions in the <u>trivalent</u> rare earths
usually occur in the vacuum ultraviolet and, therefore, have not
been studied very thoroughly.   The observed transitions in
CaF$_2$[7] are plotted <u>vs.</u> atomic number in Figure 8 along with the
various values given for the free ions by Sugar and Reader.[4]
It does not take much imagination to see where the presently
unobserved transitions must lie.   Figure 8 shows in addition to

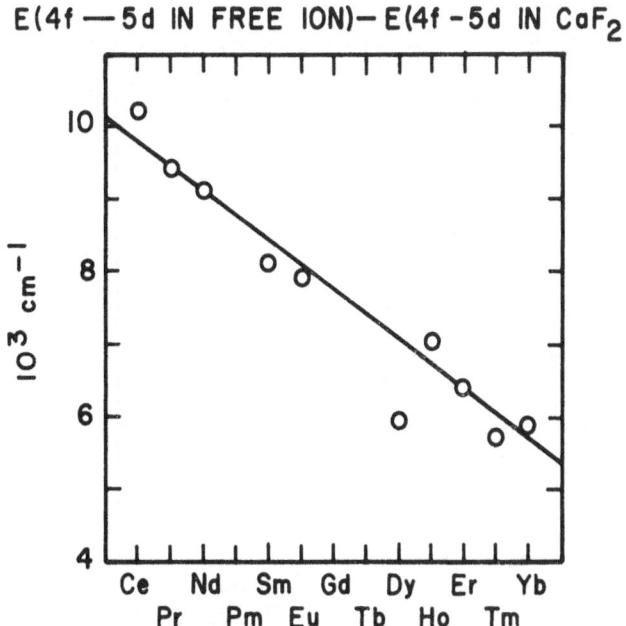

Figure 6. Energy difference between lowest $4f \rightarrow 5d$ transition in free M^{++} and in M^{++} in CaF_2. Data from refs. 4 and 5.

the points for the ions in CaF_2, some of the data for hexachloro complexes in solution.[8] The CaF_2 data again seem to show a greater shift from the free-ion values in the first half of the series compared to the second.

The detailed analysis of the $4f \rightarrow 5d$ spectra has been accomplished for Yb^{++},[9] Ce^{++} [10] and to a more limited extent for Tm^{++} [11] and Eu^{++}.[12] The crystal field calculations for even the simplest cases are formidable because of the large number of states and because the crystal field, the spin orbit coupling and the electrostatic coupling are all of the same order of magnitude. Figure 9 shows the type of results obtained for Yb^{++} in $SrCl_2$. In this case, both energies and intensities could be calculated and used for spectral identification. Although the results seem very good, they have recently been challenged,[11] and it is possible that the fit obtained is only one of many. Furthermore, there are unexplained luminescences and photo-chemical reactions which make this system an intriguing one.[13]

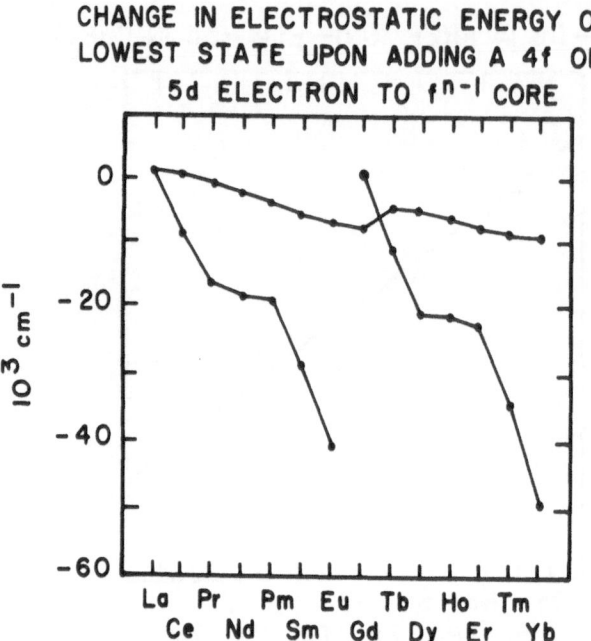

CHANGE IN ELECTROSTATIC ENERGY OF
LOWEST STATE UPON ADDING A 4f OR
5d ELECTRON TO f^{n-1} CORE

Figure 7.   Change in energy of the lowest state of a M^{+++} ion
upon adding a 4f or a 5d electron, calculated using Slater-
Condon theory (McClure and Kiss, ref. 5).

The comparison between free ion and solid state spectra
shown in Figures 5,6 is necessarily crude because the spectra
have not been analysed properly.   In the case of Yb^{++} in $SrCl_2$
where the analysis has been made, one can quote precisely how
far the centre of gravity of the fd configuration moves in going
from gas to crystal.   This is a decrease of 6670 cm^{-1} in the
$4f^{14} \rightarrow 4f^{13}5d$ interval.[9]   Fortunately, it agrees quite well
with the crude estimate of 5900 cm^{-1} made from the first bands
of vapour and crystal, and one hopes that this agreement applies
to the rest of the series.

It is interesting to compare the energies of the charge
transfer bands of the trivalent ions to the $f \rightarrow d$ energies of
the divalent ions.   For an acceptor transition, ligand $\rightarrow M^{+++}$,
equation (1) is

$$E \, (L \rightarrow M^{+++}) = B_D - I_3 - e^2/r_{12} \tag{2}$$

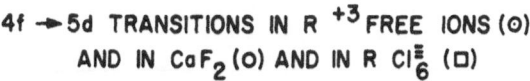

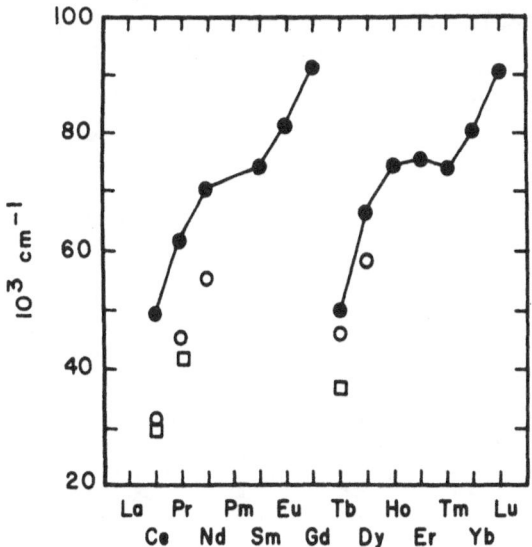

Figure 8. Comparison of 4f → 5d transition energy in free trivalent rare earth ions with data from CaF$_2$ solution or RCl$_6^=$ complexes. Data from refs. 4, 7, 8.

where the terms in E are neglected. For an f → d transition, the same equation is

$$E (f \rightarrow d, M^{++}) = I_3 - B_{5d} \qquad (3)$$

where I$_3$, the third ionization potential is the same as in equation (2), and B$_{5d}$ is the binding energy of the d-electron, shown in Figure 3. (The e^2/r$_{12}$ term is included with B$_{5d}$.) Figure 3 shows that the 5d binding energy as a function of Z is nearly monotonic. In the sum of equations (2) and (3), I$_3$ goes out and the other terms are monotonic and slowly varying with Z.

$$E (L \rightarrow M^{+++}) + E (f \rightarrow d, M^{++}) = I_D - E_d - e^2/r_{12} \qquad (4)$$

Table 2 shows some of the data of Nugent et al.[8] on the charge transfer bands of R^{III}Cl$_6^=$ in anhydrous alcohol compared in the above way to the f → d energies of the divalent ions. The sum is the same for Eu and Sm, and for Tm and Yb. The reason for

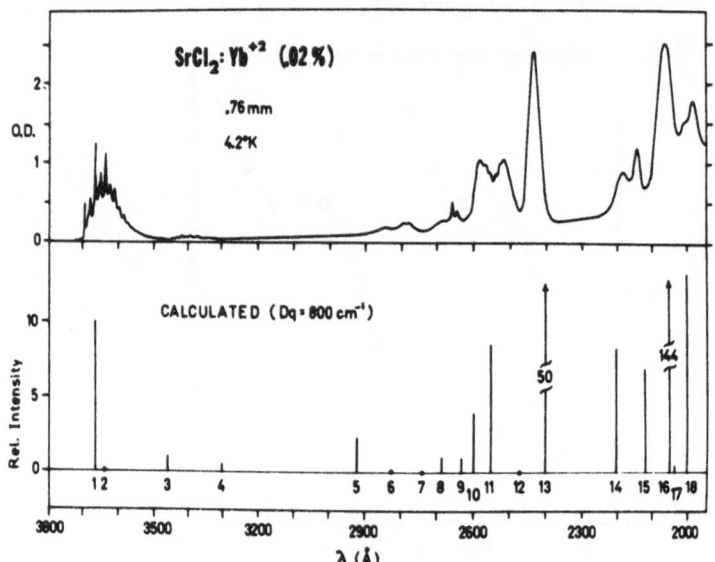

Figure 9.   Absorption spectrum of Yb^{++} in $SrCl_2$ single crystal
showing $4f \rightarrow 5d$ transitions.   The calculated state energies
and transition probabilities are shown with vertical lines
below the spectrum.   A $4f \rightarrow 6s$ contribution may be present.
Ref. 9.

Table 2.   Comparison of charge transfer energies of rare earth
hexahalide complexes, $MCl_6^=$, to $f \rightarrow d$ energies of divalent ions

	C.T.M^{+3}	f d, M^{+2}	Sum	I_3
Sm	40.3	16.4	56.7	189.0
Eu	30.3	26	56.3	199.2
Tm	44.3	17.2	61.5	191.0
Yb	35.3	27.5	62.8	201.9

the small difference, 5×10^3 cm^{-1}, between the sum for the two
groups is the difference in the f-d exchange splitting between
the first and second half of the rare earth series.   In the
second half, a state having a higher multiplicity than the ground
state is the lowest f $\rightarrow$ d state, and this weak band is not always
observed.   The quoted f-d energy is that of the first strong
band.   Therefore, it is expected that other charge transfer
bands will obey the constancy relationship shown in Table 2 and
their positions can be predicted.

Also shown in Table 2 is the third ionization potential;
it is clear that it varies in the same way as do the f $\rightarrow$ d and
CT transitions.   Indeed, it represents the binding energy of the
f-electron whose dependence on Z is the basis for almost all of
the changes with Z of the f $\rightarrow$ d and CT spectra.

The f-electron binding energy should also be the major
component of the oxidation potentials of the ions in solution.
Figure 10 shows the known values of the half reaction potentials
(against a standard hydrogen electrode) for the reaction

$$e + R \, Cl_6^{\equiv} \rightarrow RCl_6^{\equiv}$$

This reaction must involve I_3, and so it would be expected that
if the E^o values were plotted against the halogen   metal charge
transfer band energy, there would be an approximately linear
dependence both for $RCl_6^{\equiv}$ and $RBr_6^{\equiv}$.   Furthermore, since both E^o
and the CT band energy are expressed in the same units the slope
should be unity and, in fact, lines of unit slope have been
drawn in Figure 10.   The fitting only determines an intercept.
Using this line, Figure 11 shows how the reduction potentials
for the other rare earth hexachlorides can be obtained from I_3.
Some of the potentials should be low enough to be measured in
certain non-aqueous solvents.

The next orbital in order of decreasing stability in the
rare earths is the 6s orbital.   The binding energies of the 4f,
5d and 6s orbitals for divalent rare earths are shown in Figure 3.
The data are taken from the papers of Sugar and Reader.[4]   The
high binding energies of these three types of orbital suggest
that all these would have comparable stability in the solid state,
and therefore 4f $\rightarrow$ 6s should be a spectroscopically observable
transition in solids as 4f $\rightarrow$ 5d was found to be.   There is no
clear spectroscopic evidence for these transitions as yet, but
no organized attempt to identify these transitions has been made.
In the case of Yb^{++} in SrCl$_2$, certain spectral features and the
photochemical behaviour could be ascribed to the presence of the
$4f^{13}6s$ configuration in the same region of energy as the $4f^{13}5d$
configuration.

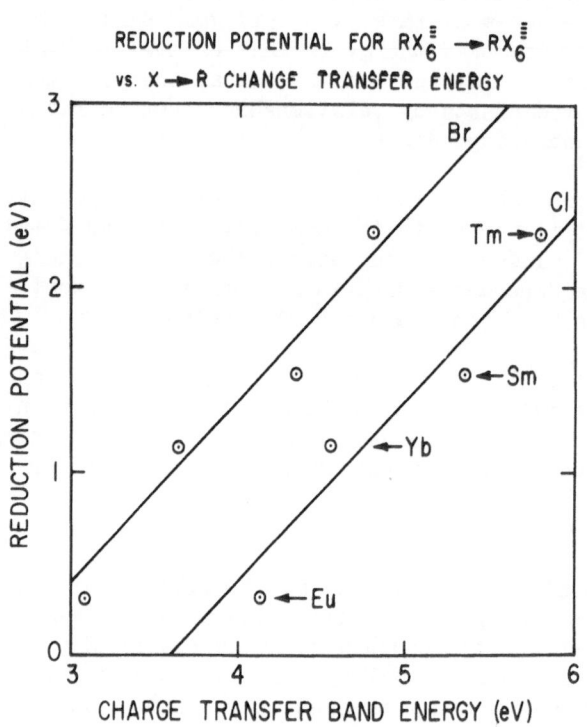

Figure 10.   Plot of reduction potential for trivalent rare
earth hexachloride and hexabromide against charge transfer energy.
Units are the same on ordinate and abscissa so the correlation
with equation (1) requires a line of unit slope as drawn in the
figure.   Ref. 8.

III. TRANSITION METAL IONS

The binding energies for 3d and 4s electrons are plotted in
Figure 12 and given in Table 3 for the first group of transition
metals.[14]

As with the 4f series the 3d binding energies show a break
at the half-filled shell.   The 4s binding energies, however,
have a much smoother curve.   Table 3 gives the values of these
binding energies based on the lowest state of the configuration.

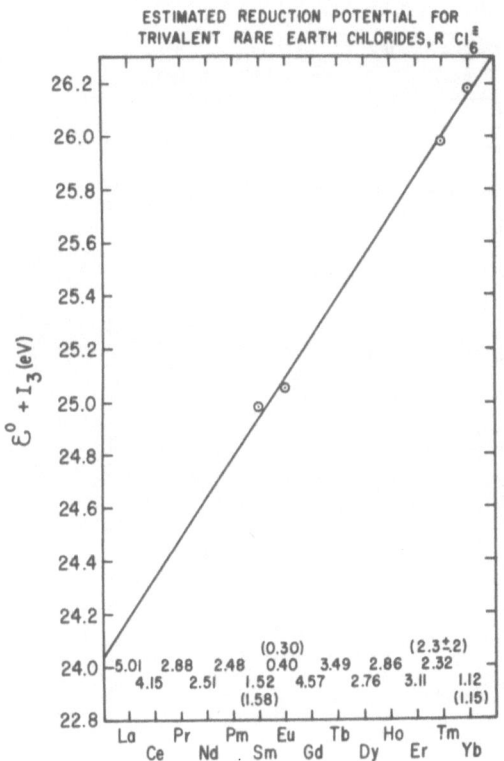

Figure 11.   Plot of ionization potential of trivalent rare
earth plus reduction potential of $RCl_6^=$ against atomic number.
The four known data points are shown and the reduction potentials
for all cases are obtained from the line drawn through them.
The values are shown at the bottom of the figure.   Measured
values are given in parentheses.   Data from refs. 4 and 8.

        Until quite recently, there had been no systematic
investigation in which $3d \rightarrow 4s$ transitions could be identified
with certainty.   Sabatini[15] however, has observed these
transitions in fluoride host crystals such as $KMgF_3$, CaF_2 and
MgF_2 doped with transition metal ions.   The kind of transition
ordinarily observed in other halogen host crystals is the ligand
to metal charge transfer.

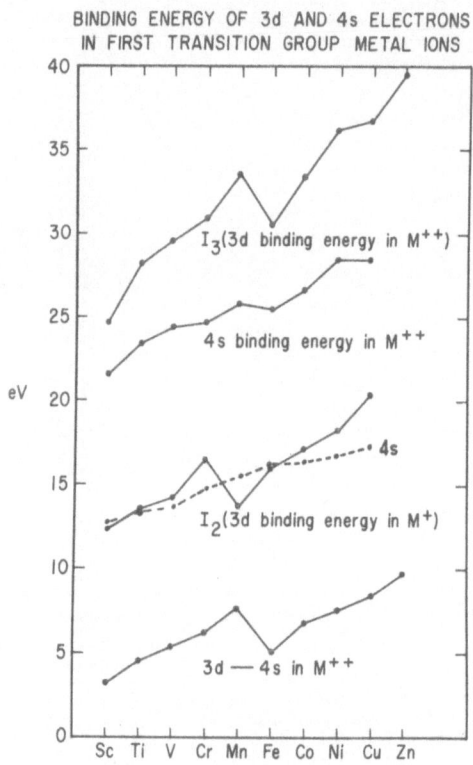

BINDING ENERGY OF 3d AND 4s ELECTRONS
IN FIRST TRANSITION GROUP METAL IONS

Figure 12.   Binding energy of 3d and 4s electrons in mono and divalent transition metal ions, defined in the same way as in Figure 3.   The 3d → 4s transition energy in M^{++} is also shown.

Figure 13 shows the Cl → metal charge transfer spectra of $CoCl_4^=$, $NiCl_4^=$ and $CuCl_4^=$ from the work of Bird and Day.[16]   Marked on these spectra are the relative band positions required by equation (1).   The fit is reasonably good.   In order to investigate the fitting any further, one would need to see the entire series of chlorides and the structure of each spectrum would have to be analysed so that comparable transitions in each ion could be used.   The crystal field corrections are not large in these examples.   The Dq values are known for divalent ions, and were taken to be one-half as large for the isoelectronic monovalent ion.

Table 3. 3d and 4s electron binding energies in transition metal ions (units: 10^3 cm^{-1})[a]

| | II (M^+) | | III (M^{++}) | | IV (M^{+++}) | |
	3d	4s	3d	4s	3d	4s
Sc	99.1	104.	199.7	174.1		
Ti	109.	110.0	227.0	188.9	348.8	268.4
V	114.6	111.7	240.	196.	391.	295.
Cr	133.0	120.7	249.7	200.2	400.	296.
Mn	111.4	126.1	271.8	209.3	I	I-112
Fe	128.0	130.5	247.2	217.1		
Co	137.6	133.1	270.2	223.2		
Ni	146.4	136.4	283.7	229.		
Cu	163.7	141.2	297.1	236		
Zn			320.3	241		

[a] Binding energies are defined as the energy from the lowest state of a configuration to the ionization energy.

Tippins[17] has studied the ligand to metal charge transfer spectra of the trivalent transition metal ions in corunderm. In this case, there is again approximate agreement with equation (1), showing that the transitions are, in fact, of the acceptor type. Tippin's spectra are plotted in Figure 14 and listed in Table 4 along with the corresponding values of $C- I_3 + \mathcal{E}_3 - \mathcal{E}_2$ where C = 286.6 kk. In this case, $\mathcal{E}_3$ and $\mathcal{E}_2$ are crystal field corrections for which Dq values are known. They were taken from ref. 18 for the +3 ions and from ref. 6 for the +2 ions, except that for Ni^{+3}, Co^{+3} and Cu^{+3} Dq = 1800 cm^{-1} was chosen and the correction given is for the spin unpaired state which is actually very close to the true ground state (paired for Ni^{+3} and Co^{+3}). For these three ions, the predicted charge transfer energies are too small, as there is no evidence in the spectra for strong bands at these energies. This disagreement must be regarded as a real breakdown of the simple correlation, the first one met with in this article.

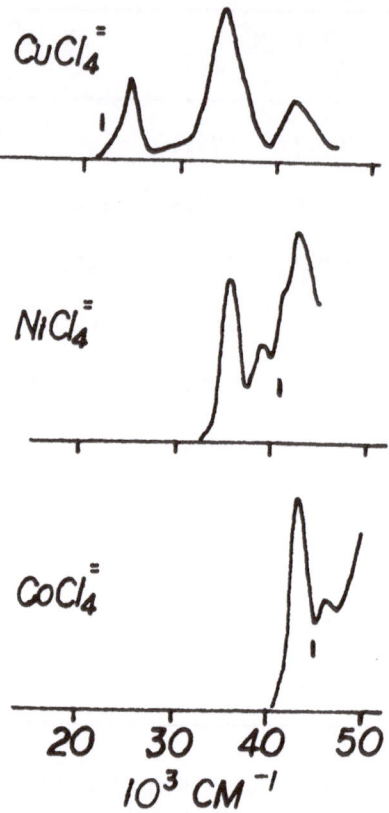

Figure 13. Chloride → metal charge transfer bands in $MX_4^=$ complexes. Marks show best fitting of energy of transitions using equation (1). Data of ref. 16.

Another type of comparison is to plot the charge transfer energies against the 2-3 oxidation potentials. This is shown in Figure 15. A 45° line is drawn through some of the points, but Co^{+++}, Ti^{+++} and Cr^{+++} do not lie on this line.

Again, we can say that the trend in the values of charge transfer energy go in the right direction as the atomic number changes, but that there are departures from quantitative agreement with equation (1), showing that there are other factors needing to be considered. That these are to be found in the spectra rather than in the oxidation potentials is shown in Figure 16.[19] Here it is seen that the E^o values follow $-I_3$ quite faithfully, except for Ti^{+++}. The E^o value for Ti^{+++} may be in error.

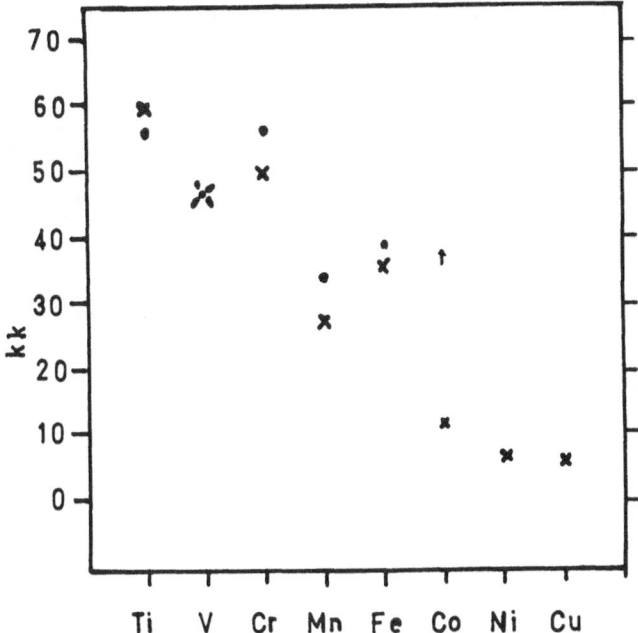

Figure 14.   The oxygen → metal charge transfer energies of trivalent transition metal ions in Al_2O_3 shown as dots: the calculated energies using equation (1) shown as crosses.   The small arrow at Co means that the observed charge transfer energy is higher than 35 kk, according to data of McClure, J. Chem. Phys. **36**, 2757 (1962).   Data from ref. 17.

It should be remarked at this point that it is by no means a simple matter to verify the carrier of an absorption band in a solid.   Even when impurities can be eliminated as a possible source of spurious absorption bands, it remains to be shown that the transition metal ion is in the correct oxidation state and that the complex one thinks he has is the one whose spectrum is being observed, or in crystal studies that a metal ion–defect cluster is not causing the spectrum.   Thus it is essential to use chemical analyses extensively on the systems being studied, and to use other physical methods such as E.S.R. spectroscopy to help establish the major components of the sample.   In the case of the $Ni^{+3}Al_2O_3$ sample of Tippins, it appears quite likely that what he interpreted as a charge transfer band of Ni^{+3} is actually a crystal field band of Ni^{+4}.[20]

Table 4.   Peaks in charge transfer spectra of trivalent
transition metal ions in Al_2O_3.   Data of H. Tippins, Phys.
Rev. BI, 126 (1970).   Peak positions compared to calculated
positions using C = 286.6.

Peak	Obs. Peak (10^3cm^{-1})	$\zeta(+3) - \zeta(+2)$		I_3	Calc. Peak (10^3cm^{-1})
Ti	55.5	7.8	− 7.3	227.0	60.1
V	46.4	14.0	−14.2	240	(46.4)
Cr	56.0	21.8	− 8.4	249.7	50.3
Mn	33.4	12.6	0.0	271.8	27.4
Fe	38.7	0.0	− 4.0	247.2	35.4
Co		7.2	− 8.0	270.2	15.6
Ni		14.4	− 9.6	283.7	7.7
Cu		21.6	− 6.0	297.1	5.1

The detailed analysis of acceptor-type charge transfer bands
must be made in terms of the molecular orbital theory of the
local cluster $MX_4^=$, $MX_6^=$ etc.   A typical molecular orbital energy
diagram for such a cluster is shown in Figure 17.   Schatz and
his coworkers[21] have done detailed studies of some of the higher
transition metals (4d, 5d) and have been able to make assignments
of bands to the transitions between the levels of Figure 17.
In this way, they have established the actual ordering of the
molecular orbital levels of octahedral complexes.

The 3d-4s transitions of the first group of transition
metals was established by a systematic study of the high energy[15]
spectra of the divalent ions in various fluoride host crystals.
There are weak bands from 35,000 to 70,000 cm^{-1} as shown in
Figure 18, and they show an energy which increases with atomic
number.   According to equation (1), these should be donor-type
transitions.   Various detailed properties of these bands show
that the 3d → 4s assignment is preferable to a 3d → conduction
band assignment.   The major one of these reasons is that the
absolute energy of the bands is almost identical to the 3d → 4s
energy of the free atoms when crystal field corrections are made.
The positions of these bands calculated on the 3d → 4s assumption,
and including crystal field corrections are shown on Figure 18.

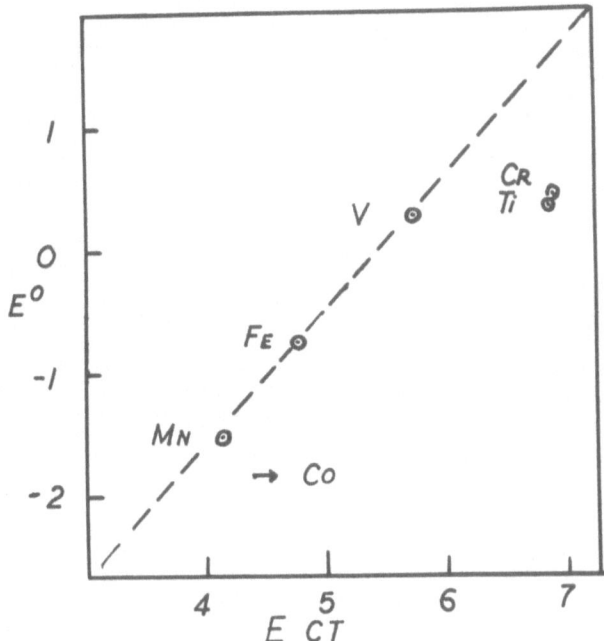

Figure 15.   Charge transfer bands of trivalent transition metal
ions in Al_2O_3 (ref. 17) <u>vs</u>. +2 → +3 oxidation potential.   The
line drawn is of unit slope.

     If these were fluoride   metal charge transfer transitions
analogous to those of Figure 13, the band energies would have
the opposite dependence on atomic number.   The question can be
asked, where are the fluoride   metal charge transfer bands?
We would expect that nickel would have the lowest energy band of
this type.   Further studies of spectra at higher energies may
reveal these bands.

     The higher atomic states have more diffuse wavefunctions
than lower ones.   There should be an increase of energy for say
4s due to the antibonding of the F-ions if the 4s function is
confined to the locale of the impurity; but the most diffuse
states must gain potential energy due to outer metal atom cores.

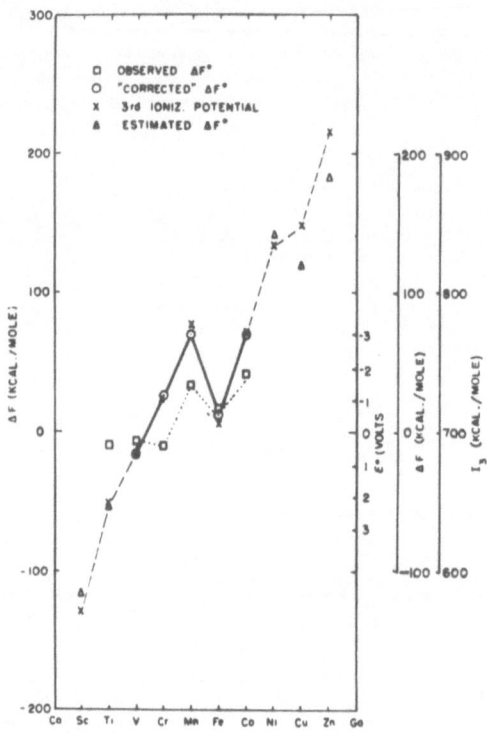

Figure 16.   The +2 and +3 oxidation potentials of divalent
transition metal ions with and without crystal field corrections.
The third ionization potential is also shown for comparison.
Ref. 19.

So, the highly excited 5s, 6s -- which in the free atoms see
less and less of the core potential, in the crystal will see
more and more of the other atom potential.   Therefore, the
excitation energy cannot rise above the energy of the conduction
band.   All such states should have about the same transition
probability from a 3d state, and in a crystal they should have
about the same energy.

    In another description of the higher non-localized orbitals,
we could call them localized excitons.   This language recalls
the localized excitons of semiconductors, and there is a close
parallel here.   Because of the lower dielectric constant, these
delocalized atomic levels will be more localized than the
corresponding orbitals around a phosphorus atom in a silicon

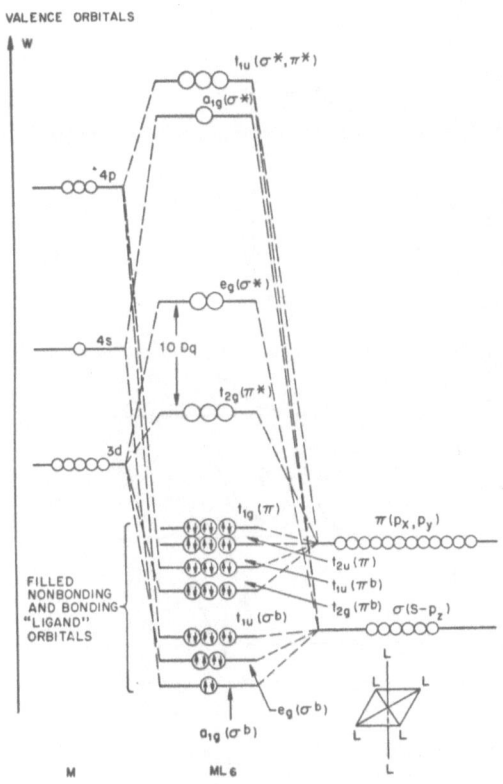

Figure 17.   A typical molecular orbital energy level diagram
of an octahedral complex (C.J. Ballhausen and H.B. Gray,
Molecular Orbital Theory, Benjamin, 1964).

crystal, for example.   They would not approach the hydrogenic
limit nearly as closely as do effective-mass excitons in semiconductors.   It would be interesting if some of the properties
of the higher atomic states could be measured in order to learn
something about electrons in these orbits of intermediate de-
localizations.

V.   CONCLUSION

We do not find any significant disagreement between the
observed and calculated positions of charge transfer in 4f → 5d
transitions in the +2 or +3 rare earth series.   Charge transfer

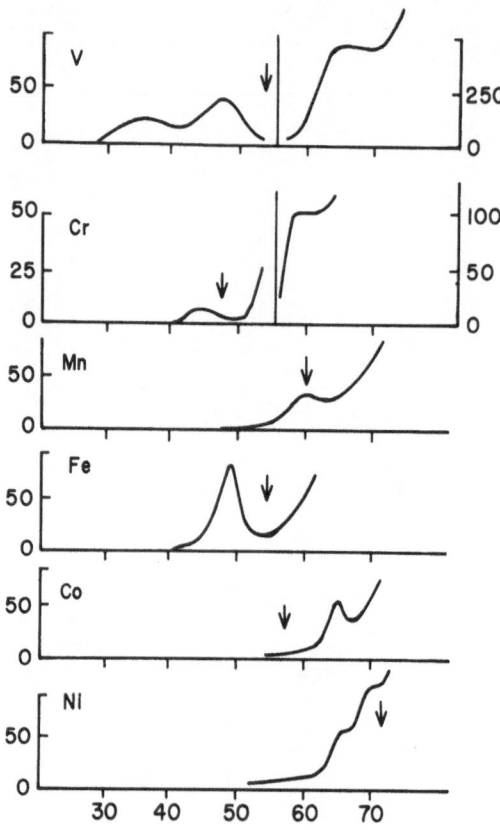

Figure 18.   The farther ultraviolet spectra of divalent
transition metal ions in $KMgF_3$ host crystal.   Arrows show
where the free atom 3d   4s transitions lie when corrected for
crystal field splittings of both upper and lower state.   This
figure is a revised version of one to be found in the thesis of
J. Sabatini, Princeton University, 1973, which will be published
shortly.

bands have only been reported for the rare earth chlorides and
bromides in which the rare earth ion has a high electron affinity.
It may be that Tb, Gd, Ce and La will not show charge transfer
bands at the normal places, since these ions in the +2 state have
$f^{n-1}d$ ground levels.   The 4f   5d transitions in the divalent
rare earths follow the energies of the free ions faithfully, and
we may conclude that the 5d orbital is not strongly perturbed by
the environment.   The bond lengths in CaF_2 where the divalent
ions have been observed are evidently large enough to provide

the space necessary for the rather extended 5d orbital (see
Figure 1).   This space may not be available in the chlorides of
the four rare earths mentioned above, so the Cl   metal transition,
ending on a 5d level, may occur at higher energies than would be
predicted from Table 2.

The question which these experiments may answer is how much
space does an atomic orbital need in order to stay at or near its
free-atom energy?

VI. EXPERIMENTAL METHODS

Much of the spectroscopy I have discussed is carried out in
the vacuum ultraviolet region.   Special techniques are needed
in order to get good data.

One of the principal problems, aside from the need for
keeping air out of the optical path, is the low intensity of the[22]
light sources.   We used microwave powered rare gas discharges.
Different rare gases provide different spectral regions.   Xenon
is the most useful from about 1900 to 1500 A.   Hydrogen
discharge sources were also used in our work.

Because of the low reflectivity of metals in the vacuum
region, the number of reflecting surfaces must be kept to a
minimum.   Nevertheless, the 2-meter McPherson spectrograph we
used has the Czerny-Turner mounting;   this uses a plane grating
and two spherical mirrors.

Since it is a single monochromator, scattered light can
become a problem, especially near the extremes of the useful
range of the source.   This problem is partly overcome by using
a 'solar-blind' photomultiplier tube as detector.   The tube we
used is insensitive to wavelengths larger than 2000 A, and thus
visible or near ultraviolet light from the source cannot
contribute to the scattered light component.

The photomultiplier was also selected so that it could be
used in a quantum counting mode.   This method of detection is
appropriate at the low light levels encountered.   By using pulse
height selecting circuits, noise pulses can be partly eliminated.
The dark noise from the phototube is extremely low because of the
high thermionic barrier of the photocathode, and is entirely due
to cosmic ray background.

The output of the quantum counter has been interfaced to a
small computer.[23]   Records can be stored in the computer, background can be subtracted and the corrected output can be plotted
on an energy scale.

Since the major problem in this research was to identify the carrier of an absorption band with certainty we have begun to apply other experimental methods. Magneto circular dichroism is one. We use a CaF_2 strain plate driven by a crystal quartz oscillator and a MgF_2 Glan-Thompson prism to produce alternating circularly polarized light. The quantum counting detection system is divided into two channels by an electronic switching and gating circuit so that right and left circular absorption can be recorded separately. This system has operated in the visible, but has not yet been tested in the vacuum ultraviolet.[23] Another type of experiment is microwave-optical double resonance. This can relate a spin-resonance signal to an optical signal and thus provide extra information to help identify an optical absorption band.[24] This method is not a panacea, however, because the microwave absorption must be partially saturated in order to modulate the optical absorption, and this is not always possible, especially with ions having low spin-orbit components such as Co^{++} and Fe^{++}.

REFERENCES

1.   J. Reader and J. Sugar, J. Opt. Soc. Am. 56, 1189 (1966).

2.   J. Reader and J. Sugar, J. Opt. Soc. Am. 60, 1421 (1970).

3.   J. Sugar and J. Reader, J. Opt. Soc. Am. 55, 1286 (1965).

4.   J. Sugar and J. Reader, J. Chem. Phys. 59, 2083 (1973).

5.   D.S. McClure and Z.J. Kiss, J. Chem. Phys. 39, 3251 (1963).

6.   O.G. Holmes and D.S. McClure, J. Ceem. Phys. 26, 1686 (1957).

7.   M. Schlesinger and J. Szazurek, Phys. Rev. B8, 2367 (1973).

8.   L.J. Nugent, R.D. Baybarz, J.L. Burnett and J.L. Ryan, Inorg. and Nucl. Chem. 33, 2503 (1971).

9.   T.S. Piper, J.P. Brown and D.S. McClure, J. Chem. Phys. 46, 1353 (1967).

10. R.C. Alig, Z.J. Kiss, J.P. Brown and D.S. McClure, Phys. Rev. 186, 276 (1969).

11. R.C. Alig, R.C. Duncan and B.K. Mokross, J. Chem. Phys. 59, 5837 (1973).

12. H.A. Weakliem, Phys. Rev. B6, 2743 (1972).

13. H. Witzke, D.S. McClure and B. Mitchell (The Photoluminescence Spectra of Single Crystals of $SrCl_2:Yb^{+2}$) in 'Luminescence of Crystals, Molecules and Solutions' (F. Williams, ed.) p. 598, Plenum Press, 1973.

14. C. Moore, Atomic Energy Levels, Vols I, II, III, NBS Circular 467.

15. J. Sabatini, Ph.D. Thesis, Princeton University, 1973. Near and Vacuum Ultraviolet Study of Metal Donor and Metal Acceptor Charge Transfer Transitions in Transition Metal Doped Fluoride Host Materials.

16. B.D. Bird and P. Day, J. Chem. Phys. $\underline{49}$, 392 (1968).

17. H. Tippins, Phys. Rev. $\underline{B1}$, 126 (1970).

18. D.S. McClure, J. Chem. Phys. $\underline{36}$, 3757 (1962).

19. P. George and D.S. McClure, Progress in Inorganic Chemistry, Vol. 1 (F.A. Cotten, ed.) Academic Press, 1959. The Effects of Inner Orbital Splitting on the Thermodynamic Properties of Transition Metal Compounds and Coordination Complexes.

20. A. Salwin, unpublished work, Princeton University.

21. S.B. Piepho, J.R. Dickenson, J.A. Spencer and P.N. Schatz, J. Chem. Phys. $\underline{56}$, 2668 (1972).

22. Techniques of Vacuum Ultraviolet Spectroscopy, James Samson, John Wiley and Sons, 1967.

23. D. Bruce Chase, Ph.D. Thesis, Princeton University, 1975.

24. Arthur Salwin, Ph.D. Thesis, Princeton University, 1974.

THEORY OF MAGNETIC CIRCULAR DICHROISM SPECTROSCOPY

P.J. Stephens

Department of Chemistry, University of Southern
California, Los Angeles, U.S.A.

In materials that are not optically active the absorption of left and right circularly polarised (CP) light is identical:

$$A_+ = A_-$$

$$A = A_- - A_+ = 0 \tag{1}$$

where A is absorbance (optical density) and + and − denote right and left CP light respectively. In the presence of a longitudinal magnetic field (that is, parallel to the light propagation direction), however, the optical properties of materials need no longer be symmetrical with respect to left and right circular polarisations and, in particular,

$$A_+ \neq A_-$$

$$\Delta A \neq 0 \tag{2}$$

This is the phenomenon of magnetic circular dichroism (MCD).

On a microscopic level, MCD relates to the perturbation of the absorbing system by the applied field − the Zeeman effect − and MCD leads to similar types of information to classical Zeeman spectroscopy. Its advantage is that it can be measured in broad absorption lines and bands whose Zeeman splittings cannot be spectrally resolved. It is, therefore, a much more general experiment.

P. Day (ed.), Electronic States of Inorganic Compounds. 141–156. *All Rights Reserved.*
Copyright © 1975 by D. Reidel Publishing Company, Dordrecht-Holland.

In these notes, we present a simple introduction to the theory and concepts of MCD. References 1-6 provide more extensive coverage of the theory and applications of MCD.

1. AN ATOMIC EXAMPLE[7]

Consider first an atom with a ^{1}S ground state and a transition to a ^{1}P excited state. The effect of the magnetic field is to split the three-fold degeneracy of the ^{1}P level. Quantitatively, the magnetic field perturbation is (to first order)

$$\mathscr{H}' = -\vec{\mu}\cdot\vec{H}$$

$$\vec{\mu} = \sum_i \frac{e}{2mc}(\vec{1}_i + 2\vec{s}_i) = -\beta(\vec{L} + 2\vec{S}) \tag{3}$$

where $\vec{\mu}$ is the magnetic dipole moment operator, $\vec{H}$ is the magnetic field, e and m are the electron charge and mass, i sums over electrons, $\vec{1}_i$ and $\vec{s}_i$ are one-electron orbital and spin angular momentum operators, $\vec{L}$ and $\vec{S}$ are total angular momentum operators and β is the electronic Bohr magneton. Taking $\vec{H}$ along the z axis, the change in energy of the M sublevels of ^{1}P is given by

$$E_M = \langle{^1}PM|\mathscr{H}'|{^1}PM\rangle = \beta MH \tag{4}$$

as shown in Figure 1.

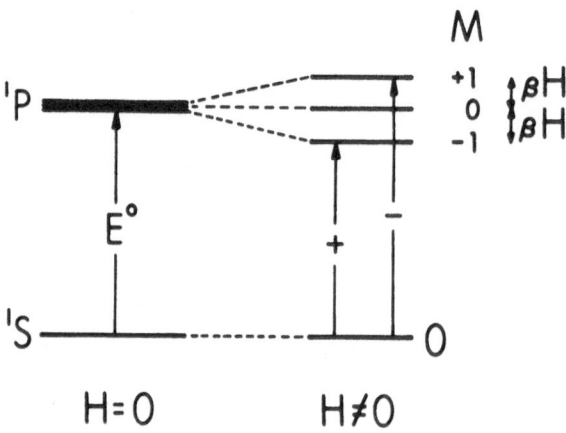

Figure 1. Energy levels and CP selection rules for a ^{1}S$\rightarrow^1$P transition with and without an applied magnetic field.

The absorbance of a transition $A \to J$ for right and left CP radiation with propagation vector $\vec{k}$ is related to the transition matrix elements of m_+ and m_- respectively, where $\vec{m}$ is the electric dipole moment operator $\sum_i e \vec{r}_i$ and $M_\pm = \frac{1}{\sqrt{2}}(m_x \pm i m_y)$.   Thus

$$A_\pm(A \to J) = \gamma \left(\frac{N_A}{N}\right) |\langle A | m_\pm | J \rangle|^2 f(E_{JA}, E)$$

$$\gamma = \left\{ \frac{\mathcal{N} \pi^2 \log_{10} e}{250 \, \hbar c} \right\} Cl; \qquad \int_0^\infty f(E_{JA}, E)/_E \, dE = 1$$

(5)

where N_A/N is the fraction of atoms in state A, $f(E_{JA}, E)$ is the line shape of the transition, $\mathcal{N}$ is Avogadro's number, C is the concentration of atoms in moles/liter and l is the optical path-length.

Atomic theory leads to the transition moments and selection rules

$$\langle {}^1S | m_\pm | {}^1P \mp 1 \rangle = m$$

Right CP: $\Delta M = -1$
Left CP: $\Delta M = +1$

(6)

Thence, with equation (5)

$$A_\pm({}^1S \to {}^1P) = \gamma m^2 f(E^\circ \mp \beta H, E)$$

(7)

as illustrated in Figure 2.

The MCD of the system is then derived by subtracting A_+ from A_- and this is shown in Figure 2 for the two extreme cases: (i) the Zeeman splitting is much smaller than the line width and (ii) the Zeeman splitting is much larger than the line width.   In case (i) one obtains a derivative (S) - shaped MCD curve. Quantitatively, from equation (7)

$$\Delta A = \gamma m^2 \left[f(E^\circ + \beta H, E) - f(E^\circ - \beta H, E) \right]$$

(8)

and, expanding $f(E^\circ \pm \beta H, E)$ in a Taylor series expansion:

$$f(E^\circ \pm \beta H, E) = f(E^\circ, E) \mp \beta H \left[\partial f(E^\circ, E)/\partial E \right] + \ldots$$

(9)

we find

$$\Delta A = - 2 \gamma m^2 \beta H \left[\partial f(E^\circ, E)/\partial E \right]$$

(10)

arising from the zero-field absorption

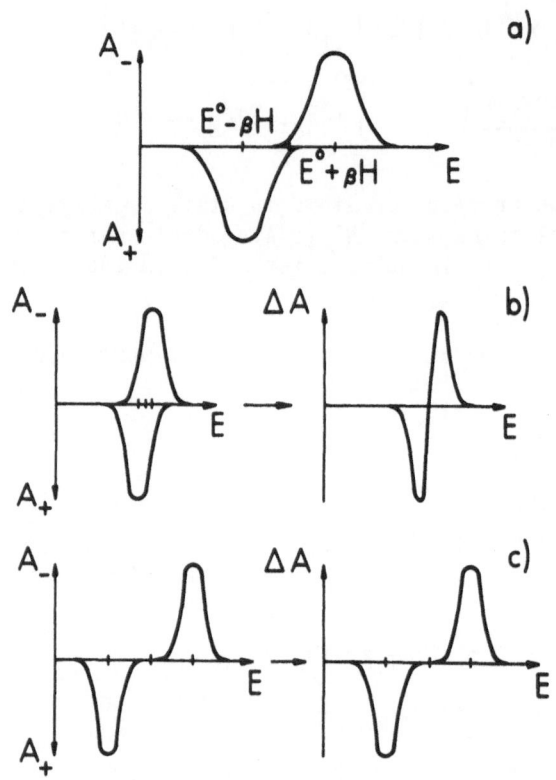

Figure 2.   CP absorption spectra and MCD of a $^1S \rightarrow {}^1P$ transition.
In Figures b and c the Zeeman splitting is respectively much
smaller and much larger than the line width.

$$A^O = A_+(H=0) = A_-(H=0) = \gamma m^2 f(E^O, E) \tag{11}$$

In order of magnitude

$$(\partial f/\partial E)_{max} \sim f_{max}/\Delta \tag{12}$$

where Δ is the line width, whence

$$[\Delta A]_{max} / A^O_{max} \sim \beta H/\Delta \tag{13}$$

The MCD is here linear in H and the order of magnitude of
$[\Delta A]_{max}/A^O_{max}$ is given by the splitting/line width ratio.   In case

(ii) A_+ and A_- exhibit the CP longitudinal Zeeman effect with fully resolved splitting and the MCD does so too.

In case (i) A_+ and A_- are only very slightly different and the Zeeman splitting is not observable in a direct Zeeman experiment.   The MCD, which derives from the Zeeman splitting, therefore provides additional information.   In case (ii), on the other hand, the MCD provides no information not observable in a direct Zeeman experiment (with CP light).

Let us now turn the system around and make ^1P the ground state and ^1S the excited state (Figure 3).   The Zeeman pattern is just as before and the transition probabilities are obtained from equation (6):

$$\langle \,^1P \pm 1 \, | \, m_{\pm} \, | \,^1S \rangle = m$$

$$(14)$$

Right CP: $\Delta M = -1$
Left CP: $\Delta M = +1$

Now, however, there is the added feature that the populations of the M sublevels in the presence of the field are unequal. According to Boltzmann

$$\frac{N_M}{N} = \frac{\exp\{- \langle \,^1PM | \,\mathcal{H}' \, | \,^1PM \rangle /kT \}}{\sum\limits_M \exp\{- \langle \,^1PM | \,\mathcal{H}' \, | \,^1PM \rangle /kT \}}$$

$$(15)$$

As T decreases the population of the M = -1 level increases until at $0°K$ only this state is occupied.   Substituting equations (14)

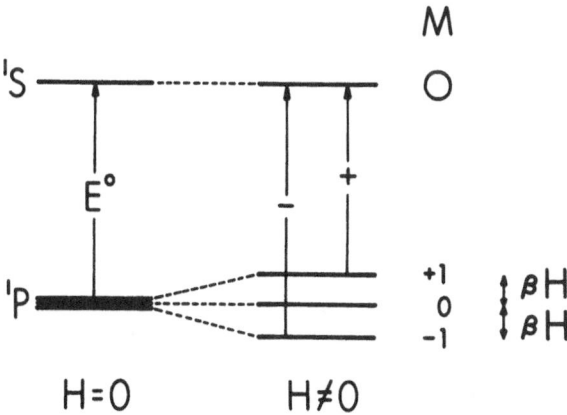

Figure 3.   Energy levels and CP selection rules for a $^1P \to {}^1S$ transition with and without an applied magnetic field.

and (15) in equation (5) gives for A_+ and A_-

$$A_{\pm}(^1P \to {}^1S) = \frac{\mathfrak{y}\exp\{\mp\beta H/kT\}}{\displaystyle\sum_{M=-1}^{+1} \exp\{\beta HM/kT\}} \; m^2 f(E^\circ \mp \beta H, E) \tag{16}$$

Equation (16) is illustrated in Figure 4, as is also the MCD deriving therefrom.   We again exhibit the two extremes of (i) unresolved and (ii) resolved Zeeman splittings and further add the effect of varying T.   The MCD in case (i) is similar to that for the S → P transition but the S-shaped dispersion is now asymmetrical, the asymmetry increasing as T decreases.   Quantitatively,

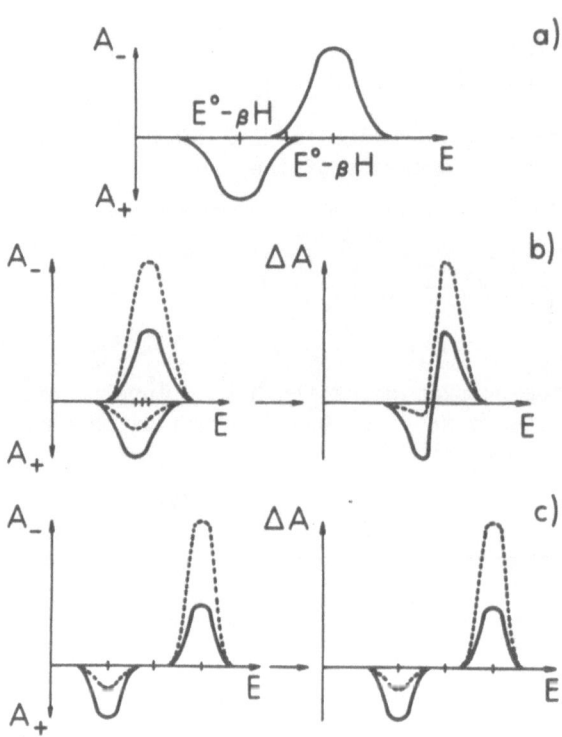

Figure 4.   CP absorption spectra and MCD of a $^1P \to {}^1S$ transition. In Figures b and c the Zeeman splitting is respectively much smaller and much larger than the line width.   The solid and dotted curves are for high and low temperatures respectively.

at 'high' T (where kT is much greater than the Zeeman splitting)
this curve can be resolved into the sum of a derivative-shaped
component and a component of identical shape to the absorption
line.   Thus, using equation (9) and simplifying equation (15)
when $\langle {}^1\text{PM} | \mathcal{H}' | {}^1\text{PM} \rangle \ll kT$ to

$$\frac{N_M}{N} = \frac{1}{3} \left\{ 1 - \frac{\beta HM}{kT} \right\} \tag{17}$$

equation (16) gives

$$\Delta A = \frac{2\gamma m^2}{3} \left\{ -(\beta H) \frac{\partial f(E^0, E)}{\partial E} + \frac{\beta H}{kT} f(E^0, E) \right\} \tag{18}$$

The first term of equation (18) is T-independent;   the second
term varies as $1/T$.   The first term arises from the Zeeman
splitting of the transition;   the second originates in the differing populations of the $M = +1$ and -1 levels.   Both terms are
linear in H.   In order of magnitude these terms have
$[\Delta A]_{max} / A^0_{max}$ ratios of $\beta H/\Delta$ and $\beta H/kT$ respectively.

Equation (18) is not valid when the Zeeman splitting is not
small compared to kT.   The MCD in this region is shown qualitatively in Figure 4, and is given quantitatively by equation (16).
Linearity in H and $1/T$ is lost and the effects of the Zeeman
splitting and the population imbalance are no longer separable.
This also obtains in case (ii).

Figure 4 shows that the MCD provides information unresolved
in a CP Zeeman experiment when the Zeeman splitting is much
smaller than the line width and kT, while there is no gain when
the effects of the field are large (Zeeman splitting large
compared to line width or kT).

2.   BROAD ELECTRONIC ABSORPTION BANDS

We turn now to the real life situation of an ion or molecule
in a crystalline solid.   Electronic transitions are here
accompanied by changes in vibrational level and appear as broad
absorption bands, rather than sharp lines.   Zeeman splittings
are rarely resolvable and MCD finds its prime application.

In this section we treat MCD by direct extension of the
discussion of Section 1 - the approach often referred to as
'rigid-shift' (RS) theory.   In Section 4 we shall comment on the
limits of the approach.

Let us consider the electronic transition $A \to J$ where A and J have degeneracies d_A and d_J at the ground state equilibrium nuclear configuration. In the magnetic field, the degeneracies are removed by Zeeman splitting. If the wave-functions diagonalizing $\mathcal{H}'$ are written ψ_{A_α} and ψ_{J_λ}, the Zeeman energies are

$$\langle \psi_{A_\alpha} | \mathcal{H}' | \psi_{A_\alpha} \rangle = \langle \psi_{A_\alpha} | L_z + 2S_z | \psi_{A_\alpha} \rangle \beta H$$

$$\langle \psi_{J_\lambda} | \mathcal{H}' | \psi_{J_\lambda} \rangle = \langle \psi_{J_\lambda} | L_z + 2S_z | \psi_{J_\lambda} \rangle \beta H \tag{19}$$

$$(\alpha = 1 - d_A; \quad \lambda = 1 - d_J)$$

The populations of the split ground state components are given by

$$\frac{N_\alpha}{N} = \frac{\exp \left\{ -\langle \psi_{A_\alpha} | L_z + 2S_z | \psi_{A_\alpha} \rangle \beta H/kT \right\}}{\sum_\alpha \exp \left\{ -\langle \psi_{A_\alpha} | L_z + 2S_z | \psi_{A_\alpha} \rangle \beta H/kT \right\}} \tag{20}$$

which at 'high' T reduces to

$$\frac{N_\alpha}{N} = \frac{1}{d_A} \left\{ 1 - \langle \psi_{A_\alpha} | L_z + 2S_z | \psi_{A_\alpha} \rangle \frac{\beta H}{kT} \right\} \tag{21}$$

Then, from equation (5), A_+, A_- and A are

$$A_\pm (A \to J) = \gamma \sum_{\alpha,\lambda} \left\{ \frac{\exp \left\{ -\langle \psi_{A_\alpha} | L_z + 2S_z | \psi_{A_\alpha} \rangle \beta H/kT \right\}}{\left[\sum_\alpha \exp \left\{ -\langle \psi_{A_\alpha} | L_z + 2S_z | \psi_{A_\alpha} \rangle \beta H/kT \right\} \right]} \right\}$$

$$\times | \langle \psi_{A_\alpha} | m_\pm | \psi_{J_\lambda} \rangle |^2$$

$$\times f_{\alpha\lambda} \left(E_{JA}^0 + [\langle \psi_{J_\lambda} | L_z + 2S_z | \psi_{J_\lambda} \rangle - \langle \psi_{A_\alpha} | L_z + 2S_z | \psi_{A_\alpha} \rangle] \beta H, E \right) \right\} \tag{22}$$

$$\Delta A = \gamma \sum_{\alpha,\lambda} \left\{ \frac{\exp \left\{ -\langle \psi_{A_\alpha} | L_z + 2S_z | \psi_{A_\alpha} \rangle \beta H/kT \right\}}{\left[\sum_\alpha \exp \left\{ -\langle \psi_{A_\alpha} | L_z + 2S_z | \psi_{A_\alpha} \rangle \beta H/kT \right\} \right]} \right.$$

$$\times \left[| \langle \psi_{A_\alpha} | m_- | \psi_{J_\lambda} \rangle |^2 - | \langle \psi_{A_\alpha} | m_+ | \psi_{J_\lambda} \rangle |^2 \right]$$

$$\times f_{\alpha\lambda} \left(E_{JA}^0 + [\langle \psi_{J_\lambda} | L_z + 2S_z | \psi_{J_\lambda} \rangle - \langle \psi_{A_\alpha} | L_z + 2S_z | \psi_{A_\alpha} \rangle] \beta H, E \right)$$

where $f_{\alpha\lambda}$ is the band shape of the $A \to J$ Zeeman component, and is assumed to be the same as that at zero-field, $f(E^o_{JA}, E)$, shifted by the $A_\alpha \to J_\lambda$ Zeeman shift. Making the excellent approximation that the Zeeman shift is small compared to the band width, we can expand $f_{\alpha\lambda}$:

$$f_{\alpha\lambda} (E^o_{JA} + [\langle \Psi_{J_\lambda} | L_z + 2S_z | \Psi_{J_\lambda} \rangle - \langle \Psi_{A_\alpha} | L_z + 2S_z | \Psi_{A_\alpha} \rangle] \beta H, E) =$$

$$f(E^o_{JA}, E) - [\langle \Psi_{J_\lambda} | L_z + 2S_z | \Psi_{J_\lambda} \rangle - \langle \Psi_{A_\alpha} | L_z + 2S_z | \Psi_{A_\alpha} \rangle] \times \beta H. \qquad (23)$$

$$\times \partial f(E^o_{JA}, E)/\partial E$$

At 'high' T, with equation (21) also, ΔA can then be reduced to

$$\Delta A = \gamma \sum_{\alpha, \lambda} \frac{1}{d_A} \left\{ \left[|\langle \Psi_{A_\alpha} | m_- | \Psi_{J_\lambda} \rangle|^2 - |\langle \Psi_{A_\alpha} | m_+ | \Psi_{J_\lambda} \rangle|^2 \right] \times \right.$$

$$\left[[\langle \Psi_{J_\lambda} | L_z + 2S_z | \Psi_{J_\lambda} \rangle - \langle \Psi_{A_\alpha} | L_z + 2S_z | \Psi_{A_\alpha} \rangle] (\frac{-\partial f(E^o_{JA}, E)}{\partial E}) \right. \qquad (24)$$

$$\left. - \langle \Psi_{A_\alpha} | L_z + 2S_z | \Psi_{A_\alpha} \rangle f(E^o_{JA}, E)/kT \right] \beta H$$

which we rewrite as

$$\Delta A = \gamma \left\{ -\alpha_1 \frac{\partial f(E^o_{JA}, E)}{\partial E} + \frac{C_o}{kT} f(E^o_{JA}, E) \right\} \beta H \qquad (25)$$

where

$$\alpha_1 = \frac{1}{d_A} \sum_{\alpha, \lambda} \left[\langle \Psi_{J_\lambda} | L_z + 2s_z | \Psi_{J_\lambda} \rangle - \langle \Psi_{A_\alpha} | L_z + 2S_z | \Psi_{A_\alpha} \rangle \right] \times$$

$$\left[|\langle \Psi_{A_\alpha} | m_- | \Psi_{J_\lambda} \rangle|^2 - |\langle \Psi_{A_\alpha} | m_+ | \Psi_{J_\lambda} \rangle|^2 \right]$$

$$(26)$$

$$C_o = \frac{1}{d_A} \sum_{\alpha, \lambda} \langle \Psi_{A_\alpha} | L_z + 2S_z | \Psi_{A_\alpha} \rangle \times \left[|\langle \Psi_{A_\alpha} | m_- | \Psi_{J_\lambda} \rangle|^2 - |\langle \Psi_{A_\alpha} | \right.$$

$$\left. m_+ | \Psi_{J_\lambda} \rangle|^2 \right.$$

For comparison, the zero-field absorption is

$$A^O = \gamma \frac{1}{d_A} \sum_{\alpha,\lambda} |\langle \psi_{A_\alpha} | m_\pm | \psi_{J_\lambda} \rangle|^2 f(E^O_{JA}, E) = \gamma D_o f(E^O_{JA}, E) \qquad (27)$$

The two terms in equation (25), labelled $\mathcal{A}$ and $\mathcal{C}$, arise from the Zeeman splitting of the transition and the population redistribution in the ground state due to its Zeeman splitting. Their origin is illustrated in Figure 5 for the specific example of a doublet → doublet transition.

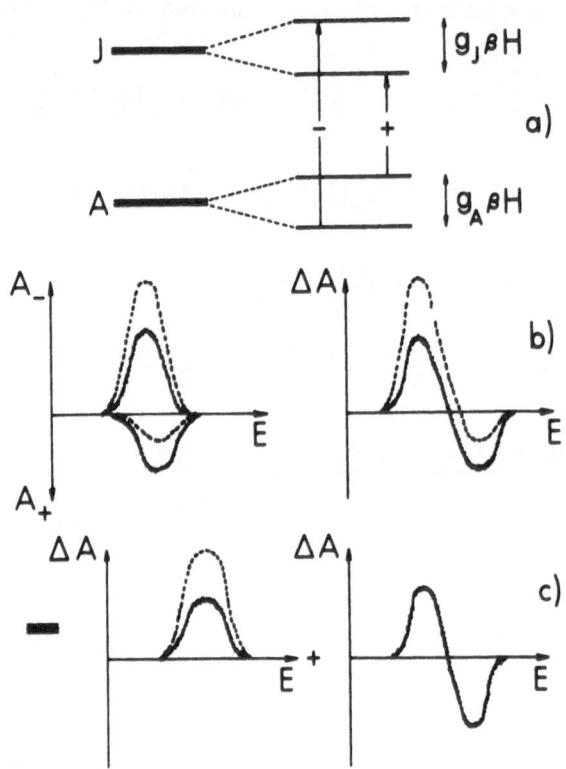

Figure 5. $\mathcal{A}$ and $\mathcal{C}$ terms for a doublet → doublet transition. The solid and dotted curves apply to high and low temperatures respectively. In this example

$$\mathcal{A}_1 = \frac{m^2}{2}(g_J + g_A); \quad \mathcal{C}_o = \frac{m^2}{2} g_A; \quad D_o = \frac{m^2}{2}; \quad \mathcal{A}_1/D_o = g_J + g_A; \quad \mathcal{C}_1/D_o = g_A,$$

where m is the magnitude of the allowed transition matrix elements of m_+ and m_-.

The relation of this general discussion to the specific examples of Section 1 is hopefully transparent. Evaluation of a_1 and C_0 using the matrix elements of equations (4), (6) and (14) and substitution in equation (25) leads to the specific equations (10) and (18) for ΔA earlier obtained.

Equation (25) becomes inadequate when kT is lowered sufficiently to become comparable with the ground state Zeeman splitting. Equation (20) is then needed in place of equation (21). Also, if structure sharp enough exists in the absorption band, equation (23) can become invalid. In both cases ΔA is no longer linear in H.

A more basic deficiency of equation (25) lies in our implicit assumption that the wave-functions ψ_{A_α} and ψ_{J_λ} are correct in the presence of the field. This is not generally true, H' coupling different zero-field electronic states together, and we should put

$$\psi_{A_\alpha}(H) = \psi_{A_\alpha}^0 + \sum_{K \neq A} \psi_K^0 \frac{\langle \psi_K^0 | H' | \psi_{A_\alpha}^0 \rangle}{W_A^0 - W_K^0} + \dots \tag{28}$$

and similarly for $\psi_{J_\lambda}(H)$, where zeros refer to H = 0. This perturbation of the wave-function does not affect the first order Zeeman energies of equation (19), but does add a term linear in H to the transition probability. If this is included, equation (25) becomes

$$\Delta A = \gamma \left\{ - a_1 \frac{\partial f}{1 \partial E} + \left(B_0 + \frac{C_0}{kT} \right) f \right\} \beta H \tag{29}$$

where

$$
\begin{aligned}
B_0 = \frac{2}{d_A} \sum_{\alpha, \lambda} \text{Re} \Bigg\{ &\sum_{K \neq J} \big[\langle A_\alpha | m_- | J_\lambda \rangle \langle K | m_+ | A_\alpha \rangle - \\
&\langle A_\alpha | m_+ | J_\lambda \rangle \langle K | m_- | A_\alpha \rangle \big] \times \langle J_\lambda | L_z + 2S_z | K \rangle / W_J^0 - W_K^0 \\
&+ \sum_{K \neq A} \big[\langle A_\alpha | m_- | J_\lambda \rangle \langle J_\lambda | m_+ | K \rangle - \langle A_\alpha | m_+ | J_\lambda \rangle \langle J_\lambda | m_- | K \rangle \big] \\
&\times \frac{\langle K | L_z + 2S_z | A_\alpha \rangle}{W_A^0 - W_K^0} \Bigg\}
\end{aligned}
\tag{30}
$$

The new $\mathcal{B}$ term thus adds a T-independent effect having the same dispersion as the $\mathcal{C}$ term.

The relative orders of magnitude of the $\mathcal{A}$, $\mathcal{B}$ and $\mathcal{C}$ terms in ΔA are

$$\mathcal{A} : \mathcal{B} : \mathcal{C} \sim \frac{1}{\Delta} : \frac{1}{\Delta W} : \frac{1}{kT} \qquad (31)$$

where ΔW is the order of magnitude of electronic state separations. For a 'typical' system at room temperature we could put $\Delta \sim 1000$ cm^{-1}, $\Delta W \sim 10,000$ cm^{-1}, $kT \sim 200$ cm^{-1} when

$$\mathcal{A} : \mathcal{B} : \mathcal{C} \sim 10^{-3} : 10^{-4} : 5 \times 10^{-3} \qquad (32)$$

Examples of $\mathcal{A}$, $\mathcal{B}$ and $\mathcal{C}$ terms are given in Figures 6–9. The $A_{1g} \rightarrow T_{1u}(^3P)$ transition of Tl^+/KI [8] and the $^1A_1 \rightarrow {}^1T_2(1)$ transition of MnO_4^- [9] exhibit $\mathcal{A}$ terms due to excited state degeneracy, as shown in Figures 6 and 7. The $^2T_{2g} \rightarrow {}^2T_{1u}(1)$, $^2T_{2u}(1)$ and $^2T_{1u}(2)$ charge transfer transitions of $Fe(CN)_6^{3-}$ [10] exhibit 1/T-dependent $\mathcal{C}$ terms, shown in Figure 8, due to the ground state degeneracy. The non-degenerate $n \rightarrow \pi^*$ transition of cyclobutanone [11] shows just a $\mathcal{B}$ term in Figure 9.

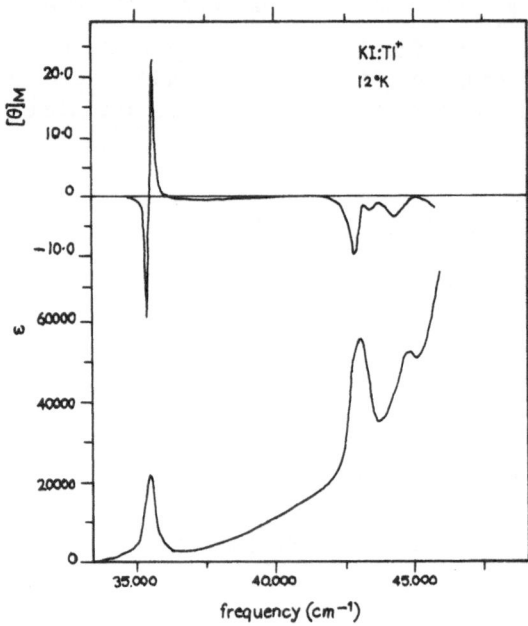

Figure 6.   MCD and absorption spectrum of Tl^+/KI at 12°K.   The 35–36,000 cm^{-1} band is due to the $A_{1g} \rightarrow T_{1u}(^3P)$ transition.

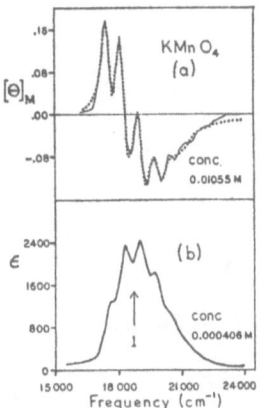

Figure 7. MCD and absorption spectrum of the $A_1 \rightarrow T_2(1)$ transition of MnO_4^- in aqueous solution at room temperature.

3. INFORMATION CONTENT

Let us now examine what information the MCD provides, within the framework of the RS equation (29). Firstly, α, $\mathcal{B}$ and $\mathcal{C}$ terms are experimentally distinguishable via the dispersion (E-dependence) and T-dependence and the α_1, $\mathcal{B}_0$ and $\mathcal{C}_0$ parameters can thus be determined individually. Second, α terms exist only when either ground or excited state is degenerate; $\mathcal{C}$ terms exist

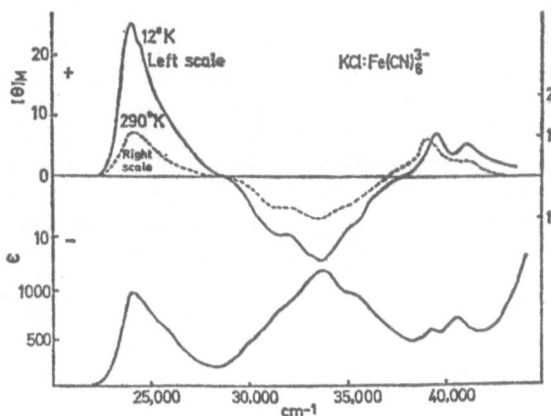

Figure 8. MCD and absorption spectrum of the lowest charge-transfer transitions of $Fe(CN)_6^{3-}$ in KCl.

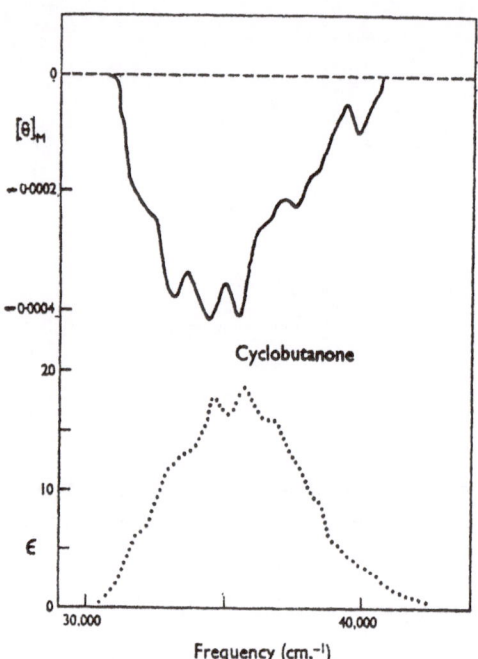

Figure 9.   MCD and absorption spectrum of the n→π* transition
of cyclobutanone.

only when the ground state is degenerate.   A non-zero ℭ term
thus proves the presence of ground state degeneracy.   A non-
zero 𝒶 term when the ground state is non-degenerate proves excited
state degeneracy.   Third, the quantitative magnitudes of 𝒶 and
ℭ terms depend on the magnetic moments (g values) of ground and
excited states and on the circularly polarised selection rules
for the transition, which in turn relate to the symmetries and
electronic natures of ground and excited states.   If the nature
of the transition is known magnetic moments can be deduced; if
magnetic moments are known (or can be calculated) the nature of
the transition can be elucidated.   Lastly, we note that little
can be said about 𝓑 terms.   They are always present (in principle);
their calculation is hard since $𝓑_0$ involves a sum over <u>all</u> electronic states of the system and usually there is no very good
reason for truncating this sum down to a few terms.

4.   COMMENTS ON THE RIGID-SHIFT APPROXIMATION

We have made many assumptions above in deriving equation (29).   Some of these are simple to remove;   others are not.   In our concluding comments we focus on the major approximation made: the rigid-shift (RS) approximation.   This consists in assuming that the absorption band associated with a Zeeman component of the electronic transition shifts rigidly – i.e. without change in shape – with the Zeeman shift.   To examine the validity of this assumption one must treat the band shape properly and include explicitly the ground and excited vibrational states and their contribution to the Zeeman effect and to the transition probabilities.   So doing, one finds that the RS approximation can be derived if the Born-Oppenheimer and Franck-Condon approximations both obtain, when the ground and excited vibronic functions are of the form

$$\Psi_a = \psi_{A_\alpha} \chi_a \qquad \Psi_j = \psi_{J_\lambda} \chi_j \tag{33}$$

where ψ_{A_α} and ψ_{J_λ} are the electronic wave functions used above and χ_a and χ_j are vibrational functions.   On the other hand, in the event that a Jahn-Teller effect exists the vibronic states are more complex, the Zeeman effect varies with vibrational state and the RS approximation is not valid.

This creates a dilemma, in that the major interesting MCD effects arise from degenerate states which are in principle always susceptible to Jahn-Teller effects (excepting Kramers doublets).   In practice, one either assumes optimistically that Jahn-Teller effects can be ignored – which, experience shows, often works quite successfully – or looks to more complex theory which includes Jahn-Teller effects explicitly.   In the latter category lies the method of moments, in which the moments of the MCD are calculated and obtained experimentally.   This method of analysis in general allows the same $\mathcal{A}_1$, $\mathcal{B}_0$ and $\mathcal{C}_0$ parameters to be obtained – and the same system information to be derived – while being more rigorous than the RS approximation.   Unfortunately, we do not have space to enter into this in detail here.

REFERENCES

1.   A.D. Buckingham and P.J. Stephens, Ann. Rev. Phys. Chem. <u>17</u>, 399 (1966).

2.   P.N. Schatz and A.J. McCaffery, Quart. Rev. Chem. Soc. <u>23</u>, 552 (1969) (Err: <u>24</u>, 329 (1970)).

3.   C. Djerassi, E. Bunnenberg, D.L. Elder, Pure Appl. Chem. 25, 57 (1971).

4.   B. Briat in Fundamental Aspects and Recent Developments in ORD and CD, ed. F. Ciardelli and P. Salvadori (1973).

5.   P.J. Stephens, J. Chem. Phys. 52, 3489 (1970).

6.   P.J. Stephens, Ann. Rev. Phys. Chem. 25, 201 (1974).

7.   This example is discussed erroneously in references 1 and 2: see Erratum to reference 2.

8.   P.N. Schatz et al., Symp. Far. Soc. No. 3, p. 14 (1969).

9.   P.N. Schatz, A.J. McCaffery, W. Suetaka, G.H. Henning, A.B. Ritchie and P.J. Stephens, J. Chem. Phys. 45, 722 (1966).

10. R. Gale and A.J. McCaffery, J.C.S. Chem. Comm. 832 (1972).

11. A.J. McCaffery et al. Chem. Comm. 520 (1966).

THE TECHNIQUE OF MAGNETIC CIRCULAR DICHROISM

R.G. Denning

Inorganic Chemistry Laboratory, University of Oxford,
South Parks Road, Oxford

1. MEASUREMENT OF OPTICAL ACTIVITY

A general expression for the electric field vector of a
circularly polarised light beam with angular frequency $\omega\,(=2\pi\nu)$
travelling in the +z direction is

$$\vec{E}_{\pm}(z,t) = E_o e^{i\omega(t-\hat{n}_{\pm}z/c)}(\vec{i}\pm i\vec{j})$$

where + and − specify right and left circularly polarised light
respectively and e.g. $\hat{n}_{+} = n_{+}-ik_{+}$ is a complex refractive index
referring to the right circularly polarised beam. The imaginary
part of the complex refractive index can be shown to be a simple
absorption coefficient representing the attenuation of the light
intensity as a function of z, i.e.

$$I(z) = I(o)\, e^{-2\omega kz/c}$$

The two parts of the complex refractive index are however
determined by the same molecular transition moments, so that if
we seek to measure these molecular quantities it is, in principle,
possible to measure either (a) the difference in the real
refractive indices for the two circular components

$$\phi = (n_{+}-n_{-})\pi/\lambda$$

by measuring ϕ the rotation of the plane of polarised light (in
radians per unit length) or (b) the difference in the absorption
coefficients

$$\Delta k = k_{-}-k_{+}$$

P. Day (ed.), Electronic States of Inorganic Compounds. 157–176. All Rights Reserved.
Copyright © 1975 by D. Reidel Publishing Company, Dordrecht-Holland.

which is called the circular dichroism.   The frequency dispersion
of the two phenomena are, however, completely different.   They
are related in a complex manner expressed by the Kramers-Kronig
transformation.   For example the rotation ϕ at a frequency ω_0
is given by

$$\phi(\omega_0) = \frac{\omega_0^2}{2\pi} \int_0^\infty \frac{\theta(\omega)d\omega}{\omega\,(\omega^2-\omega_0^2)}$$

In other words a complete knowledge of the ellipticity
$\theta = (k_- - k_+)\pi/\lambda$, over the whole electromagnetic spectrum is
necessary to determine a single value of ϕ .   In practice, if
an absorption band is well separated from others in the spectrum,
a reasonably good transformation between the two phenomena may
be obtained over a limited wavelength range.   Either measurement
is technically possible although circular dichroism is now almost
universally favoured.

Because the ellipticity or dichroism occurs only at wavelengths where the sample absorbs, the contributions of numerous
overlapping transitions are much more readily separated than in
the optical rotatory dispersion spectrum.   This is because the
dispersion of the rotation is complex and extends well outside
the regions of absorption.   The ORD experiment is particularly
inappropriate in the measurement of magnetically induced optical
activity because all the elements in the magnetic field, i.e. the
cell, the solvent and the optical components, contribute to the
rotation although they are not absorbing.   The principal
advantage of the rotation experiment is that it can detect the
contribution to the rotation of an electronic transition that
lies outside the wavelength range of the instrument.   The main
applications are in natural optical activity.

2.   THE SPECTROPOLARIMETER

A brief description of the features of this instrument is
worthwhile so that its sensitivity can be compared with a CD
spectrometer.

The principle of all spectropolarimeters is the detection of
the condition in which a polariser and an analyser are exactly
crossed.   In general-either the polariser or the analyser is made
to oscillate through a small angle on either side of the crossed
condition, which is registered by a symmetrical distribution of
the transmitted light intensity with respect to the sense of the
displacement.   The rotation of a sample placed between the two
polarisers is compensated by a servo-driven rotation of the non-
oscillating component.   A typical system is shown in Figure 1.

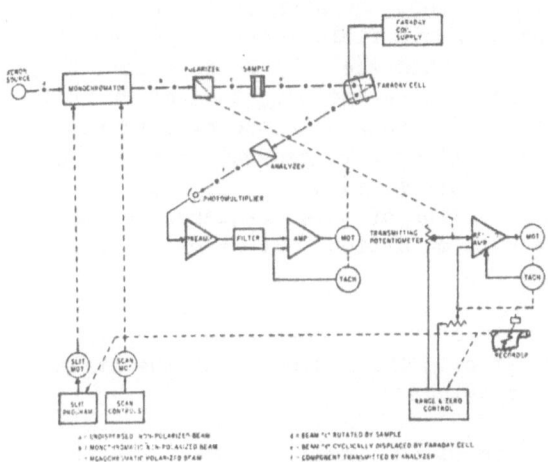

Figure 1.   The principle of a spectropolarimeter.   The Faraday
Cell provides an angular modulation of the plane of polarisation.
A frequency selective amplifier detects the fundamental frequency
of the oscillation and its output rotates the polariser to null
this signal.

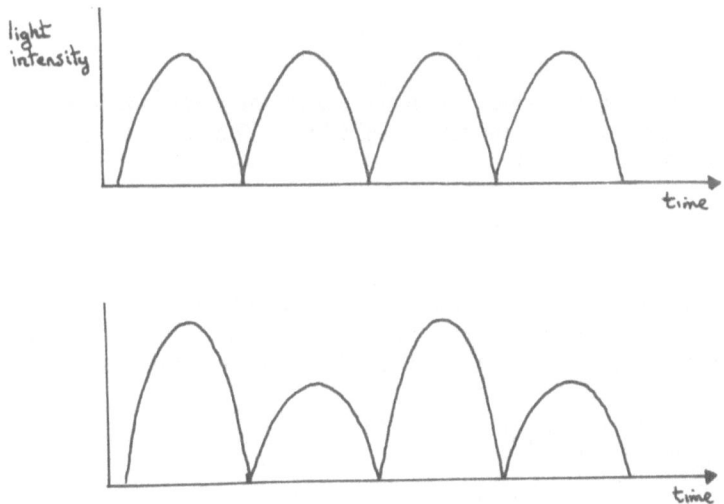

Figure 2.   Signals at the detector of a spectropolarimeter:
(a) in-balance waveform (DC + 2w);   (b) out-of-balance waveform
(DC + w + 2w).

In this system the oscillation of the plane of polarisation is achieved by means of the Faraday rotation in a cell surrounded by an AC solenoid.   Since the sign of the Faraday rotation is reversed when the magnetic field is reversed the effect of the internal reflection within the cell is to double the oscillation amplitude.   Figure 2 shows the output waveform in the crossed and the uncrossed conditions.   If the oscillation frequency is ω the uncrossed condition will generate a signal with frequency ω , whose phase is determined by the sign of the rotation. Phase sensitive detection of this frequency is used to drive the polariser to the null condition.   A recorder is coupled to the polariser drive.

The sensitivity of the spectropolarimeter is limited by its signal to noise ratio.   The intensity of the light beam as a function of time is given by

$$I(t) = I_o e^{-A} \sin^2(\phi_o + \alpha_o \sin \omega t)$$

where A is the absorption coefficient, ϕ_o is the actual rotation and α_o is the amplitude of the polariser oscillation.   Since both ϕ_o and α_o are small this becomes

$$I(t) = I_o e^{-A}(2\phi_o \alpha_o \sin \omega t + \alpha_o^2 \sin^2 \omega t) \qquad (1)$$

The signal current, due to the first term in equation (1), after rectification at the fundamental frequency is then

$$i_{signal} = d I_o e^{-A}.4\phi_o \alpha_o/\pi$$

where d is the photocathode quantum efficiency and we have ignored the gain of the photomultiplier.   The much larger background current due to the second term in equation (1) is

$$i_{background} = d I_o e^{-A}.\alpha_o^2/2$$

In well designed equipment the principal noise source is the 'shot' noise generated solely by the photon flux statistical fluctuations.   In such a case the RMS noise current is related to the mean background current by

$$i_{noise}^2 = d I_o e^{-A} \alpha_o^2.k\Delta f/2$$

where k is a constant characterising the photomultiplier and Δf is the instrumental bandwidth.   It follows that the signal-to-noise ratio is given by

$$S/N = \frac{4\phi_o \cdot \sqrt{[\delta \, I_o e^{-A}]}}{\sqrt{[k \, \Delta f/2]}} \qquad (2)$$

and is independent of the instrumental modulation amplitude.
To achieve high sensitivity long time constants are used to give
small values of Δf.   This makes the measurement of ORD (and
for the same reasons, CD) a slow procedure.   Equation (2) also
shows that photomultipliers of the highest quantum efficiency
must be used.   The main controlling feature which is amenable
to careful design is the light intensity.   Commonly intense
sources such as a 500 watt high pressure xenon arc are used.
The intensity is actually less important than the brightness of
the source.   The brightness depends on the size of the source
and the constraint on the parallelism of the light passing
through a polariser implies that more flux may be collected from
sources which are small and bright.   The size of a bright high
pressure xenon arc source is about 3 mm. by 2 mm.   The dependence
on intensity also requires good optical design in the monochromator
and other optics.   We return to this aspect later.

In practice a good commercial ORD instrument has a
sensitivity of about 1×10^{-3} degrees.   The experiment has one
major advantage in sensitivity over CD.   Equation (2) shows that
the signal to noise ratio can be increased by increasing ϕ_o as
long as the absorption coefficient, A, is zero.   This can be
done simply by increasing the concentration of the solution.
Naturally the spectroscopic information available from such a
measurement is limited, but it may, none the less, provide data
about a weakly optically active chromophore beyond the wavelength
range of the instrument.

The limitations to the accuracy of the ORD experiment
depend on (a) the mechanical stiffness of the instrument bed
and (b) the presence of spurious rotation introduced by strain
in the sample.   Serious difficulties arise from the strain often
present in solid samples at low temperatures.

3.   CIRCULAR DICHROISM

The principal advantages of the circular dichroism technique
are its simple relation to absorption spectra and its relative
independence of stray anisotropies in the refractive index
(birefringence).   CD spectrometers all revolve around a device
for generating circularly polarised light.   When plane-polarised
light is incident on a birefringent optical element such that the
polarisation plane lies at 45° to the principal axes of the
refractive index, the relative retardation of one component can
lead to a phase shift of half a wavelength between the two

components.   Figure 3(a) shows that a $\frac{1}{2}$ wavelength shift of
the horizontally polarised component will lead to plane-polarised
light polarised at 90° to the incident radiation.   If such an
element were placed between crossed polarisers the radiation
would be transmitted without attenuation.   If the phase shift
is $\lambda/4$ as shown in Figure 3(b) the resultant amplitude vector
follows a helical path which would appear as clockwise when
viewed with the radiation shining into the eye of the observer.
This is conventionally called right circularly polarised light.

One of the simplest devices for measuring CD was first
introduced by Rosenheck and Doty.[2]   Suppose that at a wavelength
λ_0 the retardation between the two components of a birefringent
plate is 250 $_0$.   To a first approximation the actual retardation
in time is independent of wavelength so that radiation of wavelength 2 $_0$ will suffer a phase shift of 125 wavelengths.   At
0.999 λ_0 the retardation will be 250.25 wavelengths and
circularly polarised light will be generated.   At 0.998 λ_0 the
retardation will be 250.50 wavelengths, giving plane polarised
light and at 0.997 λ_0 circularly polarised light with the
opposite sense to that at 0.999λ_0 will emerge.   In other words
the sense of the circular polarisation will oscillate rapidly
as a function of wavelength.   To measure CD identical solutions
and plane polarisers are placed in the two beams of a spectrophotometer and the highly chromatic retardation plate is inserted
in the sample beam.   Figure 4 shows the nature of the measured
effect and its relation to the true CD spectrum.   Although a
good illustration of principle this technique is very limited.
The sensitivity of most absorption spectrophotometers is
insufficient to measure the small magnitude of most circular
dichroism and there are obvious difficulties in reconstructing
the CD spectrum especially when the linewidth of the spectrum
approaches the phase shift of the plate.

All circular dichroism instruments generate circularly
polarised light by modulating the birefringence of an essentially
isotropic plate.   In the Pockels Cell[1] the plate is a single
crystal of KH_2PO_4 or an isomorphous salt.   It is cut so that
the optic axis of the tetragonal crystal is coincident with the
light beam.   If electrodes are applied to the faces of the
crystal an applied voltage induces an orthorhombic distortion
and with it the necessary birefringence. KH_2PO_4 is used because
it is a ferroelectric material with very large electro-optic
coefficients.   The ferroelectric property depends on the
hydrogen bonding between phosphate tetrahedra.   In the absence
of an electric field the hydrogen bonds between phosphate
tetrahedra are symmetrical.   Figure 5 shows that associating
the hydrogen atoms with two of the phosphate oxygen atoms creates
OH groups which define one orthorhombic direction while reversing
the polarity of the field creates OH groups arranged in the

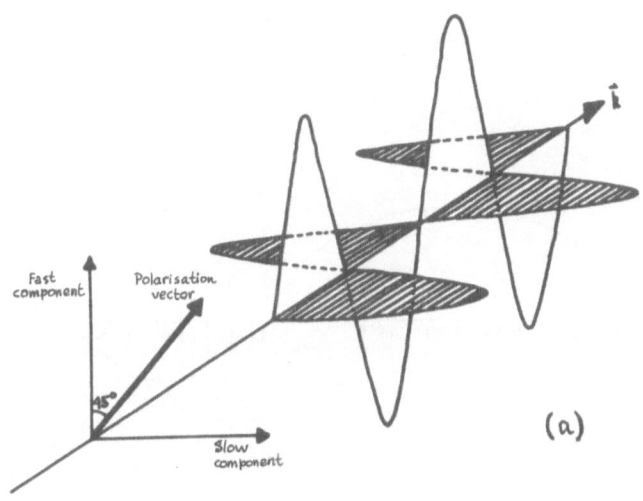

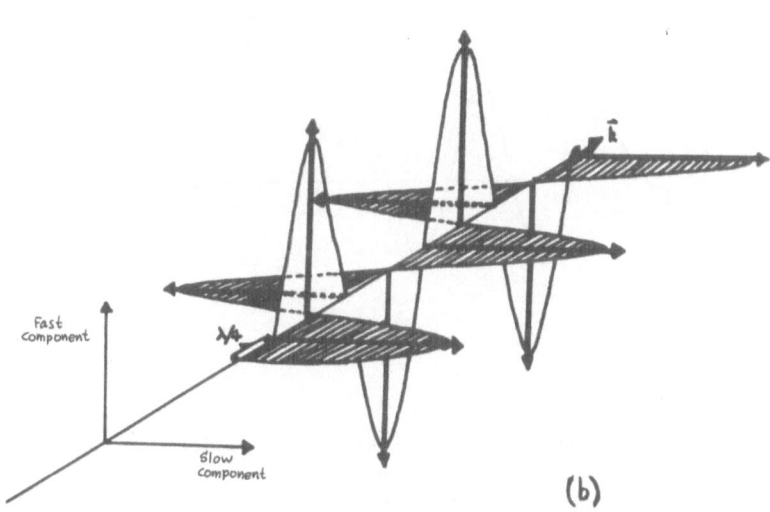

Figure 3. Effect of a birefringent plate on plane polarised radiation: (a) incident plane-polarised radiation; (b) after $\lambda/4$ retardation.

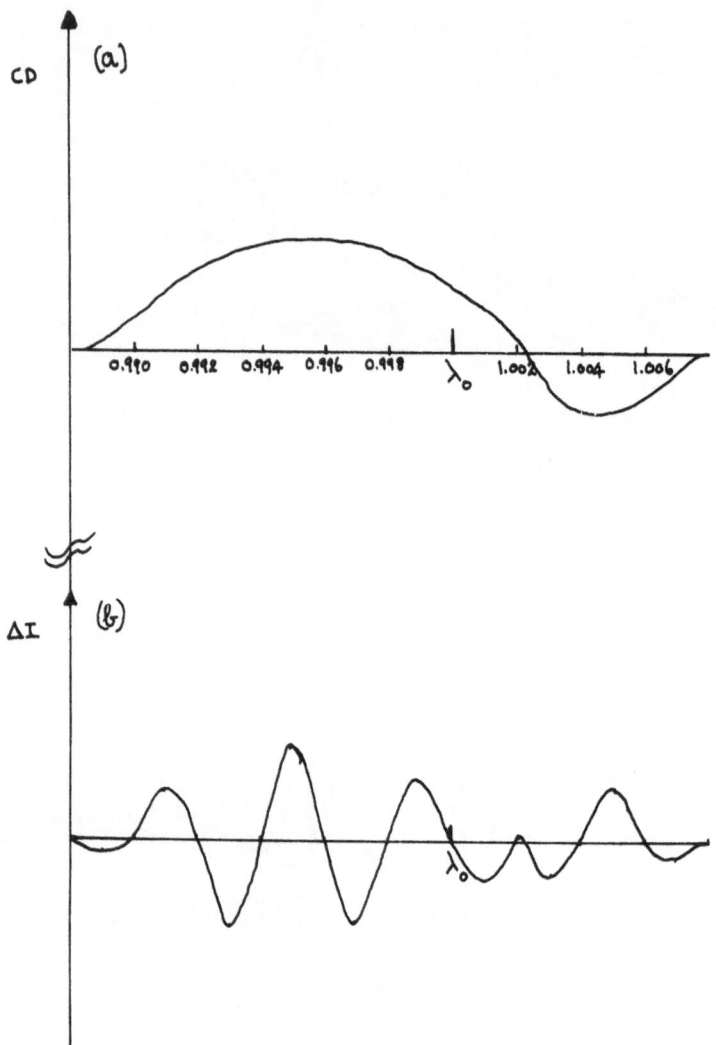

Figure 4.   The relationship between (a) the circular dichroism
spectrum and (b) the signal obtained with a highly chromatic
retardation plate, having a retardation of $250\lambda_o$, in a double
beam spectrophotometer.

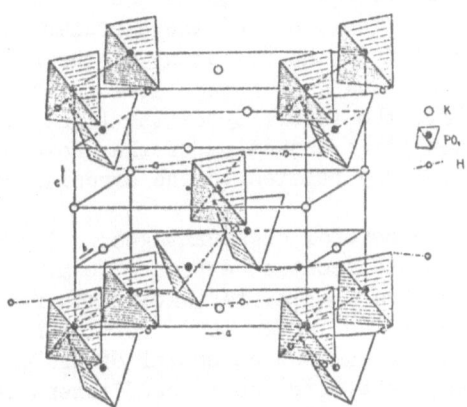

Figure 5. The structure of KH_2PO_4. Note that all the 'top'
edges of the tetrahedra are approximately parallel to the (110)
plane, and all the 'bottom' edges are nearly parallel to the
(110) plane. Applying a potential shifts the hydrogen atoms in
the symmetrical hydrogen bonds so that they are associated either
with the 'top' or 'bottom' edges of the tetrahedron, depending
upon the sign of the potential.

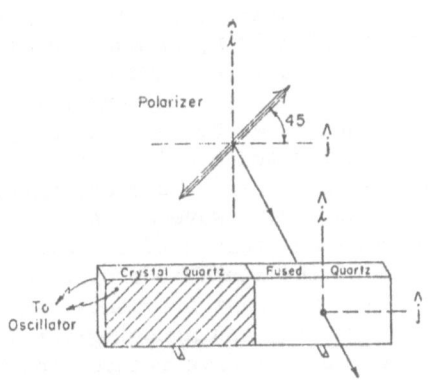

Figure 6. The resonant Acousto-optic Modulator resting upon
knife-edges. Gold electrodes have been vacuum-deposited on the
crystalline section. The strain wave has a node at the joint
which need not therefore be strong. There is an anti-node in
the region which the light beam traverses.

orthogonal direction.   Application of a.c. to the electrodes
therefore induces a birefringence whose sense alternates with
the applied potential.   The sense of the circular polarisation
is consequently modulated at the frequency of the applied a.c.
CD in a sample which follows such a modulator in the optical
path will then modulate the intensity of the light at the same
frequency.   Phase sensitive detection of the photomultiplier
output then gives a voltage related to the circular dichroism.

Pockels cells have several disadvantages.   The crystals
are expensive, fragile and decay on exposure to air.   The
potential needed to generate a quarter wavelength retardation
is about 5kV and wavelengths longer than one micron require
fields in excess of the breakdown potential of the material.
Deuteration increases the electro-optic coefficient but raises
the expense.   The most reliable electrode structure is an
evaporated gold grid which lowers the transmission of the device.
The uniaxial nature of the crystal means that it must be very
carefully oriented with respect to the incident beam to avoid
a residual birefringence in the absence of the potential.
Similarly the optical angular aperture is restricted to a solid
angle of less than about 3° because off-axis rays pass through
birefringent material and lose their polarisation.

All the instruments developed in the last few years have
used acousto-optic modulators in which the birefringence is
introduced by means of mechanical strain in the form of a
standing acoustic strain wave.   In the device invented by
Jasperson and Schnatterly and separately by Kemp[3] the driving
element is a single crystal quartz oscillator (Figure 6) to which
the isotropic fused quartz polarising element is attached by glue.
The isotropic section is carefully cut and polished so that its
natural resonant frequency is exactly in tune with that of the
driver.   The driving element must be cut from a special crystal
section so that the acoustic resonance is a pure mode developed
along the long axis of the element.   This arrangement ensures
that there is a node in the strain wave at the joint and
consequently an efficient transmission of acoustic power which
is virtually independent of the acoustic properties of the glue.
The power conversion at the resonant frequency can be extremely
efficient.   Only about 20 volts and a negligible current is
necessary to generate the quarter wave retardation at 5000 A.
Any isotropic material can be used in the polarising section
providing that it is acoustically tuned to the driver.   Schnepp[8]
has used fluorite for work in the vacuum ultra-violet, while
Stephens has used infrasil for the near infra-red.[5]

Billardon et al.[4] prefer a modulator in which the driving
elements are piezo-electric ceramic discs (Figure 7).   The
resonant condition is established by positive feedback from a

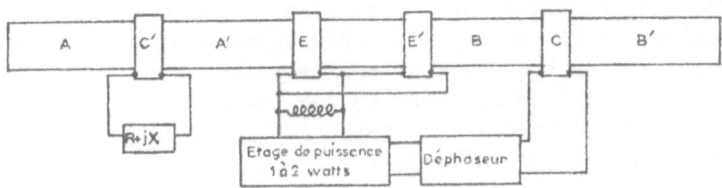

Figure 7. The forced oscillation modulator. The central
section is the optical element, which may be fused silica,
fluorite, or KRS-5 for the infrared. E and E' are the driving
piezo-electric ceramic discs. The piezo-electric disc, C,
provides positive feedback to establish the oscillation while C'
monitors the level of the oscillation. A, A', B, B' are sections
of steel which, by increasing the length, lower the oscillation
frequency and ensure a uniform strain amplitude over the optical
section.

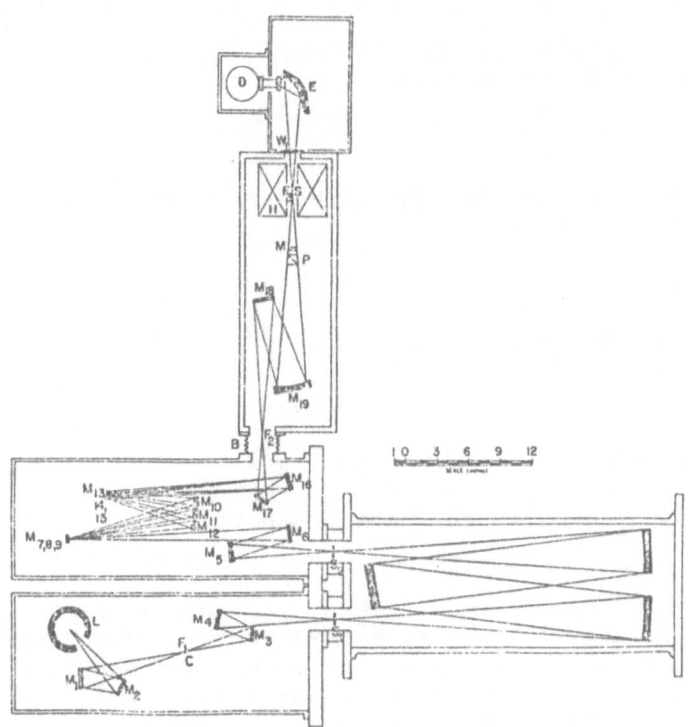

Figure 8. The infrared sensitive CD instrument of Osborne et al.
C is a chopper. The Czerny-Turner monochromator is followed by
the image slicer (M5-M17). P is the polariser, M the modulator
(infra-sil), s the sample and D the photovoltaic, InSb, detector.

sensing piezo-electric disc incorporated in the assembly.
Besides the polarising element pieces of steel are incorporated
to lengthen the resonant bar. The strain wave then has an
almost uniform amplitude over the polarising element , and a
larger aperture can be used than in the resonant device. Alkali
metal salts can be used for work in the infra-red. Holzwarth
and his colleagues[6] have used a germanium polarising element for
measurements in the far infra-red.

The resonant frequency of the modulator is determined by
its length and is usually about 50 kHz. The principal advantages
of acousto-optic modulators are (1) their robust nature, (2) their
wide wavelength range and low operating voltages, and (3) their
indifference to off-axis rays in convergent light beams. Because
the modulating material is entirely isotropic off-axis rays do not
travel through birefringent material (as in the Pockels cell) and
a much higher modulating efficiency is possible.

4.   SENSITIVITY IN THE CIRCULAR DICHROISM MEASUREMENT[5,8,9]

The retardation, δ , in a sinusoidal modulator varies with
time according to the relation

$$\delta = \delta_o \sin \omega t$$

Defining the dichroism $\Delta A = A_1 - A_r$, the transmitted intensity
consists of an unmodulated part

$$I_{dc} = \tfrac{1}{2} I_o \left[10^{-A_1} + 10^{-A_r} \right]$$

and the modulated part

$$I_{ac} = \tfrac{1}{2} I_o \left[10^{-A_1} - 10^{-A_r} \right] \sin \left[\delta_o \sin \omega t \right]$$

Taking the ratio of these two signals gives

$$I_{ac}/I_{dc} = \tanh \left[\frac{\ln 10}{2} \Delta A \right] \sin \left[\delta_o \sin \omega t \right]$$

This quantity is independent of fluctuations in the intensity
of the source and of the detector sensitivity. Even though the
CD instrument is a single beam spectrometer sufficient information
is carried on the beam to eliminate these variables. The ratio
is a function of only the dichroism and the retardation. If the
a.c. component is rectified at the fundamental frequency it can
be shown that the output is a maximum when $\delta_o = 1.84$ radians,
i.e. a little greater than $\lambda/4$. This is due to the contribution
of the third and higher odd harmonics. Integration over a cycle
then gives a ratio

$$I_{rect}/I_{dc} = 0.74 \ \tanh(1.15 \, \Delta A)$$

The retardation δ_o must be increased linearly with wavelength so that it maintains the value of 1.84 radians throughout the whole spectrum. When ΔA is less than 0.1 optical density units (which is usually the case)

$$\tanh(1.15 \, \Delta A) \sim 1.15 \, \Delta A$$

Larger values of ΔA are only likely to occur in MCD spectra at very low temperatures when the differential population of Zeeman components in the ground state is large. If the output is collected digitally the spectrum can easily be corrected to give the true ΔA values.

It is now possible to compare the sensitivity of the CD instrument with the spectropolarimeter. When the CD signal is small the rectified output is

$$i_{signal} = \delta I_o e^{-A} . \ 0.85 \, \Delta A$$

while the RMS noise current is

$$i_{noise} = \sqrt{[\delta I_o e^{-A} . k \, \Delta f]}$$

whence the signal-to-noise ratio is given by

$$S/N = \sqrt{[\delta I_o e^{-A} . \ 0.85 \, \Delta A / k \Delta f]} \qquad (3)$$

When equation (3) is compared to that for the equivalent polarimeter it is possible to compare the sensitivity of the two experiments. Assuming that the same light source, monochromator, focussing system and photomultiplier are used we find

$$\frac{(S/N)_{CD}}{(S/N)_{ORD}} = 0.47 \, \Delta A/\phi$$

This relationship can be used to compare the possibility of detecting a single isolated feature of the spectrum which has a Gaussian CD lineshape. In this case it is possible to show that the CD maximum is related to the peak-to-trough value of the ORD anomaly by the expression

$$\Delta A = 1.42 \, \phi$$

where ΔA is in optical density units and ϕ is in radians. Obviously there is little intrinsic difference between the sensitivities of the two experiments.

Equation (3) suggests that there is an optimum value for the optical density which will maximise the signal to noise ratio in the CD experiment. Rewriting equation (3)

$$S/N = k'gAe^{-A/2}$$

where k' is a constant and $g = \Delta A/A$ is the Kuhn anisotropy factor. Differentiation with respect to A gives a maximum signal to noise ratio when $A = 0.87$.

Knowing the photocurrent which is obtainable from a good monochromator at acceptable optical bandwidths it is possible to use equation (3) to calculate the absolute sensitivity of the CD experiment. In general the noise level of most modern instruments is about 1×10^{-5} optical density units with time constants of about 10 secs.

5.   OPTICAL DESIGN

The usual way of specifying the flux gathering power of a monochromator is to determine its 'f' number. This measure is related to the photographic exposure in a spectrograph. If a weak line is to be detected on a film the important quantity is the flux incident per unit area, which will determine the plate blackening for a given exposure. Increasing the width of the entrance slit merely makes the image more diffuse and does not therefore effect the exposure. On the other hand the flux received by a photoelectric detector can be increased without detriment to the optical bandpass by either increasing the slit height, or by increasing the grating dispersion; which in turn allows greater slit-widths to be used. While both of these measures increase the total flux (which is the important quantity for a photoelectric detector) neither of them increase the flux per unit area (which is important in photographic detection). The most suitable monochromator for the CD experiment therefore uses the largest aperture consistent with the aberation limit, the largest possible dispersion consistent with size and economy and the largest slit heights consistent with other optical constraints. Often double monochromators are in use. The full slit height of the monochromator can seldom be used in MCD because the demagnification necessary to form an image on a small crystalline sample gives light which is too convergent (a) to be satisfactorily polarised by a Rochon prism and (b) to meet the angular aperture requirements of a cryostat containing a super-conducting solenoid. For this reason Osborne, Cheng and Stephens[5] have used a beam slicer in their design for an infra-red sensitive circular dichroism apparatus (Fiuure 8). The slit image is dissected into three parts which are separately focussed to be superimposed at the sample.

6.   IMPERFECTIONS

The most serious difficulty in CD measurements arises from strain birefringence occurring between the modulator and the photomultiplier. The source may be the cryostat windows or even the sample itself. The result is a spurious signal whose amplitude varies with wavelength. It arises in the following way. In a perfect system the retardation oscillates symetrically between the limits $- \lambda/4$ and $+ \lambda/4$. If a static retardation plate (simulating strain birefringence) with a shift of $\lambda/4$ is interposed the oscillation will occur between the limits of zero and $\lambda/2$ retardation, i.e. between two linear polarisations. Unfortunately most photomultipliers have a sensitivity to linear polarisation. At least part of this sensitivity is due to rays which are not normal to the photo-multiplier window. They will have different reflection coefficients for the components polarised parallel and perpendicular to the reflection plane.

The best way to overcome this difficulty is to insert a depolarising element in front of the photomultiplier, although some improvement can be achieved by mechanical adjustments to the photomultiplier. Nevertheless stray birefringence means that the baseline linearity of a CD spectrum is seldom better than 1×10^{-4} optical density units. The errors are particularly serious in the ultra-violet where a given retardation corresponds to a larger proportion of a wavelength. In the Cary commercial CD instrument a multipotentiometer is used to correct for these baseline variations. Fortunately the baseline error is seldom a problem in the MCD experiment since the spectra with and without the magnetic field can be subtracted from one another. It is a more serious problem when natural CD is to be measured.

7.   DARK CURRENT

When the spectral bandpass of the monochromator must be small and when wavelengths are used at which the quantum efficiency of the photomultiplier is low the dark current of the tube may make a sizeable contribution to the anode current. In this case the ratio of a.c. and d.c. components will be smaller than the true circular dichroism.

This difficulty is more severe when photovoltaic or photoconductive detectors are used in the infra-red because they have a d.c. output in dark conditions which is liable to drift and cannot be effectively offset. The solution involves chopping the incident beam mechanically so that a dark period is included in the output. In the infra-red sensitive CD instrument of

Osborne et al.[5] this chopper runs at 540 Hz. The circular
dichroism is obtained by dividing the signal rectified at 50 kHz
by that rectified at 540 hz (Figure 9). The analogue divider
must be capable of working over a wide dynamic range to
accommodate the change in the 540 Hz component which occurs for
example when the instrument scans through an absorption band.
With this type of double modulation Holzwarth and his colleagues[6]
have been able to measure electronic circular dichroism in the
1500 cm[-1] region of an optically active cobalt(II) complex.
The accuracy of instruments with photomultiplier detectors would
also benefit from double modulation.

8.   COMPARISON WITH ABSORPTION SPECTROSCOPY

The analysis of MCD often depends on a careful comparison
with absorption spectra. In particular the moments analysis
of a feature in MCD may depend critically on the value taken for
the centre of gravity of the absorption band. Instrumentally
this is made difficult because the wavelength calibration and
spectral bandpass of the absorption spectrophotometer will differ
from that of the CD instrument. Ideally one would measure both
absorption and CD spectra simultaneously on the same instrument.
Figure 10 shows an instrument developed in this laboratory[7] which
accomplishes this. The chopper has two blades occupying about
10% of the circumference; one which is blackened, to provide
a reference period during which the dark current may be subtracted, the other carrying a mirror which directs the light around
the sample so that its transmittance may be measured. The
purpose of the small blade size is to avoid interrupting the
data collection and integration time in the CD experiment
because of the importance of small bandwidths in this measurement.
The timing pulses which are derived from the position of the
chopper allow the separation of the measurements from a photomultiplier output. A typical result on a single crystal
specimen of Cs_3CoBr_5 at 4.2K is shown in Figure 11. Both
spectra were measured in a field of 4.75T and were recorded
simultaneously with a two-pen recorder.

9.   SAMPLE MOUNTING

Measurements on solutions at room temperature present
little difficulty in MCD provided that precautions are taken to
avoid strain in the cell windows. On the other hand the great
majority of measurements are made at lower temperatures with the
objective of both enhancing the resolution of the spectrum and
identifying any temperature dependent contributions to the MCD.
The sample phases are therefore either single crystals or some
sort of glass. In either case the sample will be mounted in an

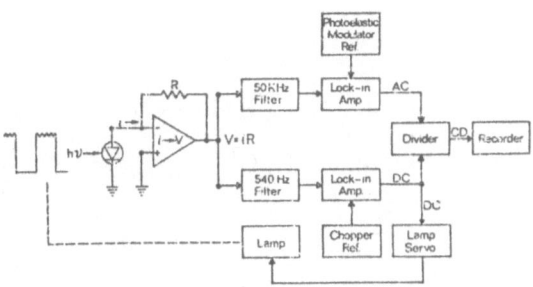

Figure 9. Electronic layout for the double modulation instrument in Figure 11. The CD is modulated at 50 kHz and the complete light beam is chopped at 540 Hz.

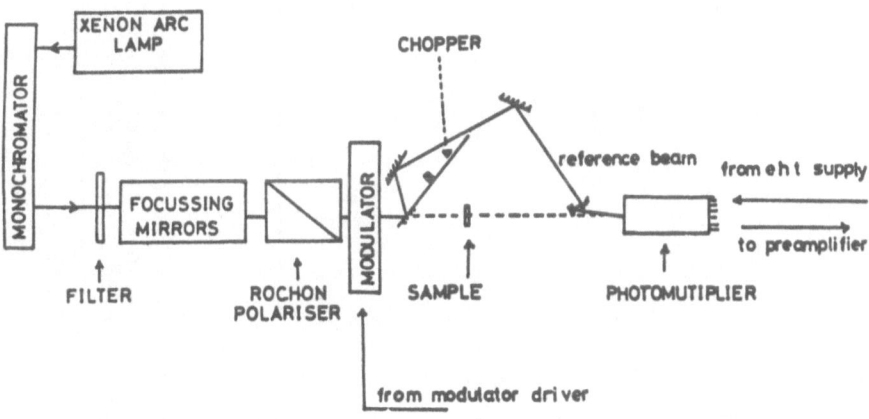

Figure 10. Optical layout for a simultaneous CD/Absorption Spectrophotometer. The chopper carries one mirror blade to provide the reference beam and one blackened blade to define the 'dark' condition in the detector.

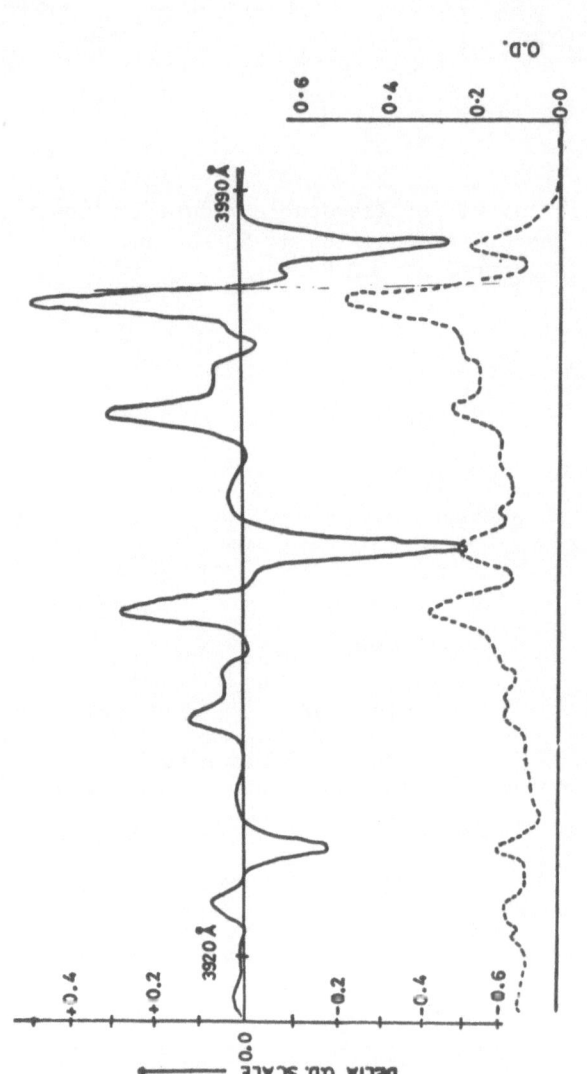

Figure 11. The simultaneous absorption/MCD spectrum of the $^4A_2 \longrightarrow {}^2E(^2D)$ transition in a single crystal of Cs_3CoBr_5 at 4·2 K and 4·75 T. The first two main features in the absorption spectrum are separated by 38 cm^{-1}. The progression in the totally symmetric stretching mode is clearly visible.

optical cryostat which also contains a superconducting solenoid.
Occasionally small permanent magnets with fields of about 0.8T
have been used.

Suitable cryostats are available from many manufacturers
and are arranged with a horizontal solenoid axis.   Optical
windows provide for the propagation of light parallel to this
axis.   Some systems have compact tail sections which permit
their insertion into the cell compartment of commercial CD
spectrometers.   Some care must be taken to avoid the stray
magnetic field from extinguishing the arc source of the spectrometer.   In some optical cryostat applications single crystals
of sapphire are used for the cold windows because of their high
thermal conductivity.   The sapphire is usually cut in a
birefringent plane and they must therefore be avoided in MCD
measurements.   Optical silica is used and the window mounts
must be carefully designed to avoid strain birefringence which
may develop on cooling.   When an 'exchange gas' method of
temperature regulation is employed the cold windows can be under
a considerable pressure differential which in turn leads to
strain.

It should be obvious that only cubic or uniaxial specimens
are suitable for the measurement of MCD.   The intrinsic birefringence of biaxial materials will depolarise the radiation.
Of course an optic axis exists in biaxial crystals but its
direction is not symmetry controlled and will in general vary
with wavelength.   It is usually very difficult to find the axis,
let alone to cut a face perpendicular to it.   When uniaxial
crystals are used their orientation can be critical and it is
useful to be able to adjust the sample mounting with respect to
the optical axis.

The samples must be strain free if possible.   Crystals
grown from the melt are sometimes badly strained and prolonged
annealing may be necessary.   Similarly crystals should not be
glued rigidly to a copper mount.   A dab of grease at one corner
of the sample is a useful technique.   Glasses are often strained
upon formation but polymer films are strain free.   The effect
of strain birefringence can be assessed by measuring the natural
CD spectrum of an optically active solution placed after the
cryostat and before the detector.   Serious birefringence will
reduce the amplitude of the measured natural CD.

Temperature control and measurement is essential in the
analysis of low temperature MCD spectra.   The best regulation
is achieved with 'exchange gas' cooling and compensatory heating.
Good regulation is also possible with systems which rely on the
flow of helium gas at atmospheric pressure to provide cooling.
The latter type also has the great advantage that samples may be

changed without the loss of the main vacuum. Difficulties in the control of temperature of polymer films have been found, probably because of their poor conductivity.

REFERENCES

1. L. Velluz, M. Legrand and M. Grosjean 'Optical Circular Dichroism' (Weinheim:Verlag Chemie, 1965).

2. K. Rosenheck and P. Doty, Proc. Nat. Acad. Sci. 47, 1775 (1961).

3. J.C. Kemp, J. Opt. Soc. Amer. 90, 950 (1969); S.N. Jasperson and S.E. Schnatterly, Rev. Sci. Inst. 40, 1136 (1969).

4. M. Billardon and J. Badoz, Comptes Rend. 262, 1672 (1966).

5. G.A. Osborne, J.C. Cheng and P.J. Stephens, Rev. Sci. Inst. 44, 10 (1973).

6. E.C. Hsu and G. Holzwarth, J. Amer. Chem. Soc. 95, 6902 (1973).

7. J.C. Collingwood, P. Day, R.G. Denning, P.N. Quested and T.R. Snellgrove, J. Phys. (E) (in press).

8. O. Schnepp, S. Allen and E.F. Pearson, Rev. Sci. Inst. 41, 1136 (1970).

9. M. Billardon, J.C. Rivoal and J. Badoz, Rev. Phys. Appl. 4, 353 (1969).

APPLICATIONS OF MAGNETIC CIRCULAR DICHROISM TO CHARGE TRANSFER AND LIGAND FIELD SPECTRA

B. Briat

Laboratoire d'Optique Physique, EPCI, 10 rue Vauquelin, 75231 Paris, Cedex 05

1. INTRODUCTION

The first MCD spectra appeared in the literature in the middle sixties. For ·technical reasons, they were obtained from solutions and were run at room temperature. Meanwhile, there appeared the theoretical paper most often quoted in MCD reports.[1] In it, the authors defined a set of three parameters and expressed them in terms of molecular quantities; they also offered a comprehensive interpretation for the existing literature and stimulated further extensive work.

During the past ten years, MCD has become increasingly popular among inorganic chemists for two reasons. On the one hand, there have been important advances in technique,[2,3] while on the other hand, MCD has been found especially fruitful for those systems having cubic or uniaxial symmetry.

A large number of diamagnetic and paramagnetic transition metal ions of the d and f series have been investigated over the years, magnetic solids being also considered recently. Most interesting applications of MCD have been of a spectroscopic nature and have arisen from low-temperature work. In the course of this work, the original parameters derived by Buckingham and Stephens have been slightly rearranged and put in a form more suitable for computational purposes. Also, it has been found necessary to make an extensive use of the irreducible tensor method[4-6] so as to derive theoretically MCD and absorption parameters.

P. Day (ed.), Electronic States of Inorganic Compounds. 177–221. *All Rights Reserved.*
Copyright © 1975 by D. Reidel Publishing Company, Dordrecht-Holland.

There is no point here in making a review of all the applications of MCD since this has been well done in several places.[7],[8]   I shall do my best also not to duplicate a previous personal contribution to a summer school on the same topic.[9] I have chosen rather to discuss in detail the specific evaluation of MCD parameters in many situations which, I believe, are those encountered most often for chemical systems.   I shall do my best to present theoretical derivations – especially those concerning spin- and/or parity-forbidden bands in the shortest (although complete) and simplest possible form.   My hope is that this contribution may eventually save time and energy for future workers in the field.   The formalism used throughout is that agreed upon by a large number of people on a summer evening at Oxford.

2.   FORMALISM

In order to define the parameters of interest to us, we write the absorbance A_γ of a material for a polarization γ and a transition between degenerate levels I and J as:[10]

$$A_\gamma = \eta E f(E) \sum_{i,j} Ni \left| \langle J_j | \hat{m}_\gamma | Ii \rangle \right|^2 \Big/ \sum_i Ni \qquad (1)$$

Here η is a constant factor, i and j stand for the sub-states of the ground (I) and excited (J) states respectively, E is the energy (hereafter expressed in cm^{-1}) and $f(E)$ is a shape function chosen so as to satisfy $\int f(E)dE = 1$; $\hat{m}$ is the electric dipole moment operator and Ni is the population of the i substate.   It follows from the definition of $f(E)$ that:

$$\langle A_\gamma \rangle_0 = \int \frac{A_\gamma}{E} \, dE = \eta \sum_{i,j} Ni p_\gamma^2 \Big/ \sum_i Ni \qquad (2)$$

where $p_\gamma^2 = \left| \langle J_j | m_\gamma | Ii \rangle \right|^2$ and $\langle A_\gamma \rangle_0$ is the zeroth order moment of the A_γ/E function.

In the absence of a magnetic field, then $Ni/\sum Ni = 1/d$ where d stands for the degeneracy of the ground state.   One has therefore:

$$\langle A_\gamma \rangle_0 = \frac{\eta}{d} \sum_{i,j} p^2 \qquad (3)$$

Choosing z as the direction of propagation of the light beam, the operators to be associated with left and right circularly polarized light in the above equation are $\hat{m}_+$ and $\hat{m}_-$ respectively, where $\hat{m}_\pm = \mp (i/\sqrt{2})(m_x \pm im_y)$.   Thus, for example,

$$\langle A \rangle_o = \frac{\eta}{2d} \sum_{i,j} |\langle J_j | \hat{m}_+ | Ii \rangle|^2 + |\langle J_j | \hat{m}_- | Ii \rangle|^2 \qquad (4)$$

for a cubic material, the absorbance being independent of the polarization in this particular case. This means that the absorption spectrum run in zero magnetic field with linearly polarized light will be just the same as that obtained with right or left circularly polarized light.

We now apply a magnetic field B along z, i and j being the Zeeman sublevels of the ground and excited states respectively. Then $Ni = \exp(-x)$ where $x = \langle i | Lz + 2Sz | i \rangle \mu_B B/kT$; kT and $\mu_B B$ are both expressed in wave numbers, u_B is the Bohr magneton and equals 0.465 when B is in Teslas ($1T = 10^4$ gauss). It follows from equation (2) that the zeroth order moment of MCD may be written:

$$\langle \Delta A \rangle_o = \langle A_+ \rangle_o - \langle A_- \rangle_o = \eta \sum_{i,j} (p_+^2 - p_-^2) \exp(-x) / \sum_i \exp(-x) \quad (5)$$

This equation is valid for any value of x and it does not require a knowledge of the shape function of the absorption curve nor the assumption that p_+^2 and p_-^2 are field independent. When this latter condition is satisfied, then:

$$\langle \Delta A \rangle_o = \eta \sum_{i,j} (p_+^2 - p_-^2) shx / \sum_i chx \qquad (6)$$

This equation should be used for the interpretation of MCD measurements at large x values. In the limit where x is small, $shx \rightarrow x$ and $chx \rightarrow 1$ and:

$$\langle \Delta A \rangle_o = \eta \sum_{i,j} (p_+^2 - p_-^2) \; x/d \qquad (7)$$

Therefore a parameter c_o can be defined, satisfying:

$$\langle \Delta A \rangle_o / \langle A \rangle_o = c_o \; (\mu_B B/kT) \qquad (8)$$

where:

$$c_o = -\sum_{i,j} \langle i | L_z + 2S_z | i \rangle (p_+^2 - p_-^2) / \sum_{i,j} (p_+^2 + p_-^2)/2 \qquad (9)$$

If we consider that p_+^2 and p_-^2 may be field dependent, a second parameter b_o has to be introduced in equation (8) besides c_o/kT. This occurs to take account of the mixing of $|Ii\rangle$ and (or) $|Jj\rangle$ with some other state $|Kk\rangle$ via the Zeeman Hamiltonian.

We now define n^{th} order moments of A/E and $\Delta A/E$ relative to an origin $\bar{E}$ as:

$$\langle A \rangle_n = \int \frac{A}{E} (E - \bar{E})^n \, dE \tag{10}$$

$$\langle \Delta A \rangle_n = \int \frac{\Delta A}{E} (E - \bar{E})^n \, dE \tag{11}$$

These can be used to gain additional spectroscopic information regarding the states involved in the transition. Although MCD moments of order 2 and 3 have been considered on a few occasions, the most commonly employed is certainly $\langle \Delta A \rangle_1$. Choosing $\bar{E}$ as the baricentre of the absorption band ($\bar{E}$ then satisfies $\langle \Delta A_1 \rangle = 0$), a new parameter a_1 can be defined such that, in most practical situations,[8,11] and independently of electron-lattice interaction:

$$\langle \Delta A \rangle_1 / \langle A \rangle_0 = a_1 \, (\mu_B B) \tag{12}$$

where

$$a_1 = c_0 + \sum_{i,j} \langle j | L_z + 2S_z | j \rangle (p_+^2 - p_-^2) \Big/ \sum_{i,j} (p_+^2 + p_-^2)/2 \tag{13}$$

In practice, c_0 (b_0) and a_1 parameters are therefore best obtained through a moment analysis of the experimental data, whenever this is feasible. The rigid shift approximation may be very crude sometimes. It should be realised that it only provides a rough estimate for the molecular quantities of interest, using the following equations:

$$\frac{A}{E} = \eta' \mathcal{D}_0 f \tag{14}$$

$$\mathcal{D}_0 = (3/2d) \sum_{i,j} (p_+^2 + p_-^2) \tag{15}$$

$$\frac{\Delta A}{E} = \eta' \left\{ a_1 \mathcal{D}_0 \left(-\frac{\partial f}{\partial E} \right) + \left(b_0 + \frac{c_0}{kT} \right) \mathcal{D}_0 f \right\} \mu_B B \tag{16}$$

c_0 and a_1 have been called c and $-a$ in all our previous work in this laboratory (e.g.[12,13]). Compared with the parameters suggested by P.J. Stephens in Oxford:

$$a_1 = \mathcal{A}_1 / \mathcal{D}_0 \qquad c_0 (b_0) = \mathcal{C}_0 (\mathcal{B}_0) / \mathcal{D}_0$$

a_1 and c_0 are dimensionless, whereas b_0 is expressed in cm^{-1}; $\eta' = \eta/3 = 10^3 \, cl/9.18$ where c and l stand for the concentration (mole.l^{-1}) and length (cm) of the investigated material. $\mathcal{D}_0$ is the dipole strength to be expressed in (Debye)^2 ($D \sim 3.335640 \times 10^{-30}$ cm).

Referring again to our previous work on sharp lines in rare earths,[13] E in equations (15) and (16) may be considered as approximately constant within the line and we may write:

$$A = A_m F \qquad (17)$$

$$\Delta A/A_m = \left\{ a_1 \left(-\frac{\partial F}{\partial E} \right) + \left(b_0 + \frac{c_0}{kT} \right) F \right\} \mu_B B \qquad (18)$$

where F is a Lorentzian or Gaussian shape function.   Choosing
for example a Lorentzian and writing $F = 1/1 + \chi^2$ where $\chi = (E - E_0)/\gamma$, then A_m is the maximum absorbance at the centre E_0
of the line, γ standing for the half-width at half-height.
Comparing equations (17) and (14) leads to the approximate
result:[*]

$$\mathcal{D}_0 \sim 9.18 \times 10^{-3} \; A_m \gamma \pi / E_0 c l \qquad (19)$$

since

$$\int F dE = \gamma \int dx/1 + \chi^2 = \gamma \Big[\tan^{-1} x \Big]_{-\infty}^{+\infty} = \gamma \pi$$

It follows also from equation (18) that $\Delta A/A_m$ at the centre
of the line is a good approximation for $\langle \Delta A \rangle_0 / \langle A \rangle_0$.

3.   QUALITATIVE APPLICATIONS OF MCD

In this section, I wish to illustrate the specificity and
sensitivity of MCD as an analytical or spectroscopic tool.   I
shall also comment upon those numerous applications which do not
require elaborate computational work since they rely essentially
on rather simple group theoretical arguments.

3.1   Specificity of MCD

It is clear from equation (18) that a MCD spectrum contains
much more information than an absorption spectrum.   To each
electronic transition, one may associate three parameters instead
of one, these being either positive or negative.

I should mention at this stage that obtaining meaningful
parameters from MCD experiments sometimes presents some difficulties.   When the material in question has an orbital or (and)
spin-degenerate ground state, one should run spectra over a wide
range of temperatures in order to figure out the number of
transitions which are actually involved in the spectral range
considered and to associate them with well-characterised MCD

---

[*] In the case of a Gaussian, γ and π should be replaced by δ and
$\sqrt{\pi}$ respectively, δ being the halfwidth at $1/e$;   then $F = \exp(-y^2)$ where $y = (E-E_0)/\delta$.

parameters.   This can be exemplified with low-spin iron(III)
species of biological interest,[14] ferricytochrome b_2 being a
representative member.   We observe in Figure 1 (ΔA_M stands for
$\Delta A/u_B B$) that its room temperature spectrum in the 6000-5000 Å
range has little, if anything, in common with the low-temperature
data obtained on a polyvinyl alcohol polymer film.   The former
shows an S-shaped curve which would undoubtedly be erroneously
interpreted as an 'a$_1$ term' in the absence of additional experiments.   On the contrary, the latter unambiguously demonstrates
the presence of four 'c$_0$ terms'.   These are likely to be associated with the predominantly x and y polarized components of the
0-0 (α) and 0-1 (β) Q bands of the molecule.

An obvious qualitative use of MCD work is therefore to serve
as a detector of overlapping bands which are hidden in the
absorption spectrum.   A second example of this sort is given in
Figure 2.[14]   Reduced spinach ferredoxin is an optically active
iron-sulfur protein which contains both iron(III) and iron(II) in
their high-spin states.   In contrast to the absorption[15] and CD
spectra[14] which show broad and rather structureless features even
at very low temperature, the MCD spectrum is much more informative
and may thus serve to number and locate the various transitions

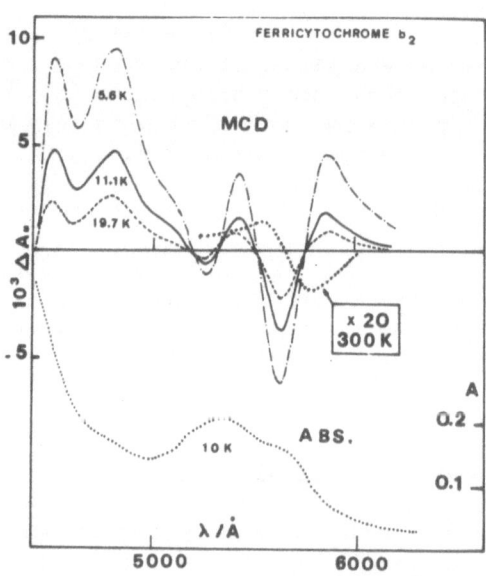

Figure 1.   Absorption and MCD of ferricytochrome b_2 through its
Q bands. ΔA_M stands for $\Delta A/\mu_B B$.

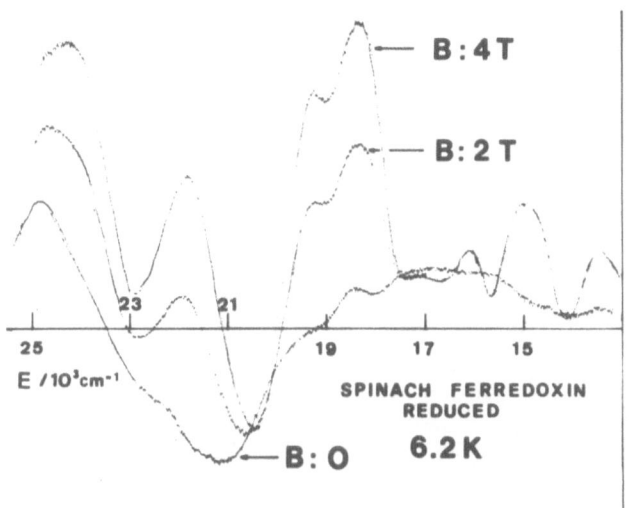

Figure 2.   CD spectra of reduced spinach ferredoxin at 6.2 K,
in the absence (B=0) or presence of a magnetic field.   The
spectrum at B=2T covers onlythe region 17000-25000 cm^{-1}.

which are in this case associated with bell-shaped MCD components
('c$_0$ terms').

     A similar situation is represented in Figure 3 for ruthenium
acetylacetonate.[14]   The MCD varies approximately as $(kT)^{-1}$ in
the temperature range considered.   Even a very crude gaussian
analysis of the data leads to the unambiguous conclusion that at
least six absorption components are present in the spectral range
600-300 nm.   This is what one would expect as a result of the
combined action of a trigonal field and spin-orbit coupling on
orbital triplet electronic states for a d^5 system.

     MCD spectroscopy is also of great qualitative value and
complementary to other physical techniques when a material may
contain several insufficiently characterised species.   Choosing
an example in the field of solid state chemistry, it has been
experimentally proven that the doping of yttrium-gallium (or
aluminum) garnets with ruthenium or iridium leads to samples of
different colours, depending upon the preparative conditions.
A bunch of green, blue and orange crystals was, for example,
obtained in the case of YGaG/Ru.   These colours may be attributed to different valencies of the ruthenium ions or else, to
different site symmetries associated with a given valency.

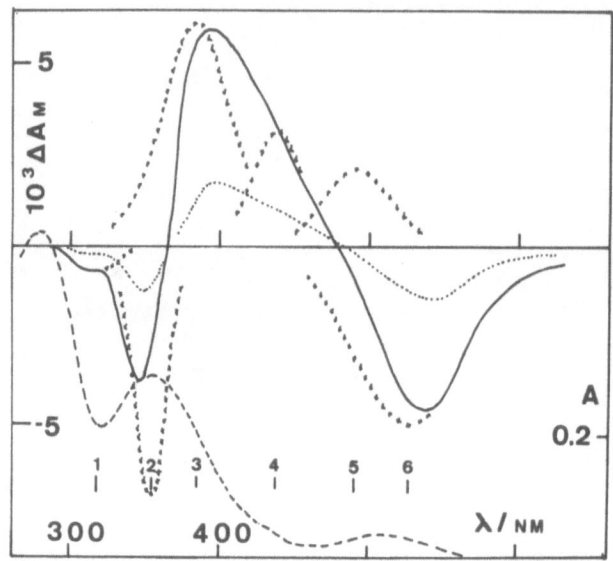

Figure 3.   Absorption (---30 K) and MCD (——— :6 K; ... :18 K) of
ruthenium acetylacetonate in a cellulose acetate/CH_2Cl_2 polymer
film;   xxx: crude gaussian analysis.

Figure 4 shows the absorption and MCD data for a green sample
which is known from EPR measurements to contain at least Ru^{3+}
ions at an approximately octahedral site.   The absorption
spectrum shows two broad bands around 630 nm and 440 nm.   MCD
measurements at two temperatures demonstrate that these are
associated with diamagnetic and paramagnetic ions respectively.
We therefore assign band I (400–540 nm region) to Ru^{3+}, band II
being due to a diamagnetic species which we have shown to be
largely predominant in our blue crystals.

In the course of a MCD study on MnX_4^{2-} ions (X = Cl, Br, I)
we have encountered a situation similar in many respects to that
of the ruthenium garnets and this will be commented on as a last
example in this section.   Figure 5 shows the absorption and MCD
data for the n-butylammonium derivative of MnI_4^{2-} in a cellulose
acetate/acetonitrile polymer film.   It proved difficult to get
rid of free iodine in the film and this impurity lead to a broad
absorption band which might have completely masked the weak spin-
forbidden bands of MnI_4^{2-}.   In fact, low-temperature MCD experiments

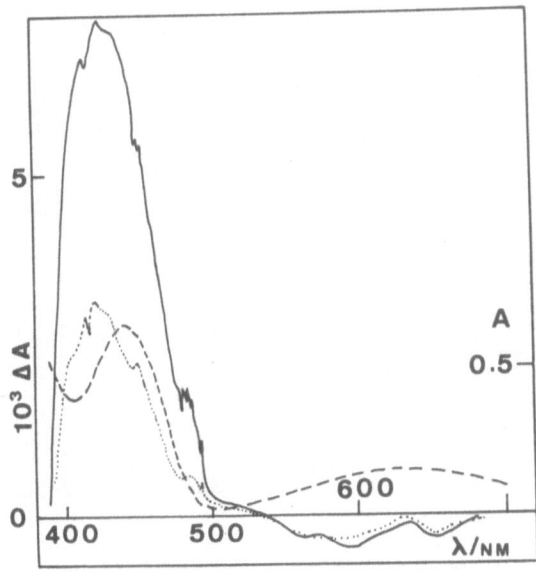

Figure 4.   Absorption (--- 15 K) and MCD (— 5 K;   ... 19 K) of
a green ruthenium gallium garnet.[14]

solved the problem very easily since MnI_4^{2-} is paramagnetic, while
I_2 is diamagnetic, temperature dependent signals being thus
associated with the former.

     One may finally wonder how to run low-temperature spectra
when no cubic or uniaxial crystal is available.   McCaffery et al.[16]
have used doped polystyrene thin films deposited on a glass window
and rigid polymethylmetacrylate matrices.   These offer the
possibility of being cut and polished to the appropriate size
and they prove free from residual birefringence down to liquid
helium temperature.   However, their use is limited to certain
compounds and they absorb in the ultraviolet region.   In our
laboratory, we have preferred the thin film technique which can
be applied to most practical problems, as long as a solvent can
be found for the investigated material.   Sample and polymer are
first dissolved in an appropriate solvent and then a drop of the
mixture is evaporated on a silica window, or else a small amount
of the solution is left to evaporate at the bottom of a Petri dish,
the film thus obtained being then lifted out and cut.   For most
solvents in frequent use, we have selected the polymer which is

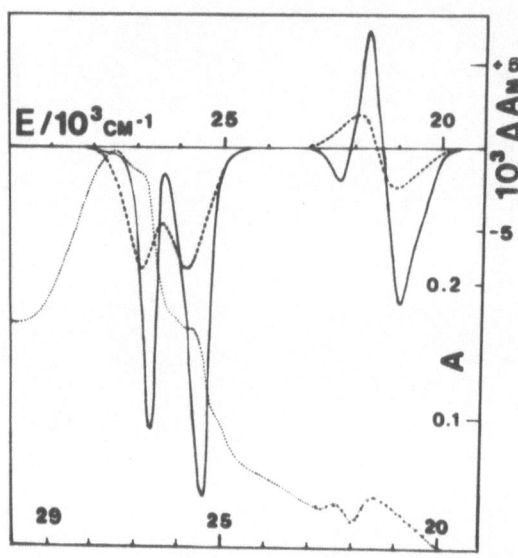

Figure 5.   Absorption and MCD (——) spectra of MnI_4^{2-} in a CA/CH$_3$CN polymer film at 10.8 K.   The RT MCD (- - -) is also shown for comparison (enlarged scale).

the least absorbing in the ultraviolet region, appropriate polymer/solvent combinations being given in Table 1.

Table 1.   Polymer/solvent mixtures appropriate for low-temperature MCD work

Polymer	Solvent
Polyvinyl alcohol	Hot water
Polyoxyethylene	Hot ethanol Hot methanol
Polyvinyl acetate	Dioxane
Cellulose acetate	Acetonitrile Acetone Methylene chloride Nitromethane

When water must be used as a solvent and for those chromophores which are only weakly absorbing (e.g. iron haemoproteins
in the near infrared region) we obtained good results by freezing
very quickly a solution saturated with sucrose in a gas flow
cryostat (using helium gas), a clear glass being obtained under
these conditions.

3.2 Sensitivity of MCD

c_0 and a_1 parameters as expressed by equations (9) and (13)
are typically of the order of unity. b_0 in equation (16) can be
written $b_0 = b'_0/\Delta W$ when the K state is not appreciably populated
at the temperature of experiment, b'_0 being also of the order of
unity and W standing for the difference in energy between
states I and K or J and K.   Assuming, for example, a gaussian
shape for the band, equation (18) can be written:

$$\Delta A/Am = \left\{ \frac{a_1}{\delta} \ G + \left(\frac{b'_0}{\Delta W} + \frac{c_0}{kT} \right) F \right\} \mu_B B \qquad (20)$$

with $G = 2y \exp(-y^2)$, the peak to peak value of G being 1.7.
The various contributions to $\Delta A/Am$ will therefore be as $(Width)^{-1}$,
$(\Delta W)^{-1}$ and $(kT)^{-1}$ for the a_1, b_0 and c_0 'terms' respectively.
Thus, even if we consider a transition between well-isolated
levels ($b_0 \sim o$), then the shape of the MCD signal for a paramagnetic chromophore will strongly depend upon the width of the
line as well as upon the temperature.   Broad bands (e.g. ions
of the d series in solution) will essentially show bell-shaped
signals ('c_0 terms') even at room temperature.   In contrast 'a_1
terms' are likely to be observable at 300 K when the lines are
sharp (e.g., rare earths in solution or in crystals) and it will
then often be necessary to cool the sample to get accurate c_0
parameters.

Most MCD experiments on paramagnetic materials are run using
an electromagnet $(B \sim 1$ T) or a superconducting magnet $(B \sim 4$ T)
which both allow the use of cryogenic devices.   Considering the
former situation and a doublet doublet transition originating
from a ground state having $g_Z = 2$, then the 'paramagnetic'
contribution to $\Delta A/A_m$ will be 10^{-1} for $kT \sim 10$ cm^{-1}.   Keeping
in mind that an MCD machine is capable of detecting a difference
in absorbance of 10^{-6}, this means that it may act as a very
sensitive spectrophotometer.   It is therefore capable of
detecting traces of highly absorbing species and/or helping in
the study of forbidden bands which are hardly seen in absorption
spectrophotometry.

The former argument has recently been illustrated in a study
of matrix isolated mercury atoms.[17]   The repeated appearance of
a distinct 'a_1 term' at 2467 Å in the MCD spectra of various

species isolated in argon was demonstrated to be due to atoms arising from untrapped mercury manometers.

Rare earth ions have long attracted the attention of early MCD (or MORD) experimentalists since they exhibit large $\Delta A/A_m$ ratios for most of their lines. This is due to their sharpness (even at room temperature), and to the occurrence of a paramagnetic ground state (and quite often large g_z values) for many of them. Considering also that the lines are reasonably well isolated from one ion to another, we have taken advantage of the above arguments to propose the use of Faraday effect spectroscopy for the qualitative and quantitative analysis of lanthanide ions mixtures.

The increased sensitivity of MCD versus absorption is illustrated in Figure 6, which shows the MCD of a solution containing the Eu^{3+} ion. Characteristic MCD features can be accurately determined, whereas the absorption bands of the same sample, as run on a commercial spectrophotometer, are barely discernable (see, for example, regions 1 and 3 in Figure 6).

The possibility of MCD studies on chromophores at low concentrations is highly attractive in cases where ion pairing may occur or when a broadening of the lines and therefore a loss of resolution result from the increase of concentration.

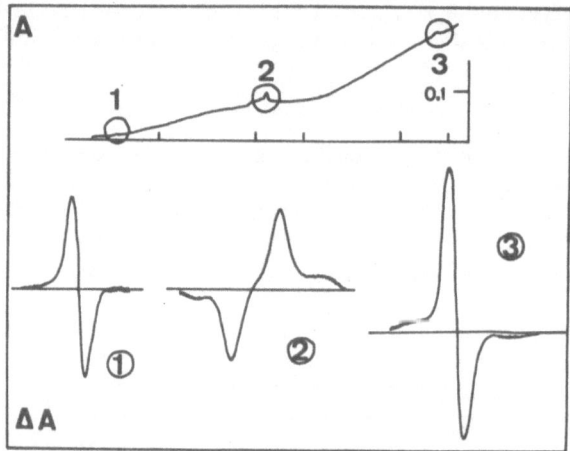

Figure 6. RT absorption (top) and MCD (bottom) spectra of the $EuCl_6^{3-}$ ion in CH_3CN. $B = 5T$; $c = 10^{-2}M$; $l = 1$ cm. The noise level is indicated in the MCD spectra.

A counterpart of MCD sensitivity is the possible interference between features attributable to weak lines (e.g., parity and spin-forbidden transitions) of a 'concentrated' ion and those associated with the more intense lines (e.g., parity-forbidden but spin-allowed) of an impurity ion.   This point is illustrated in Figure 7 taken from Reference 18.   It shows that most of the MCD features of a 'KMgF$_3$:Co' crystal in the 415–440 nm region are actually to be attributed to a Ni impurity present in the crystal. We experienced the same kind of difficulty in the course of a preliminary study of the tetraalkyl ammonium salts of FeBr$_4^-$ and FeCl$_4^-$ in the red spectral region; the spin-allowed $^4A_2 \to {}^4T_1$ bands of CoX$_4^{2-}$ (Co^{2+} is found in commercial iron trichloride or tribromide) overlap with certain sextet quartet transitions of the tetrahaloferrate ions.   Especially in low temperature work on paramagnetic ions one should therefore be aware of the fact that some of the richness of an MCD spectrum may possibly be due to artifacts caused by undesired species.

One of the most interesting applications of MCD spectroscopy is certainly the precise location and assignment of the spin and parity forbidden d-d bands of octahedral compounds, since these are very weak and difficult to study through absorption.   Such work has been carried out on, e.g., Co^{2+}, Ni^{2+} and Cr^{3+} in various environments.   Figure 8 illustrates such proposed assignments[18] in the case of the Co^{2+} ion in KMgF$_3$ (Co/Mg ~ 2.1 x 10^{-2});

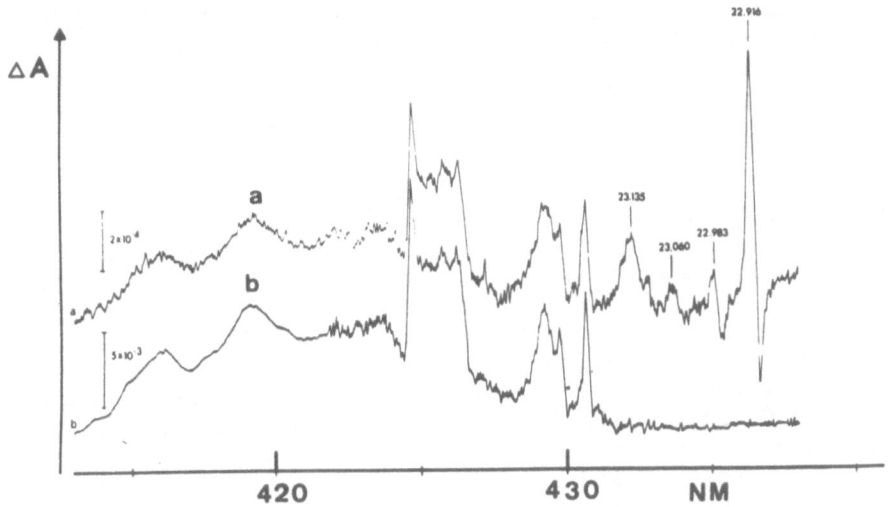

Figure 7.   MCD spectrum of the 24,000 cm^{-1} region of (a) KMgF$_3$: Co at 16 K; (b) KMgF$_3$: Ni at 11 K.   B = 4.5 T for both experiments.   The data are taken from Reference 18.

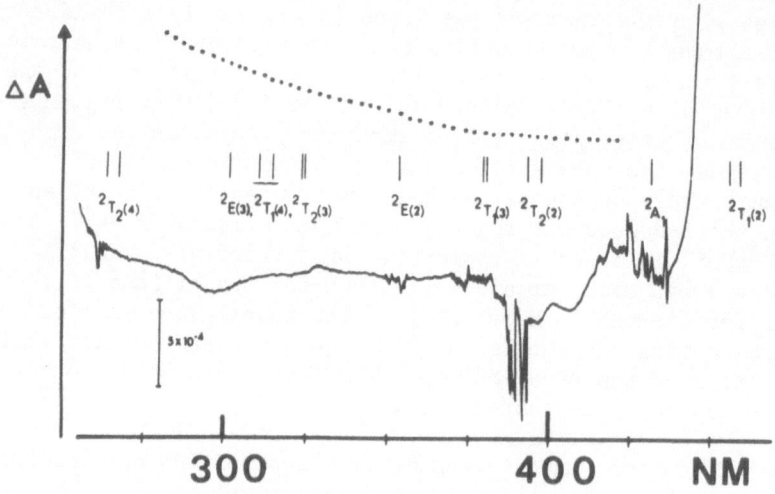

Figure 8.   Proposed assignment of the doublet lines in
$Co^{2+}/KMgF_3$.[18]   A survey MCD spectrum is shown in full line.
The dotted line indicates the absorption.

even in these survey spectra, MCD is obviously, once again, much
more informative than absorption.   Similarly in Figure 9, the
$^4A_{2g} \rightarrow {}^2T_{2g}$ transitions of Cr^{3+} in alum, although hardly seen in
the 16 K absorption spectrum, are easily detected in the low-
temperature MCD spectra;   this arises since c_0 has, a priori, no
reason to be even approximately the same for a transition to $^2T_{2g}$
or, e.g., to $^4T_{2g}$.   The theoretical aspects of such MCD work
will be discussed under a forthcoming heading.   Let me just
state here that the information available from these studies is
of prime importance for checking individual sets of crystal field
parameters.

      Lastly, in this section, I wish to discuss a few other
aspects of MCD work at low temperature and/or high magnetic field.
Considering* first a well isolated ground state doublet, then to
a o_+ transition originating from one of the ground state sublevels
will correspond a o-transition of equal probability from the other
sublevel.   Equation (5) then leads to

$$\langle \Delta A \rangle_o = \eta p^2 th(g\mu_B B/2kT) \tag{21}$$

_____
*The arguments given here apply to an ordered or isotropic dis-
ordered system.

where $p^2 = p_+^2 = p_-^2$ and g is the parallel Lande factor in the
ground state.   The MCD will therefore saturate at high B (or g)
or low T.   We now assume that a non-degenerate K state lies near
the ground state I and that I and K may mix under the action of
the magnetic field.   When $E_K - E_I$ is much less than the line
width of the absorption bands corresponding to I→J and K→J a
second, 'temperature dependent b_o' term will contribute to $\langle \Delta A \rangle_0$,
this being expressed now in the more general form

$$\langle \Delta A \rangle_0 = \alpha \, th(g\mu_B B/2kT) + \beta \mu_B B \, th[(E_K - E_I)/2kT] \qquad (22)$$

where α and β are field and temperature independent.   Therefore,
the second contribution as well as the first one will vary as
$1/kT$ when $kT \gg E_K - E_I$.   At low temperature, the 'β term' will
remain linear in B and begin to saturate when $kT \sim E_K - E_I$, in
contrast with the 'α term' which will begin to saturate when
$kT \sim g\mu_B B$.   We thus understand that temperature dependent b_o
terms are distinguishable from genuine c_o terms in view of their
different field dependence at very low temperatures.   Furthermore, the zero-field splitting in the ground state can be derived
from this kind of experiment.   More generally, if K is also a
degenerate state,the MCD is a sum of the contributions of
individual states, each weighted by an appropriate Boltzmann
factor.

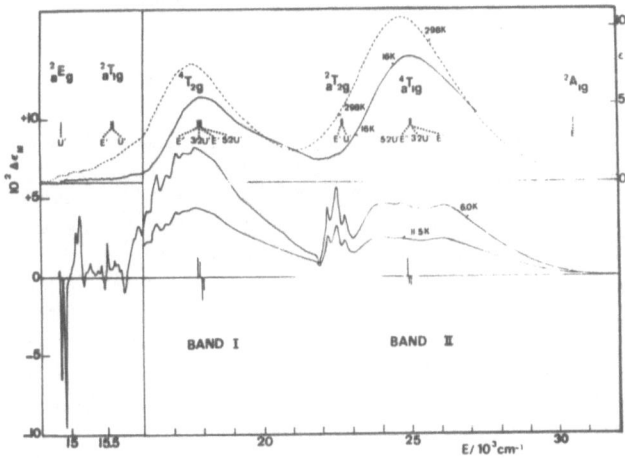

Figure 9.   Absorption and MCD spectra of Cr^{3+} in alum.

3.3 Symmetry arguments in the interpretation of MCD data

3.3.1 Wigner-Eckart theorem

This theorem is most useful since it serves (via the V coefficient) to find the selection rules and relative intensities for transitions between the various Zeeman sublevels of given electronic states.   Following Griffith[5] and using complex orbital components, we express it as:

$$\langle a\alpha | g_\gamma^c | b\beta \rangle = [-1]^{a+\alpha} V \begin{pmatrix} a & b & c \\ \alpha & \beta & \gamma \end{pmatrix} \langle a \| g^c \| b \rangle \tag{23}$$

α and β are the complex tetragonal (or trigonal) components of states transforming as the a and b irreducible representations of a given group; g is an operator (e.g., electric or magnetic dipole moment) transforming as c; γ is a complex component of this operator.

$[-1]^{\jmath} = 1$ if $\jmath$ is A_1, A_2 or E (or their components)

$= -1$ if $\jmath$ is T_1 or T_2

$= (-1)^{\jmath}$ if $\jmath$ is one component of T_1 or T_2.

$\langle a \| g^c \| b \rangle$ is a so-called reduced matrix element, whose value is independent of α, β and γ.

The above equation is valid when the direct product b x c contains a only once, i.e., as long as U' states are not involved. The reader should refer to reference 6 for more detailed information on this point.

3.3.2 Low-spin d^5 ions in an O_h crystal field

I shall choose the example of iron(III) first, for historical reasons, since the MCD study of potassium ferricyanide in solution[21] beautifully illustrated the value of this kind of measurement for the precise assignment of charge-transfer bands. Bands were broad under the experimental conditions and 'c$_0$ terms' were expected to dominate the spectrum even at room temperature.

Figure 10 shows a simplified molecular orbital diagram for this compound.   The ground state corresponds to a hole in the t_{2g} orbital (configuration t_{2g}^5) and is therefore $^2T_{2g}$.   The electric dipole moment transforms as T_{1u} and transitions are allowed to $^2A_{1u}$, ^2E_u, $^2T_{1u}$ and $^2T_{2u}$ since $T_2 \times T_1 = A_1 + E + T_1 + T_2$. Among these, $^2T_{1u}$ and $^2T_{2u}$ are certainly the best candidates as the two lowest excited states (irrespective of their relative energy at this moment) since they are obtained by promoting an

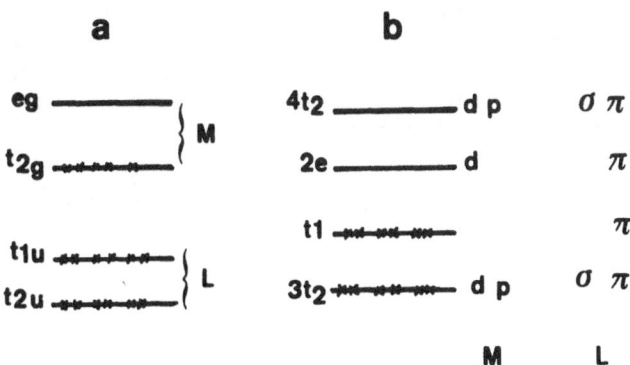

Figure 10.   Simplified molecular orbital diagram for (a) potassium ferricyanide;   (b) tetrahedral d^0 compounds.

electron from one of the highest filled bonding or non-bonding molecular orbitals to the hole in t_{2g}.   The problem in 1966 was to figure out the relative energy of $^2T_{1u}$ and $^2T_{2u}$.

Using the W-E theorem and tetragonal components chosen to diagonalize the Zeeman Hamiltonian (B directed along a C_4 axis) it is easy to demonstrate that the expected MCD is opposite for the $^2T_{2g} \rightarrow {}^2T_{1u}$ transition on the one hand and the $^2T_{2g} \rightarrow {}^2T_{2u}$ transition on the other.   This is done by asking ourselves which is the polarization of the transition(s) originating from, for example, the $|T_2 1\rangle$ ground state component in the two situations. In the former (T_{1u} excited state), we need to consider

$$ V \begin{pmatrix} T_1 & T_2 & T_1 \\ -M_{\Gamma} & +1 & \gamma \end{pmatrix} $$

where we operate on the ground state first and γ stands for the components of the dipole moment operator transforming as T_1 ($\gamma = -1$ for σ-light).   Inspection of Table C 2.3 in Reference 5 shows that only one σ transition is allowed to the $T_1 0$ substate for $\gamma = +1$.   In the latter situation (T_{2u} excited state) we have to change T_1 for T_2 in the first row, first column of the above V coefficient;   the same Table tells us now that one σ transition is allowed to $|T_2 0\rangle$ for $\gamma = -1$.   Had we chosen $|T_2 -1\rangle$ as the starting ground sublevel, then the polarisations would have been inverted, the absolute value of V being $1/\sqrt{6}$ in the four situations.

The sign of c_0 being positive or negative according to whether a σ_+ or σ_- transition originates from the lowest Kramers doublet in the ground state, we have thus proved that c_0 is opposite for the two transitions, irrespective of the relative energy of $|T_{2g}1\rangle$ and $|T_{2g}-1\rangle$. The actual sign to be associated with each of the $^2T_{1u}$ and $^2T_{2u}$ excited states requires, however, the knowledge of this ordering. This can be done simply using, for example, equation 7.14 of reference 5.[22]

$$\langle t_2^5 \, ^2T_{2g} \, 1/2 \, 1 \, | \, \hat{L}z \, | \, t_2^5 \, ^2T_{2g} \, 1/2 \, 1 \rangle = \langle t_2 1 \, | \, \hat{l}_z | t_2 1 \rangle g^5_{t_2 t_2}$$

$$(t_2 t_1) = -1 \tag{24}$$

since $g^5_{t_2 t_2}(t_2 t_1) = (-1)^{t_1+1} g^1_{t_2 t_2}(t_2 t_1) = 1$

and $\langle t_2 1 \, | \, \hat{1}z \, |t_2 1\rangle = -i \langle t_2 1 | \hat{1}_0 | t_2 \rangle = -1$

Therefore $|^2T_{2g}1\rangle$ is the lowest sublevel and we can now state that the MCD associated with the $^2T_{2g} \rightarrow \, ^2T_{1u}$ and $^2T_{2g} \rightarrow \, ^2T_{2u}$ transitions will be positive and negative respectively.

The numerical evaluation of c_0 requires a little more labour, since we need (see equation (9)) the relative intensities of all σ_+ transitions originating from $^2T_{2g}$, i.e., also those from $|^2T_{2g}0\rangle$. Using again the W.E. theorem, then one arrives at the Zeeman diagram shown in Figure 11. Thus, from our equation (9):

$$c_0 = +1 \text{ for } ^2T_{2g} \rightarrow \, ^2T_{1u}$$

$$c_0 = -1 \text{ for } ^2T_{2g} \rightarrow \, ^2T_{2u}.$$

Comparison of these predictions with experimental data therefore leads to the definite conclusion that the bands around 24500 cm^{-1} and 34000 cm^{-1} correspond to the $^2T_{1u}$ and $^2T_{2u}$ excited states respectively.

An interesting point is worth emphasis here. If we consider a free ion, it is well known that a σ_+ transition is associated with an increase of the angular momentum component M_L or M_J by one unit. Although 'intuition' would suggest the same rule to apply if we consider M_Γ instead of M_L (or M_J), this is clearly not the case for, e.g., a $T_2 \rightarrow T_1$ transition, since σ_+ is associated with a decrease of M_Γ by one unit in this case.

In cubic crystals when the electronic levels involved in the transition are separated from other levels by an amount considerably greater than the Zeeman splittings, one may avoid the task of calculating V (or Clebsch-Gordan) coefficients, since this has

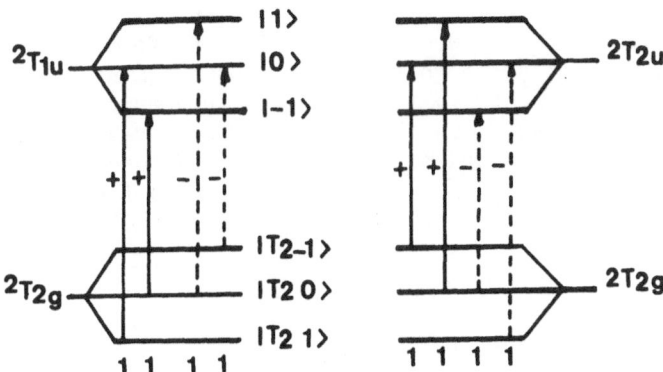

Figure 11.   Zeeman diagram for the $^2T_{2g} \rightarrow {}^2T_{1u}$ transitions of potassium ferricyanide.   Transition probabilities are expressed in units of $|\langle \Gamma \| m \| T_2 \rangle|^2 / 6$ where $\Gamma = T_{1u}, T_{2u}$.

already been done for all types of transitions (except $U' \rightarrow U'$) for an orientation of the magnetic field along either $[001]$ $(O_h \rightarrow C_{4h})$, $[111]$ $(O_h \rightarrow S_6)$ or $[110]$ $(O_h \rightarrow C_{2h})$.   For C_{4h} and S_6, group theory gives an immediate answer concerning circular polarisation since in these groups, the m_+ components transform according to one of the irreducible representations of the group.[25]   Choosing again the example of $Fe(CN)_6^{3-}$ and using Koster's notations,[25] $\Gamma_5(O_h) \rightarrow \Gamma_2 + \Gamma_3 + \Gamma_4$ under C_{4h}, Γ_4 transforming as $|T_2 + 1\rangle$ and having thus the lowest energy in the ground state.   The m_+ operators transforming as $\Gamma_3(m_+)$ and $\Gamma_4(m_-)$ respectively, a σ_+ transition is allowed from Γ_4 to the Γ_1 component of Γ_4 while a σ_- transition is allowed to the Γ_2 component of Γ_5 (in agreement with Figure 11).

Similar arguments can be used if we now consider a model taking account of first order spin-orbit coupling in low-spin d^5 compounds.   Such situations are expected in practice in the ground state of $4d^5$ and $5d^5$ derivatives since the spin-orbit coupling constant of the metal is then much larger than that for Fe^{3+}.[26]   They will also occur in the excited states when ligands such as Br^- are chosen.   ($\zeta_{Br^-} \sim 2460$ cm^{-1} while $\zeta_{Cl^-} \sim 590$ cm^{-1}).   Under the O_h^* double group, $\Gamma_4 \rightarrow \Gamma_6 + \Gamma_8$ and $\Gamma_5 \rightarrow \Gamma_7 + \Gamma_8$, Γ_7 being known from EPR to lie lowest in, for example, iridium(IV) compounds.   When the magnetic field is applied along a C_4 axis, $O_h \rightarrow C_{4h}$ and $\Gamma_7 \rightarrow \Gamma_7 + \Gamma_8$, $\Gamma_8 \rightarrow \Gamma_5 + \Gamma_6 + \Gamma_7 + \Gamma_8$ ($\Gamma_7 \rightarrow \Gamma_6$ is forbidden in O_h^* symmetry since $\Gamma_7 \times \Gamma_4 = \Gamma_7 + \Gamma_8$).

The transition probabilities between the various Zeeman sublevels as given in Reference 23 are reported in Table 2. If we refer to the notation in O_h^*, it appears clearly that (i) the MCD will be of an opposite sign for $\Gamma_7 \to \Gamma_7$ and $\Gamma_7 \to \Gamma_8$; (ii) the absolute value of c_0 is twice as great for the former than for the latter. In order to go further, we must again figure out which among Γ_7^+ and Γ_8^+ (C_{4h}) is the lowest in energy in the ground state. We find that it is the former, since a basis function for Γ_7^+ is $\phi(3/2, -3/2)$ according to Koster.[25] We can also look for Griffith's expression of Γ_7 ($E''\alpha''$) in terms of a linear combination of $|M_\Gamma, M_S\rangle$[26]

$$|E''\alpha''\rangle = \frac{1}{\sqrt{3}} |0\ 1/2\rangle - \frac{\sqrt{2}}{\sqrt{3}} |1\ -1/2\rangle$$

Thus $\langle E''\alpha''|L_z + 2S_z|E''\alpha''\rangle = \frac{2}{3} \langle t_2^5\ {}^2T_2\ 1|\hat{L}_z|t_2^5\ {}^2T_2\ 1\rangle - \frac{1}{3} = -1$

We finally come to the conclusion that $c_0 = -2$ for $\Gamma_7 \to \Gamma_7$ and $c_0 = +1$ for $\Gamma_7 \to \Gamma_8$. These predictions have been used to interpret fully the visible spectra of, e.g., $IrBr_6^{2-}$ in various environments.

Table 2.   Relative intensities of transitions between the various Zeeman sublevels of d^5 low-spin compounds in the presence of spin-orbit coupling.[25]   G and E stand for the ground and excited states respectively.   M transitions are allowed for the $\hat{m}_0 = i\hat{m}_z$ component of the electric dipole moment operator. $B // [001]$.

G \ E	Γ_7		Γ_8			
C_{4h}	Γ_7	Γ_8	Γ_8	Γ_7	Γ_6	Γ_5
Γ_7 {	π_1	σ_1^-	σ_1^-	π_4	σ_3^+	
Γ_8 {	σ_1^+	π_1	π_4	σ_1^+		σ_3^-

Table 3. π and σ polarizations for S_4 symmetry

G	Γ_5	Γ_6	Γ_7	Γ_8
E				
Γ_5		σ'_-	π	σ_+
Γ_6	σ_+		σ'_-	π
Γ_7	π	σ_+		σ'_-
Γ_8	σ'_-	π	σ_+	

3.3.3 Tetragonal compounds with an odd number of electrons

We consider a properly oriented uniaxial crystal (e.g., CuX_4^{2-}). The combined action of the tetragonal field and spin-orbit coupling on the cubic (strong or intermediate field strength) or free ion terms (weak field) results in Kramers doublets whose degeneracy is only lifted by a magnetic field. In D_{2d} symmetry, they correspond to the $\Gamma_{5,6}$ and $\Gamma_{7,8}$ irreducible representations. When the field is applied, $D_{2d} \to S_4$;[25] σ_+ transform as Γ_4 (σ_+) and Γ_3 (σ'_-) respectively and this leads to the selection rules given in Table 3.

In CuX_4^{2-} compounds, the ground state is known from EPR to be $\Gamma_{7,8}$ (D'_{2d}) originating from $\Gamma_4(D_{2d})$ and $\Gamma_5(T_d)$, Γ_8 transforming like ϕ (1/2, -1/2) and being thus the lowest in energy. With this information and Table 3, we find that c_0 is positive for a $\Gamma_{7,8} \to \Gamma_{5,6}$ transition and negative for $\Gamma_{7,8} \to \Gamma_{5,6}$.

The one aspect of the above situation which is new with respect to the cubic case is that we do not even have to bother about the relative intensities of σ_+ pairs for the various doublets, as long as their energy separation is large compared to individual band widths.

The last condition is strictly or approximately respected in the case of tetragonal Nd^{3+} centres[29] and CuX_4^{2-} ions[12] respectively. Figure 12 shows the MCD and absorption spectra of a polymer film containing the ethylammonium derivative of $CuBr_4^{2-}$. We observe two groups of three components in the 15000–33000 cm^{-1} spectral region. Comparison of these data with the theoretical predictions concerning the expected 'c_0 terms' leads us to the energy level diagram shown in Figure 13. MCD served here in particular to assess which of the Γ_6 and Γ_7 spin-orbit components of the ^2E states lies lowest in energy and to estimate through

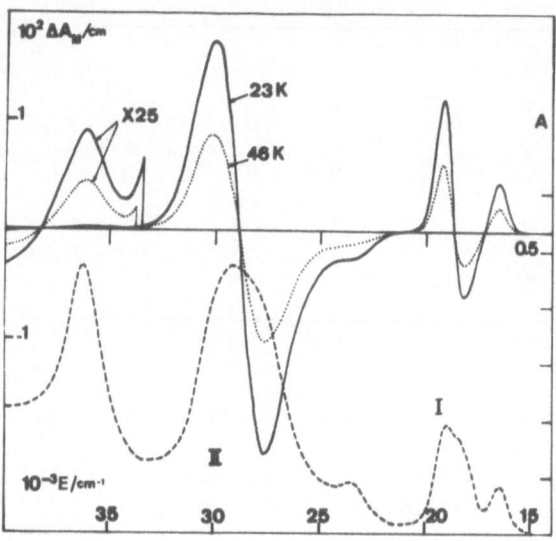

Figure 12. Low temperature absorption (bottom curve) and MCD
(topcurves) spectra of $CuBr_4^{2-}$ in a polymer film.

the first order moment the spin-orbit coupling corresponding to
$^2T_1$ (t_1^5 $4t_2^6$) and $^2T_2$ ($3t_2^5$ $4t_2^6$). We recall here that the quantitative analysis of the MCD for randomly oriented tetragonal
centres differs from that for a properly oriented uniaxial crystal
and refer to reference 12 for more detailed information on this
point.

4.   MCD STUDIES OF CHARGE TRANSFER BANDS

4.1 Limit to the use of symmetry arguments

     In the preceding section, I have already considered the above
problem for paramagnetic compounds in situations where the
separation between excited electronic states was large compared
to the individual band widths.   The use of symmetry arguments
together with a knowledge of the symmetry and wavefunctions in
the ground state, thus lead to the determination of the symmetry
of the excited states.   Nothing else could be derived about them
since we observe only 'c$_0$ terms' in the spectra and the theoretical estimate of c_0 is independent of the relative ordering of
the Zeeman excited components.

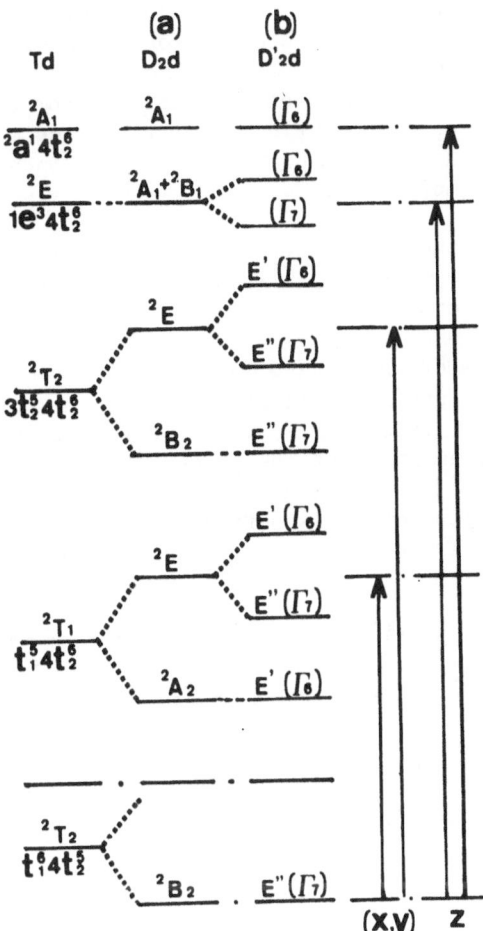

Figure 13.   Energy level diagram for $CuBr_4^{2-}$.   (a) role of the tetragonal field on the cubic terms;   (b) role of spin-orbit coupling.

   I shall now consider the case of (i) cubic diamagnetic compounds; (ii) cubic paramagnetic materials where spin-orbit is effective but results only in weakly (or not at all) structured absorption bands.   In both situations information will emerge from MCD, concerning the orbital angular momentum properties of excited states.   These will in turn serve to elucidate their origin and to gain chemical information.

4.2 Use of 'a$_1$ terms'

Figure 14 shows such terms for ReS_4^-,[30] a member of the tetrahedral d⁰ series.   In the ground state, metal orbitals are empty while bonding (or non-bonding) orbitals of essentially ligand character are filled, leaving an $^1A_1$ ground state (see Figure 10).   The promotion of an electron from one of the upper field orbitals to, for example, 2e will provide a $^1T_2$ excited state to which an electric dipole transition is allowed from $^1A_1$ since m transformed as T_2.

The use of the W.E. theorem and our equation (13) leads to $a_1 = 2l$ where $l = \langle ^1T_2\ 1|L_z\ |\ ^1T_2 1\rangle$ is the orbital angular momentum of the excited state.

In the case of the O_h and T_d symmetry l has been expressed in terms of M.O. coefficients for a number of excited configurations of various ions.[21,30]   It is therefore only necessary to recall the various steps of the process here.

In my example, excited configurations can be written a^5b, where a and b are two molecular orbitals.   Using Griffith's book[5] we have:

$$l = -i\langle a^5b\ T_2\ 1|\hat{L}_o|a^5b\ T_2\ 1\rangle = k_1\ \langle a^5b\ T_2\ \|\hat{L}\|\ a^5b\ T_2\rangle$$

$$= k_2\ \langle a\ \|\hat{l}\|\ a\rangle + k_3\ \langle b\ \|l\|\ b\rangle$$

$$= k_4\ \langle a_x|\hat{l}_z|a_y\rangle + k_5\ \langle b_x|\hat{l}_z|b_y\rangle$$

where $k_1 \ldots k_5$ stand for factors which may be determined easily.

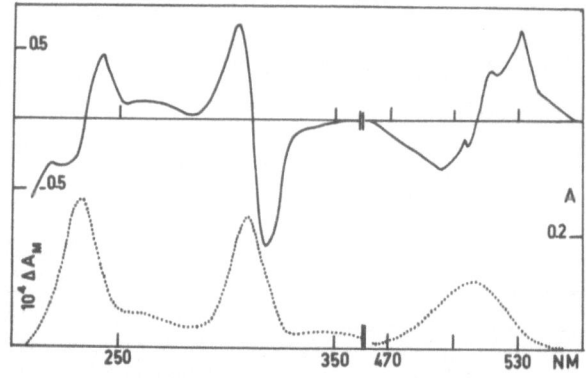

Figure 14.   Room temperature absorption (...) and MCD (——) spectra of ReS_4^-.[30]

We now write a particular component δ (x or y) of a or b as a linear combination of ligand and metal orbitals in the form:

$$a_\delta = \sum_i \rho_i L_i + \sum_j \rho_j M_j$$

and arrive at the final result:

$$1 = \sum_i \pi_i \rho_i^2 + \sum_j \pi_j \rho_j^2 + \sum_{i,j} \pi_{ij} \rho_i \rho_j$$

By using a M.O. diagram and appropriate eigenvectors, one is therefore able to choose, among various plausible assignments, (i) the true origin of excited states when various propositions have been made and (ii) the ordering of molecular orbitals. Although the relative ordering of the states arising from various excited configurations does not necessarily parallel the difference in energy between the corresponding molecular orbitals in the ground state, it is generally assumed that the actual model is not too far from the above one.

We have seen that the method of moments provides a means of extracting 1. I must recall here that A_1/A_0 for a diamagnetic material is independent of linear electron-lattice coupling, spin-orbit coupling and static distortions which do not mix appreciably neighbouring states. Therefore, as long as summations are taken over a whole set of excited vibrational states, the first order moment will give a direct measure of orbital reduction factors. This has been done, for example, for a series of iron(II) phthalocyanines and the results obtained were found to be in good agreement with expectations based on chemical intuition.[31]

4.3 MCD of paramagnetic ions with an orbitally non-degenerate
    ground state

Let me choose the case of tetrahedral d^7 compounds such as CoX_4^{2-} (X = Cl, Br, I) and ignore the possibility of interaction between successive Russell-Saunders terms belonging to different excited configurations.

The ground term is $^4A_2$ and transitions are electric dipole allowed to $^4T_1$ excited state. Now the representation associated with a 3/2 spin is U' and first-order spin-orbit coupling will result in three spinor states arising from $^4T_1$ and characterized by the representations E', $U'_{3/2}$, E" + $U'_{5/2}$ of the double group.[27] Griffith gives also the associated wavefunction $|ShJt\tau\rangle$, where t = 3/2 or 5/2 serves to differentiate among different U' components, in terms of a linear combination of $|ShM_\Gamma M_s\rangle$. For example:

$$|^4T_1 \ U'_{3/2} \ k\rangle = (3/5)^{1/2}|^4T_1 \ 0 \ 3/2\rangle - (2/5)^{1/2}|^4T_1 \ 1 \ 1/2\rangle$$

$$|^4T_1 \ U'_{3/2} \ \lambda\rangle = (2/5)^{1/2}|^4T_1 \ -1 \ 3/2\rangle + (1/15)^{1/2}|0 \ 1/2\rangle -$$
$$(8/15)^{1/2}|1-1/2\rangle$$

From these expressions, we can evaluate the relative energies of the Zeeman components. Thus, for example,

$$\langle U'_{3/2} \ k|L_z|U'_{3/2} \ k\rangle = (2/5)1 + 11/5$$

$$\langle U'_{3/2} \ \lambda|L_z|U'_{3/2} \ \lambda\rangle = (2/15)1 + 11/15$$

where $1 = \langle ^4T_1 \ 1|\hat{L}_z|^4T_1 \ 1\rangle$. The ν and μ states of U' have energies opposite those of k and λ.

We now use the W.E. theorem to find the selection rules in polarized light between the spin components $|0,M_s\rangle$ of $^4A_2$ and the Zeeman sublevels of individual excited spin-orbit components:

$$p_\gamma = \langle ^4T_1 M_r|\hat{m}_\gamma|^4A_2 0\rangle = [-1]^{T_1+M} \ v\begin{pmatrix} T_1 & A_2 & T_2 \\ -M_r & 0 & \gamma \end{pmatrix}\langle T_1\|\hat{m}\|A_2\rangle$$

Inspection of Griffith's Table C 2.3[5] immediately tells us that: (i) σ_+ transitions are allowed to $M_r = \pm 1$ states; (ii) $[-1]^{T_1+M}_r \ v$ is $+1/\sqrt{3}$ for all σ and π transitions. Therefore if we consider $|U'_{3/2} \ k\rangle$, a σ_+ transition with $p_+^2 = 2/5$ will be allowed from the $|1/2\rangle$ spin state in $^4A_2$; $|U'_{3/2} \ \lambda\rangle$ is accessible from $|3/2\rangle$ with $p_-^2 = 2/5$ and from $|-1/2\rangle$ with $p_+^2 = 8/15$. The $\Delta M_s = 0$ selection rule always applies and tells us the ground state component from which transitions originate. The relative values of p_γ for all possible transitions are given in Table 4.

Considering that $\langle ^4A_2 \ M_s|2\hat{S}_z|^4A_2 \ M_s\rangle = 2M_s$ in our model, we have all the information needed to evaluate c_0 for the various transitions. Using our equation (9), we thus find:

E' state $c_0 = - (-3 \times 15 - 1 \times 5)/10 = +5$

E" $c_0 = - (3 \times 5 + 1 \times15)/10 = -3$

$U'_{3/2}$ $c_0 = - (-3 \times 12 - 1 \times 4)/20 = +2$

$U'_{5/2}$ $c_0 = - (3 \times 22 - 1 \times 6)/20 = -3$

Therefore the MCD on the red side of the spectrum will be positive or negative according to whether the E' state lies lowest or highest in energy.

Table 4. Transition amplitude p_γ from the spin-components of $^4A_2$ to the Zeeman components of spin-orbit states issued from $^4T_1$. p_γ is expressed in terms of $\langle ^4T_1 \parallel \hat{m} \parallel ^4A_2 \rangle /(3 \times 30)^{1/2}$.

Excited / Ground		+3/2 σ_+	+3/2 π	+3/2 σ_-	+1/2 σ_+	+1/2 π	+1/2 σ_-	-1/2 σ_+	-1/2 π	-1/2 σ_-	-3/2 σ_+	-3/2 π	-3/2 σ_-
E'	α'			$+(15)^{1/2}$		$-(10)^{1/2}$		$+(5)^{1/2}$					
	β'						$+(5)^{1/2}$			$-(10)^{1/2}$	$+(15)^{1/2}$		
E^a	α'	$+(5)^{1/2}$								$-(15)^{1/2}$	$-(10)^{1/2}$		
	β'		$-(10)^{1/2}$		$-(15)^{1/2}$								$+(5)^{1/2}$
3/2 U'	κ		$+(18)^{1/2}$		$-(12)^{1/2}$								
	λ			$+(12)^{1/2}$		$+(2)^{1/2}$		$-(16)^{1/2}$					
	μ						$+(16)^{1/2}$		$-(2)^{1/2}$		$-(12)^{1/2}$		
	ν									$+(12)^{1/2}$		$-(18)^{1/2}$	
5/2 U'	κ		$-(2)^{1/2}$		$-(3)^{1/2}$								$-(25)^{1/2}$
	λ			$+(3)^{1/2}$		$+(18)^{1/2}$		$+(9)^{1/2}$					
	μ						$-(9)^{1/2}$			$-(18)^{1/2}$	$-(3)^{1/2}$		
	ν	$+(25)^{1/2}$								$+(3)^{1/2}$			$+(2)^{1/2}$

'a$_1$ terms' are also easy to estimate for individual spin-orbit components, in terms of $l = \langle ^4T_1 \, 1|\hat{L}_z|^4T_1 \, 1\rangle$. Figure 15 shows the Zeeman scheme for the $A_2 \to {}^4T_1 \, U'_{3/2}$ transition. From this and our equation (13), we find:

$$a_1 = +4 \,(1 + 13)/15$$

A simple procedure has been given in reference 12 to express also the individual 'b$_0$ terms' as a function of l. It is too lengthy to be reported here.

In cases where spin-orbit components would be fully resolved in the spectra, the above predictions could be compared directly to the experimental data so as to provide detailed information about the various excited states. In practice, however, this might not be the case and it is such a situation which I wish to consider now.

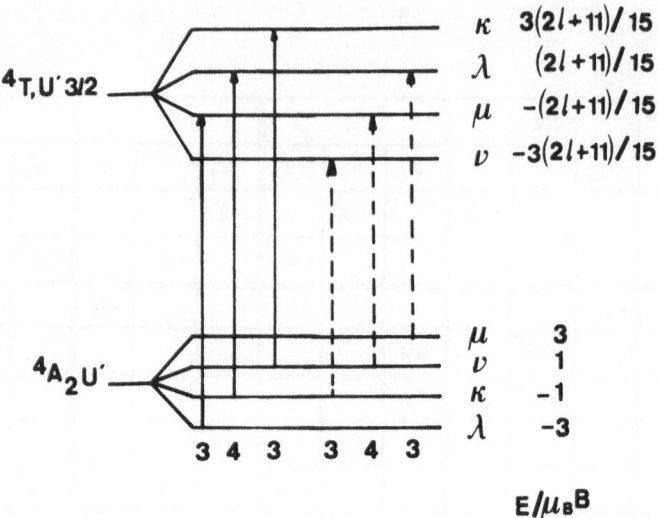

Figure 15.   Zeeman scheme for the $^4A_2 \to {}^4T_1U'3/2$ transition in CoBr$_4^{2-}$.   Full line: σ_+; dotted line: σ_-.

If we consider a transition to the ith spin-orbit state, whose baricentre is at $\bar{E}i$, then from our equations (8) and (12):

$$\langle \Delta Ai_M \rangle_{0\ \bar{E}i} = (b_o + \frac{c_o}{kT})_i \langle Ai \rangle_{0\ \bar{E}i}$$

$$\langle \Delta Ai_M \rangle_{1\ \bar{E}i} = (a_1)_i \langle Ai \rangle_{0\ \bar{E}i}$$

Now taking moments with respect to the baricentre $\bar{E}$ of all the spin-orbit components, one has:[32]

$$\langle \Delta Ai_M \rangle_{0\ \bar{E}} = \langle \Delta Ai_M \rangle_{0\ \bar{E}i}$$

$$\langle Ai \rangle_{0\ \bar{E}} = \langle Ai \rangle_{0\ \bar{E}i}$$

$$\langle \Delta Ai_M \rangle_{1\ \bar{E}} = \langle \Delta Ai_M \rangle_{1\ \bar{E}i} + (\bar{E}i - \bar{E})\langle \Delta Ai_M \rangle_{0\ \bar{E}i}$$

Integrating over the whole set of components finally leads to:

$$\langle \Delta A_M \rangle_0 \underset{\bar{E}}{=} /\langle A \rangle_0 = \sum_i (b_0 + c_0/kT)_i \langle Ai \rangle_0 / \sum \langle Ai \rangle_0 \tag{25}$$

$$\langle \Delta A_M \rangle_1 \underset{\bar{E}}{=} /\langle A \rangle_0 = \sum_i (a_1)_i \langle Ai \rangle_0 + (\bar{E}i - \bar{E})(b_0 + c_0/kT)_i \times$$

$$\langle Ai \rangle_0 / \sum \langle Ai \rangle_0 \tag{26}$$

Once the MCD parameters for the individual lines are known, it is easy to check the general result that $\langle \Delta A_M \rangle_0 = 0$ when the ground state is a pure spin state and second order spin-orbit coupling is also ignored between $|Jt\tau\rangle$ states arising from different excited cubic terms. Regarding $\langle \Delta A_M \rangle_1$, it will obviously contain information about the orbital angular momentum of $^4T_1$ via $(a_1$ and $b_0)_i$, and also about the total spread of spin-orbit components via $\bar{E}i - \bar{E}$. For ions with an orbitally non-degenerate ground state, the general result is:[9,12]

$$\langle \Delta A_M \rangle_1 / \langle A \rangle_0 = 21 - \frac{4S(S+1)}{3(2S+1)} \cdot \frac{\Delta}{kT} \tag{27}$$

where S stands for the spin in the ground state and Δ is taken as positive when the level associated with the largest 'J' value lies higher in energy.

In the limit where Δ is zero, then $\langle \Delta A_M \rangle_1 / \langle A \rangle_0 = 21$ as expected since the spin components need not here to be considered explicitly. In general, however, the first order moment will contain two contributions and measurements at several temperatures will be necessary to separate them. The MCD curve will show an S shape corresponding to a pseudo 'a_1 term'.

The concrete theoretical evaluation of   in terms of M.O. coefficients is beyond the scope of this lecture. Let me just state that the procedure follows[12,33] that outlined for the estimate of 1. Δ depends both upon the spin-orbit coupling constant of the ligand and that of the metal. For complexes of the 3d series, however, it is found to be dominated by the spin-orbit coupling constant of the ligand.

The great merit of MCD here is to provide two parameters with a sign which serves as a stringent test to proposed spectral assignments and sets of eigenvectors.

Equation (26) is independent of linear electron-lattice interaction and therefore provides no information about it. Actually, this information is contained in the second order moment of absorption and the third order moment of MCD.[32] These moments are reliable only when cubic terms are well isolated from each

other and integrations can be taken over the appropriate spectral
range.   They serve to differentiate among the contributions of
cubic (a_{1g}) and non-cubic (e_g and t_{2g}) modes to the broadening
of the lines.   This kind of analysis is of course especially
interesting in the case of broad bands since it offers the
possibility of extracting detailed information which is hidden
in the absorption spectrum.   It should certainly be more often
considered in the future, especially in the case of crystals for
which stress linear dichroism may further be used to differentiate
the e_g and t_{2g} contributions unambiguously.

5.   MCD STUDIES ON LIGAND FIELD BANDS

If we consider spin and parity allowed bands, the parametric
expression of MCD terms and moments as a function of the magnetic
moments in G and E of Δ in E is similar to that for charge-transfer
bands between states of similar symmetry.   This applies, for
example, to the LF $^4A_2 \rightarrow {}^4T_1$ band of CoX_4^{2-} in the red and the
CT $^4A_2 \rightarrow {}^4T_1$ of the same ion in the near ultraviolet.   Of course
1 and Δ will have a different meaning in the two situations since
$^4T_1$ wavefunctions rely solely on the metal orbitals in the former
and essentially on the ligand orbitals in the latter.   Equation
(26) can still be used in various situations including the
presence of tetragonal or trigonal distortions.   Relevant to
this point are the data on $VO(H_2O)_5^{2+}$ of C_{4v} symmetry.[34]   Its MCD
in the red shows a pseudo 'a$_1$ term' which has a different sign at
room and low temperature, due to two opposite contributions to
$\langle \Delta A_M \rangle_1$ in equation (26).   One should therefore be very careful
not to derive erroneous conclusions based on insufficient exper-
imental evidence.

In this section I shall exemplify the derivation of MCD
parameters in two frequently encountered situations which have
not yet been considered: (i) spin-forbidden but parity-allowed
bands; (ii) parity-forbidden bands.

5.1 Spin-forbidden bands in tetrahedral complexes

Considering again CoX_4^{2-} and assuming T_d symmetry, transitions
from $^4A_2$ to ^2E, $^2T_1$ and $^2T_2$ doublet spin states are forbidden in
the absence of S.O. coupling.   When this is considered, however,
(the associated operator transforms as T_1) transitions will become
allowed to ^2E, $^2T_1$ and $^2T_2$ via the mixing of these states with
$^4T_1$ since $T_1 \times T_1 = A_1 + E + T_1 + T_2$.   We note that the origin
of $^4T_1$ need not be specified at this stage.

Again using Griffith's nomenclature,[5] spin-orbit coupling
between the doublets and $^4T_1$ can be obtained through:

$$\langle a^m b^n \, ^4T_1 Jt\tau | \mathcal{H}_s | a^p \, b^q \, ^2\Gamma \, J't'\tau' \rangle =$$

$$\delta(t_2 t') \, \delta(\tau,\tau') \Omega_{JJ'} \begin{pmatrix} 3/2 & 1/2 & T_1 \\ \Gamma & T_1 & t \end{pmatrix} \langle a^m b^n \, ^4T_1 \| \sum \hat{s}\hat{u} \| \, ^4T_1 \rangle \quad (28)$$

The Ω coefficients as given in Table E 1 of Griffith's book[5] have been slightly rearranged and put in Table 5.

Table 5. Ω coefficients (all multiplied by $(360)^{1/2}$) for the quartet-doublets mixing in CoX_4^{2-}

$^4T_1$		E'	E''	$U'_{3/2}$	$U'_{5/2}$
^2E	U'			$-(3)^{1/2}$	$-(27)^{1/2}$
$^2T_1$	E'	$+(10)^{1/2}$			
	U'			$+(25)^{1/2}$	
$^2T_2$	E''		$+(30)^{1/2}$		
	U'			$+(3)^{1/2}$	$-(12)^{1/2}$

Using first order perturbation theory leads to, for example,

$$|^2EU'\tau\rangle = |^2EU'\tau\rangle + \alpha |^4T_1 \, U'_{3/2}\tau\rangle + \beta |^4T_1 \, U'_{5/2}\tau\rangle$$

where $|^2EU'\tau\rangle$ stands for the corrected wavefunction and:

$$\alpha = - \frac{\langle ^2EU'\tau | \mathcal{H}_s | ^4T_1 U'_{3/2}\tau\rangle}{W_{2EU'} - W_{4T_1 U'_{3/2}}} \quad , \quad \beta = \frac{\langle ^2EU'\tau | \mathcal{H}_s | ^4T_1 U'_{5/2}\tau\rangle}{W_{2EU'} - W_{4T_1 U'_{5/2}}}$$

Thus:

$$P_\gamma(^4A_2 \, M_S \to ^2EU'\tau) = \alpha P_\gamma (^4A_2 \, M_S \to ^4T_1 U'_{3/2}\tau) + \beta P_\gamma \times$$
$$(^4A_2 \, M_S \to ^4T_1 U'_{5/2}\tau) \quad (29)$$

Making now the very crude assumption that mixing occurs with only one $^4T_1$ cubic term and that the energy denominators ΔW_1 and ΔW_2 in α and β are roughly equal $(\Delta W_1 \sim \Delta W_2 = \Delta W)$, one arrives at the transition moments p_γ shown in Table 6 for $^4A_2 \to ^2EU'$ and $^4A_2 \to ^2T_2U'$. The p_γ^2 are given in Table 7 and may now be used to determine c_0 (or a_1) with our equation (9):

Table 6. Relative transition moments p_γ for the $^4A_2 M_S \rightarrow {}^2EU'$ and $^2T_2U'$ transitions in CoX_4^{2-}

$^4A_2 \{$	M_s	+3/2			+1/2		
	γ	σ_+	π	σ_-	σ_+	π	σ_-
$^2EU'$	κ		$-(54)^{1/2}$ $+(54)^{1/2}$		$+(36)^{1/2}$ $+(81)^{1/2}$		
	λ			$-(36)^{1/2}$ $+(81)^{1/2}$		$-(6)^{1/2}$ $-(486)^{1/2}$	
	μ						$-(48)^{1/2}$ $+(243)^{1/2}$
	γ	$-675^{1/2}$					
$^2T_2U'$	κ		$+(54)^{1/2}$ $+(24)^{1/2}$		$-(36)^{1/2}$ $+(36)^{1/2}$		
	λ			$+(36)^{1/2}$ $-(36)^{1/2}$		$+(6)^{1/2}$ $-(216)^{1/2}$	
	μ						$+(48)^{1/2}$ $+(108)^{1/2}$
	γ	$-(300)^{1/2}$					

Table 7. Relative transition probabilities p_γ^2 for the $^4A_2 M_S \rightarrow {}^2EU'$ and $^2T_2U'$ transitions in CoX_4^{2-}. The p_γ^2 are expressed in terms of $(\langle {}^2\Gamma \| \sum \hat{s}\hat{u} \| {}^4T_1 \rangle \langle {}^4T_1 \| \hat{m} \| {}^4A_2 \rangle / \Delta)^2 / 2 \times 6^3$.

$^4A_2 \{$	M_s	+3/2			+1/2		
	γ	σ_+	π	σ_-	σ_+	π	σ_-
$^2EU'$	κ				3		
	λ			3		8	
	μ						1
	γ	9					
$^2T_2U'$	κ		2				
	λ					2	
	μ						4
	γ	4					

$$^2EU' \quad c_0 = -(3 \times 6 + 1 \times 2)/8 = -5/2$$

$$^2T_2U' \quad c_0 = -(3 \times 4 + 1 \times -4)/4 = -2$$

For those excited states ($^2T_1E'$, U' and $^2T_2E''$) which mix with only one component of $^4T_1$ we obtain the same c_0 as for that component.

An important point is emphasised in Table 6. Certain Zeeman transitions (e.g., σ_- from $|^4A_2 + 3/2\rangle$ to $|^2T_2U'\lambda\rangle$) are forbidden although both the α and β corfficients are non-zero. Actually, a general means of checking a table of transition probabilities is to ensure that $\sum p_\sigma^2 = \sum p_+^2 = \sum p_-^2$ when considering the whole set of transitions from G to a given spin-orbit component of the excited state.

In realistic situations, more than one $^4T_1$ state will contribute to the intensity borrowing mechanism and $\Delta W_1 \neq \Delta W_2$ for a given $^4T_1$. Tables 6 and 7 will thus have to be rearranged, ΔW_1 and ΔW_2 now entering as parameters. Also the symmetry may be tetragonal D_{2d} instead of tetrahedral; this, however, does not prevent the use of our tables.

Comparison of MCD experimental data with the new predictions (or deviations from the previous ones) will therefore provide precious information regarding those $^4T_1$ states and U' components which contribute most to the intensity gaining mechanism of quartet-doublets transitions. Such a kind of work has recently been done in the case of $CoBr_4^{2-}$ [35] and has served to demonstrate (i) that coupling to the CT states was the most effective; (ii) the U' parentage of several doublets.

5.2 MCD study of d-d bands in octahedral complexes

5.2.1 Magnetic dipole transitions

There is no special problem in that both the spin allowed and spin-forbidden bands must be treated in the same way as those for tetrahedral complexes, using now the magnetic dipole operator $\hat{L}$ instead of $\hat{m}$. $\hat{L}$ transforms as T_{1g} and all bands are Laporte allowed.

5.2.2 Electric-dipole transitions

The parity-forbiddenness in centrosymmetrical compounds is relaxed via the influence of an odd parity vibrational mode. By vibration, I mean here the motions within an isolated complex molecular ion. Adopting the standard adiabatic vibronic theory[37] and calling $Q_{s\mu}$ an asymmetrical vibrational coordinate of species s (t_{1u} or t_{2u}) we can expand the crystalline field potential in this coordinate:

$$\mathcal{H} = \mathcal{H}_0 + Q_{s\alpha} U_{s\alpha}$$

where α is one component of the s_{th} ground state normal coordinate and $U_{s\alpha} = (\partial\mathcal{H}/\partial Q)_0$. Keeping the same notations as for parity-allowed transitions, we thus write:[37]

$$
\begin{aligned}
P_\gamma &= \langle J_j v_j | \hat{m}_\gamma | \Gamma_i v_I \rangle \\
&= \sum_{s,\alpha} \Bigg[\sum_k \frac{\langle J_j | \hat{m}_\gamma | K\kappa \rangle \langle K\kappa | \hat{U}_{s\alpha} | I_i \rangle}{W_I - W_k} \\
&\qquad \sum_{k'} \frac{\langle J_j | \hat{U}_{s\alpha} | K'k' \rangle \langle K'k' | \hat{m}_\gamma | I_i \rangle}{W_J - W_{k'}}
\end{aligned}
\tag{30}
$$

We assume (i) that the excited states have equilibrium nuclear configurations of identical symmetry to that in the ground state; (ii) that we are working at low temperature and no ground-state vibrations are excited. Therefore v_I is totally symmetric a_{1g} and $Q_{s\alpha}$ must transform as v_J. Since $\mathcal{H}$ transforms like the totally symmetric representation and $Q_{s\alpha}$ like v_J, we get that $U_{s\alpha}$ must also transform like v_J. Finally, $|Kk\rangle$ must be even since the electric dipole moment is odd.

Let us apply equation (30) in the case of the $^4A_{2g} \rightarrow {}^4T_{1g}E'$ transition of d^3 ions, assisted by a t_{1u} vibrational mode. We ask ourselves which are the polarizations and relative intensities of transitions from $|^4A_{2g}0M_s\rangle$ to $|^4T_{1g}E'\alpha'\rangle$ which may be written[27]

$$|^4T_1E'\alpha'\rangle = \frac{1}{\sqrt{2}} |-1\ 3/2\rangle - \frac{1}{\sqrt{3}} |0\ 1/2\rangle + \frac{1}{\sqrt{6}} |1\ -1/2\rangle \quad (31)$$

$A_2 \times T_1 = T_2$; therefore $|K\rangle$ and $|K'\rangle$ are necessarily $^4T_2$. Assuming now for simplicity that only one $^4T_{2g}$ state contributes to p_γ, then the summations in equation (30) reduce to:

$$
\begin{aligned}
P_\gamma = \sum_{\alpha, j, k} \Bigg[&\frac{\langle T_1 j | T_1\gamma | T_2 k \rangle \langle T_2 k | T_1\alpha | A_2 0 \rangle}{W_{A_2} - W_{T_2}} \\
&+ \frac{\langle T_1 j | T_1\alpha | T_2 k \rangle \langle T_2 k | T_1\gamma | A_2 0 \rangle}{W_{T_1} - W_{T_2}} \Bigg] \langle t_1\alpha | t_1\alpha | a_1 \rangle
\end{aligned}
\tag{32}
$$

where α, γ, j and k stand for tetragonal components.

$\langle t_{1\alpha} | t_{1\alpha} | a \rangle$ is independent of the α component considered and will therefore be dropped for convenience. Using the W.E. theorem and ranging the components in the V coefficients according to Table C 2-3 of Griffith[5] then leads to:

$$P_{\gamma} \simeq \sum_{\alpha,j,k} [-1]^{T_1+T_2+j+k} \; V_1 \begin{pmatrix} T_1 & T_1 & T_2 \\ \gamma & -j & k \end{pmatrix} V_2 \begin{pmatrix} A_2 & T_1 & T_2 \\ 0 & \alpha & -k \end{pmatrix}$$

$$+ \sum_{\alpha,j,k} [-1]^{T_1+T_2+j+k} \; V_3 \begin{pmatrix} T_1 & T_1 & T_2 \\ \alpha & -j & k \end{pmatrix} V_4 \begin{pmatrix} A_2 & T_1 & T_2 \\ 0 & \gamma & -k \end{pmatrix}$$

which we write symbolically:

$$P_{\gamma} \simeq V_1 V_2 + V_3 V_4. \tag{33}$$

We now consider separately the three orbital components M_Γ in the expression of $|^4T_1E'\alpha'\rangle$:

(i) $j = 0$ implies $\gamma = \alpha = \pm 1$ in V_1 and V_3. However, V_1 and V_3 both change their sign for $\gamma = \alpha = + 1$ or $- 1$ while V_2 and V_4 keep their sign when reversing that of α or γ. Therefore $p_\gamma = 0$ to $j = 0$; keeping in mind the selection rule $\Delta M_S = 0$, this means that no π or σ transition is allowed from $|^4A_{2g}0\,1/2\rangle$ to $|^4T_{1g}E'\alpha'\rangle$.

(ii) $j = 1$ together with $\gamma = \pm 1$ for $\sigma^{\pm}$ transitions implies $\gamma = -1$ and $\alpha = 0$ in V_1. Thus V_3 is zero and only $V_1 V_2$ contributes to p_γ for the amount: $-(1/\sqrt{6})(1/\sqrt{3})$.

(iii) $j = -1$ implies $\gamma = +1$ and $\alpha = 0$ in V_1 and $V_1 V_2$ is again the only contribution $(+1/\sqrt{18})$.

Our conclusion is therefore that $\sigma\pm$ transitions are allowed to the $M_\Gamma = \mp 1$ components of $|^4T_{1g}E'\alpha'\rangle$. $(V_1 V_2)^2$ being the same for $j = \pm 1$, inspection of equation (31) tells us that the σ_+ and σ_- transitions originate from the $|3/2\rangle$ and $|-1/2\rangle$ spin levels of $^4A_{2g}$, their respective intensities being $1/2$ and $1/6$ (or 3 and 1). Remembering that transitions occur in pairs, we do not even have to draw the complete Zeeman scheme to find

$$c_0 = -(1 \times 1 + 3 \times 3)/4 = -5/2$$

To obtain c_0 (and also a_1) for other spin-orbit components of $^4T_{1g}$ now presents no difficulty if we use the above selection rules. $(c_0 = -1$ and $+3/2$ for $U'3/2$ and $E'',U'5/2$ respectively).[38]

These data may in turn be used with our equations (25) and (26) to obtain the zeroth and first order moments of MCD when spin-orbit components are not fully resolved.

Due to the symmetry properties of V coefficients, it can be easily established that a t_{2u} vibrational mode will allow $\sigma+$ transitions to the $M_\Gamma = \pm\ 1$ components of $^4T_{1g}$. All MCD parameters will thus change their sign when t_{1u} goes to t_{2u}.

The important conclusions which emerge from this discussion are therefore that MCD may serve either (i) to figure out the dominant t_{1u} or t_{2u} contribution to the intensity stealing mechanism when crystal field assignments are firmly established or (ii) to ascertain these when one vibrational mode is known from the temperature dependence of $\langle A\rangle_o$ to contribute most efficiently.

The selection rules (spatial part) given here for $\sigma+$ polarizations to a T_{1g} excited state are equally valid if the ground state is $^3A_{2g}$ as encountered for Ni^{2+} in a weak or strong field. If we now consider T_{2g} instead of T_{1g}, then $\sigma+$ will change into $\sigma-$ and vice-versa, MCD parameters changing their sign.

In all the above situations, where G is non-degenerate, MCD parameters are independent of reduced matrix elements which occur in equation (32). This is not so if we consider a ground state having an orbital angular momentum, for example, $^4T_{1g}$ for Co^{2+} in a weak crystal field. For a t_{1u} vibration, then $^1T_1xT_1 = A_1+E+T_1+T_2$ and the summations over K and K' states will have to be made in equation (30).[39]

The basic information required to treat spin-forbidden bands in O_h symmetry is, at least in principle, contained in the preceding pages (5.1 and 5.2). Each p_γ generally contains several contributions and great care should be taken concerning the phases of coupling coefficients. Once the tedious labour is completed, one should certainly be very wise, anyhow, to check that his total intensity for and σ_+ (or σ_-) components for a given transition is the same! The case of d^5 complexes has recently been treated in our laboratory.[40]

6.   MCD OF MAGNETIC MATERIALS AND PAIRS

A large amount of work has already been accumulated on anti-ferromagnetic compounds of transition metal ions. I shall concentrate on $KNiF_3$[41] which is cubic above and below $T_N = 246$ K, and more precisely in the spectral region covering the $^3A_{2g} \rightarrow\ ^1E_g$ of the free ion.

Spin-orbit coupling transforms the $^3\Gamma_2^+ \to {}^1\Gamma_3^+$ transition in the O_h group into the $\Gamma_5^+ \to \Gamma_3^+$ transition of the double group. In the simplest model, each ion is considered as being isolated and submitted to an effective molecular field created by the other ions. A low external magnetic field H directed along, e.g. a C_4 axis, serves essentially to orient the spins antiparallel along to that direction in the two sublattices, lowering the symmetry to C_{4h} in each of them.

The Zeeman diagram for the two sublattices is shown in Figure 16a. In C_{4h}, $\Gamma_5^+ \to \Gamma_2^+ + \Gamma_3^+ + \Gamma_4^+$ and $\Gamma_3^+ \to \Gamma_1^+ + \Gamma_2^+$, the degeneracy of these being not lifted, the energy of the two $M_S = \pm 1$ spin levels is $\pm \mu_B H_{exch}$. The exchange field is large and only the lowest spin level in each sublattice is appreciably populated at low temperature. The external field B acts further as a small perturbation, Γ_3^+ in A being lowered by $g\mu_B H$ while Γ_4^+ in B is raised by the same amount.

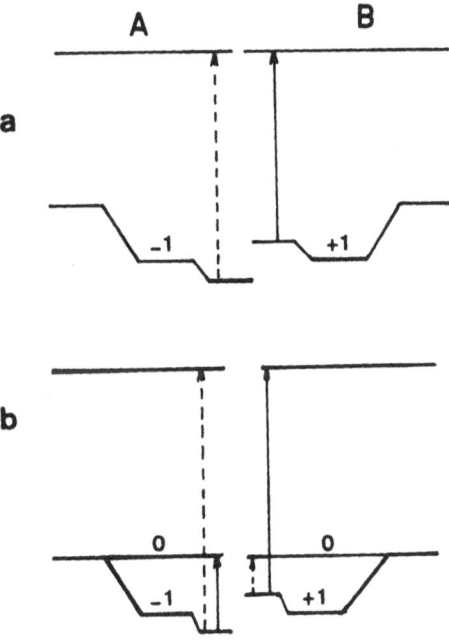

Figure 16. Zeeman scheme for the $^3A_{2g} \to {}^1Eg$ transition in KNiF$_3$. (a) pure exciton line; (b) exciton-magnon line. The full and dotted lines correspond to σ_+ and σ_- transitions respectively.

The $\Gamma_5^+ \to \Gamma_3^+$ transition is parity-forbidden for $\hat{m}$ but allowed for $\hat{L}(\Gamma_4^+)$. Group theory for C_{4h} tells us that the $\Gamma_3^+(\Gamma_4^+) \to \Gamma_1^+$ + Γ_2^+ transitions are allowed for both σ_+ (Γ_3^-) and σ_- (Γ_4^-) polarizations. In order to find the dominant one, we consider that $^1E\Gamma_3^+$ gains some spin triplet character via its mixing with $^3T_{1g}\Gamma_3^+$. Therefore, the polarizations for the $^3A_{2g} \to {}^1E_g$ transition are the same as those for $^3A_{2g} \to {}^3T_{1g}\Gamma_3^+$; these are indicated in Figure 16a. The two sublattices being independent of each other, the ground states in A and B have the same population. Therefore, our model predicts an 'a$_1$ term' (a$_1$ = -2g) for the magnetic dipole line. This is indeed what is observed at low temperature at the shortest energies in the investigated spectral range.[41]

Our model is oversimplified in that one should talk in terms of excitons, i.e., excitations which propagate in the crystal and are associated with a particular wave vector $\vec{k}$. The zero-phonon transition which we have just considered is an exciton at the centre ($\vec{k}$=0) of the Brillouin zone. This improved description, however, does not modify the answer reached in the molecular field approximation.

Magnons are analogous to excitons which imply a change in the spin component in the ground state of an ion. For example, we shall create a magnon in the A sublattice when inducing a transition from -1 to 0 or in the B sublattice for 1 → 0. An exciton-magnon transition corresponds to the simultaneous excitation of an exciton and a magnon on opposite sublattices. In our simple model (Figure 17b), it appears that σ_+ and σ_- now occur at the same energy, the 'a$_1$ term' being thus zero. This kind of negative proof has indeed been used a few times to ascertain the location of such transitions.

It appears from recent work that MCD is a very promising way of studying pair transitions. Figure 17 illustrates this point by showing a selected portion of the absorption and MCD spectra of a $Cs_3Cr_2Cl_9$ single crystal at two temperatures.[14] The spectral region covers the $^4A_2$-$^4A_2 \to {}^4A_2$-^2E transitions of the pairs and the absorption is that expected in a pair model; the MCD should serve in particular to chose among several mechanisms possibly responsible for the transitions. These experiments are also of great value to determine the isotropic and anisotropic parts of the exchange interaction in both the ground and excited states of the pair.

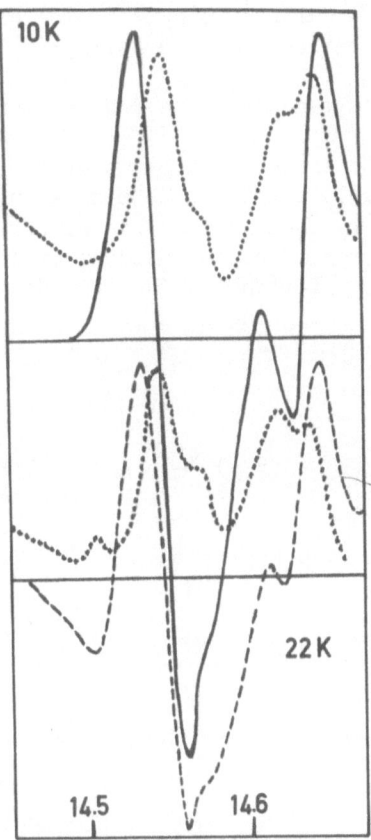

Figure 17.   Selected absorption and MCD data for $Cs_3Cr_2Cl_9$ in the red region, at 10 K (top) and 22 K (bottom).

    Double exciton lines are also found in concentrated or pure materials.   These are allowed for the exchange electric dipole moment operator.   Their MCD is predicted to be zero (e.g., $^4A_2 - ^4A_2 \rightarrow {}^2E - {}^2E$) but experimental work is needed in various situations so as to check the existing theory.

7.   CONCLUDING REMARKS ON MLD

In a Voigt-Cotton-Mouton effect experiment, the magnetic field is applied perpendicular (along Z) to the direction of propagation X of the light beam and we define:

$\Delta A = A_Z - A_Y$ as the magnetic linear dichroism.

In order to clarify the broad aspects of a MLD spectrum, I take the case of a cubic compound, Z being along a C_4 axis. I further assume that (i) the rigid shift approximation is valid, (ii) electric dipole moments are independent of the magnetic field, (iii) the Zeeman splitting in the ground state is small compared to kT.   Then, using the notations of section 2:

$$N_i = 1 - \frac{y}{kT} + \frac{1}{2}\left(\frac{y}{kT}\right)^2$$

$$f(E_{IiJj}, E) = f(E_{IJ}, E) - (x-y)\frac{\partial f}{\partial E} + \frac{1}{2}(x-y)^2\frac{\partial^2 f}{\partial E^2} , \text{ where}$$

$$x = \left\langle Ii | L_z + 2S_z | Ii \right\rangle \mu_B B$$

$$y = \left\langle Jj | L_z + 2S_z | Jj \right\rangle \mu_B B$$

These equations lead to the following expression for the MLD:

$$\frac{\Delta A_L}{E} = \eta'\left\{ \frac{\mathcal{L}_0}{(kT)^2}\mathcal{D}_0 f + \frac{\mathcal{L}_1}{kT}\mathcal{D}_0\left(-\frac{\partial f}{\partial E}\right) + \mathcal{L}_2\,\mathcal{D}_0\left(\frac{\partial^2 f}{\partial E^2}\right) \right\}(\mu_B B)^2 \quad (33)$$

where:

$$\mathcal{L}_0 = \frac{1}{2}\sum_{i,j}\left\{ p_0^2 - (p_+^2 + p_-^2)/2 \right\}x^2 / \sum_{i,j} p_0^2 \qquad (34)$$

$$\mathcal{L}_1 = -\sum_{i,j}\left\{ p_0^2 - (p_+^2 + p_-^2)/2 \right\}y(x-y) / \sum p_0^2 \qquad (35$$

$$\mathcal{L}_2 = \frac{1}{2}\sum_{i,j}\left\{ p_0^2 - (p_+^2 + p_-^2)/2 \right\}(x-y)^2 / \sum p_0^2 \qquad (36)$$

The three MLD parameters can, in principle at least, be extracted from the experimental data via a moment analysis, since:

$$\langle \Delta A \rangle_o / \langle A \rangle_o = \mathcal{L}_o \left(\mu_B B / kT \right)^2$$

$$\langle \Delta A \rangle_1 / \langle A \rangle_o = \mathcal{L}_1 \left(\mu_B B \right)^2 (kT)^{-1}$$

$$\langle \Delta A \rangle_2 / \langle A \rangle_o = 2\mathcal{L}_2 \left(\mu_B B \right)^2 + \left(\langle A \rangle_2 / \langle A \rangle_o \right) \left(\langle \Delta A \rangle_o / \langle A \rangle_o \right)$$

From equation (33) we conclude that the first term ('$\mathcal{L}_o$ term') may only occur for paramagnetic compounds. Assuming $B \sim 1T$ $kT \sim 10$ cm^{-1} and $x \sim y \sim 1$, then the MLD/MCD ratio is predicted to be 0.1 if $\mathcal{L}_o$ is non-zero. The '$\mathcal{L}_2$ term' is expected when the ground and/or excited state is degenerate; its contribution is proportional to $(x-y)^2/(\text{width})^2$ and thus expected to be large for sharp lines, even at room temperature. Quite generally, the relative contributions of the three MLD terms depend upon the ratio kT/width.

In practice we have obtained reasonably large MLD signals at pumped helium temperature on several paramagnetic compounds, using an electromagnet ($B \sim 1T$). A split coil superconducting magnet may also be used in less favourable circumstances.

To a certain ordering of excited states there will correspond different sets of CD signs in the Voigt and Faraday configurations. MCD and MLD can therefore be used as complementary tools for spectroscopic assignments.

A most important property of MLD is that it may be anisotropic even in the case of cubic compounds. Choosing d^5 tetrahedral complexes as an example, $\mathcal{L}_o$ is predicted to be invariant upon rotation of the crystal around a C$_4$ axis when the excited state is $^4A_2$ or $^4T_2$; it varies as $\cos 4\phi$ when the excited state is $^4T_1$ or ^4E. This is illustrated in Figure 18 for the MnBr$_4^{2-}$ ion. This property is clearly of very great potential value for precise assignments. Similar results have been obtained theoretically for tetragonal centres and used to interpret a series of experiments on Nd^{3+} in CaWO$_4$.[42]

ACKNOWLEDGMENTS

This contribution owes much not only to all the workers in our laboratory, but also to a number of people with whom fruitful collaboration has been established on parity-forbidden bands (Dr. Harding), d^5 compounds (Dr. Vala), iron sulfur proteins (Dr. Thomson), porphyrins (Dr. Risler) and garnets (Drs. Hodges and Krishnan). I wish to express my deep thanks to all of them.

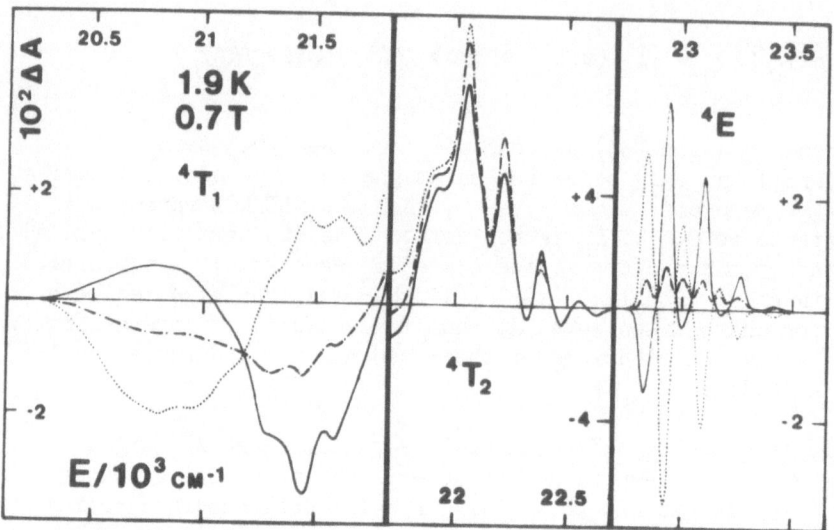

Figure 18.   Anisotropy of the MLD spectrum of a $(NEt_4)_2$ $MnBr_4$ crystal.   The light propagates along a C_4 axis and the crystal is rotated around that direction.   ——: $\phi = 0$;   -.-.: $\phi = 30^\circ$, ...: $\phi = 45^\circ$ (see text for notations).

REFERENCES

1.   A.D. Buckingham and P.J. Stephens, Ann. Rev. Phys. Chem. 17, 399 (1966).

2.   J. Badoz in 'Fundamental aspects and recent developments in ORD and CD', F. Ciardelli and P. Salvadori Ed., Heyden (1973).

Among the latest technological developments, I would like to mention (i) the photoelastic modulator which has extended the available spectral range of MCD machines towards the red; (ii) the use of very dispersive monochromators for the study of sharp lines; (iii) measurements at pumped helium temperatures under a high or low magnetic field (superconducting or electromagnet respectively) so as to populate selectively the sublevels of a ground state and eventually help in the interpretation of a spectrum; (iv) MCD spectroscopy on matrix isolated species.

3.   R.G. Denning, this book.

4.   B.R. Judd, 'Operator techniques in atomic spectroscopy',
     McGraw Hill (1963).

5.   J.S. Grifith 'The irreducible tensor method for molecular
     symmetry groups', Prentice Hall, Englewood Cliffs, N.J.
     (1962).   The method is invaluable for evaluating matrix
     elements of multideterminental wavefunctions and sums of
     matrix elements.

6.   P.A. Dobosh, Phys. Rev. A 5, 2376 (1972).   Griffith's
     work[5] did not extend to the double-group representations,
     those which are important in odd-electron systems when
     spin-orbit coupling is a major element in the molecular
     Hamiltonian.   In Dobosh's paper, the irreducible tensor
     method is extended to the double O^* group.

7.   P.N. Schatz and A.J. McCaffery, Quart. Rev. Chem. Soc. 23,
     552 (1969) (and 24, 229 (1970)).

8.   P.J. Stephens, Ann. Rev. Phys. Chem. 25, 201 (1974).

9.   B. Briat in 'Fundamental aspects and recent developments in
     ORD and CD', F. Ciardelli and P. Salvadori Ed., Heyden
     (1963).

10. A_γ/E is proportional to the extinction index k which occurs
     in the definition of the complex refractive index $\tilde{n}$ of
     the material: $\tilde{n} = n-jk$ with $j : \sqrt{-1}$.

11. P.J. Stephens, J. Chem. Phys., 52, 3489 (1970).

12. J.C. Rivoal and B. Briat, Mol. Phys. 27, 1081 (1974).

13. See, for example, J. Badoz, M. Billardon, A.C. Boccara and
     B. Briat, Symp. Faraday Soc., 3, 27 (1969).

14. Unpublished results from our group.

15. J.C. Sutherland, I. Salmeen, A.S.K. Sun and M.P. Klein,
     Biochim. Biophys. Acta, 263, 550 (1972).

16. See, for example, M.D. Rowe, A.J. McCaffery, R. Gale and
     D.N. Copsey, Inorg. Chem. 11, 3090 (1972).

17. I.N. Douglas, R. Grinter and A.J. Thomson, Mol. Phys. 28,
     1377 (1974).

18. A.J. Mann and P.J. Stephens, Chem. Phys. 4, 96 (1974).

19.   M.J. Harding and B. Briat, Mol. Phys. 27, 1153 (1974).

20.   J.C. Collingwood, P. Day and R.G. Denning, J. Chem. Soc.
      Faraday Trans. II, 69, 591 (1973).

21.   P.N. Schatz, A.J. McCaffery, W. Suetaka, G.N. Henning,
      A.B. Ritchie and P.J. Stephens, J. Chem. Phys. 45, 722
      (1966).

22.   We should in principle consider also the spin part $(2\hat{S}z)$
      of the Zeeman Hamiltonian.   In the absence of spin-orbit
      coupling, this will actually lift the spin degeneracy of
      each of the orbital components in Figure II but the MCD
      result will remain unaffected.   We note further that the
      result in equation (24) rests upon the fact that the matrix
      elements of $\hat{L}z$ within T_2 are the same as those of $-\bar{L}z$ within
      a P state.   Finally, in a more rigorous treatment, one
      should consider the possible reduction of orbital moments
      (and thus c_o) via covalency.

23.   B.P. Zakharchenya and I.B. Rusanov, Opt. Spectrosc. 19, 207
      (1965).

24.   D.F. Johnston, S. Marlow and W.A. Runciman, J. Phys. C 1,
      1455 (1968).

25.   G.F. Koster, J.O. Dimmock, R.G. Wheeler and H. Statz,
      Properties of the thirty-two point groups, M.I.T. Press,
      Cambridge (1963).

26.   T.M. Dunn, Trans. Faraday Soc. 57, 1441 (1961).

27.   J.S. Griffith, The theory of transition metal ions,
      Cambridge University Press (1961), see Table A 20.

28.   J.R. Dickinson, S.B. Piepho, J.A. Spencer and P.N. Schatz,
      J. Chem. Phys. 56, 2668 (1972) and references therein.

29.   A.C. Boccara, J. Ferré and J. Badoz, Phys. Status Solidi,
      36, 601 (1969).

30.   R.H. Petit, B. Briat, A. Müller and E. Diemann, Mol. Phys.
      27, 1373 (1974).

31.   M.J. Stillman and A.J. Thomson, J.C.S. Faraday II, 805
      (1974).

32.   C.H. Henry, S.E. Schnatterly and C.P. Slichter, Phys. Rev.
      A137, 583 (1965).

33. B.D. Bird and P. Day, J. Chem. Phys. 49, 392 (1968).

34. D.J. Robbins, M.J. Stillman and A.J. Thomson, J.C.S.,
     Dalton Trans. 813 (1974).

35. P.N. Quested, R.J. Tacon, P. Day and R.G. Denning, Mol.
     Phys. 6, 1553 (1974).

36. C.J. Ballhausen, Introduction to Ligand Field Theory,
     McGraw Hill, N.Y., (1962), chapter 8.

37. P.J. Stephens, J. Chem.Phys. 44, 4060 (1966).

38. M.J. Harding and B. Briat, Mol. Phys. 27, 1153 (1974).

39. M.J. Harding and B. Briat, Mol. Phys. 25, 745 (1973).

40. M. Vala, J.C. Rivoal and J. Badoz, in press.

41. R.V. Pisarev, J. Ferré, R.H. Petit, B.B. Krichevtsov and
     P.P. Syrnikov, J. Phys. C, Solid State Phys. 7, 4143
     (1974).

42. A.C. Boccara and N. Moreau, Phys. Status Solidi 45, 573
     (1971).

MCD SPECTRA OF CHARGE-TRANSFER TRANSITIONS: OCTAHEDRAL Ir^{4+}

P.N. Schatz

Department of Chemistry, University of Virginia,
Charlottesville, Virginia 22901, U.S.A.

1. INTRODUCTION

The utility of magnetic circular dichroism (MCD) spectroscopy
is now well recognized, and its application to the octahedral
transition metal hexahalides provides an excellent example of the
power of the technique.   In this short report, attention will be
focussed on one 'simple' example, the Ir^{4+} ion, in preference to
an attempt to survey the considerable number of hexahalides that
have now been studied in detail.   A study of Ir^{4+} illustrates
many of the features of MCD spectroscopy both at low and high
resolution as well as demonstrating important characteristics of
the lower energy ligand-to-metal charge-transfer (C.T.)
transitions in hexahalides.

A schematic molecular orbital (M.O.) energy level diagram
for an octahedral transition metal hexahalide is shown in Figure 1.
We note that in the ground state all 'ligand' M.O.'s are filled,
and the t_{2g} (and e_g) 'metal' M.O.'s contain a total of n electrons
for a metal ion of configuration d^n.   Ligand-to-metal charge-
transfer transitions then involve the excitation of a 'ligand'
electron into a 'metal' M.O.   The octahedral Ir^{4+} ion which has
a d^5 configuration is a particularly simple case since the ground
configuration $(t_{2g}{}^5)$ involves a single hole in the metal t_{2g} M.O.
whereas the low-lying C.T. states involve a single hole in a
ligand M.O.   There is then a one-to-one correspondence between
configurations and states, i.e. interelectronic repulsions can
produce no additional splittings.   An energy level diagram for
some of these C.T. states is shown in Figure 2.

P. Day (ed.), Electronic States of Inorganic Compounds. 223–240. *All Rights Reserved.*
Copyright © 1975 by D. Reidel Publishing Company, Dordrecht-Holland.

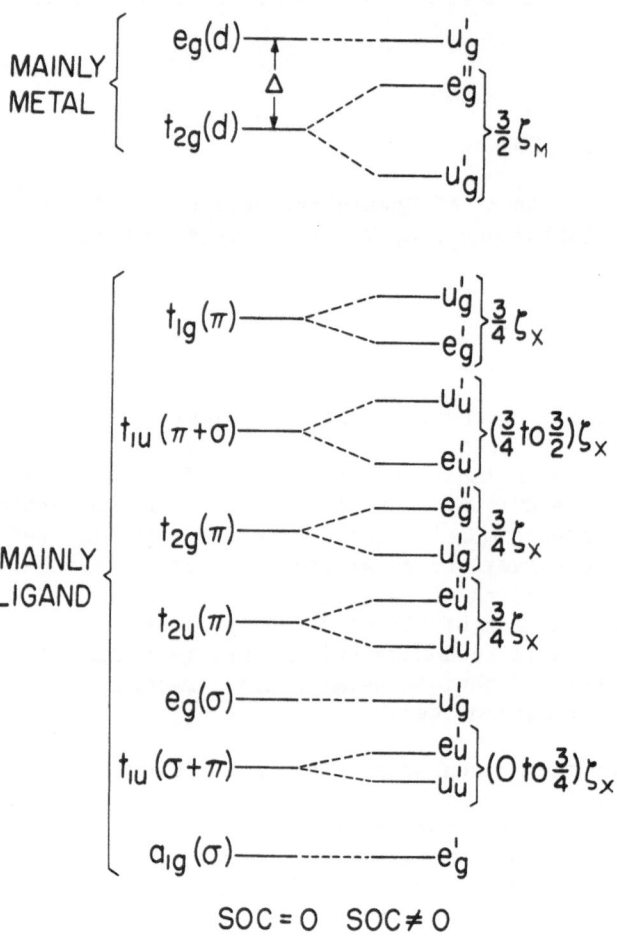

Figure 1.   Schematic M.O. energy level diagram in the absence
and presence of spin-orbit coupling (S.O.C.).   In the ground
state of Ir^{4+} (M = Ir), all levels through U'_g ($t_{2g}(d)$) are filled
and e_g ($t_{2g}(d)$) contain one electron.   The double group notation
is that of Griffith with U' levels holding a maximum of four and
e' and e'' levels holding a maximum of two electrons.   The spin-
orbit splittings have been calculated using the approximation
described in Piepho et al., Mol. Phys. 19, 781 (1970).

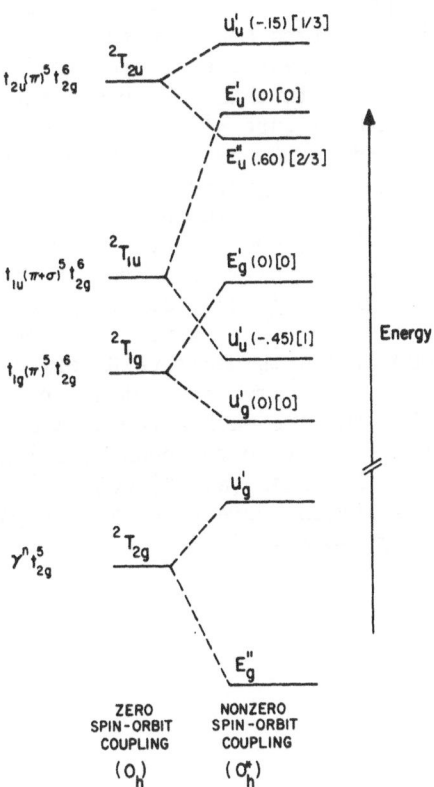

Figure 2. Energy level diagram for states of $\gamma^n t_{2g}{}^5$ and low-lying $\gamma^{n-1} t_{2g}{}^6$ configurations. The Griffith double group notation is used. The energy level spacings are roughly to scale and the order corresponds to our assignments in IrBr$_6^{2-}$. The numbers in parentheses are the theoretically calculated C values (using an orbital reduction factor of 0.85), and the numbers in square brackets are the corresponding theoretical dipole strengths (D), C/D being in units of Bohr magnetons.

2. LOW RESOLUTION (SOLUTION) RESULTS

The $^2T_{2g}$ ground state of Ir^{4+} is split by spin-orbit coupling (S.O.C.) into an E$_g''$ (Kramers doublet) lower component and a U$_g'$ upper component, (Figure 2) and this splitting to first order is calculated to be $1.5 \, \zeta \text{Ir}^{4+} \sim 3600$ cm^{-1}. (Experimentally,[1,2] the splitting is observed to be 5000 cm^{-1}, the increased value resulting from configuration interaction between the E$_g''$ ground state and low-lying E$_g''$ excited states.)[2] Clearly, only the E$_g''$

ground state will be occupied at room or low temperature.   Since
this state is paramagnetic with a large magnetic moment $(g \sim 1.75)$,
the room temperature MCD spectrum is expected to be dominated by
C terms[3] whose predicted signs and magnitudes are shown in Figure
2 along with the predicted relative dipole strengths of the
transitions.   In accord with this, the room temperature solution
spectrum of $IrCl_6^{2-}$ shows (Figure 3) two intense, low energy
transitions (20,000 and 23,000 cm^{-1}) with negative and positive
C terms respectively.[4]   (A negative C term corresponds to
positive MCD and vice versa.)[3]   These signs demonstrated[4]
unambiguously that the lower and higher energy transitions
correspond respectively to excitations from the $t_{1u}(\pi + \delta)$ and
t_{2u} (π) ligand M.O.'s (into $t_{2g}(d)$), contrary to the order that
had generally been accepted previously.   The considerably weaker
transition at 32 000 cm^{-1} with negative C term corresponds to
the t_{1u} $(\delta + \pi) \rightarrow t_{2g}$ (d) excitation (not shown in Figure 2 –
see Figure 3, Ref. 5), and the much stronger transition at
$\sim 42,000$ cm^{-1} almost certainly corresponds to a ligand $\rightarrow e_g$ (d)
excitation.   The weak shoulder (at $\sim 17,000$ cm^{-1}) on the red side
of the first C.T. transition corresponds to the parity-forbidden
C.T. excitation, t_{1g} $(\pi) \rightarrow t_{2g}$ (d).

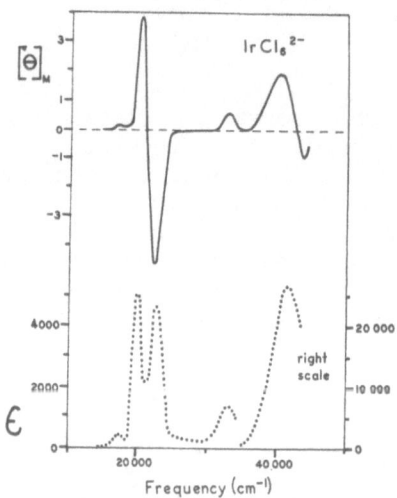

Figure 3.   Absorption spectrum and MCD of $IrCl_6^{2-}$ in dichloro-
ethane. $[\theta]_M$ is the molar ellipticity (defined as in natural
optical activity in degrees.deciliter decimeter^{-1}.mole^{-1}) per
gauss in the direction of the light beam. ε is the molar
extinction coefficient.

This IrCl$_6^{2-}$ MCD spectrum bears a striking resemblance to that of Fe(CN)$_6^{3-}$,[6] and in fact if ligand spin–orbit coupling is neglected, one predicts similar C/D ratios.[4,6]   Such an assumption is justified for IrCl$_6^{2-}$ in solution since the observed bandwidth of the second allowed transition is ~2,000 cm^{-1} whereas the expected excited state spin–orbit splitting is $(3/4)\mathcal{J}_{Cl}$ ~440 cm^{-1}.

The effect of halogen S.O.C. becomes clearly apparent even in solution when Br is substituted for Cl ($\mathcal{J}_{Br}$ ~2460 cm^{-1} vs $\mathcal{J}_{Cl}$ ~590 cm^{-1}), as is seen in Figure 4.   The clear separation of the E$_u''$(^{2}T$_{2u}$) and U$_u'$(^{2}T$_{2u}$) components (~ 16,500 and 18,000, 18,600 cm^{-1}) is apparent with the latter further split (~600 cm^{-1}) into two Kramers doublets perhaps by solvent interactions.   The U$_u'$(^{2}T$_{1u}$) excited state is also split (Figure 2) and corresponds to the two lowest energy absorption bands in Figure 4 (~14,500 cm^{-1}), but the third band in that region which actually has the largest E$_{max}$ has been clearly shown[7,8] to be a forbidden (vibronically allowed) transition since it decreases markedly in intensity at low temperatures[7] and is built on a forbidden origin[8] (see below).   The E$_u'$(^{2}T$_{1u}$) state is orbitally forbidden

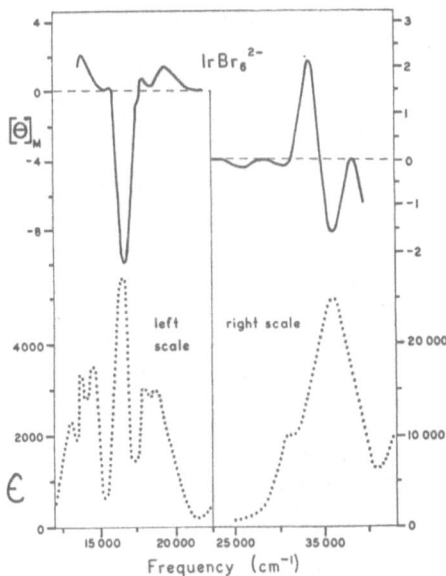

Figure 4.   Absorption spectrum and MCD of IrBr$_6^{2-}$ in dichloro-ethane.   [Θ]$_M$ and ε are defined as in Figure 3.   The MCD further to the red, to about 12,000 cm^{-1}, closely mirrors the absorption spectrum.

and is not apparent in the solution spectrum. The C term signs
for the three allowed C.T. components are seen to agree with the
theoretical predictions (Figure 2) if the $^2T_{1u}$ state is placed
lowest. The strong band at about 35,000 cm^{-1} again almost
certainly corresponds to a ligand $\rightarrow e_g(d)$ excitation.

3. HIGH RESOLUTION (DOPED CRYSTAL) RESULTS

One anticipates much greater spectral detail if one studies
the Ir^{4+} ion at low temperature since features are expected to
become sharper and individual vibronic transitions may be resolved.
In this event, A terms may also become apparent since their
amplitudes vary as the inverse second power of the line (or band)
width.   They thus gain enormously in prominence as lines become
sharp even though the C terms which vary as T^{-1} also become much
larger at low temperature.

These expectations are fully realized when Ir^{4+} is doped at
low concentration into the cubic hosts, Cs$_2$ZrX$_6$ (X = Cl, Br).
Ir^{4+} substitutes for Zr^{4+} at a site of O$_h$ symmetry, and a wealth
of vibronic detail is revealed.   These spectra are shown in
Figures 5-12, and a comparison with Figures 3 and 4 shows the
order of magnitude difference between the room temperature solution
spectrum and the low temperature spectrum of the same moiety doped
into a favourable host.   These high resolution spectra have been
discussed in considerable detail elsewhere,[8-11] and only some of
the main features will be outlined here.

Examining first the simpler IrBr$_6^{2-}$ spectrum, we see the
strong U$_u'$($^2T_{1u}$) band at ~13,000 cm^{-1} (bands 3,4 - Figure 5) and
the E$_u''$ and U$_u'$ spin-orbit components of $^2T_{2u}$ (bands 6 and 8
respectively - Figures 6, 8, 9).   Band 5 (Figures 5, 7) is the
vibronic transition referred to earlier which had $\mathbf{\varepsilon}_{max}$ larger
than bands 3,4 in solution.   That it is built on a forbidden
origin is clear from the hot band structure (Figure 7) and its
marked decrease in intensity at low temperature and its very sharp
vibronic structure is notable.   This band is assigned as the
upper (E$_g'$) spin-orbit component of $^2T_{1g}$, and bands 1 and 2 are
assigned to the lower (split) component, U$_g'$($^2T_{1g}$).   The
appearance, throughout the spectrum of prominent A terms
associated with much of the sharp, vibronic structure is clear
and is an important aid in making detailed assignments.[8]
Significant vibronic interactions occur in bands 3,4 and 6,7.
In the former, a Jahn-Teller effect appears to be operative,[8]
and in the latter there is strong intensity mixing.   Band 6 is
formally assigned to the allowed E$_u''$($^2T_{2u}$) state and band 7 to the
orbitally forbidden E$_u'$($^2T_{1u}$) state, but in fact band 7 contains
at least half of the total intensity, presumably due to strong
vibronic interaction with band 6.

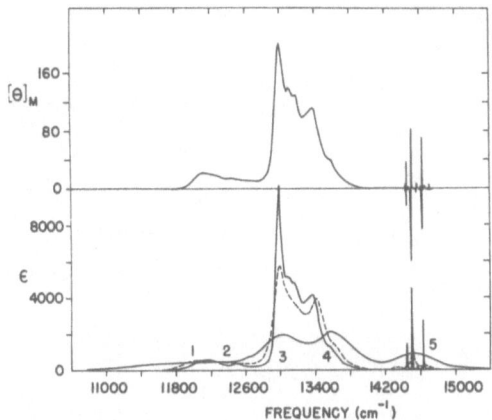

Figure 5.   Absorption and MCD spectrum of Cs$_2$ZrBr$_6$:Ir^{4+}. $[\theta]_M$
is the MCD in molar ellipticity units (defined as in nstural
optical activity in degrees.deciliter decimeter^{-1}.mole^{-1}) per
gauss in the direction of the light beam. $[\theta]_M = 3.30 \times 10^3$
$(\epsilon_l - \epsilon_r)/H$.   is the molar extinction coefficient.   The MCD and
sharpest absorption spectrum (solid line) were run at 8°K, the
dashed absorption spectrum was rnn at liquid nitrogen temperature,
and the broad solid one at room temperature.   The bands are
numbered in accordance with our discussion in the text.

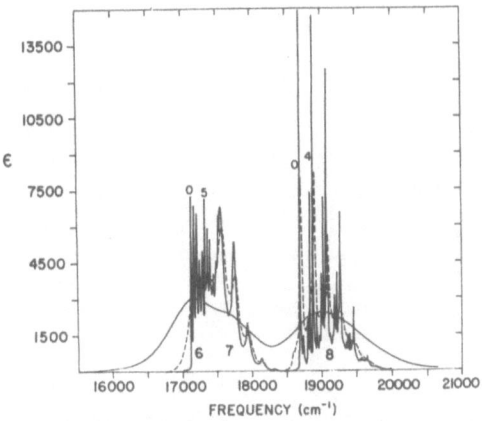

Figure 6.   Absorption spectrum of Cs$_2$ZrBr$_6$:Ir^{4+}.   Units,
notation, and temperatures are as in Figure 5.

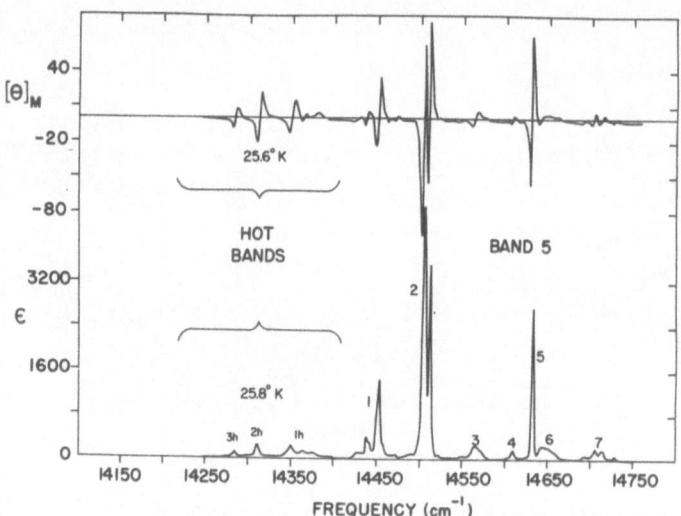

Figure 7.   Absorption and MCD spectrum of $Cs_2ZrBr_6:Ir^{4+}$
including hot bands.   Units and notation are as in Figure 5.
The cold spectrum was run at 8°K.   The hot bands have been
blown up along the ordinate for clarity the factors being 10 and
2 for $[\theta]_M$ and ε, respectively.

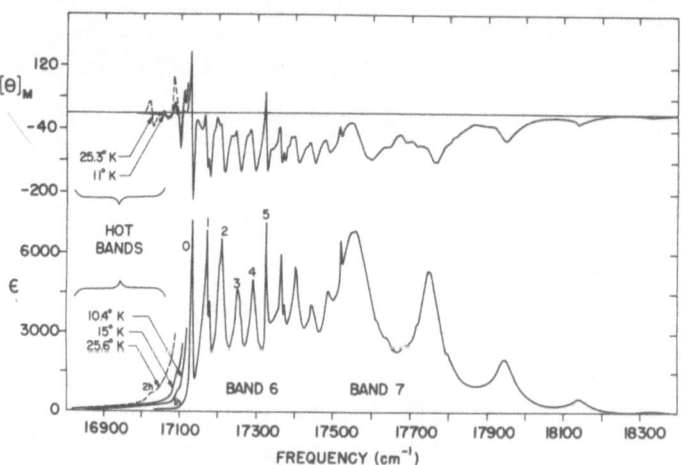

Figure 8.   Absorption and MCD spectrum of $Cs_2ZrBr_6:Ir^{4+}$ including
hot bands.   Units and notation are as in Figure 5.   The cold
spectrum was run at 8°K.   The hot bands hvve been blown up along
the ordinate for clarity the factors being 40 and 3.75 for $[\theta]_M$
and ε, respectively.

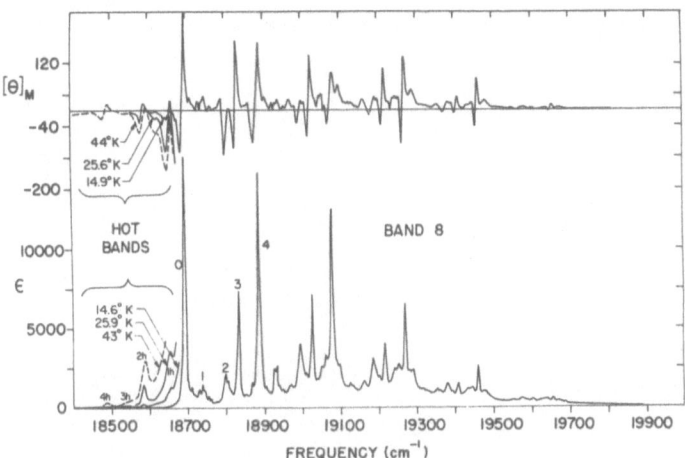

Figure 9. Absorption and MCD spectrum of Cs$_2$ZrBr$_6$:Ir^{4+} including hot bands. Units and notation are as in Figure 5. The cold spectrum was run at 8°K. The hot bands have been blown up along the ordinate for clarity the factors being 20 and 4.17 for $[\theta]_M$ and ε, respectively.

It is quite illuminating to compare the IrCl$_6^{2-}$ spectrum (Figures 10-12) with that of IrBr$_6^{2-}$. The obvious differences (aside from the well-known 5000 cm^{-1} blue shift of corresponding C.T. baricentres) relate to the decreased ligand spin-orbit splitting ($\jmath_{Cl}$ 590 cm^{-1} vs $\jmath_{Br}$ 2460 cm^{-1}). The two strong bands, at 19,000 and 23,000 cm^{-1} are assigned as in solution, to the allowed C.T. states $^2T_{1u}$ and $^2T_{2u}$, but there is no indication of the very clear spin-orbit splitting of the $^2T_{2u}$ state observed in the bromide (Figures 6,8,9), a point which will be discussed in some detail in section 4.

Lines 1-3 (Figure 10) are associated with the parity-forbidden $^2T_{1g}$ C.T. state, lines 1 and 2 being assigned to the U$_g'$ component and line 3 to the E$_g'$ component. The reasons for the 240 cm^{-1} interval between lines 1 and 2 and the abnormally large spin-orbit splitting between U$_g'$ and E$_g'$ ($\sim$700 cm^{-1} vs $\frac{3}{4}\jmath_{Cl}$ 440 cm^{-1}) are not clear.

The strong band at 19,500 cm^{-1} is undoubtedly dominated by the U$_u'$($^2T_{1u}$) C.T. state,[9] but the details of the vibronic structure are unclear. Originally,[9] the structure (lines 6-13, Figure 10) was associated with both the allowed U$_u'$($^2T_{1u}$) and the orbitally forbidden E$_u'$($^2T_{1u}$) components perhaps interacting through a Jahn-

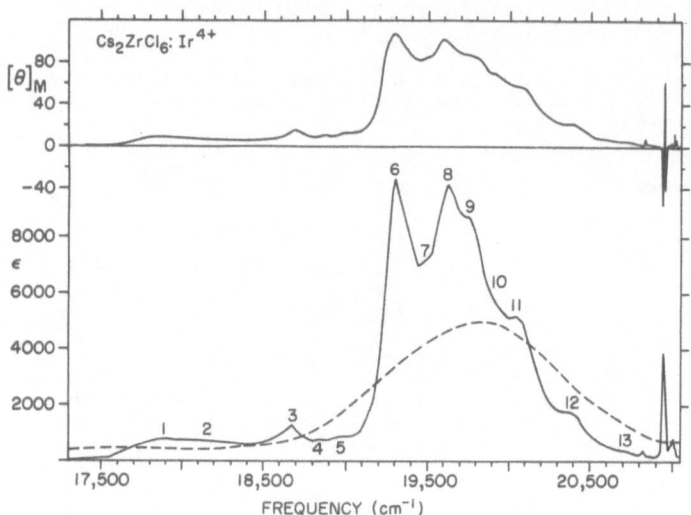

Figure 10.   Absorption and MCD spectrum of $Cs_2ZrCl_6:Ir^{4+}$. $[\theta]_M$
is the MCD in molar ellipticity units (defined as in natural
optical activity in degrees deciliter decimeter^{-1} mole^{-1}) per
gauss in the direction of the light beam. $[\theta]_M = 3.30 \times 10^3$
$(\epsilon_l - \epsilon_r)/H$. ϵ is the molar extinction coefficient.   The solid
curves were run at 11°K and the dashed curve at room temperature.

Teller effect.   The highly structured band (Figure 11) starting
just to the blue ($\sim$20,800–22,800 cm^{-1}) was then assigned to the
parity-forbidden $t_{2g}(\pi) \rightarrow t_{2g}(d)$ C.T. excitation.[9]   Recently,
on the basis of careful comparisons of the spectra of $IrCl_6^{2-}$,
$OsCl_6^{2-}$ and $ReCl_6^{2-}$, Collingwood has argued[12] that in fact the
structured band starting at 20,800 cm^{-1} should be associated
with the orbitally-forbidden $E_u'(^2T_{1u})$ state even though this
requires an <u>effective</u> ligand spin-orbit splitting of $\sim$1500 cm^{-1}
<u>vs</u> a maximum calculated value of 880 cm^{-1} ($\frac{3}{2}\zeta_{Cl}$ assuming
substantial $\delta-\pi$ mixing in the t_{1u} ($\pi+\delta$) orbital).   A splitting
this large can be rationalized[12] if it is assumed that appreciable
($\sim$10%) metal 6p orbital is mixed into the t_{1u} ($\pi+\delta$) ligand M.O.,
since the former is associated with a very large spin-orbit
coupling constant.[13]   Adopting the Collingwood argument, one then
moves the t_{2g} (π) $\rightarrow$ $t_{2g}(d)$ excitation to higher energy
($>$25,000 cm^{-1}).   A detailed explanation of lines 6–13 (Figure 10)
is still required, but rationalization through a Jahn-Teller
effect in the $U_u'(^2T_{1u})$ state seems possible.

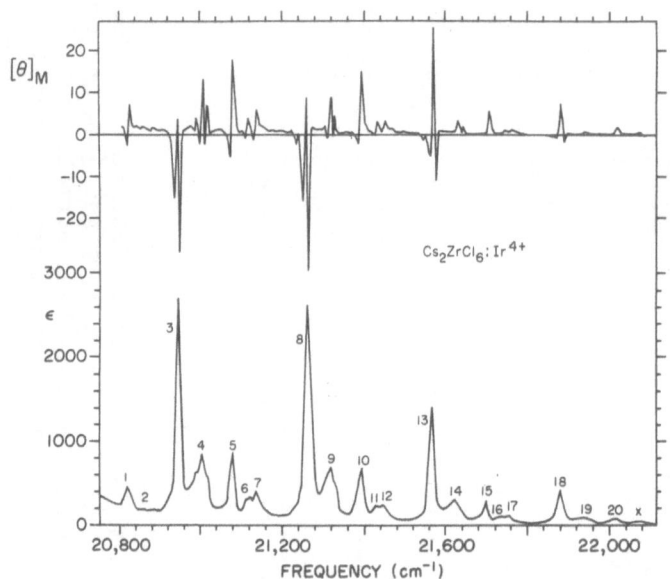

Figure 11.   Absorption and MCD spectrum of $Cs_2ZrCl_6:Ir^{4+}$.
Units, notation and temperatures are as in Figure 10.

4.   HAM (JAHN–TELLER) EFFECT IN THE $^2T_{2u}$ BAND OF $IrCl_6^{2-}$

The $^2T_{2u}$ band of $IrCl_6^{2-}$ (Figure 12) is very interesting
because there now seems little doubt that a large Jahn–Teller
effect occurs which quenches the spin-orbit splitting of the
E_u'' $(^2T_{2u})$ and U_u' $(^2T_{2u})$ no-phonon lines by a factor of the order
of 100.[10,11]   In $IrBr_6^-$, the two no-phonon lines (line 0, band 6
and line 0, band 8 – see Figure 6) are separated by about
1560 cm^{-1}, a value of the magnitude expected (0.75ζ_{Br} ~1845 cm^{-1}).
In $IrCl_6^{2-}$ one therefore anticipates two (overlapping) bands with
no-phonon lines separated by 0.75ζ_{Cl} ~440 cm^{-1}.   Though the
band in question exhibits extensive resolved vibronic structure
(Figure 12), one finds no evidence of the expected pattern.[9,10]
However, there is a shoulder on the blue side of line 0 which is
most clearly indicated in the MCD by the fact that the line 0
A term recrosses the axis on the high energy side.   Detailed
analysis[10,11] shows that this shoulder is the $U_u'(^2T_{2u})$ no-phonon
line separated by only about 5 cm^{-1} from the $E_u''(^2T_{2u})$ no-phonon

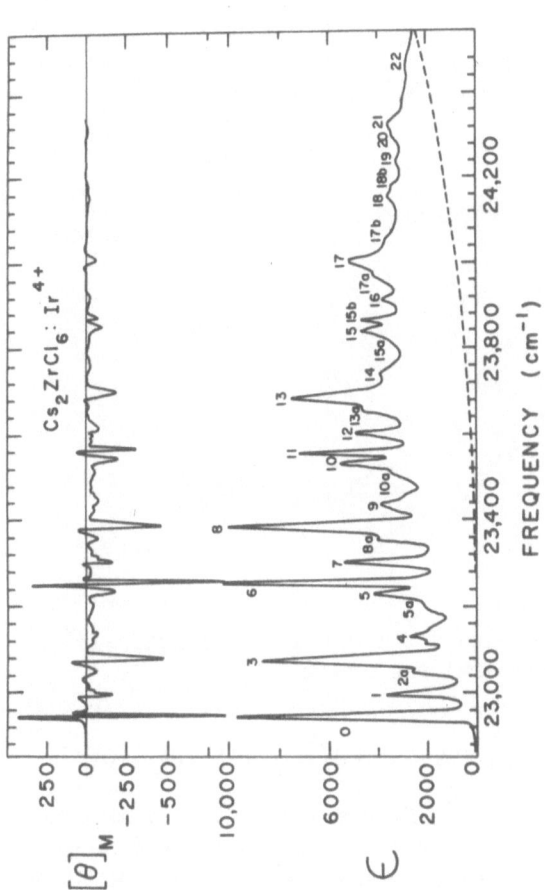

Figure 12.   Absorption and MCD spectrum of $Cs_2ZrCl_6:Ir^{4+}$.   Units and temperature are as in Figure 10; however, ε and $[\theta]_M$ are rough ($\pm$ 25%) <u>absolute</u> values, whereas they have been normalized to larger solution values in Figures 10 and 11.9   The sharpest lines are not fully resolved (cf. Figure 14), but all main features are shown. The dashed curve is the estimated baseline.

line (line O).   Thus the expected spin-orbit splitting of these
lines is quenched by a factor of about (440/5).   It is found
that this quenching can be accounted for by a dynamic Jahn-Teller
(Ham) effect involving coupling of the $^2T_{2u}$ electronic state with
a t_{2g} vibrational mode.[10]   Such a model is able to rationalize
quite reasonably the complex vibronic structure throughout the
band as well as account semi-quantitatively for the behaviour of
the no-phonon region as a function of magnetic field and
temperature.[10,11]   We briefly discuss this latter aspect here.

Figure 13 shows the predicted and observed absorption and
MCD spectrum of the no-phonon region as a function of field and
temperature.[10]   The theoretical integrated absorption intensity
has been normalized to match the experimental value at zero field,
and the absorption line shapes have been fitted by a linear
combination of two gaussians.   The shape and absolute magnitude
of the MCD as a function of field and temperature is then
predicted by the Jahn-Teller treatment.[10]   The agreement is seen
to be quite good.   The most serious discrepancy (Figure 13) in
the theoretical treatment is the somewhat enlarged central
component predicted for theMCD and absorption at high field.
(The smaller amplitudes observed in the experimental vs
theoretical A term at the higher temperatures is not of serious
concern since it is almost certainly due to a slight increase in
experimental line width with temperature, an effect not included
in the theoretical calculation.)   The quite reasonable
correspondence between theory and experiment can be regarded as
strong support for the Jahn-Teller analysis.

After the completion of this work, a paper by Massuda and
Dorain appeared[14] claiming on the basis of a high field Zeeman
analysis that line O (our proposed $E_u''(^2T_{2u})$ no-phonon line) could
not be no-phonon and in fact must be a vibronic line arising from
a parity-forbidden transition coupled to an ungerade vibration.
To examine this assertion, we carried out a careful Zeeman study
as a function of field and temperature using circularly polarized
light.[11]   The results are summarized in Figures 14-16.

Figure 14 shows the experimental results, and Figure 15 shows
the selection rules and energy level behaviour predicted as a
function of field on the basis of our Ham effect no-phonon model.
Comparison (Figure 16) of experiment (solid lines) with theory
(solid dots) clearly indicates quantitative agreement.   It should
be noted that though Zeeman effects involving U' states in cubic
systems are in general non-isotropic, this is not the case for
the transitions discussed here.   It can in fact be shown
analytically[15] in the present case that the Zeeman pattern
predicted both for the energies and polarization intensities in
the same for any arbitrary orientation of the applied magnetic
field with respect to the crystal axes.

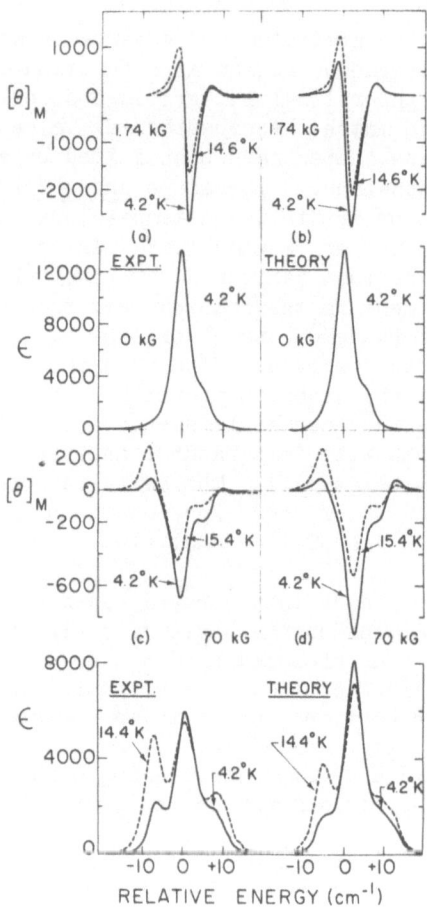

Figure 13.   Comparison of theory and experiment for the no-
phonon region as a function of temperature and field with <u>all</u>
experimental features fully resolved.   Units are as in Figure
12.

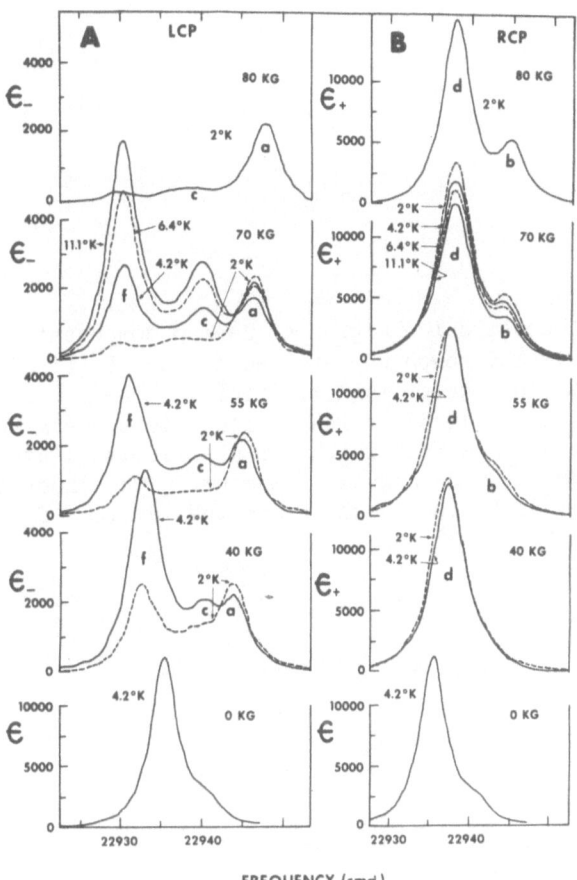

Figure 14.   Absorption spectra as a function of temperature and field (kilogauss) for left (A) and right (B) circularly polarized light propagated parallel to a magnetic field applied along the (111) axis of a dilute crystal of Cs$_2$ZrCl$_6$:Ir^{4+}.   The lines are labelled a-f in accordance with Figure 15. ϵ, ϵ_+, ϵ_- are the molar extinction coefficients for unpolarized right (+) and left (-) circularly polarized light respectively.

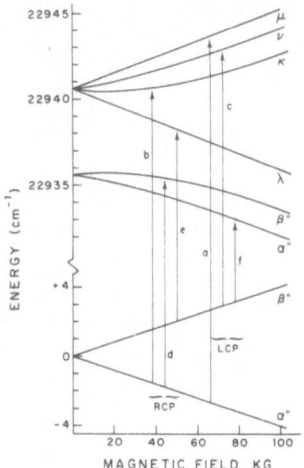

Figure 15. Theoretical energies of Zeeman components of ground state $E_g''(^2T_{2g})$ and no-phonon excited states $E_u''(^2T_{2u})$ and $U_u'(^2T_{2u})$ as a function of magnetic field. The ground state calculation uses g = 1.755. The excited state energies are calculated for a Ham effect which for simplicity assumes complete quenching of the orbital angular momentum. The theoretical polarization of each line is shown and the labels a–f accord with Figures 14 and 16.

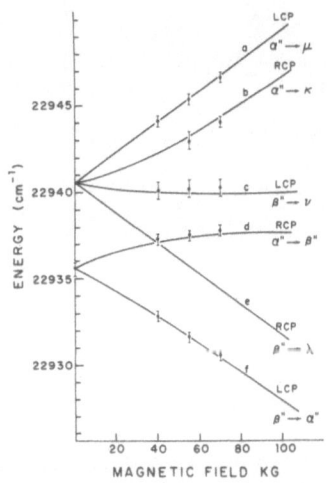

Figure 16. Experimental and theoretical transition energies between Zeeman components. The dots are the experimental points (Figure 14) with attached error bars. The solid curves are the theoretical transition energies obtained from Figure 15 as a function of magnetic field.

A careful analysis[11] of the Massuda-Dorain work[14] indicates that their erroneous conclusions are a result primarily of a misassignment of Zeeman components and the neglect of non-linear effects (field-induced mixing). Their results are also inconsistent with the observed hot band data.[11]

The splittings reported by Massuda and Dorain[14] in higher members of the a_{1g} progression of line O cannot be accounted for by their fourth order vibronic treatment. Several alternate explanations are possible,[11] the most plausible being an isotope splitting arising from the 3/1 natural abundance of $^{35}Cl/^{37}Cl$, as recently suggested by Tacon, Day and Denning.[16]

5. ACKNOWLEDGEMENTS

The work reported has been supported by the National Science Foundation, and this report was prepared while the author was on leave of absence supported by a Sesquicenntenial Research Associateship of the University of Virginia and by a Fellowship from the John Simon Guggenheim Memorial Foundation.

REFERENCES

1. G.C. Allen, R. Al-Morbarak, G.A.M. El-Sharkawy and K.D. Warren, Inorg. Chem. 11, 787 (1972).

2. J.L. Slater et al. - to be published.

3. For a review of MCD theory and a discussion of nomenclature and related matters, see for example, P.N. Schatz and A.J. McCaffery, Q. Rev. Chem. Soc. 23, 552 (1969); 24, 324 (erratum).

4. G.N. Henning, A.J. McCaffery, P.N. Schatz and P.J. Stephens, J. Chem. Phys. 48, 5656 (1968).

5. A.J. McCaffery, P.N. Schatz and T.E. Lester, J. Chem. Phys. 50, 379 (1969).

6. P.N. Schatz, A.J. McCaffery, W. Suetaka, G.N. Henning, A.B. Ritchie and P.J. Stephens, J. Chem. Phys. 45, 722 (1966).

7. A.J. McCaffery, J.R. Dickinson and P.N. Schatz, Inorg. Chem. 9, 1563 (1970).

8. J.R. Dickinson, S.B. Piepho, J.A. Spencer and P.N. Schatz, J. Chem. Phys. 56, 2668 (1972).

9.   S.B. Piepho, J.R. Dickinson, J.A. Spencer and P.N. Schatz,
     J. Chem. Phys. <u>57</u>, 982 (1972).

10. W.C. Yeakel and P.N. Schatz, J. Chem. Phys. <u>61</u>, 441 (1974).

11. W.C. Yeakel, J.L. Slater and P.N. Schatz, J. Chem. Phys.
     (in press).

12. J.C. Collingwood – to be published.

13. C.K. Jørgensen, 'Absorption Spectra and Chemical Bonding in
     Complexes', Pergamon Press, London, 1962; Chapter 9,
     p. 158.

14. R. Massuda and P.B. Dorain, J. Chem. Phys. <u>59</u>, 5652 (1973).

15. W.C. Yeakel – private communication.

16. R.J. Tacon, P. Day and R.G. Denning, J. Chem. Phys. <u>61</u>, 751
     (1974); reply by P.B. Dorain, J. Chem. Phys. <u>61</u>, 753
     (1974).

MAGNETIC CIRCULAR DICHROISM SPECTROSCOPY OF MATRIX ISOLATED SPECIES

A.J. Thomson

School of Chemical Sciences, University of East Anglia, Norwich NR4 7TJ, Norfolk

The technique of matrix isolation spectroscopy was first used by Pimentel[1] for the trapping in an inert gas of unstable reaction intermediates at low temperature. This enabled a leisurely analysis of their structures to be carried out by infrared spectroscopy. Since then the electronic,[2] e.s.r.,[3] Mössbauer[4] and Raman spectra[5] of matrix isolated species have been recorded.

For the study of electronic spectra, inert gas matrices give the advantages of a low temperature specimen, that is, lack of hot bands and, in most cases, absence of rotational structure. (There may, however, be fine structure introduced by the matrix due to effects such as aggregation and local site splitting.) The rare gases generate matrices which are transparent into the vacuum ultraviolet. Because of these advantages there are now many examples in the literature of the electronic spectra of matrix isolated species. However, no attempt has been made to test assignments either by the use of polarised light or with Zeeman studies. The measurement of the magnetic circular dichroism (m.c.d.) spectroscopy of matrices would provide this sort of information and, in addition, could supply data about the ground state magnetic properties.[6] At present, if the e.s.r. spectrum is undetectable, magnetic properties remain unknown since measurement of susceptibilities of matrices are not possible owing to the small amount of total material.

This article describes an apparatus designed to measure the m.c.d. spectra of condensed gas matrices. Experiments demonstrating that reliable spectra can be obtained are presented together with a short discussion of the gases likely to be suitable matrix

P. Day (ed.), Electronic States of Inorganic Compounds. 241–253. All Rights Reserved.
Copyright © 1975 by D. Reidel Publishing Company, Dordrecht-Holland.

solvents for m.c.d. spectroscopy.   Finally some examples are
described to illustrate the scope of the technique.

The matrix isolation apparatus, Figure 1, was constructed
around a standard liquid helium dewar (A) fitted with a 4.8 T
superconducting solenoid (D) (Oxford Instruments Limited).   At
the centre of the solenoid is mounted a quartz window (E) in
thermal contact with the bore of helium can via a copper tube (F).
The temperature of this window can be varied by a small resistive
heater wound around the copper mount.   The temperature of the
window is monitored by means of an Allan-Bradley resistor.

Matrices are deposited on this window by firing a beam along
the magnet bore from the room temperature jacket.   This rotatable
under high vacuum, the seal (C) consisting of a pair of piston
O-rings and a roller bearing.   After deposition of the matrix

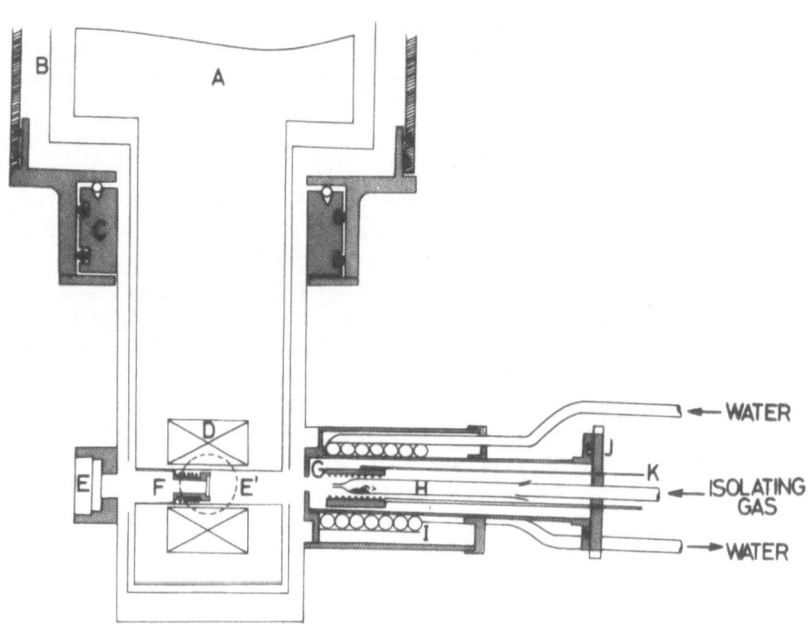

Figure 1.   Cross section of the tail-piece and furnace.   A:
liquid helium can;   B: liquid nitrogen shield;   C: rotatable
joint;   D: superconducting magnet;   E,E': quartz windows;   F:
target window holder and heater;   G: furnace;   H: quartz furnace
tube;   I: water-cooled heat shield;   J: flange supporting the
furnace;   K: tungsten electrical connections.

the outer jacket can be rotated through 90° about the vertical
axis of the dewar to bring a pair of silica viewing windows (E')
into line with the axis of the solenoid bore. The measuring
beam of circularly polarised light can then be passed through
the matrix.

The furnace (G) for the preparation of beams of involatile
materials is mounted in a water cooled compartment on the outer
vacuum jacket.   It consists of a 1 cm diameter fused silica tube,
4 cm long, within which is wound a tungsten wire resistive heater.
This is mounted from the back flange (J) by two tungsten rods,
which also provide electrical connections to the windings.   A
detachable quartz tube (H) drawn down to a jet is directed towards
the isolating window.   Material to be volatilized is placed in
(H).   The isolating gas is passed directly through (H) over the
solute material.   An auxiliary gas supply, not shown in the
Figure, by-passing the furnace can be used to spray pure gas on
to the target.   The complete disposition furnace is cooled by
water flowing through copper coils.   The direct heat of the
furnace is prevented from reaching the isolating window by the
stop at the exit of the furnace.   Experiments have shown that
it is important to maintain the diameter of the hole as small as
possible to prevent appreciable heat input to the helium bath.
Isolating gas is supplied by a standard vacuum line from one
litre bulbs, the flow of the gas being controlled by a needle
valve.

M.c.d. spectra are recorded with a Cary 61 dichrograph.
This instrument has two special advantages for this type of work.
The sample compartment is commodious and the absorption spectra
of the matrices can be measured by recording the dynode voltage
of the photomultiplier.   This latter advantage can be important
since an apparatus constructed around a superconducting magnet
is likely to be too bulky to fit into the sample compartment of
an orthodox absorption spectrophotometer.

In order to obtain a satisfactory m.c.d. spectrum from a
sample it must be isotropic in the direction of propagation of the
circularly polarised light.   This requires that a solid specimen
be either cubic or uniaxial and virtually free from scattering
centres.   Since many gases crystallize in one of the close-
packed phases, either cubic or hexagonal, these might be expected
to provide matrices suitable for m.c.d. spectroscopy.   A list of
such gases is provided in the Table.   A further consideration,
however, is the temperature and the speed at which matrices are
deposited since these often have a pronounced effect upon the
transparency of a solid film.   For example, xenon forms an
opalescent, highly scattering matrix in our apparatus and we have
not succeeded in obtaining m.c.d. spectra with it.   It has been
claimed that a reasonably transparent matrix is formed by xenon

TABLE.   Properties of matrix solvents

		m.p. (K)	b.p. (K)	T_d (a) (K)	T_{vap} (b) (K)	Crystal structure (c)
I	Ne	24.6	27.1	10	11	c.c.p. (d)
	Ar	83.3	87.3	35	39	c.c.p.
	Kr	115.8	119.8	50	54	c.c.p.
	Xe	161.4	165.0	65	74	c.c.p.
II	N_2	63.2	77.4	30	34	c.c.p. 0→35.6 K h.c.p. (>35.6 K)
	CO_2	216.6	194.6(sub)	63	106	c.c.p.
	CH_4	90.7	111.7	45	48	c.c.p. from 4.2–75 K
III	CO	68.1	81.6	35	38	Two phases isomorphous with N_2 but transition T at 61.6 K.
	NH_3	195.3	239.5	–	103	c.c.p. (e)
	N_2O	182.4	184.7	–	99	c.c.p.

(a) Temperature at which diffusion first becomes appreciable – this is a rough guide only.
(b) Temperature at which vapour pressure is 10^{-3} mm.
(c) c.c.p. = cubic close packed; h.c.p. = hexagonal close packed.
(d) Although the stable form of argon is c.c.p., h.c.p. phase is metastable below 84K in high purity crystalline argon.   On adding impurities, 1-2% of air, or CO, argon crystallises with h.c.p. structure.   As temperature is lowered more impurity is needed to stabilize the h.c.p. phase.   At 20K more than 50% of N_2 is required to stabilize the h.c.p. phase.
(e) I. Olovsson and D.H. Templeton, Acta Cryst. 12, 832 (1959). Data taken from 'Vibrational Spectroscopy of Trapped Species', edited by H.E. Hallam, Wiley, 1973.   Chapter 2 by H.E. Hallam and G.F. Scrimshaw.

if deposited at 66 K.[8]   High rates of deposition can cause
'splash patterns' to be frozen in and these can be a source of
bi-refringence.   However, a method of deposition developed by
Rochkind[9] for the deposition of a volatile solute mixed with the
isolating gas, advocates the deposition of a pulse of, say, 10 ml
volume at 200 mm pressure of Hg.   Annealing of each pulse occurs
on impact with the window and consequently the method yields
matrices free from scatter and from multiple site effects.   This
technique promises to be especially useful for the preparation of
matrices for use in m.c.d. spectroscopy.

    All the gases listed in the Table crystallize in close
packed structures and are therefore potentially of value.   Two
groups, in addition to the noble gases, are given.   Group II
consists of relatively inert gases which, like the noble gases,
will not react with solute species.   However, those in Group III
may prove to be useful for the preparation of novel chemical
species in situ.   For example, it should be possible to isolate
sodium atoms in an ammonia gas matrix thereby generating the
solvated electron.   The magneto-optic properties can be studied
and it may prove possible to use this matrix as a highly reducing
one for the preparation of unusual oxidation states of other
solute molecules.

    The other properties listed in the Table refer to the
temperature range over which the matrices are likely to be stable.
In order to disentangle m.c.d. C terms from A and B terms, it is
necessary to show that the signal is temperature dependent.
Therefore the temperature range over which a matrix is stable is
of particular interest for m.c.d. work.   The Table lists the
temperature (T_{vap}) at which a gas reaches a vapour pressure of
10^{-3}mm.[10]   This is a suitable index of the upper temperature to
which a matrix may be raised before it is boiled off into the
vacuum space.   Clearly gases such as argon, krypton and nitrogen
provide a suitable working range for temperature dependent
studies.   However, it is well to be aware that even before the
matrix begins to boil away diffusion of small solute molecules
may take place within the matrix and this may lead to changes,
possibly irreversible, in the spectrum due to aggregation.   The
temperature, T_d, has been suggested as an index of the onset of
diffusion, although this is only a rough guide since this factor
will depend, amongst other things, on the size and shape of the
solute molecules.

    Figure 2 compares the m.c.d. spectra of metal-free phthalo-
cyanine measured at room temperature in a solution of 1-chloro-
naphthalene[11] and isolated in argon at 15 K.[7]   The matrix was
prepared using a furnace temperature of 360°C and the argon was
flowed for 2-3 hours.   Agreement between the two spectra provides
a good test of the technique.   Although the solute is a highly

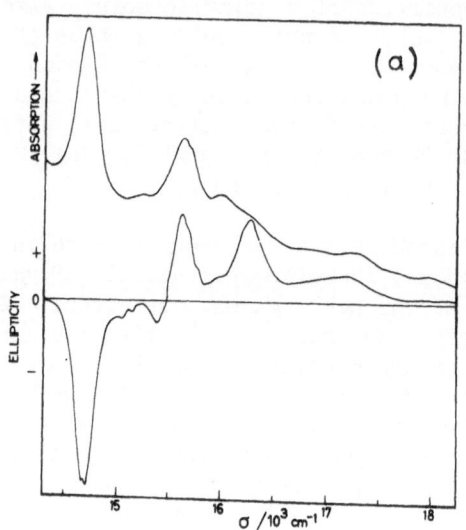

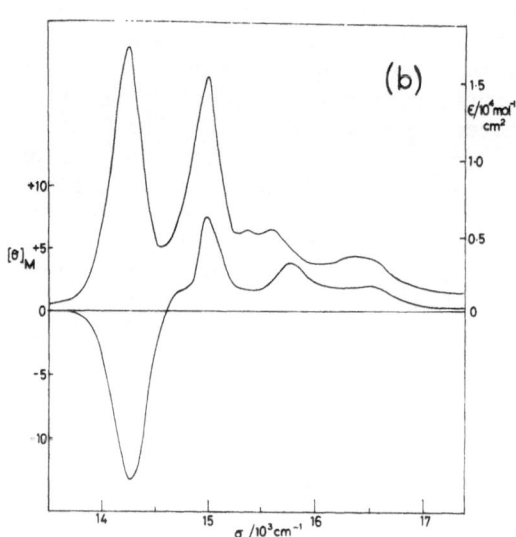

Figure 2.   The m.c.d. (lower) and absorption (upper) spectra of
metal-free phthalocyanine.   (a) isolated in an argon matrix at
20 K;   (b) in a solution of 1-chloronaphthalene.

anisotropic absorber no linear dichroism of the matrix was detec-
table in zero magnetic field.   We have not so far detected any
linear dichroism in clear matrices although there are reports of
preferential orientation of solutes being detected by E.P.R.
spectroscopy.[3]   It is a wise precaution to check each matrix for
such effects.   One disadvantage of a flow through furnace design
as used here is that small particles of material can be blown by
the isolating gas on to the window.   This will give rise to
linear dichroism.   A small piece of asbestos wool in the tip of
the jet overcomes this problem.   Some depolarisation of the
circularly polarised light may occur even though the matrix transparency is apparently high.   The extent of depolarisation can be
assessed by measuring the circular dichroism spectrum of a
naturally optically active sample placed after the matrix isolation dewar in zero field.   We have checked many matrices in
this way and find that the signal strength can be reduced by depolarisation to between 50 and 80% of the true value.   However,
no wavelength dependence of this effect has been seen and, for
quantitative work, it should be possible to correct adequately
for this effect.

We have obtained reliable m.c.d. spectra for matrices of
argon, krypton and nitrogen.   As mentioned earlier, xenon did
not give success, neither did oxygen.   The latter, which is
known to crystallise in the monoclinic structure, gave large
linear dichroism in the region of the oxygen transitions.   It is
possible that annealing of matrices would improve their quality.
We have carried out some experiments to anneal but have not been
able to obtain a marked improvement.

The first example chosen is the m.c.d. spectrum of mercury
atoms in argon and in nitrogen, Figure 3, (a) and (b), respectively.[7]   The spectra shown were measured at a mercury to solvent
atom ratio of about 1:3000.   However, it has proved possible to
detect m.c.d. spectra down to levels of $1:1.6 \times 10^5$, at which
concentration the absorption spectrum is undetectable.   Previously
the limit of detection of typical absorption experiments has been
$1:10^4$.   This illustrates well the superior sensitivity of the
m.c.d. experiment provided that the excited (or ground) state is
degenerate.

The transition arises from the configurational jump
$6s^2$ $6s^1 6p^1$ and the excited state at this energy is $^3P_1$.   No
absorption to the $^3P_0$ level can be detected and the $^1P_1$ state is
out of range of our dichrograph at < 185 nm.   The line is expected to be sharp and single.   The splitting observed in the
absorption spectrum has given rise to a great deal of discussion
concerning its origin.[12,13,14]   Two explanations have been
advanced.   One suggestion is that the different lines correspond
to mercury atoms at different solvent sites experiencing slightly

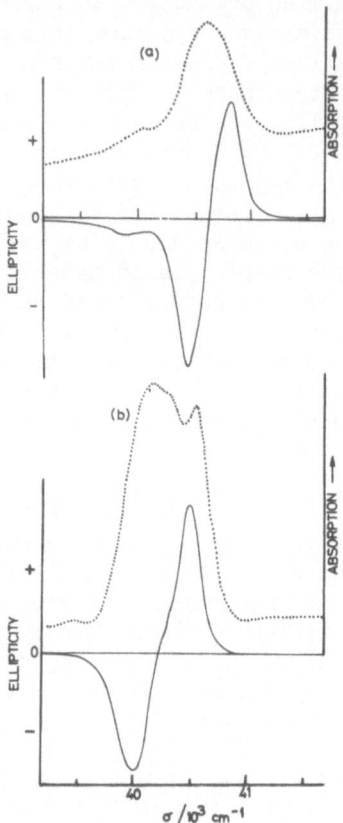

.Figure 3.   The m.c.d. (lower) and absorption (upper) spectra of
mercury isolated in (a) argon and (b) nitrogen.

different solvent shifts.   The second explanation proposes that
the site symmetry of the mercury atom has dropped below cubic,
the degeneracy of the $^3P_1$ state being lifted.   This may be due
either to asymmetric distortion of the lattice by substitution of
a large atom for a solvent atom or to lattice vacancies present
at the solute site.   Again, dimer formation (which is well known
for mercury in the gas phase) would lead to a lowering of site
symmetry.

M.c.d. spectroscopy can help in deciding between these
alternative proposals.   At a dilution of 1:1.6 x 10^6, mercury in

argon shows a perfect A term with no trace of the low energy
shoulder.   This suggests that at low concentration mercury atoms
take up a perfectly cubic site in argon.   This strongly suggests
that some form of aggregation is responsible for the band splitting.   By contrast, the spectrum of mercury in nitrogen remains
invariant down to 1 part in 1.6×10^5.   This would suggest that,
in this case, the curious splitting pattern is not due to
aggregation effects but rather to some lower symmetry component
at the mercury site, possibly due to distortion of the lattice
by the mercury atom.   Indeed, one might have supposed that a
spherical atom would have been accommodated rather asymmetrically
in a lattice of diatomic molecules.

It should, of course, be possible to extend this type of
investigation by studying these changes as a function of atom size
and by varying the symmetry of the excited state being investigated by choice of suitable atoms.

The second example illustrates nicely the considerable
simplification which can result when hot bands are frozen out.
Figure 4 compares the m.c.d. spectrum of benzene in the gas phase[15]
with the spectrum of benzene isolated in nitrogen.[16]   Attention
is drawn to some features of these interesting but puzzling
spectra.   The transition is $^1A_{1g} \rightarrow {}^1B_{2u}$ which is vibronically
allowed by an e_{2g} vibrational mode.   There are four such modes
in benzene with the following frequencies in the ground state.

608 cm^{-1} (ν_6) C-C bend

1596 cm^{-1} (ν_8) C-C stretch

1178 cm^{-1} (ν_9) C-H bend

3056 cm^{-1} (ν_7) C-H stretch

The gas phase absorption spectrum has its intensity almost entirely built upon one quantum of ν_6 as a false origin followed by a
progression in the totally symmetric mode (a_{1g}, 923 cm^{-1} in the
ground state).   This progression is clearly seen in m.c.d. but
in addition two further prominent progressions based on two lines
labelled I and II are seen.   The positions of these origins
correspond well with the expected false origins based on ν_9 and
ν_8, respectively.

The matrix spectrum is considerably simpler, one progression
being seen in absorption, based upon ν_6 as false origin.   (The
origin of the splitting in each band is not known, but could be a
matrix effect.)   However, in the m.c.d. spectrum three false
origins are clear with associated a_{1g} progressions built on each.
Why the vibrations ν_8 and ν_9, the latter being a C-H band, should

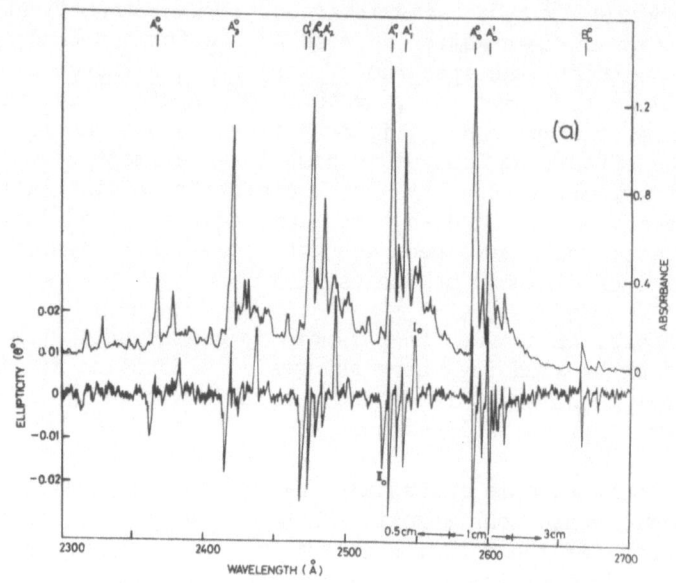

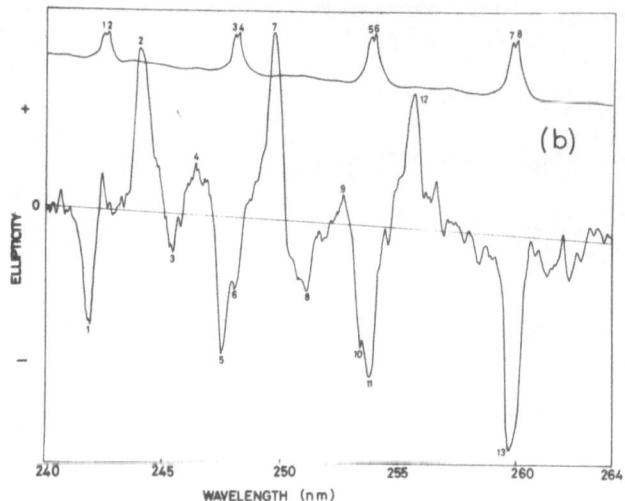

Figure 4.(a) The m.c.d. (lower) and absorption spectra of the $^1A_{1g} \rightarrow {}^1B_{2u}$ transition in benzene vapour.   The path lengths used in the different region of the m.c.d. spectrum are indicated. The absorption spectrum was measured with a path length of 1.0 cm. (b) The m.c.d. (lower) and absorption spectra of benzene isolated in nitrogen at 15 K.

be so effective in inducing magneto-optical activity into the m.c.d. spectrum is a mystery.   Also the change in sign of the m.c.d. due to 9, on the one hand, and 6 and 8, on the other, is equally puzzling.

However, it is clear that matrix isolation spectroscopy will allow the m.c.d. spectra to be obtained of many organic compounds to be determined under high resolution, low temperature conditions. In view of the paucity of highly symmetrical single crystal organic hosts this is a valuable extension.

Finally, a number of other examples which have been studied are briefly mentioned.   It has proved possible to measure the ultraviolet m.c.d. spectra of 20% oxygen in 80% nitrogen, Figure 5.[17]   This illustrates the value of temperature dependent studies.   The m.c.d. spectrum is clearly temperature dependent, the spectrum being due to C-terms caused by spin-orbit splitting in the excited state.   It can be shown quite unambiguously that this transition is $^3\Sigma_g^- \rightarrow {}^3\Delta_u$, confirmation of an assignment suggested first by Finkelnburg and Steiner[18] and supported by Herzberg.[19]   This example illustrates dramatically the sensitivity of the technique especially for paramagnets at low temperatures. Herzberg required a gaseous sample 800 metres in path length at a pressure of 2-3 atmospheres in order to pick up this band photographically.

The anhydrous linear dihalides, $NiCl_2$ and $CoCl_2$, have been studied isolated in argon.[7]   Their isolation requires a furnace reaction at 500°C. $CoCl_2$ is clearly shown to be paramagnetic as expected.   The transitions detected so far are the intense charge-

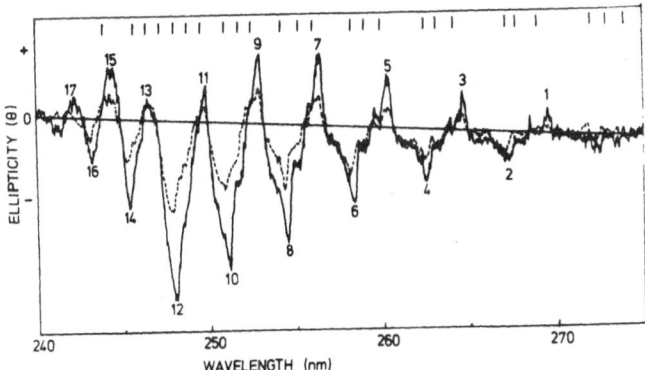

Figure 5.   The m.c.d. spectra of atmospheric oxygen: —T = 19 K, ---T = 48 K.   Above the spectra are the positions of the triple-headed bands observed by Finkelnburg and Steiner (reference 3).

transfer bands.   It will, perhaps, be possible with the superior
sensitivity of m.c.d. spectroscopy to detect the d-d bands of
matrix isolated dihalides.   Previously the long paths of gas
phase cells were needed to detect the spectra.[20]   But they are
rather broad and featureless at the temperature required for such
measurements.

This brief list of examples serves to demonstrate the range
of stable and unstable molecules which are amenable now to the
study by m.c.d. spectroscopy.

REFERENCES

1.   E. Whittle, D.A. Dows and G.C. Pimentel, J. Chem. Phys.
       22, 1943 (1954).

2.   J.W. Hastie, R.H. Hauge and J.L. Margrave, 'Spectroscopy in
       Inorganic Chemistry', ed. Rao and Ferraro, Acad. Press
       (1970) vol. 1, p. 57.

3.   W. Weltner Jr., Advances in High Temperature Chemistry,
       2, 85 (1969).

4.   T.K. McNab, H. Micklitz and P.H. Barrett, Phys. Rev. B4,
       3787 (1971).

5.   See 'Vibrational Spectroscopy of Trapped Species', ed.
       H.E. Hallam, Wiley 1973, chapter 9 by G.A. Ozin.

6.   For a recent review see P.J. Stephens, Ann. Rev. Physical
       Chem. 25, 201 (1974).

7.   I.N. Douglas, R. Grinter and A.J. Thomson, Mol. Phys. 28,
       1377 (1974).

8.   E.D. Becker and G.C. Pimentel, J. Chem. Phys. 25, 224
       (1956).

9.   M.M. Rochkind, Anal. Chem. 39, 567 (1967); 40, 762 (1968).

10.   D. Meyer, 'Low Temperature Spectroscopy', Elsevier, 1971.

11.   M.J. Stillman and A.J. Thomson, J. Chem. Soc. Far. Trans.
       II, 70, 805 (1974).

12.   W.W. Duley, Proc. Phys. Soc. 90, 263 (1970).

13.   L. Brewer, B. Meyer and G.D. Brabson, J. Chem. Phys. 43,
       3973 (1965).

14. M. McCarty Jr., J. Chem. Phys. 52, 4973 (1970).

15. I.N. Douglas, R. Grinter and A.J. Thomson, Mol. Phys. 26, 1257 (1973); Ibid (to be published).

16. I.N. Douglas, R. Grinter and A.J. Thomson (unpublished data).

17. I.N. Douglas, R. Grinter and A.J. Thomson, Chem. Phys. Lett. 28, 192 (1974).

18. W. Finkelnburg and W. Steiner, Z. Physik. 79, 69 (1932).

19. G. Herzberg, Can. J. Phys. 30, 185 (1952); Ibid 31, 657 (1953).

20. C.W. DeKock and D.M. Gruen, J. Chem. Phys. 49, 4521 (1968).

ACKNOWLEDGEMENTS

The work described here was carried out in close collaboration with Dr. R. Grinter, University of East Anglia, and with the able assistance of Dr. I.N. Douglas, now at the Australian National University, Canberra. The author wishes to record his debt to them.

CIRCULARLY POLARISED EMISSION SPECTROSCOPY

A.J. McCaffery

School of Molecular Sciences, University of Sussex,
Brighton BN1 9QJ.

This paper concerns optical emission spectroscopy at optical
frequencies and therefore we shall be concerned with eigenstates
of the neutral molecule.   As is well known, emission spectra
give information on low-lying electronic states, ground
vibrational levels and on states such as organic triplets, which
are often not accessible by direct absorption processes.   In
this contribution I shall describe the measurement of circular
polarisation of emission, a phenomenon which may occur in a number
of circumstances.   Those I shall be concerned with are:-

1.   Magnetically induced circular emission; the emission
analogue of MCD in which the magnetic parameters give spectroscopic information on the ground and the emitting states, and

2.   Circular emission induced by excitation with circularly
polarised light.   In this case we may obtain information on
energy transfer process arising from elastic and inelastic
collisions in the gas phase, or by coupling with elementary
excitations in the solid state.

Figure 1 shows some of the elementary processes in the
molecule-photon reaction sequence which leads to emission of a
second photon.   In absorption spectroscopy the individual excited
vibronic levels E_0, E_1, E_2 are displayed as a function of frequency
as the spectrometer is scanned.   Whatever the excitation frequency,
in condensed phases emission generally takes place only from the
lowest excited level following rapid intra-molecular energy
conversion processes, vibrational relaxation, internal conversion
and intersystem crossing.   The emission parameters therefore
provide a probe into these energy conversion processes in excited

P. Day (ed.), Electronic States of Inorganic Compounds. 255–265. *All Rights Reserved.*
Copyright © 1975 by D. Reidel Publishing Company, Dordrecht-Holland.

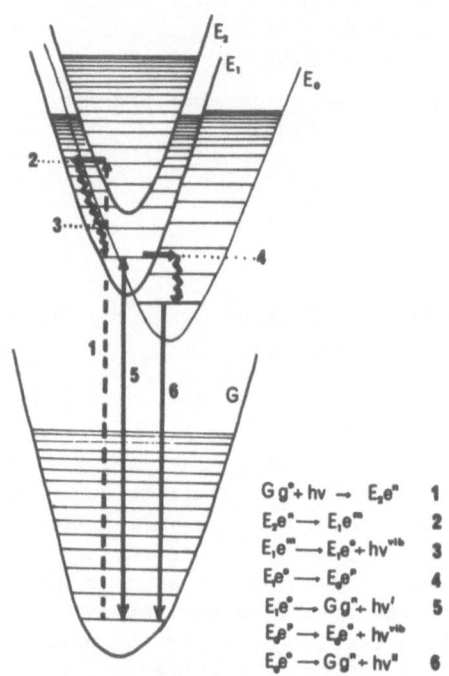

$$Gg^\circ + h\nu \rightarrow E_2e'' \quad \mathbf{1}$$
$$E_2e'' \rightarrow E_1e''' \quad \mathbf{2}$$
$$E_1e''' \rightarrow E_1e'^\circ + h\nu^{vib} \quad \mathbf{3}$$
$$E_1e'^\circ \rightarrow E_1e'^\circ \quad \mathbf{4}$$
$$E_1e'^\circ \rightarrow Gg'' + h\nu' \quad \mathbf{5}$$
$$E_1e'^\circ \rightarrow E_1e'^\circ + h\nu^{vib} \quad \mathbf{6}$$
$$E_1e'^\circ \rightarrow Gg'' + h\nu'' \quad \mathbf{6}$$

Figure 1.   Intramolecular energy conversion processes following
absorption of a photon (1).   These may be internal conversion (2),
vibrational relaxation (3), intersystem crossing (4) and may lead
to fluorescence (5) and/or phosphorescence (6).

species which is not available from absorption studies.   Emission
spectra have two distinct functions therefore.   Firstly the
spectroscopic identification of emitting and final electronic
states, ground vibrational levels, etc.   Secondly, and of great
current interest, are the intramolecular processes described above.

        The molecule-photon reaction sequence of Figure 1 is conven-
tionally written in terms of energy states of photons and molecule
in which energy is conserved in each step.   A more precise
description may be obtained by prescribing the angular momentum
quantum numbers of each component.   In addition to utilising
energy (and lifetime) parameters to probe the intramolecular
processes therefore we may use angular momentum quantum numbers.
Experimentally this involves exciting the molecule with photons
of well-defined angular momentum and determining the angular
momentum of those emitted.   Thus we excite with circularly
polarised radiation and measure the circular polarisation of
emission.

The same experiment also gives a more intimate picture of
the excited states of gas phase molecules which exhibit the
phenomenon of resonance fluorescence, i.e. emission from the level
directly excited.   This will be described in more detail below.

EXPERIMENTAL

The emission instrument[1] (Figure 2) is based on Spex 1406
0.85 metre double grating monochromator and the sample is excited
either by an argon ion laser or by a Xenon arc lamp followed by a
Spex 'Minimate' monochromator.   The exciting radiation may be
circularly polarised by means of an achromatic quarter-wave plate
or by a stressed quartz plate.   The emission is collimated and
passes through a photo-elastic modulator set for $\lambda/4$ retardation
and then through a calcite linear polariser.   This converts the
polarisation modulation to an intensity modulation at 50 k Hz.
The photomultiplier detects the signal, following wavelength
analysis, and its anode current consists of a d.c. and an a.c.
component.   The former is amplified by a d.c. amplifier and is
proportional to the total emitted intensity $(I_+ + I_-)$.   The
latter is amplified by a phase-sensitive detector locked into the
modulation frequency and is proportional to the difference in
emitted intensity for right and left circularly polarised light
$(I_+ - I_-)$.   The two amplifier outputs are displayed simultaneously
on a double pen chart recorder.   A number of points are worth

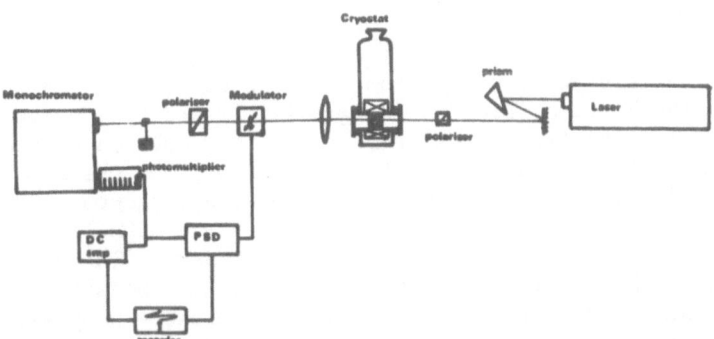

Figure 2.   Apparatus for measuring magnetically induced circular
emission.   The sample in the cryostat is excited by an argon ion
laser.   The emitted radiation is wavelength analysed by a Spex
0.85 M spectrometer and polarisation analysed using a photoelastic
modulator.   Both emission and circular polarisation are recorded
simultaneously.

mentioning about this apparatus.   First, it is possible to record
absorption and dichroism spectra on this instrument by passing
white light through the sample and analysing the transmitted
frequencies.   Secondly, the final polariser plays a very important role by limiting the monochromator input to one polarisation.
This is due to the variation in efficiency of gratings as a
function of incident polarisation.   Thirdly, it is straightforward
to obtain time resolution using the technique of chronospectroscopy
first suggested by Mollenauer.[2]   The exciting radiation is
chopped and by a combination of varying the chopping frequency
and phase selection on the psd, it is possible to 'phase in' or
'phase out' of the spectrum features having different lifetimes.
Examples of this technique will be shown below.

RESULTS

1.   Magnetically induced circular emission

     Figure 3 displays typical conditions for observation of mce.
They are very similar to those required for med.   A degenerate
excited state will give rise to a Faraday C term, the sign and
magnitude of which depends on the magnetic moment of the emitting
state and the sign of the Zeeman splitting.   A thermal distribution among the Zeeman components leads to circular polarisation
of emission.   An example of this type of behaviour is found in
PdOEP (Figure 4).   Metalloporphyrins generally exhibit fluorescence from ^1Eu and phosphorescence from ^3Eu excited states.   The
broad phosphorescence in PdOEP shows a very strong mce whose
magnitude obeys a $^1/T$ law from 25-65°K.   The magnetic moment for
this state is 2.1 μ_B .[3]

     Systems containing Cr^{3+} often show strong luminescence.
Emission takes place usually from the ^2E state and we see what are
known as the R lines, i.e. the origin - followed by a number of
ground state vibrational features. Cr^{3+}:MgO shows a strong R
line (Figure 5) together with strong emission from charge-
compensated sites[4] (tetragonal and rhombic) and phonon structure
(Figure 6).   The R lines are easily split by a magnetic field and
here, differential techniques are not necessary.   The dotted
curve is the R line in the absence of the field and this splits
as shown with high circular polarisation ratios.   Ruby also has
a strong emission.   The spectrum of 0.25% Cr^{3+} in Al_2O_3 shows
numerous features associated with pairs of Cr^{3+} ions together with
the single ion spectrum, these normally being superimposed.

     We can now illustrate the advantage of 'chronospectroscopy'
described earlier.   We can phase-in to the pair events and phase
out the phonon structure (i.e. put the R line in antiphase).
This removes the centre of R line and the phonon structure.

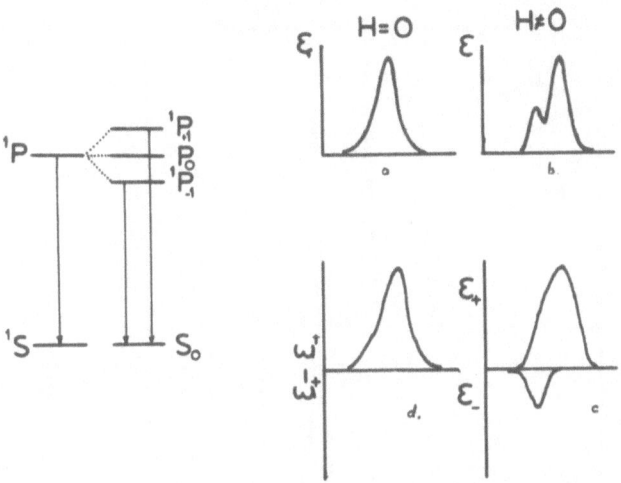

Figure 3.   Conditions for emission of circularly polarised
radiation.   In the absence of a field the emission from $^1P \rightarrow {}^1S$
is unpolarised and may be split by a magnetic field as in b.   In
the presence of a magnetic field the Zeeman components emit
circularly polarised radiation, as shown in c, a Boltzmann
distribution of populations being responsible for the relative
magnitudes.   The difference spectrum is displayed in d.

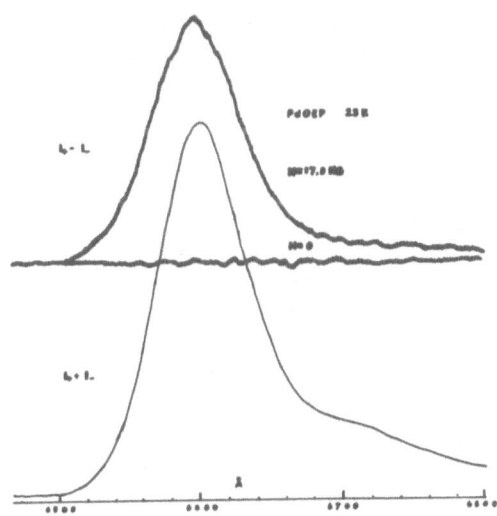

Figure 4.   Magnetically induced circular emission (MCE) of
palladium octaethylporphyrin (PdOEP) at 20°K.   The upper curve
is the emission intensity and the lower the MCE.

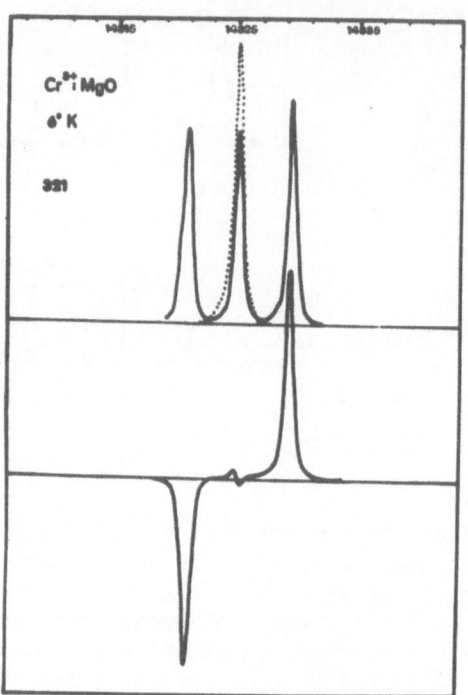

Figure 5.   Emission spectrum (upper curve dotted) of the
$^2E \rightarrow {}^4A_2$ band of Cr^{3+} in MgO showing the splitting in a magnetic
field.   The circular polarisation is shown in the lower curve.

This is illustrated in Figure 7.

ZERO-FIELD EXPERIMENTS

We move away from experiments where a thermal distribution
among M_J states gives circularly polarised emission to zero-field
experiments.   Here we force a non-Boltzmann distribution of M_J
states usually by exciting with circularly polarised radiation.
Two main areas will be described: (1) resonance fluorescence from
gas phase molecules and (2) non-resonant emission from molecules
in condensed phases.

1.   Resonance fluorescence of iodine – a study of gas-phase
       energy transfer processes

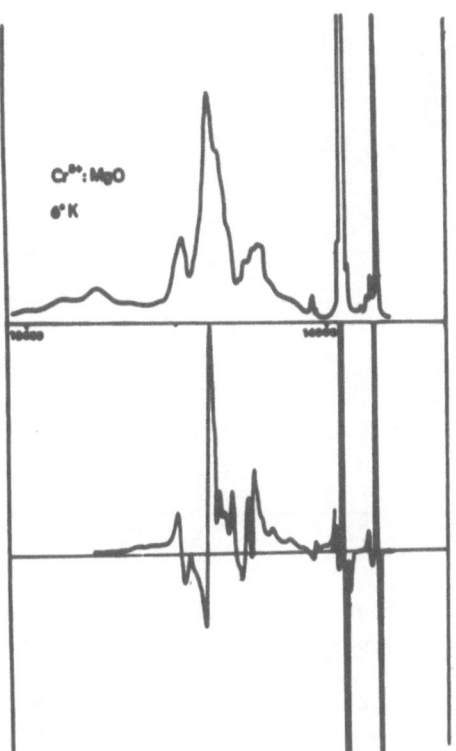

Figure 6.   Emission spectrum (upper curve) and MCE of Cr^{3+}:MgO
showing emission from the cubic and distorted lattice sites and
to excited phonon states.

The 5145Å Ar^+ laser line excites the $\nu'' = 43$, $\nu' =$ n progression in I_2 vapour (Figure 8).   If we excite with circularly
polarised laser radiation, the rotationally resolved emission is
also circularly polarised.   Figure 9 shows the rotational
structure of one of the vibrational lines (43,2).

The main features[5] are three strong components, strongly
circularly polarised.   The 5145Å line excites

$J'' = 13 \rightarrow J' = 12$
$J'' = 15 \rightarrow J! = 16$

coincidentally.   These return directly back and give the central
component.   In addition since the rigid rotor selection rule
allows $\Delta J = \pm 1$, the excited states can relax back radiatively
as follows:-

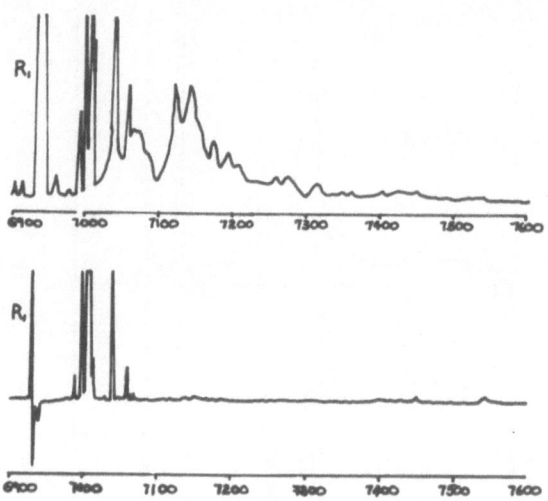

Figure 7.   Chronospectroscopic resolution of the pair lines and single ion emission from Cr^{3+} in Al_2O_3.   The lower spectrum is obtained by placing the R line centre in antiphase thus leaving the pair line spectrum displayed.   Note thatthe R line wings have different temporal characteristics to the centre giving rise to the unusual line shape shown.

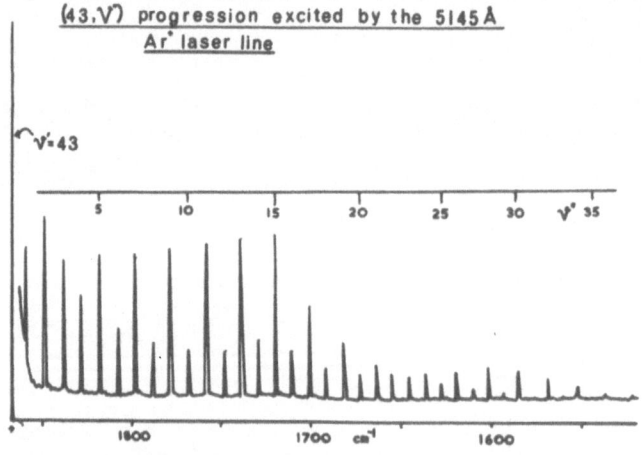

Figure 8.   Vibrational progression in iodine vapour excited by the 5145Å line of the argon laser.

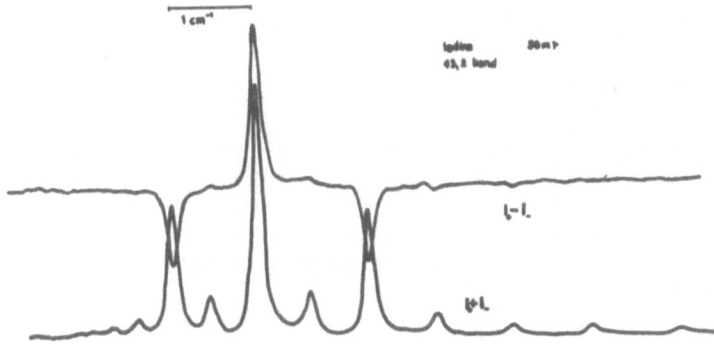

Figure 9.   Rotationally resolved emission spectrum of I_2 excited
by circularly polarised radiation.   The circular polarisation is
the upper curve.

    J' = 12   J" = 11 high energy
    J! = 16   J" = 17 low energy

These are circularly polarised because the circularly polarised
light forces a particular distribution of population of M_J states
- mainly high values of M_J (for J' = 12, J' = 16).   These then
emit circularly polarised radiation.   However, the lifetime of
the excited state is ~3 x 10^{-6} sec., time for a number of rotations
and, more important, a number of collisions.   Elastic collisions
may destroy the coherence of the excited state and the preferred
orientations are lost.   We therefore get 'level crossing' and
consequently depolarisation of emission.

    In addition to the resonance triplet we see features arising
from energy transfer in the excited state.   Inelastic collisions
cause changes in J' during the lifetime of the excited state
according to the $\Delta J = \pm$ 2n selection rule.   It is interesting to
see if orientation is preserved in this process.   Figure 9 shows
the triplet and some rotational transfer lines.   It can be seen
that there is some small circular polarisation signal, i.e. some
coherence is retained despite $\Delta J'$ change.   We are currently in
the process of enhancing these signals using repetitive scanning
techniques.

    To analyse the data a rate matrix has been set up for the
sub-level populations in which elastic, inelastic and radiative
processes all influence the population of a particular sub-level.[6]
The variation with pressure of foreign gas is then explored in

terms of parameters representing both kinds of collisions. Experimentally a start has been made using argon as a quenching gas.

2.   Non-resonant processes

In solids, excitation is usually followed by very rapid intramolecular energy conversion before emission of a photon from the lowest excited state.   The include the non-radiative steps of internal conversion, and intersystems crossing and also vibrational relaxation.   There is much interest in the mechanism of these processes.

If we excite directly into the origin of a triply degenerate state with pure right circularly polarised light say, then pure right circularly polarised light will be emitted, since we have achieved a non-Boltzmann population, provided that spin lattice relaxation does not immediately destroy our distribution.   This is the case only if spin lattice relaxation is very slow compared to the lifetime of the state and this occurs only at very low temperatures in crystal lattices.   It is essential therefore to work in the range 1.6–4.2K or information is lost.   If this facility is available, then experiments on angular momentum retention may be undertaken.   Thus excitation into the zero phonon level of an excited state should produce non-Boltzmann populations and changes in circular polarisation of emission. However, whether this is achieved on pumping into higher vibronic levels depends on angular momentum retention in the energy conversion process.   By selective optical pumping it should be possible to isolate vibrational relaxation, internal conversion and intersystem crossing.   These occur via matrix elements of the form

$$\langle SM_s \Gamma_\gamma | V_\delta | S'M_s' \Gamma'_{\gamma'} \rangle$$

V may be $T(Q)$, $H_{so}(Q)$ or some other operator and the presence (or absence) of angular momentum retention should enable us to determine experimental selection rules for the process and speculate on the nature of V.

We can also use the technique to study energy transfer between single ions and pairs, in the concentrated ruby for example, or between sublattices in a two sublattice antiferro- magnet.   We hope to describe the results of experiments of this type in the near future.

REFERENCES

1. R.A. Shatwell and A.J. McCaffery, J. Phys. (E) $\underline{7}$, 297 (1974).

2. H. Engstrom and L.F. Mollenauer, Phys. Rev. B $\underline{7}$, 1616 (1973).

3. R. Gale, A.J. McCaffery, R.A. Shatwell and K. Sichel, to be published.

4. R.A. Shatwell and A.J. McCaffery, to be published.

5. R. Clark, S.R. Jeyes, A.J. McCaffery and R.A. Shatwell, Chem. Phys. Lett. $\underline{25}$, 74 (1974).

PHOTOELECTRON SPECTROSCOPY AND ALLIED TECHNIQUES – GENERAL
INTRODUCTION

A.F. Orchard

Inorganic Chemistry Laboratory, University of Oxford

Photoelectron spectroscopy (PES) is concerned with the
analysis of the kinetic energy spectrum of the electrons
(photoelectrons) emitted when materials – gases, solids, or
even liquids – are exposed to monochromatic ionising radiation
(the photoelectric effect).   Sufficiently energetic incident
radiation may eject more than one 'species' of electron (as
distinguished by the molecular orbitals or, in the case of
solids, the frequently delocalised 'lattice orbitals', that
they originally occupied) so that, in general, a polychromatic
flux of photoelectrons is produced.   The kinetic energy E_k of
the kth species of electron is related to its ionisation energy
(or 'binding energy') I_k by the following equation
(Einstein 1905)[1]

$$E_k = h\nu - I_k \qquad\qquad\qquad (1)$$

where ν is the frequency of the ionising radiation, and $\underline{h}$
Planck's constant.   When dealing with solids (or liquids), as
opposed to gases, equation (1) requires an additional term
catering for contact potential and charging effects (vide infra).
Both UV and X-radiation are routinely employed to excite PE
spectra (see Appendix).   Incident UV photons ($h\nu \leqslant 50$ eV, say)
can eject only valence electrons while X-rays can ionise both
valence and core electrons.

The measurement of photoelectron energy distributions
provides information concerning both the ionised states of a
system and also, if the orbital approximation be assumed and one
neglects relaxation effects in the ionised system, the initial
(orbital) energies of the electrons ejected during photoionisation.

P. Day (ed.), Electronic States of Inorganic Compounds. 267–304. All Rights Reserved.
Copyright © 1975 by D. Reidel Publishing Company, Dordrecht-Holland.

The latter notion is based on what has become known as
Koopmans' theorem,[2] according to which the ionisation energy of
an electron species k may be identified with the negative of
the self-consistent field (SCF) energy eigenvalue E_k^{SCF}:

$$I_k = -E_k^{SCF} \tag{2}$$

This relation has much chemical appeal, but its approximate
nature should constantly be borne in mind: what PES actually
provides is the ionisation potential of the system, i.e. its
first ionisation energy I_1,* together with the relative energies,
$I_k - I_1$, of certain excited electronic states of the (singly)
ionised system.

The UV aspect of the technique (UV-PES), as applied to
gases, originates in experiments by Al-Joboury and Turner (1963)[3]
and, independently, Vilesov et al. (1961),[4] and owes its
development** principally to Turner and coworkers[5] (at Imperial
College, London, and at Oxford), Price and coworkers[6] (at King's
College, London) and McDowell and associates[7] (at the University
of British Colombia, Vancouver).   This rapid evolution of
molecular UV-PES depended largely on the use of the He-I
resonance line (58.4 nm = 21.22 eV photon energy),[8] which
accounts for at least 98% of the vacuum UV emission of helium
gas excited under normal conditions by a d.c.[5,8] or microwave[7a]
discharge.   The helium source can be operated in a 'window-less'
configuration by virtue of differential pumping between the
discharge and ionisation chambers.

Until quite recently,[9] He-I PES was exclusively a gas phase
technique.   But solid-state UV-PES - what physicists refer to,
somewhat unfortunately, as UV photoemission - has a long history
stretching back to Millikan (1916),[10] who performed the key
experiments verifying Einstein's theory of the photoelectric
effect.   The modern phase of solid-state UV-PES dates perhaps
from the late forties,[11] but prior to about 1970 the technique
was limited to exciting photon energies of less than 11.6 eV,
the cut-off point of the LiF windows separating the source from
the high vacuum sample chamber.   In these early experiments (as
indeed in the work of Vilesov et al.)[4] the exciting radiation
was obtained, via a vacuum monochromator, from hydrogen or noble
gas discharge lamps.   The adoption of the window-less helium

---

* I_1 is sometimes referred to as the (first) ionisation potential
of the system.   But the term second (or subsequent) ionisation
potential is often used, especially in connection with atoms, for
the production of doubly (or multiply) ionised species M^{2+} in
their ground states.
** See Phil. Trans. A, 268, No. 1184, 1970, for a perspective of
the subject in 1969.

discharge source (initially by Eastman and Cashion)[9] and, more recently, the application of synchroton radiation,[12] have dramatically extended the scope of traditional photoemission studies.

X-ray photoelectron spectroscopy (X-PES) can also be traced back to the early part of this century[13] but its dramatic development in recent times is due to Siegbahn and coworkers[14-16] (at Uppsala), who refer to the technique as 'Electron Spectroscopy for Chemical Analysis' (with the familiar acronym, ESCA).* The maturation of X-PES also owes a great deal to workers in the United States.[19] In contrast to UV-PES, X-PES was originally a **solid state** technique[15] but it has been profitably extended,[20,21,16] again largely through the efforts of the Uppsala group,[16] to encompass the study of gases. (Very recently, measurements on **liquids** in the form of high pressure jets have also been reported.)[22]

The photoelectrons are usually excited by means of Al Kα (hν = 1486.6 eV) or Mg Kα (hν = 1253.6 eV) radiation both of which have,for X-rays, relatively narrow line widths δ_X (0.9 eV and 0.8 eV, respectively) and are easy to generate in sufficient flux. A line width of the order of 1 eV does however normally impose a limit on resolution greater than that due to instrumental and other factors. The situation has recently been much improved by the use of monochromatised X-rays (with $\delta_X \sim 0.25$ eV),[23-26] but the resolution attainable in X-PES remains an order of magnitude less than is expected on a routine basis in UV-PES work on gases.

EXPERIMENTAL ASPECTS

The essential features of a photoelectron spectrometer are illustrated in Figure 1.** The determination of the kinetic energy (KE) distribution of the photoelectrons of course demands high vacuum conditions, but the study of gases is made possible by efficient differential pumping between the analyser and the sample region. It is also normally essential to have differential pumping between the radiation source and the sample chamber,

---

* Recently upgraded to 'Electron Spectroscopy for Chemical Applications'.[17] See ref. 18 for a perspective of the technique in 1971.
** Design principles in electron spectroscopy have recently been reviewed by Wannberg **et al.**, J. Phys. E (Sci. Instrum.), 1974, **7**, 149.

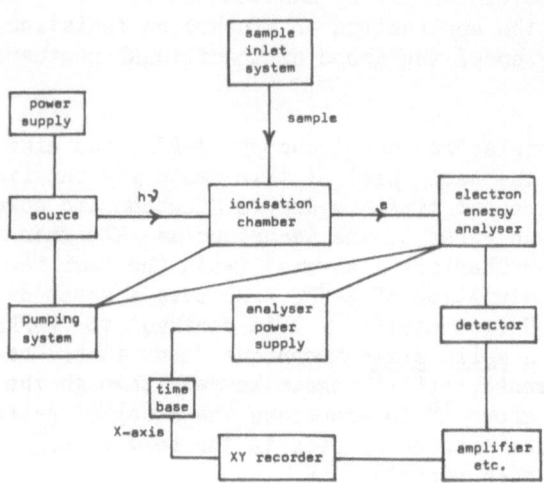

Figure 1.   Block diagram of a photoelectron spectrometer.

(a)

(b)

(c)

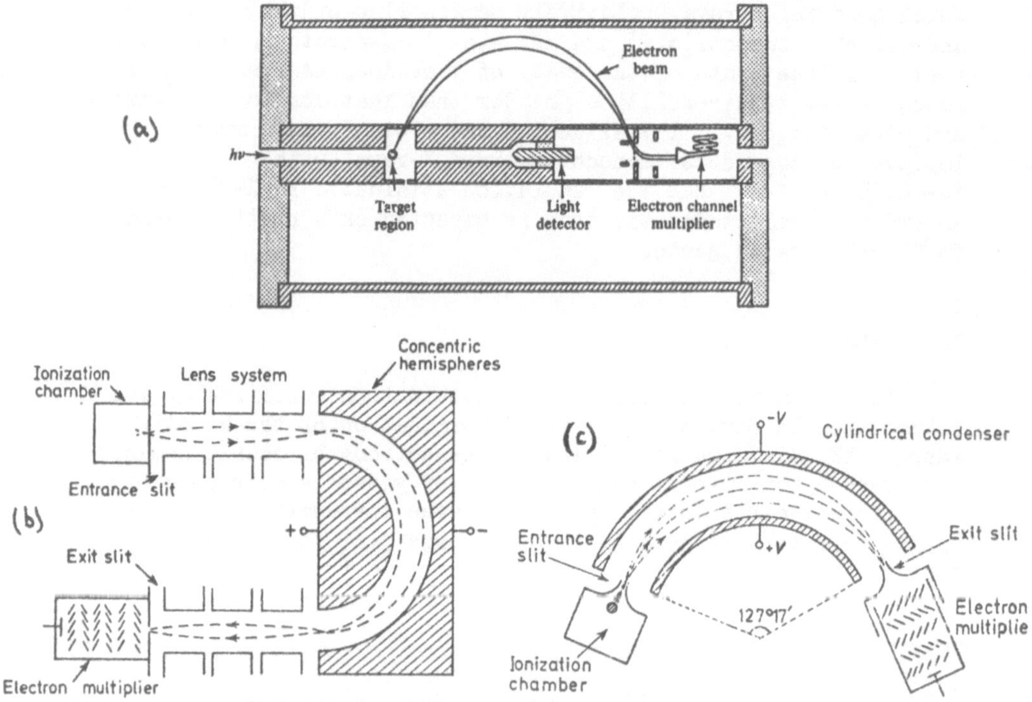

Figure 2. (a) section of a photoelectron spectrometer incorporating a cylindrical mirror analyser; (b) hemispherical electrostatic analyser (with virtual slits); (c) 127° cylindrical sector electrostatic analyser (with curved slits designed to reduce the effects of fringing fields).

either to protect the source from gaseous samples or gases
emanating from solid samples, or, in UV-PES work with window-
less discharge lamps, to minimise the flow of discharge gas
into the sample region. The analyser and sample chamber are
connected via a narrow slit which serves to define a narrow
beam of photoelectrons entering the analyser and, incidentally,
also much facilitates the maintenance of a high vacuum in the
analyser even when gaseous samples are investigated.

Energy analysis of the photoelectrons

   Most contemporary PE spectrometers employ electrostatic
dispersion analysis based on the focusing deflection of the
photoelectrons by a radial electric field.   There are   *
essentially two types of dispersive electrostatic analyser,
namely (1) simple deflection or 'prism' systems, in which the
principal electron trajectory follows an equipotential surface,
and (2) 'mirror' types, in which the electrons are reflected
by a retarding field. The most popular designs of analyser
are shown in Figure 2. The cylindrical and spherical sector
analysers belong to category (1), while category (2) is
represented by the cylindrical mirror analyser. The deflecting
electric fields in the spherical sector and cylindrical mirror
analysers operate in two-dimensions and they are, for this
reason, referred to as double-focusing systems: the cylindrical
sector analyser is, on the other hand, a single-focusing design.
In much of the pioneering work in PES the electron energy analysis
was actually carried out with deflecting magnetic fields[15,5]
(as in β-ray spectroscopy), but magnetic deflection analysers
are nowadays much less popular than electrostatic dispersion
systems.

   The energy analysis of the photoelectrons may alternatively
be achieved by retarding electric fields, a 'filtering' as
opposed to dispersive technique.   This non-dispersive method,
used in the early gas-phase UV-PES experiments,[3,8,4] and very
extensively in Auger electron spectroscopy (vide infra), has
many disadvantages but, in experienced hands, can be put to
remarkably good use.   Analysers based solely on electron
retardation (operating as low energy cut-off filters) yield a
step-function or 'integral' PE spectrum (Figure 3) whereas
deflection systems, which provide energy band resolution, yield
the 'differential' spectrum directly.   It is however possible
to obtain differential spectra by means of a combination of

_____

*
  See Lindau et al., Rev. Sci. Instrum., 1973, 44, 265, for a
useful bibliography for deflection analysers.

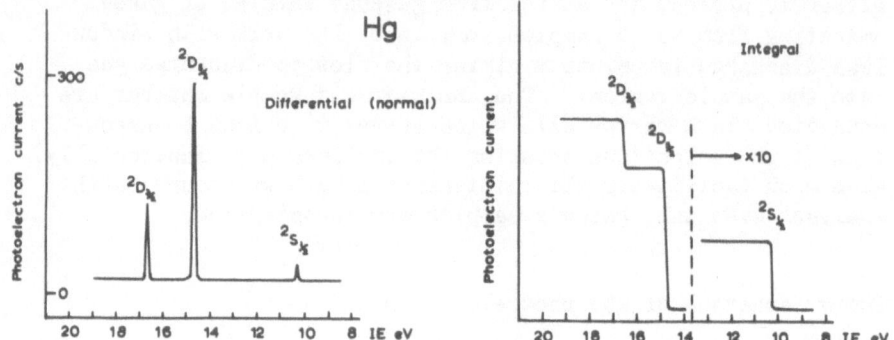

Figure 3.   Differential (normal) and integral He-I photoelectron
spectra of atomic mercury.

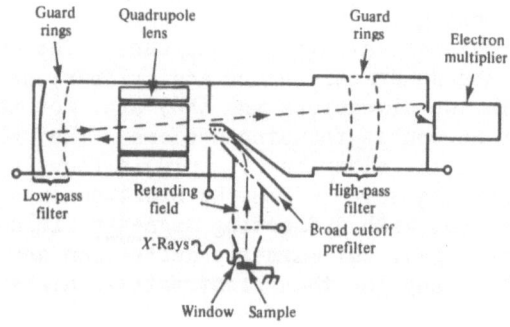

Figure 4.   The retarding field analyser used in the Du Pont
650 X-PE spectrometer.   The photoelectrons first pass through
a retarding field into a broad cut-off filter, and are then
directed through a series of low-pass and high-pass filters.
Electrons within a narrow band of kinetic energy are selected by
first removing electrons of excess energy by means of the low-
pass filter, which is a spherical electron mirror (operating in
reflection mode).   The reflected lower energy electrons subsequently approach a high-pass filter, a conventional spherical
retarding grid operating in transmission mode, which eliminates
all but those electrons of highest kinetic energy.   With correct
adjustment of the transmission characteristics of the two filter
systems, only electrons of narrowly defined kinetic energy can
reach the detector.   The quadrupole lens system serves to
optimise performance.   The photoelectron KE spectrum is scanned
by varying the initial retarding field.

retarding fields acting as low-pass and high-pass filters with a small energy overlap.[27] This ingenious technique is used with great profit in the Du Pont 650 electron spectrometer (Figure 4).

The <u>resolving power</u> of an analyser, $R = \frac{E}{\Delta E}$, where ΔE is the <u>resolution</u> attainable for electrons of kinetic energy E, depends on many design factors. These cannot be detailed here but, as regards dispersive analysers, the obvious point can be made that resolving power increases with the length of the focused electron trajectory, i.e. with the physical dimensions of the system. When dealing with high energy electrons, say E = 1000 eV, a relatively modest resolution of 1 eV requires a deflection analyser with R = 1000. The design of analysers for X-PES has thus been dominated by the consideration of high resolving power. However, in UV-PES, high resolving power is often of much less concern: thus, for E = 5 eV (a representative figure for He-I work), a resolving power of merely 100 will permit a resolution of 50 meV, adequate for many purposes. In general, moreover, the need for high resolving power can be partially circumvented by the device of initial retardation of the photoelectrons.

It is usual in X-PES to apply a fairly substantial retarding potential to the photoelectrons prior to energy analysis. On the other hand, when dealing with very slow electrons, it is sometimes necessary to arrange for a pre-acceleration of the photoelectrons. This commonly leads to a peak at zero on the electron KE scale (the so-called 'zero-energy peak') arising from various stray and multiply scattered electrons.

The photoelectrons are usually detected by means of dynode multipliers or (more commonly nowadays) by channel electron multipliers, so-called 'channeltrons'. The detector is normally situated behind a narrow adjustable slit which limits the energy band of the in-coming electrons. For deflection systems, the electron KE spectrum may be scanned in a simple <u>deflection mode</u>, which consists of varying the deflecting potential linearly with time, or in a <u>retardation mode</u>, when the analyser potentials are set so as to transmit only electrons of a certain energy E (or, strictly, electrons belonging to an energy band $\Delta E = \frac{E}{R}$) and a retardation potential imposed between the ionisation region and the analyser entrance slit is varied with time.

Scanning the deflecting potential produces an electron flux at the multiplier which, to a first approximation, is directly proportional to the electron KE - that is, the sensitivity declines towards low KE while the resolution $\left(\frac{E}{R}\right)$ tends to

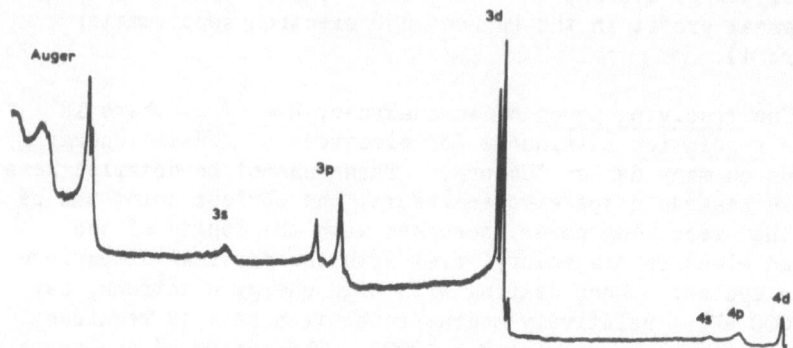

Figure 5.   Wide-scan X-PE spectrum of metallic silver (sputtered film) scanned in the retardation mode.

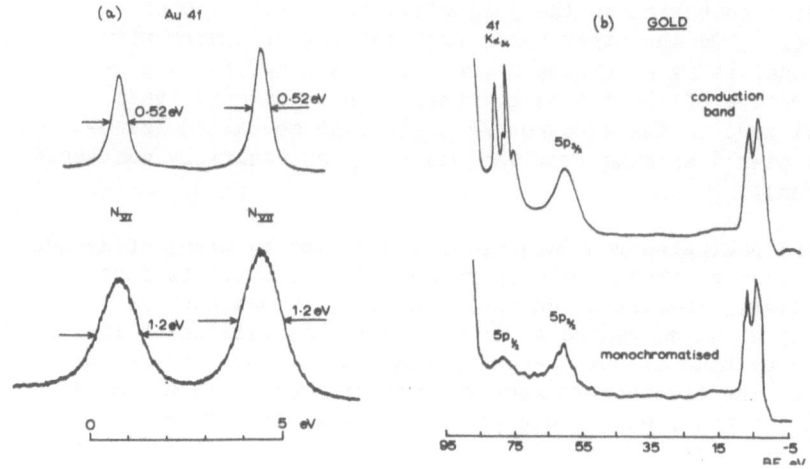

Figure 6.   (a) X-PE spectra of metallic gold illustrating the advantages of monochromatic exciting X-rays (Hewlett-Packard Ltd.); (b) low kinetic energy region of the X-PE spectrum of gold.

improve.   The retardation mode, on the other hand, can give
essentially constant sensitivity and resolution over the
operating range of electron KE.*   Both modes of scanning are
widely used and sometimes the spectra are scanned in a hybrid
mode: in the AEI ES 100/200 system, for example, the photoelectrons are continuously retarded to 1/20 of their original
kinetic energy and subsequently dispersed in a synchronised
sweep of the analyser deflection potential.

Where count rates are low the sensitivity can of course
always be improved by widening the detector slit, but only at
the expense of resolution.   A considerable gain in sensitivity,
without loss of resolution, may however be achieved by the use
of a multi-detector, a two-dimensional array of 'point'
multipliers, coupled to a multi-channel analyser.   This allows
one to count electrons simultaneously over a range (20 eV say)
of electron KE, a technique known as position-sensitive detection.
The method offers a gain in sensitivity of in principal x 100
(though apparently only x 10 in practice) compared with the
usual process of narrow band focusing onto a single channel
multiplier, and is used with considerable success in, for
example, the Hewlett-Packard 5950A X-PE spectrometer (vide infra).

Finally a word about the background signal in PE spectra.
X-PE spectra obtained (as in usually the case) with achromatic
exciting radiation suffer particularly high background counts
due to electrons ionised by continuum photons (the bremsstrahlung
radiation).   Also contributing to the background signal in
solid state PES are inelastically scattered electrons which
accumulate in flux towards low KE.   This is the origin of the
continuously rising background observed in X-PE spectra recorded
in the retardation mode (e.g. Figure 5).

Radiation sources

The conventional means of exciting photoelectrons are Al Kα
or Mg Kα X-rays, produced by electron bombardment of aluminium
or magnesium anodes, and He-I radiation from a suitable helium
discharge source.   Little use is made of harder X-ray lines,
despite their advantage in allowing us to probe more deeply into
the inner electron shells of heavy atoms, mainly because of
their large line widths: Cr Kα_1 radiation (hν = 5414.7 eV),
for example, has $\delta_X \sim$ 2 eV.   And although great interest attaches
to the potentialities of ultra-soft X-radiations such as yttrium

---

* The sensitivity characteristics of retardation instruments are
in general rather complex, however.   See, for example ref. 28.

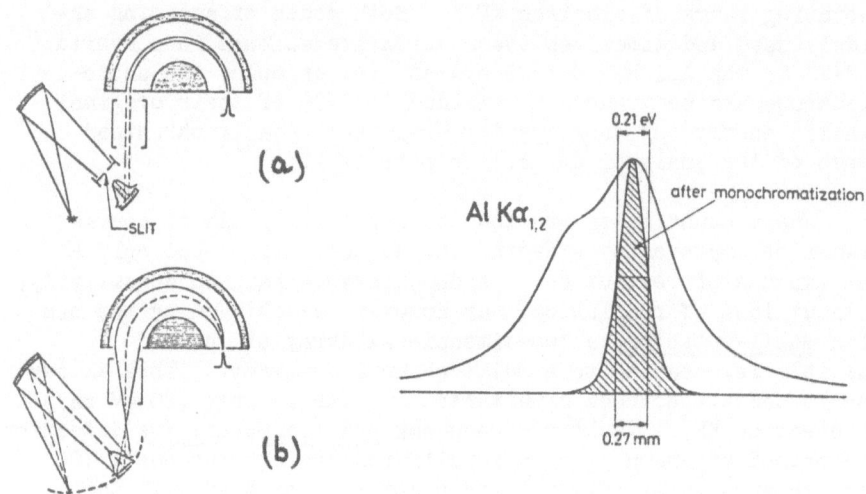

Figure 7. Monochromatisation of X-rays for photoelectron spectroscopy. (a) slit monochromatisation; (b) dispersion compensation.

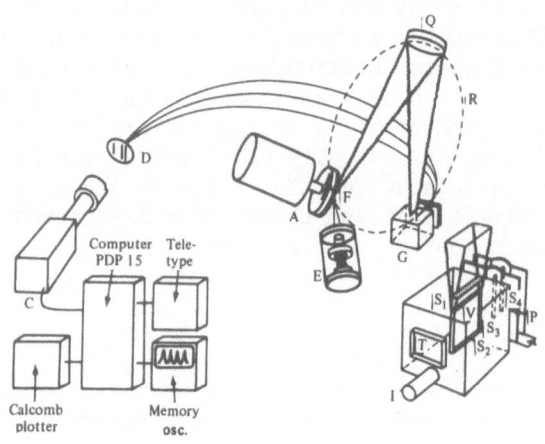

Figure 8. Sketch of the latest Uppsala X PE spectrometer, designed mainly for gas phase work. (U. Gelius and K. Siegbahn,
Faraday Disc. Chem. Soc., 54, 257 (1974). E=electron gun, A=
rotating anode, F=focal spot, Q=spherically bent quartz crystal,
R=Rowland circle, G=gas cell, S_1-S_4=slits, V=effective irradiated
gas volume, T=temperature raising device, I=gas inlet system, P=
two-stage differential pumping system with electron retardation
step, D=multichannel plate detector and C=television camera. The
line-width of the exciting radiation is determined by slit S_1.

M ζ (hν = 132.3 eV, δ_χ = 0.44 eV)[29] for the study of the 'deeper'
valence electron shells, technical difficulties (such as the
mundane problem of constructing sufficiently durable anodes)
have impeded their exploitation.   Much use has been made,
however, of the He-II line at 40.8 eV photon energy (λ = 30.4 nm)
which can be obtained (though, usually, not without difficulty)
when conventional helium discharge sources are operated at
particularly low pressure and at high current density.[6a]
Unfortunately, application of He-II radiation is complicated by
the fact that the He-I line normally still dominates the
radiative output of the source, though it is possible by
filtration techniques[30,31] to obtain a nearly pure He-II photon
flux.   The emission lines of other noble gas discharges (see
Appendix) are also occasionally employed in UV-PES.

     For many applications of X-PES it is more than desirable
to use monochromatised radiation - most obviously in order to
secure higher resolution, but also to eliminate interference
from satellite lines (such as the K$\alpha_{3,4}$ emissions when using
Al or Mg K$\alpha_{1,2}$ exciting radiation) and to reduce the background
counts due to bremsstrahlung radiation.   The Hewlett-Packard
X-PE spectrometer (see Appendix) incorporates this important
facility.   Al Kα radiation is dispersed by means of a bent
crystal system and the dominant K$\alpha_{1,2}$ component ($\delta_\chi \sim 0.9$ eV)
selected by a mechanical slit: additional virtual monochromatisation is achieved through a cleverly designed 'dispersion
compensation' system (see Appendix).   Some of the advantages that
accrue from the use of monochromatised X-rays are illustrated
by Figure 6.

     The more conventional techniques of monochromatisation,
using crystal dispersion with slit selection or fine-focussing
to cut the X-ray line-width down to (say) 0.25 eV (Figure 7)
has the obvious disadvantage of dramatically reducing the photon
flux, and therefore the ultimate PE counts.   The crystal
dispersion approach has however been most effectively employed
by Siegbahn and coworkers,[23-26] who counter the sensitivity
problem by means of a high intensity X-ray source - an Al Kα
source operating at some 8 kW power (with a water-cooled anode
rotated at 5000 r.p.m.), as opposed to the 200-500 W X-ray power
employed in ordinary X-PE spectrometers - together with position-
sensitive detection.   A sketch of this remarkable instrument is
given in Figure 8.

     Applications of synchrotron radiation for PES are also
being reported.[12,32,33]   The photon flux emerging from a
synchrotron source has a wide spectral range with a smooth
intensity distribution.   Given a sufficiently flexible
monochromator system, one has therefore a continuously tunable
radiation source, invaluable for the investigation of photo-

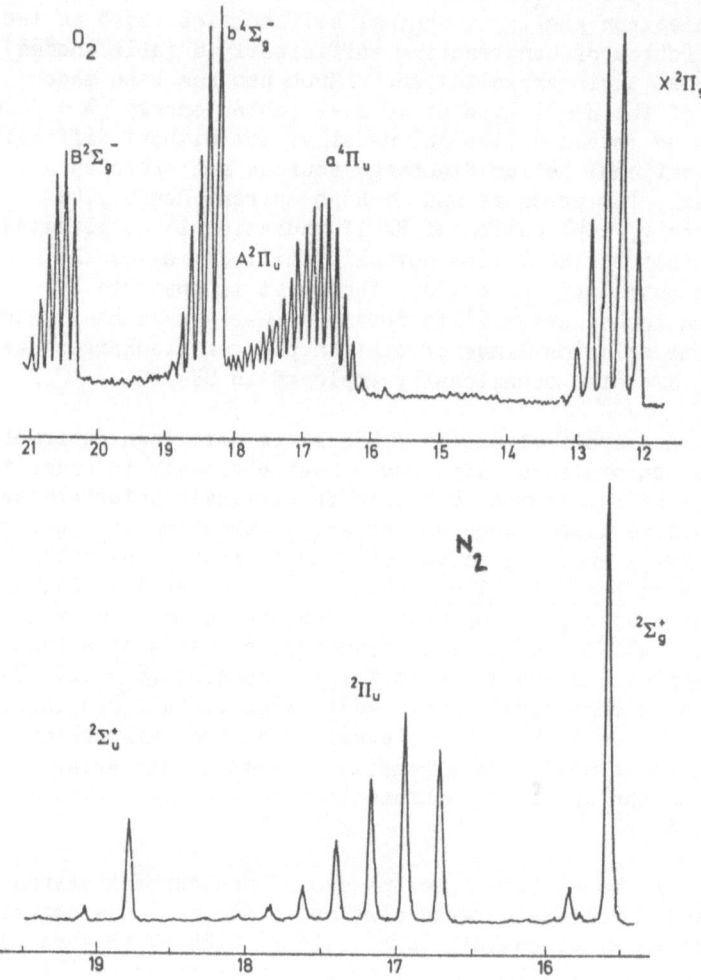

Figure 9.   He-I (21.2 eV) photoelectron spectra of molecular
nitrogen and oxygen.

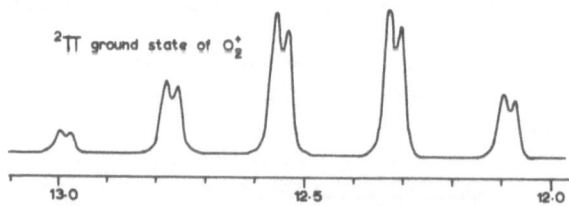

Figure 10.   High resolution scan of the first photoelectron band
of O_2 (cf. Figure 9). He-I exciting radiation. (O. Edqvist, E.
Lindholm, L.E. Selin & L. Asbrink, Physica Scripta, 1, 25 (1970).

ionisation cross-sections (and the angular distribution of the
PE flux, <u>vide infra</u>) as a function of the exciting photon energy.
Particularly useful are the photons in the 50-1000 eV energy
range, where there is a dearth of conventional line sources.

Gas phase measurements

The investigation of <u>molecular</u> electronic structure by
PES is clearly most advantageously conducted on gaseous samples.
There are however problems with relatively involatile materials,
which require special arrangements for high temperature measurements.[34]   The calibration of gas phase spectra (by reference to
signals from admixed calibrant gases) is quite straightforward,
and the spectra offer the advantage of higher resolution than can
be obtained with solid samples.

In He-I work the resolution is limited not by line width of
the ionising radiation (effectively a few meV, allowing for Doppler
broadening effects in the discharge), but by instrumental factors
which appear to contribute at least 5 meV to the half-width of
the PE signal.*   In practice, one must usually be content with a
resolution of around 30 meV (as defined by the half-width of the
argon $^2P_{3/2}$ line.   This is more than sufficient to resolve
vibrational fine structure in the case of small molecules (e.g.
Figure 9).   With higher resolution additional detail, such as the
multiplet splitting of the $^2\Pi$ terms of O_2^+ and N_2^+ (Figure 10)
or, in certain very small molecules, rotational fine structure
(e.g. Figure 10), can be observed.

Sensitivity is not normally a problem in He-I PES: for
example, an argon $^2P_{3/2}$ signal of at least 10,000 counts s^{-1},
at 30 meV resolution, can be obtained with most of the commercial
spectrometers.   However, count rates are commonly rather low in
He-II work, a high flux of the 40.8 eV photons being very difficult
to obtain.   Signal strength becomes a real problem in gas phase
X-PES where, using the ordinary instruments, one must normally
be content with count rates in the 10-100 counts s^{-1} range,**

---

* The Doppler effect in the photoionisation process also degrades
the resolution of PE signals, especially in the case of light
molecules or high temperature vapours.

** However, dramatically higher count rates - e.g. 45,000 counts
s^{-1} at 0.94 eV resolution for the 1s signal of neon, as compared
with some 1000 counts s^{-1} (at 0.8 eV resolution) using more
conventional equipment - have been reported for the proto-type
of the Du Pont X-Pe spectrometer.[27b]

the PE signals associated with ionisation of valence electrons
being much weaker than core PE signals.   The resolution of
the X-PE technique is moreover limited (with achromatic
radiation) to about 0.8 eV, which severely restricts its scope,
especially as regards valence electron studies.

Amazingly, however, Siegbahn and coworkers have succeeded
with gas phase measurements using monochromatised X-radiation
(Figure 8 above).*   The high-power Al Kα source and recourse
to position sensitive detection are of course crucial for this
application.   The line width of the exciting radiation
(200 meV) is sufficient for the resolution of vibrational fine
structure only in the case of certain small molecules but, as
shown in Figure 11, the instrument is capable of resolving quite
closely spaced bands in the low ionisation energy region.   The
ionisation of core electrons yields intrinsically broader PE
signals but, even so, the use of monochromatised X-rays can lead
to a substantial improvement in resolution.   The PE line widths
observed in some cases are given in the Appendix.   The neon 1s
line (half-width = 0.39 eV) is the narrowest yet reported in
the literature.

Solid state measurements

Solid samples necessarily yield higher count rates (x 100
perhaps) than gases but, in both UV-PES and X-PES, various solid
state effects lead to some broadening of the PE signals,
typically by about 0.2-0.3 eV for core ionisations.   The
narrowest X-PE band yet observed in the solid state is 0.5 eV
wide (Figure 6).   In addition to reduced resolution, solid
state PE spectra tend to suffer more background signal (due to
photoelectrons inelastically scattered as they make their way
out of the solid).

Calibration is also something of a problem in solid state
work, especially when dealing with poorly conducting (non-
metallic) samples which acquire a surface charge - usually a
positive charge - during the PES experiment.   This charging
effect is an equilibrium phenomenon determined by (i) the rate
at which electrons leave the sample surface, and (ii) the stray
electron current arriving at the surface (e.g. from the X-ray
window) and any limited current from earth, which neutralise the
positive holes.   All things considered, the most satisfactory
calibration procedure would appear to be the metal 'decoration'
technique,[36] in which a small (sub-monolayer) quantity of metal

---

* The dispersion compensation scheme employed in the Hewlett-
Packard instrument is not readily applicable to gaseous samples.

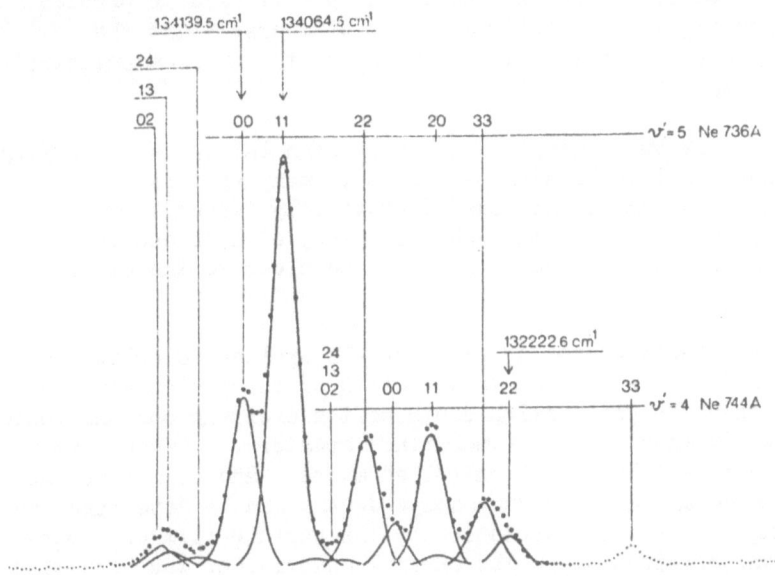

Figure 11. Rotational fine structure in the Ne-I photoelectron spectrum of H_2. The portion of the spectrum shown represents the transition to the v=5 vibrational level of H_2^+ excited by the 16.83 eV photons and the transition to the v=4 level excited by the 16.65 eV photons. (L. Asbrink, Chem. Phys. Lett. 7, 549 (1970).

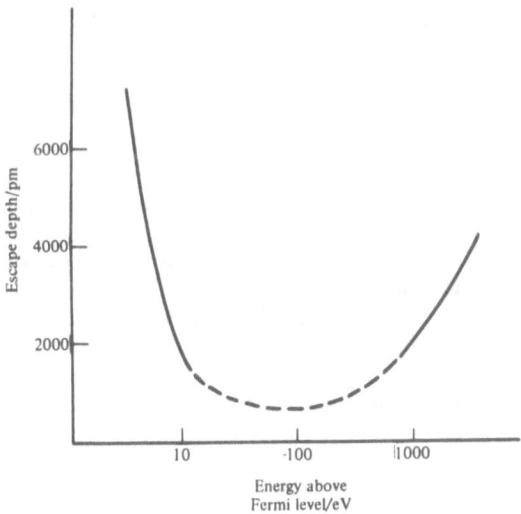

Figure 12. Photoelectron escape depth from gold vs. electron kinetic energy (log. scale). (M. Klasson, J. Hedman, A. Berndtsson, R. Nilsson, C. Nordling and P. Melnik, Physica Scripta 5, 93 (1972).

sputtered on to the surface of the sample is used to provide
reference signals, it being assumed, of course, that the
calibrant 'film' will faithfully take up the surface potential
of the sample.

In addition to displacing the PE signals, surface charging
also exerts a line-broadening effect in many systems.   This
can be eliminated, or at least dramatically reduced, by
discharging the sample through the agency of a 'flood-gun',
a source of thermal electrons, situated close to the sample
surface.[37]

The effective probing depth of PES applied to solids is
determined by the average escape depth of the photoelectrons
rather than the penetration depth of the exciting photons (which
is normally many orders of magnitude greater).   Given a means
of depositing films in a controlled manner, and of monitoring
their thicknesses, electron escape depths can be determined by
studying either the attenuation by the surface film of photoelectrons originating in the support material or the variation
in intensity of the photoelectron flux emitted by the film
itself as a function of thickness.   The photoelectron flux (at
a particular electron KE) penetrating a film of thickness $\underline{d}$ is
given by

$$I = I_0 \exp\left(-\frac{d}{\lambda}\right) \tag{3}$$

where I_0 is the signal strength of the clean backing material
and   is the mean escape depth (or attenuation length) for the
photoelectrons.   On the other hand, the intensity of a PE signal
from the film is

$$I = I_\infty \left[1 - \exp\left(\frac{d}{\lambda}\right)\right] \tag{4}$$

λ represents the depth from which $\frac{1}{e}$ of the photoelectrons
originating at that depth can escape.   (An alternative measure
of escape depth would be $\lambda\frac{1}{2} = \lambda\log_e 2$, the depth from which one
half of the photoelectrons can be observed.)   It turns out that
the escape depth is highly dependent upon the electron KE[38]
(e.g. Figure 12) and, understandably, depends also on the
material in question.   For inorganic solids, λ is typically
5-30 Å (it may often be somewhat larger for organic materials),
so that PES is very much a surface technique.   This introduces
exciting possibilities for the investigation of surface films
(e.g. adsorbed gases) but, at the same time, raises considerable
problems regarding the acquisition of data characteristic of
bulk solids.   Clearly, great care must be taken to ensure a
clean sample surface which, at least in the case of involatile
solids, demands UHV conditions (pressure $\leqslant 10^{-9}$ torr) in the
sample chamber.   Many workers in the field would feel that, at
the present time, few instruments meet the standards required.

Angular distribution studies

The photoelectron flux from both solid and gaseous samples is usually highly anisotropic and naturally much interest attaches to the characterisation of this anisotropy.   Many different techniques for studying the angular distribution of photoelectrons have been described in the literature.   The more obvious approaches to the problem involve the use of rotatable sources[39] or rotatable analyser systems.[40]

The angular dependence of the PE flux has a relatively simple general form in the case of gases, where the atoms or molecules are of course randomly oriented.   If the exciting radiation is unpolarised (as is the case for most of the experimental work so far reported) the angular distribution of photoelectrons is of the form

$$I(\theta) \quad \propto \quad 1 + \frac{1}{2}\beta\,(\frac{3}{2}\sin^2\theta - 1) \qquad (5)$$

where θ is the angle between the direction of propagation of the photons and the direction of the outgoing photoelectrons, and where β is the asymmetry parameter, which can range in value from -1 to $+2$.   (Clearly, the strongest signal is observed when $\theta = 90^\circ$, the most common observation angle in PE spectrometers.) For gaseous samples, the angular distribution is thus fully characterised by the asymmetry parameter.   This will normally vary according to the frequency of the exciting radiation and the particular orbital ionisation process involved.   There is however no thoroughgoing theory of PE angular distributions for molecular systems: only the theory for atoms is reasonably well established.

The angular dependence of photoelectrons emitted from solids can be a great deal more complicated, both phenomenologically and from the theoretical viewpoint, than in the case of gases. In addition to the obvious dependence of PE flux on the various angular coordinates involved, the spatial distribution will to some extent reflect the symmetry of the crystal lattice.   An intriguing example, concerning one aspect of the angular distribution of the Ta 5d and Se 3p photoelectrons from $TaSe_2$,[41] is shown in Figure 13.

Auger electron spectra

The core holes produced in the primary process of X-PES decay either by the emission of softer X-rays or by the ejection of secondary electrons (Figure 14) which are, of course, analysed together with the photoelectrons.   The latter decay process is known as the Auger effect.[42]   A core hole W is filled by the transition of an electron from some higher energy subshell X and

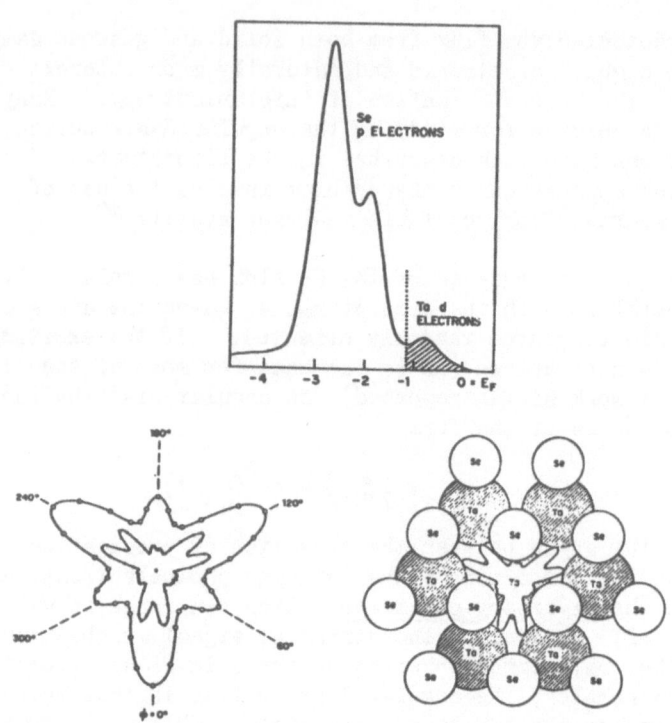

Figure 13. The angular dependence of the photoelectron flux from the layer compound 1T-TaSe$_2$. The incident radiation (hν =10.2 eV) is normal to the surface. The photoelectrons are 'viewed' at an angle θ to the normal and at an azimuthal angle ϕ. (M.M. Traum, N.V. Smith & F.J. Di Salvo, Phys. Rev. Lett. 32, 1241 (1974). (a) H. Lyman α (10.2 eV) PE spectrum of TaSe$_2$ for θ =55° and ϕ =0°; (b) radial plots of the azimuthal dependence (at θ =55°) of the total PE flux (outer contour) and, with x 5 expansion, the Ta 5d PE flux (inner contour); (c) the Ta 5d emission pattern (for θ =55°) superposed on the two uppermost atomic sheets of the TaSe$_2$ crystal lattice.

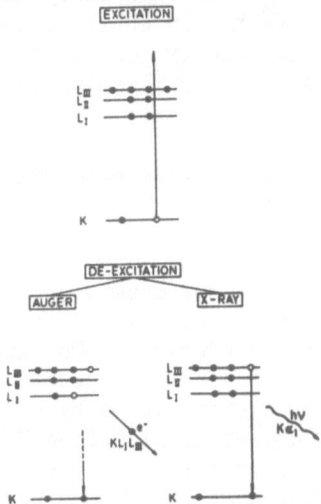

Figure 14. The processes of Auger electron emission and X-ray fluorescence by which atoms having core holes decay (after Siegbahn et al. (1967).

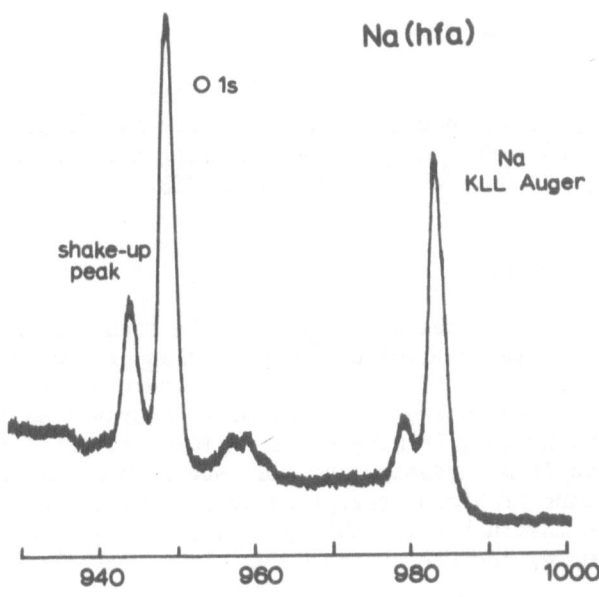

Figure 15. O 1s region of the X-PE spectrum of sodium hexafluoro-acetylacetonate showing structure due to Auger processes. (P. Burroughs, A. Hamnett and A.F. Orchard, unpublished work.)

the excess energy, $E_W - E_X$, is consumed by the simultaneous ionisation of some high energy subshell Y, leaving a doubly ionised system (with lower energy core holes X and Y).* Further Auger processes may occur leading to multiply ionised species.

The kinetic energy of an Auger electron may be expressed as

$$E(WXY) = E_W - E_X - E_Y \tag{6}$$

Notice that this energy does not depend upon the energy of the exciting X-photons (or indeed on any feature of the mechanism by which the core hole W was produced). The Auger electron signals that are frequently observed in X-PE spectra (e.g. Figure 15) may therefore readily be distinguished by changing the frequency of the X-rays (e.g. by comparing Al Kα and Mg Kα spectra): the PE signals will undergo a uniform shift in the spectrum, while the position of the Auger signals remains unchanged.

The strength of Auger signals is limited by the rate at which the primary core holes are produced, and the latter can be dramatically increased by turning to electron beam excitation. This technique, which pre-dates X-PES, is known as <u>Auger electron spectroscopy</u> (AES).[43,15,16]

APPLICATIONS OF PES

It is convenient, in the present context, to distinguish (1) the study of valence ionisations, which can be pursued by both UV-PES and X-PES, and (2) the investigation of core electron ionisations, which is the preserve of X-PES.

Valence electron PES

He-I (21.2 eV) excitation permits the study of only the higher energy valence electrons of molecules or, in the case of solids, only the 'upper' region of the valence bands. Of course, these electron levels are often the most important as regards chemical bonding but, in many cases, vital additional information can be gained from investigating also the deeper valence sub-shells by means of higher energy photons. He-II (40.8 eV) radiation is invaluable for this purpose. To take a somewhat extreme example, the He-I PE spectrum of CF_4 reveals only the

---

* Auger decay in which one of the two final vacancies remains in the same <u>principal</u> quantum shell as the primary vacancy is known as a <u>Coster-Kronig process</u>.

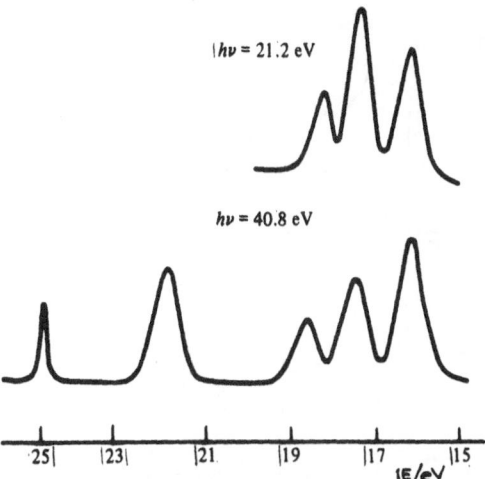

Figure 16.   Medium resolution UV photoelectron spectra of CF_4.
(W.C. Price, A.W. Potts and D.G. Streets in 'Electron Spectroscopy',
ed. D.A. Shirley (1972) p. 187.) The high resolution spectra show
some vibrational detail.

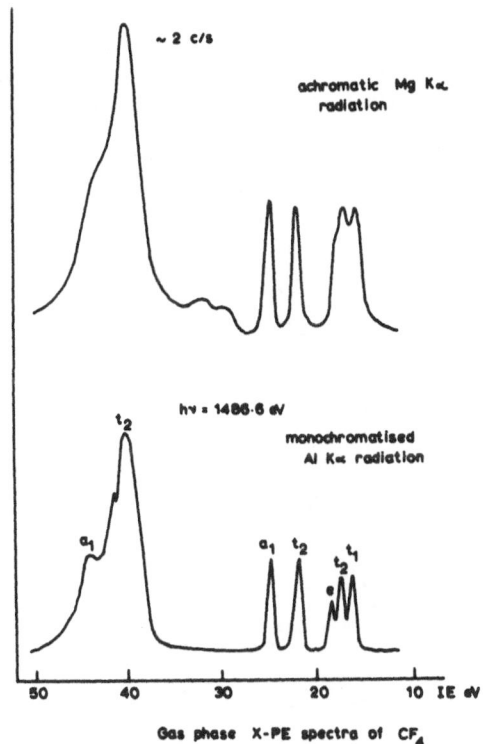

Gas phase X-PE spectra of CF_4

Figure 17.   Gas phase X-PE spectra of CF_4 excited with achromatic
and monochromatised X-radiation. (K. Siegbahn et al., 'ESCA Applied
to Free Molecules' (1969); U. Gelius, E. Basilier, S. Svensson,
T. Bergmark and K. Siegbahn, UUIP 817, Uppsala, April 1973.)

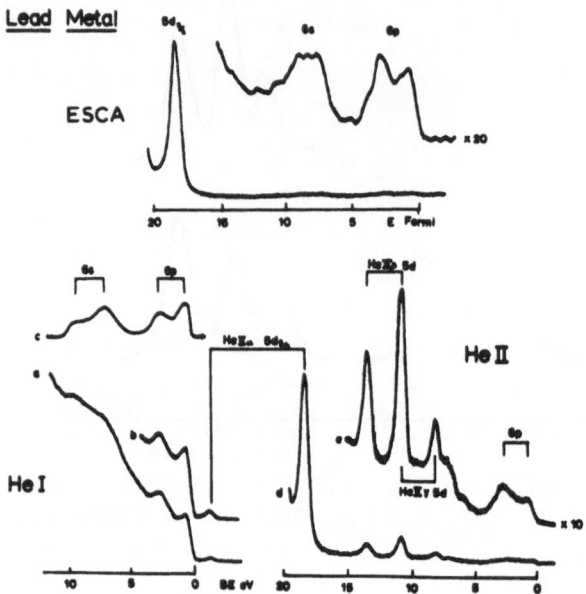

Figure 18.   X-PE and UV-PE spectra of metallic lead. (Hewlett-Packard Ltd. and S. Evans and J.M. Thomas, to be published.)   The X-PE spectrum was obtained with monochromatic X-rays.

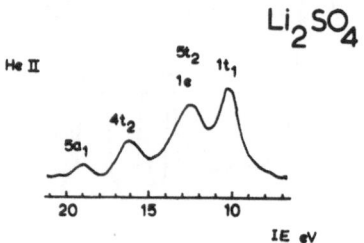

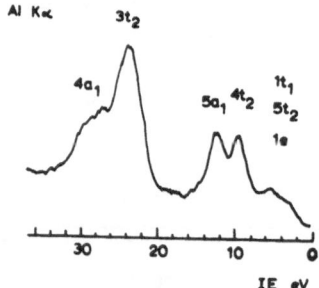

Figure 19.   He-II and Al Kα photoelectron spectra of Li₂SO₄. (J.A. Connor, I.H. Hillier, M.H. Wood and M. Barber, J. Chem. Soc. Faraday Trans. II, 70, 1040 (1974); J.A. Connor, I.H. Hillier, V.R. Saunders and M. Barber, Mol.Phys. 23, 81 (1972).

first three valence electron levels, whereas the He-II spectrum
shows a further two bands (Figure 16).[44]   These latter bands
relate to electrons having a major bonding role.   But even He-II
photons are insufficiently energetic, in the case of CF_4, to
ionise all species of valence electron.   As revealed by the gas
phase X-PE spectrum of CF_4 (Figure 17), there are further, very
tightly bound valence electrons (in molecular orbitals correlating
with the 2s orbitals on the fluorine atoms) to be considered.
The above example shows that X-PES has an important complementary
role to UV-PES, though one must not underestimate the difficulties
involved in obtaining spectra such as those reproduced in Figure
17, especially the spectrum measured with monochromatised X-
radiation.[26]   Healthier count rates are obtained in measurements
on condensed samples, but there is of course a loss of information
due to solid state broadening effects.   X-PES proves a particularly powerful complement to UV-PES in the study of solid materials,
where even the UV spectra tend to suffer rather low resolution.
Some illustrative examples are given in Figures 18 and 19.

There is a further reason why combined PE studies using more
than one radiation source are advantageous - namely the fact that
the individual orbital ionisation cross-sections, and therefore
the pattern of band intensities in the PE spectra, usually depend
(often dramatically) on the energy of the exciting photons.   This
is, in itself, an interesting, fundamental aspect of PES and one
which requires detailed attention from theoreticians: but, more
significantly, at least from the point of view of the experimentalist, the intensity data may often be interpreted in terms of quite
simple ideas regarding orbital cross-sections, the application of
which can much facilitate the essentially empirical assignment of
the different bands in the spectra.[45]

The use of X-PES, in addition to UV-PES, is particularly
important in solid state work directed at the determination of
the **density of states** function n(E) characterising the valence
electronic structure.   It would appear that a He-I spectrum does
not normally give an immediate picture of n(E) since the measured
intensity pattern depends also on the density of states for the
low energy region of the continuum, together with the transition
moment matrix connecting the two n(E) functions.[46]   But, since
the continuum should be relatively unstructured for photoelectrons
of high kinetic energy, it may well be that in most cases the X-
PE spectrum yields directly the occupied density of states,
modulated only by variations in the cross-sections of the atomic
orbitals involved.[47]   There is indeed some evidence to suggest
that even a He-II spectrum may give a reliable picture of this
density of states.[32]

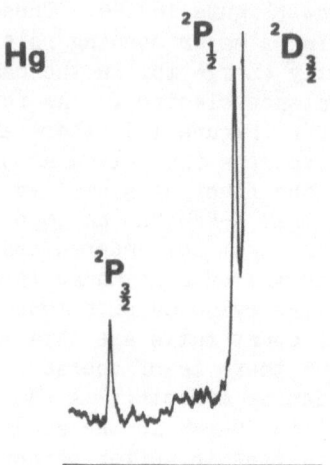

Figure 20.   Shake-up structure in the He-I photoelectron spectrum
of atomic mercury (cf. Figure 3).   (S. Evans, A. Hamnett and A.F.
Orchard, unpublished work.)

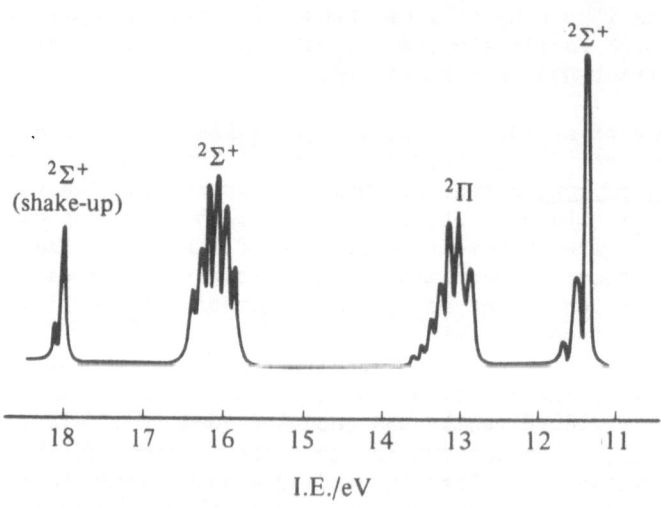

Figure 21.   He-I photoelectron spectrum of the transient species
CS.   (N. Jonathan, A. Morris, M. Okuda, K.J. Ross and D.J. Smith,
Faraday Disc. Chem. Soc., 54, 48 (1972).

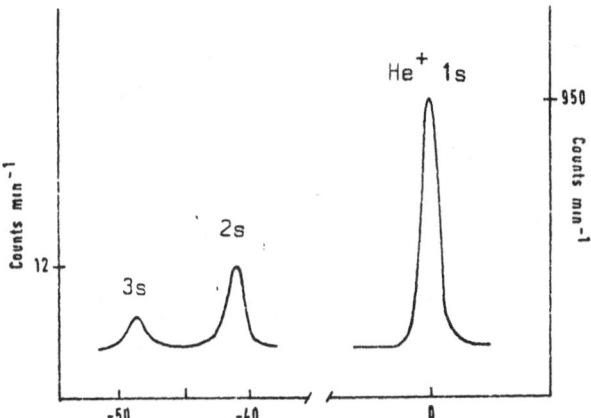

Figure 22. The photoelectron spectrum of helium excited by Al Kα radiation. Energies are referred to that of the principal peak which occurs at a kinetic energy of 1462 eV. (After T.A. Carlson, M.O. Krause and W.E. Moddeman, J. Phys. (Paris) Colloq. 32, C4-76 (1971).

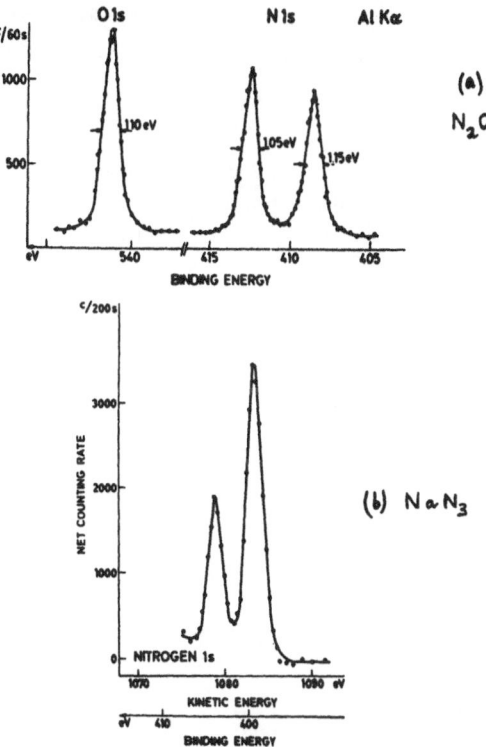

Figure 23. The nitrogen 1s region of the X-PE spectra of some inorganic compounds containing inequivalent N atoms: (a) nitrous oxide (gas phase); (b) sodium azide. (K. Siegbahn et al., (1967, 1969.)

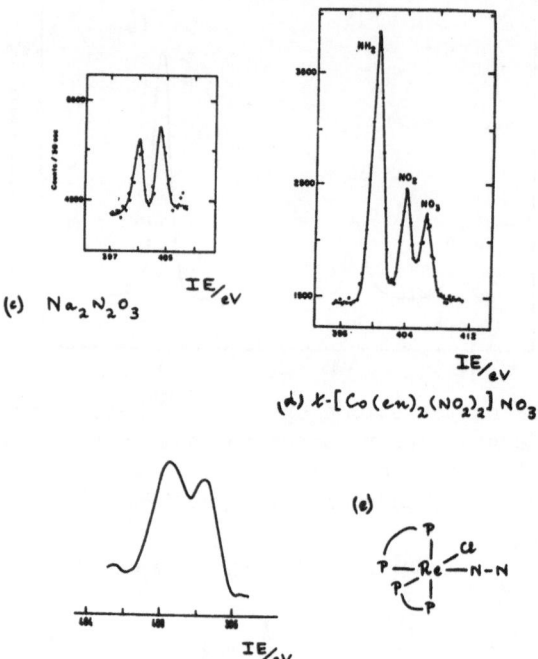

Figure 23.   (c) $Na_2N_2O_3$;   (d) <u>trans</u> $[Co(en)_2(NO_2)_2]NO_3$;
(e) $Re(diphos)_2Cl(N_2)$.   (D.N. Hendrickson, J.M. Hollander and
W.L. Jolly, Inorg. Chem. <u>8</u>, 2642 (1969); W. Bremser, Chem. Ztg.
<u>95</u>, 819 (1971).

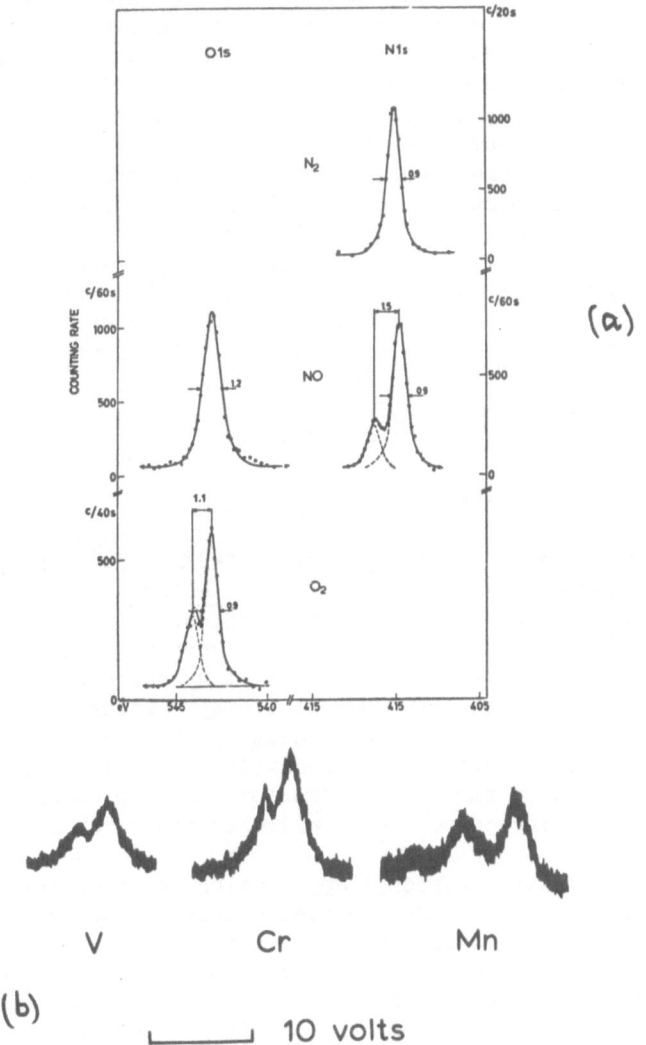

Figure 24. (a) K shell photoelectron spectra of N_2, NO and O_2 excited in the gas phase using Mg Kα radiation. (K. Siegbahn et al., 'ESCA Applied to Free Molecules', 1969.) Note the exchange splittings of the N 1s signal of NO and the O 1s signal of O_2. (b) Exchange fine structure in the 3s region of the X-PE spectra of the metallocenes $V(C_5H_5)_2$, $Cr(C_5H_5)_2$ and $Mn(C_5H_5)_2$. Achromatic Al Kα exciting radiation. (P. Burroughs, A. Hamnett and A.F. Orchard, unpublished work.) Note that the exchange splitting is greatest for manganocene, which has five unpaired spins in the outer 3d subshell.

It has been assumed thus far that the primary structure in
a PE spectrum can be understood, at least qualitatively, simply
on the basis of Koopmans' theorem, i.e. that one observes just
one band per occupied electron energy level.   However
additional structure, apparently forbidden in the one-electron
approximation, may sometimes be observed.   Figures 20 and 21
provide two simple examples.   In the case of atomic mercury
(Figure 20), the appearance of signals corresponding to the
^2P $(5d^{10}6p)$ state of Hg^+,[48] which is formally inaccessible by
ionising Hg ^1S $(5d^{10}6s^2)$, can be understood in terms of
simultaneous ionisation of a 6s electron coupled with the
excitation 6s $\rightarrow$ 6p.   Many-electron transitions of this kind
are referred to as shake-up processes.[16]   Similar processes
are frequently evident in conjunction with core electron
ionisations (e.g. Figure 22).

Core electron PES (X-PES)

The enormous interest apparent in X-PES (ESCA) has been
stimulated largely by the chemical shift effect that is usually
observed for core electron binding energies.   The core
ionisation energies of an atom normally vary with chemical
environment, correlating quite well in many cases with the
effective charge on the atom calculated by various methods.[15,16*]
When a molecule contains the same atom in chemically inequivalent
positions, it frequently proves possible to resolve more than
one PE signal for a particular core ionisation process.   Some
examples concerning compounds of nitrogen are given in Figure 23.

Apart from its occasional application to structural problems
the chemical appeal of X-PES resides mainly in the possibility,
realised in some instances, of determining the atomic charge
distributions in molecules and in 'continuous' solids from
measured core electron binding energy data.   But this represents
relatively crude information, especially in the case of ionic
materials, for which chemical shifts are frequently difficult to
interpret.   In the particular case of open-shell systems, more
subtle information concerning electronic structure may often be
obtained through the observation of exchange fine structure for
the signals of some s core subshell (e.g. Figure 24).   This
provides an indication of the localisation properties of the
unpaired valence electrons.   What may perhaps ultimately prove
of even more interest, in the general context of inorganic
electronic spectroscopy, is the study of secondary structure in

---

* See D.A. Shirley, Adv. Chem. Phys., 1973, 23, 85 and U. Gelius,
Physica Scripta, 1974, 9, 133 for a thorough review of this topic.

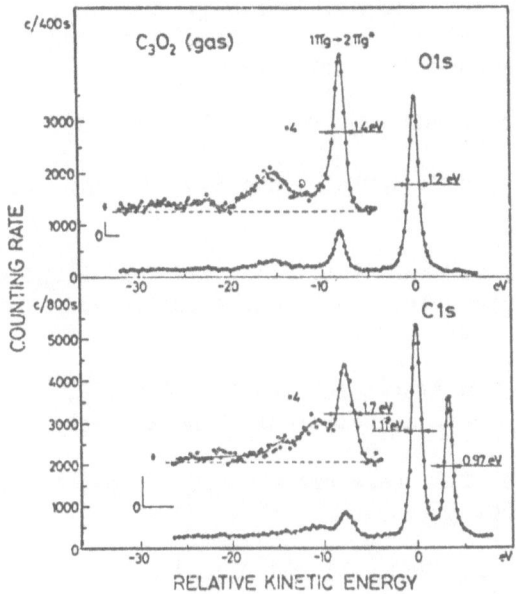

Figure 25. Satellite structure in the core regions of the X-PE spectrum of C_3O_2: Mg Kα exciting radiation. (U. Gelius, C.J. Allan, D.A. Allison, H. Siegbahn and K. Siegbahn, Chem. Phys. Letters, 11, 224 (1971).

X-PE spectra due to multi-electron processes – especially that which may be interpreted in terms of shake–up transitions. Some illustrative examples are shown in Figure 25. Shake–up structure is much more common, and usually more pronounced, in X-PE spectra than in UV-PE spectra but is, for the most part, less well understood theoretically. This secondary structure may however prove susceptible to interpretation by relatively simple models: for example, one might discover simple correlations with high energy UV absorption data.

References

1. A. Einstein, Ann. Physik, 1905, 17, 132.

2. T. Koopmans, Physica, 1934, 1, 104.

3. M.I. Al-Joboury and D.W. Turner, J. Chem. Soc., 1963, 5141.

4. F.I. Vilesov, B.C. Kurbatov and A.N. Terenin, Soviet Phys. Dokl., 1961, 6, 490.

5. D.W. Turner, C.Baker, A.D. Baker and C.R. Brundle, 'Molecular Photoelectron Spectroscopy', Wiley-Interscience, 1970 - and refs. therein.

6.(a) W.C. Price, in 'Molecular Spectroscopy', p. 221, ed. P. Hepple, Inst. of Petroleum, London, 1968.

   (b) H.J. Lempka, T.R. Passmore and W.C. Price, Proc. Roy. Soc. A, 1968, 304, 53.

7. D.C. Frost, C.A. McDowell and D.A. Vroom, (a) Phys. Rev. Letters, 1965, 15, 512; (b) J. Chem. Phys. 1967, 46, 4255; (c) Proc. Roy. Soc. A, 1967, 296, 566.

8. D.W. Turner and M.I. Al-Joboury, J. Chem. Phys., 1962, 37, 3007.

9. D.E. Eastman and J.K. Cashion, Phys. Rev. Letters, 1970, 24, 310.

   D.E. Eastman, J. Phys. (Colloques), 1971, 32, C1-293.

10. R.A. Millikan, Phys. Rev., 1916, 7, 18; 1921, 18, 236.

11. L. Apker, E. Taft and J. Dickey, Phys. Rev., 1948, 74, 1462.

12. D.E. Eastman and W.D. Grobman, Phys. Rev. Letters, 1972, 28, 1327, 1378.

13. H. Robinson and W.F. Rawlinson, Phil. Mag., 1914, 28, 277.

    H. Robinson, Proc. Roy. Soc. A, 1923, 104, 455.

    M. de Broglie, C.r. hebd. Seanc. Acad. Sci., Paris, 1921, 172, 274.

14. E. Sokolowski, C. Nordling and K. Siegbahn, Arkiv. Fysik, 1957, 12, 301; Phys. Rev., 1958, 110, 776; Arkiv. Fysik, 1958, 13, 483.

15. K. Siegbahn et al., 'ESCA: Atomic, Molecular and Solid State Structure studied by means of Electron Spectroscopy', Nova Acta Reg. Sci. Upsaliensis, ser. IV, 1967, vol. 20 (and refs. therein).

16. K. Siegbahn et al., 'ESCA Applied to Free Molecules', North Holland, Amsterdam, 1969.

17. C. Nordling, Angew. Chem. Internat. Edn., 1972, 11, 83.

18. D.A. Shirley, ed. 'Electron Spectroscopy', North Holland, Amsterdam, 1972.

19. C.S. Fadley, S.B.M. Hagstrom, J.M. Hollander, M.P. Klein and D.A. Shirley, Science, 1967, 157, 1571.

20. M.O. Krause, Phys. Rev., 1965, 140, 1845.

21. T.A. Carlson, Phys. Rev., 1967, 156, 142.

22. H. Siegbahn and K. Siegbahn, J. Electron Spectroscopy, 1973, 2, 319.

23. K. Siegbahn, D. Hammond, H. Fellner-Feldegg and E.F. Barnett, Science, 1972, 176, 245.

24. K. Siegbahn, in ref. 18.

25. U. Gelius and K. Siegbahn, Faraday Disc. Chem. Soc., 1972, 54, 257.

26. U. Gelius, E. Basilier, S. Svensson, T. Bergmark and K. Siegbahn, J. Electron Spectroscopy, 1973, 2, 405.

27. J.D. Lee, Rev. Sci. Instrum., (a) 1972, 43, 1291; (b) 1973, 44, 893; (c) R.D. Davies et al., Advan. X-ray Analysis, 1973, 16, 90.

28. R.T. Poole, R.C.G. Leckey, J. Liesegang and J.C. Jenkin, J. Phys. E, 1973, 6, 226.

29. M.O. Krause, Chem. Phys. Letters, 1971, 10, 65.

30. J.N.A. Ridyard, in 'Molecular Spectroscopy', ed. P. Hepple, Inst. of Petroleum, London, 1972.

31. A.W. Potts, T.A. Williams and W.C. Price, Faraday Disc. Chem. Soc., 1972, 54, 104.

32.   D.E. Eastman and others in 'Research Application of
       Synchrotron Radiation', proceedings of a study-symposium
       held at the Brookhaven National Laboratory, Sept. 1972,
       ed. R.E. Watson and M.L. Perlman.

33.   H. Fellner-Feldegg, U. Gelius, K. Siegbahn, C. Nordling and
       K. Thimm, University of Uppsala Institute of Physics,
       report no. UUIP-856, March 1974.

34.(a) J. Berkowitz et al., J. Chem. Phys., 1973, 59, 3645;

   (b) J.D. Allen et al., J. Electron Spectroscopy, 1973, 2,
       289;

   (c) Y.S. Khodeyev et al., Chem. Phys. Letters, 1973, 19, 16.

35.   U. Gelius et al. to be published (UUIP-860, March 1974).

36.   D.J. Hnatowich, J. Hudis, M.L. Perlman and R.C. Ragaini,
       J. Appl. Phys., 1971, 42, 4883.

37.   D.A. Huchital and J.D. Rigden, Appl. Phys. Letters, 1972,
       20, 158; M.F. Ebel and H. Ebel, J. Electron Spectroscopy,
       1974, 3, 169.

38.   e.g. M. Klasson, J. Hedman, A. Berndtsson, R. Nilsson,
       C. Nordling and P. Melnik, Physica Scripta, 1972, 5, 93.

39.   e.g. T.A. Carlson and A.E. Jonas, J. Chem. Phys. 1971,
       55, 4913.

40.   e.g. D.C. Mason, A. Kuppermann and D.M. Mintz, in ref. 18.

41.   M.M. Traum, N.V. Smith and F.J. Di Salvo, Phys. Rev.
       Letters, 1974, 32, 1241.

42.   P. Auger, J. Phys. Radium, 1925, 6, 205.

43.   P.W. Palmberg, in ref. 18.

44.   A.W. Potts, H.J. Lempka, D.G. Streets and W.C. Price,
       Phil. Trans., 1970, A 268, 59.

45.   W.C. Price, A.W. Potts and D.G. Streets, in ref. 18.

46.   D.E. Eastman, in ref. 18.

47.   C.S. Fadley and D.A. Shirley, J. Res. Nat. Bur. Stand.,
       sect. A, 1970, 74, 543.

48.   S. Evans, A. Hamnett and A.F. Orchard, unpublished work.

BIBLIOGRAPHY[*] – REVIEWS & MONOGRAPHS

General

'Photoelectron Spectroscopy', by A.D. Baker and D. Betteridge, Pergamon Press 1972

'Photoelectron Spectroscopy', in 'Electronic Structure and Magnetism of Inorganic Compounds', Chem.Soc. Specialist Periodical Report: A.F. Orchard and A. Hamnett vol. 1, 1972; S. Evans and A.F. Orchard, vol. 2, 1973; A.Hamnett and A.F. Orchard vol. 3, 1974; A.F. Orchard vol. 4, 1975.

A.D. Baker, C.R. Brundle and M. Thompson, Chem. Soc. Rev., 1972, 1, 355

R.G. Albridge, in 'Physical Methods in Chemistry', part IIID, p. 308 (Techniques of Chemistry, vol. I), ed. A. Weissberger and R.W. Rossiter, Wiley-Interscience, 1972

UV-PES

'Photoelectron Spectroscopy', by J.H.D. Eland, Butterworths, 1974

R.L. DeKock and D.R. Lloyd, Adv. Inorg. Chem. Radiochem., 1974, 16, 66

'Molecular Photoelectron Spectroscopy', by D.W. Turner, A.D. Baker, C. Baker and C.R. Brundle, Wiley, 1970

X-PES

'ESCA: Molecular and Solid State Structure Studied by means of Electron Spectroscopy', by K. Siegbahn et al., Nova Acta Reg. Soc. Sci. Upsaliensis, ser. IV, 1967, vol. 20

'ESCA Applied to Free Molecules', by K. Siegbahn et al., North Holland, 1969

W.L. Jolly, Coord. Chem. Rev., 1974, 13, 47

D.A. Shirley, Adv. Chem. Phys., 1973, 23, 85

U. Gelius, Physica Scripta, 1974, 9, 133

D.M. Hercules, Analyt. Chem., 1972, 44, 106R

---

[*] A selection from the recent literature.

D.M. Hercules and J.C. Carver, Analyt. Chem., 1974, <u>46</u>, 133R

Conference Proceedings

'<u>Electron Spectroscopy</u>', ed. D.A. Shirley, North Holland, 1972 (Proc. of Internat. Conf. held at Asilomar, California, Sept. 1971).

<u>Faraday Disc. Chem. Soc.</u>, 1972, no. 54 (Proc. of a meeting on 'The Photoelectron Spectroscopy of Molecules', held at University of Sussex, U.K., Sept. 1972).

<u>J. Electron Spectroscopy</u>, to be published (Proc. of a meeting held in Namur, Belgium, April 1974).

APPENDIX

A. Ionising radiations employed in photoelectron spectroscopy

UV photoelectron spectroscopy

He I	21.22 eV photon energy	Ar I	11.62 (2) 11.83 (1)
He II	40.81	Ar II	13.30 (2) 13.48 (1)
Ne I	16.67 (1) 16.85 (7)		
		H Lyman α	10.20
Ne II	26.81 (1) 26.91 (1)		

The relative intensities of the doublet components are given in parentheses.

The emission from a gas discharge always contains many different lines. Helium discharge lamps give varying proportions of the following lines:

He I α	21.22 eV	(100)
He I β	23.09	($\sim$2)
He I γ	23.74	($\sim$0.5)
He II α	40.81	($<$1)
He II β	48.37	
He II γ	51.015	
He II δ	52.24	

Relative intensities typical for a capillary discharge under normal He-I conditions are shown in parentheses. Common contaminant lines include H Lyman α (10.20 eV), N I (10.93 eV) and Ne I α (16.85 eV).

X-ray photoelectron spectroscopy

Listed below are some of the sources that have been employed in X-PES:

Y	Mζ	132.3 eV	(0.44)	Cr	Kα_1	5414.7	($\sim$2)
Zr	Mζ	151.4	(0.84)	Cu	Kα_1	8047.8	($\sim$3)
Mg	Kα	1253.6	(0.8)				
Al	Kα	1486.6	(0.9)				

The half-widths (FWHM) of the lines are given in parentheses.
The figures for Al Kα and Mg Kα radiation relate to the dominant
K$\alpha_{1,2}$ components of the X-ray emission.   The latter X-rays have
satellite lines, the K$\alpha_{3,4}$ and K$\alpha_{5,6}$ emissions, from the doubly-
and triply-ionised atoms, respectively.   The relative intensities
of these satellites, expressed as a percentage of the intensity
of the K$\alpha_{1,2}$ lines, are as follows:

	K$\alpha_{3,4}$	K$\alpha_{5,6}$
Mg	14.0	1.7
Al	10.8	1.1

B.   **Dispersion compensation – the Hewlett-Packard 5950A
     Spectrometer**

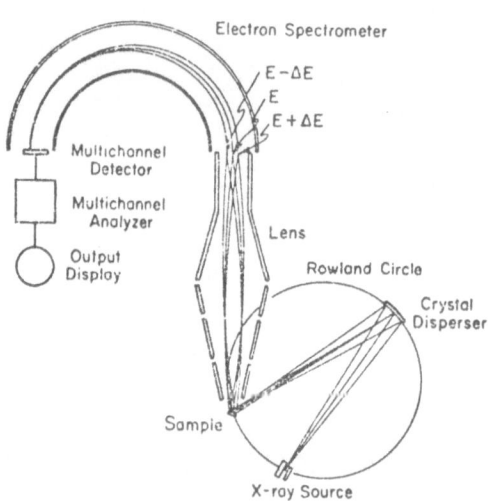

Dispersion compensation for high resolving power as
used in the design of the Hewlett-Packard ESCA instrument.
In the first approximation the electron linewidth is independent of both the sample size and the inherent width of the
exciting radiation.

Dispersion compensation scheme for the virtual monochromatisation of X-rays

In the Hewlett-Packard 5950A ESCA spectrometer Al $K\alpha$ radiation is dispersed in the usual way by means of a bent (quartz) crystal. The sample is situated so that it is irradiated only by the Al $K\alpha_{1,2}$ radiation, but the full width (ca. 0.9 eV) of this line is in fact used. The Al $K\alpha_{3,4}$ satellite radiation is eliminated, together with most of the bremsstrahlung continuum radiation which otherwise makes a major contribution to the background PE signal. Within the limits set by the resolving power of the crystal dispersion system, each point on the sample surface is exposed to monochromatic radiation, the wave-length of which varies across the sample. The photo-electrons ejected from a particular energy level will therefore have kinetic energies which depend systematically on the point on the sample surface from which they are emitted. The photo-electrons pass through a multi-component lens system designed to focus the image of the sample at the entrance to the analyser (preserving the spatial energy distribution), and to magnify the image to match the dispersion of the analyser, a hemispherical system (15.5 cm central radius). The dispersion of the analyser compensates for the energy spectrum of the photoelectron species in question, all the electrons being focussed on to a line perpendicular to the plane of the figure on the previous page. The beauty of this scheme is that it effectively monochromatises the exciting radiation without the introduction of mechanical slits which occasion a large loss in sensitivity. Apart from the natural line-width of the electrons the observable PE band-width is limited only by electron optical aberrations ($\leqslant$0.2 eV) and the energy resolution of the monochromator (ca. 0.3 eV).

The Hewlett-Packard instrument also utilises position-sensitive detection, parallel two-dimensional detection over a 10 eV window being achieved by means of an array of 250 x 60 high-gain electron multipliers.

C.   The line-widths (FWHM in eV) of some core PE signals
observed for gaseous samples with the new Uppsala X-PE spectrometer using monochromatised Al K$\alpha_{1,2}$ radiation.   (Gelius et al.,
UUIP-817, Uppsala, April 1973)

Ne 1s	0.39	C 1s (CO)	0.65
N 1s (N_2)	0.46	C 1s (CF_4)	0.52
O 1s (O_2)	0.46[*]	C 1s (C_6H_6)	0.57
O 1s (CO)	0.52	C 1s (CH_4)	0.72
F 1s (CF_4)	1.30		
S 2p (COS)	0.52[*]		

---

[*] Resolved multiplets

ULTRAVIOLET PHOTOELECTRON SPECTROSCOPY (UV-PES) OF MOLECULES IN
THE GAS PHASE

D.R. Lloyd

Chemistry Department, University of Birmingham

I.   PRINCIPLES

    Much of the material of these two lectures is covered in
greater detail in the recent book by Eland,[1] and this can be
recommended for background reading.   As has been indicated in
the previous lecture, the main photon sources used for UV-PES are
the He-I and He-II resonance emissions, though occasionally Ne-I
is useful.   Most of these two lectures will be concerned with
studies made at room temperature or slightly above, but studies
at high temperatures (~ 1000 K or greater) are potentially very
interesting for inorganic chemists.   Although not much work at
high temperatures has yet been reported, it is certain that the
literature of high temperature work will expand, and I shall make
occasional references to some of the particular problems of such
work.

    As a preliminary to the chemistry, some aspects of the basic
physics are important.   A relatively trivial point is that for a
free molecule the energy E_k in the Einstein relationship

$$E_k = h\nu - I_k$$

is the total energy released, which is shared between the emitted
electron and the recoiling ion.   Since momentum is conserved, i.e.

momentum of ion p_k = momentum of electron p_e

then the ratio of the kinetic energies of the electron to that of
the ion is

$$\frac{p_e^2}{2m_e} \cdot \frac{2m_k}{p_k^2} = \frac{m_k}{m_e}$$

where m_k and m_e are the masses of the ion and electron. Thus even for the lightest molecular ion H_2^+ the error in equating E_k with the electron kinetic energy, rather than the total energy, is only 1 in 3600, and in the subsequent discussion we ignore this correction. However, so far we have assumed the molecule before ionization to be at rest, and this is unlikely in a gas at room temperature: we have to consider the momentum which the electron carries as a result of the molecular motion. Since there is a distribution of molecular velocities we have correspondingly a distribution of electron energies, and hence a limitation on energy resolution ΔE_k. If we assume that an additional momentum p_m from the molecular motion adds to or subtracts from p_e, then the measured electron energy E lies within the range:

$$\frac{(p_e \pm p_m)^2}{2m_e} \simeq \frac{p_e^2 \pm 2p_e p_m}{2m_e} = E_k \pm \frac{E_k^{\frac{1}{2}} p_m}{(2m_e)^{\frac{1}{2}}}$$

i.e. the energy resolution ΔE_k depends on $E_k^{\frac{1}{2}}$. If a value for p_m appropriate to the r.m.s. velocity of the molecules is inserted, then[2]

$$\Delta E_k \simeq 0.7 \left(\frac{E_k T}{M}\right)^{\frac{1}{2}} \text{ meV}$$

where M is the molecular weight of the molecule, T is in degrees Kelvin and E_k is in eV. Graphs of this function for pairs of the three variables are given in reference 2. Typically at room temperature for M = 32 and E_k = 10 eV (e.g. for the first band in the He–I spectrum of O_2) ΔE_k is about 7 meV; for He–I ionization of H_2 ΔE_k rises to over 20 meV. The line width due to this thermal motion of the target molecules could be reduced by cooling or by restricting the molecular motion to one dimension by using a molecular beam and observing electrons emitted perpendicular to the beam. Neither of these has been achieved in practice. However, reduction of E_k will also reduce ΔE_k, and this has been very successful in obtaining very high resolution for H_2 (see below). For work at high temperatures higher values of ΔE_k will be encountered; thus for 10 eV electrons emitted from BF at 2000 K, ΔE_k will be about 20 meV. Such work will almost certainly be done using molecular beams, and proper design of the analyser geometry allows this thermal motion to be reduced. However, some analysers, particularly the cylindrical mirror analyser, are inherently unsuitable for this since there is no axis for the molecular beam which is perpendicular to the directions of observation of electrons.

There are several other sources of line width.   The He-I
photon source itself has been shown to have a line width varying
from about 1 to 7 meV, depending on the discharge conditions.[2]
There is also the effect of analyser slit width;   if the instrument is scanned in the dispersion mode (see previous AFO I
lecture) then ΔE decreases as E_k decreases.   Thus higher
resolution can often be obtained by working with lower photon
energies, and this is one of the major uses of the Ne-I source.
Several studies have used this to improve the resolution of
photoelectron bands at low IE.[3,4]   The disadvantage of using
this source is that there are two closely spaced photon emissions,
that at lower energy having about 1/4 the intensity of the other,
so the structure is complicated by the additional 'shadow' bands
due to the weaker emission line.

In practice, resolution is often limited by stray electrostatic fields, particularly those from insulating deposits which
build up around the slit at the ionisation region.   Inevitably
many compounds of interest to inorganic chemists are somewhat
corrosive, and obtaining high resolution with these compounds is
very much a matter of patience in cleaning the instrument
frequently.

STATES WHICH CAN BE OBSERVED IN UV-PES

(a)   Electronic states of the molecular ions

In this course we are interested in electronic states, but
in optical spectroscopy electronic information can often be
obtained from vibrational and rotational structure, so it is
important to know what corresponding information might be
available from photoelectron work.   As far as the electronic
transitions are concerned we can observe any state which can be
derived by removing one electron from any occupied orbital
(provided, of course, that there is sufficient energy available
in the photon).   In this sense there are no selection rules
comparable with those which govern single electron transitions
in absorption or emission spectroscopy.   With the He-I source,
production of 2^+ ions is not possible on energetic grounds for
most molecules, and even with He-II it is not very likely that
this will be observed in a photoelectron spectrometer since the
two electrons are produced with continuum energy distributions,
though the doubly charged ions can be detected with a photoionisation mass spectrometer.[5]   However, two-electron processes
in which one electron is ionised and the other is excited to a
higher level are possible energetically.   Such transitions are
well known in X-PES and are often referred to as 'shake-up'
processes.   Several instances of shake-up effects have been
reported in He-II work, though many require filtered He-II

radiation free of He-I to observe them.[6]   Such shake-up states
seem to be rare in He-I spectra, but they have been reported for
CS,[7] CS_2 and CSe_2,[8] and there may be a related but more complicated
process in butatriene.[9]

If the electron is removed from a degenerate orbital set
then the PE band intensity is usually higher than for a non-
degenerate orbital ionisation; when orbitals have similar
character (e.g. AO composition) then there is often an approximate proportionality between intensity and degeneracy.   Bands
due to ionisation from degenerate orbitals often show splittings
which can arise from the operation of either the Jahn-Teller
effect or spin-orbit coupling.   As a simple example of spin-
orbit coupling, the He-I spectra of the inert gases Ar-Xe are all
doublets (Figure 1).   The configuration of the ion produced is
... np^5, and coupling of orbital and spin angular momentum in
the np electron hole gives states $^2P_{1/2}$ and $^2P_{3/2}$, which have
degeneracies in the ratio 2:1.   As the nuclear charge increases
down the group, so does the spin-orbit coupling, and the energy
separation of the states increases from less than 0.2 eV for Ar
to 1.3 eV for Xe.   In very small molecules we normally have only
2-fold orbital degeneracy, e.g. π orbitals of linear molecules,
so the effect of spin-orbit coupling is to split bands representing ionisation of degenerate sub-shells into two approximately
equal components (e.g. Figure 2).   In cubic symmetry, however,
we can have 3-fold degeneracy, and ^2T bands split into two
components with unequal intensity ratio (theoretically 2:1) can
sometimes be observed: see Figure 3.   The observation of
multiplet fine structure in polyatomic molecules usually depends
on the presence of a heavy atom,* and it is sometimes possible
to obtain useful information about the localisation properties
of degenerate MOs from the magnitude of the splitting.   However,
in all but very small molecules, one needs rather heavy atoms in
order to resolve multiplet splittings: thus, although one
commonly observes splittings for bromides it is unusual to observe
such fine structure in the PE spectra of chlorides.

According to the Jahn-Teller theorem, which is discussed in
Professor Stephens' lectures, there must be some distortion of
molecular geometry whenever an electron is removed from a degenerate set of orbitals.   In practice band splittings or asymmetry
which can be ascribed to this effect are usually observable only
in the spectra of hydrides (e.g. the first band of CH_4 and the
second band of NH_3 in Figure 4).   In non-hydrides it is very
rare to observe any effects of this type, though the spectrum

---

* Quite small multiplet splittings (e.g. ca. 120 cm^{-1} in the
ground state of O_2^+) may however be observed in the case
of very simple molecules where the associated vibrational fine
structure is fully resolved. (cf. AFOI, Figure 10a).

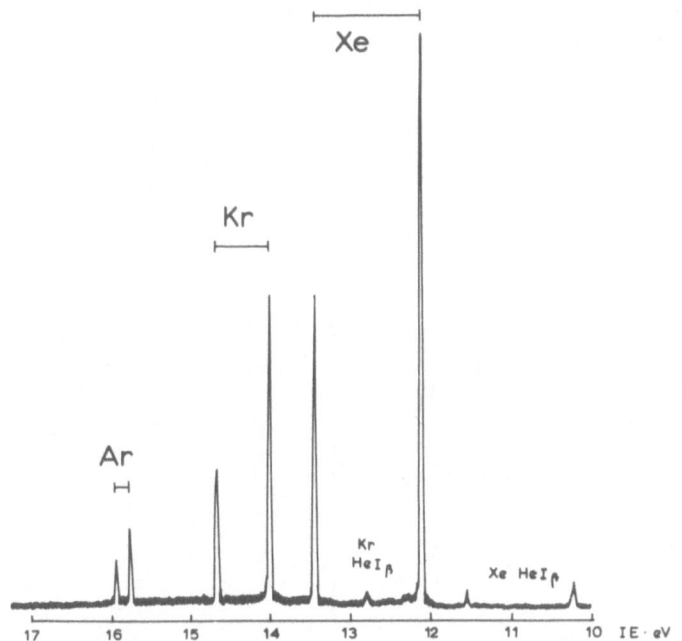

Figure 1.   He-I spectrum of an inert gas mixture.

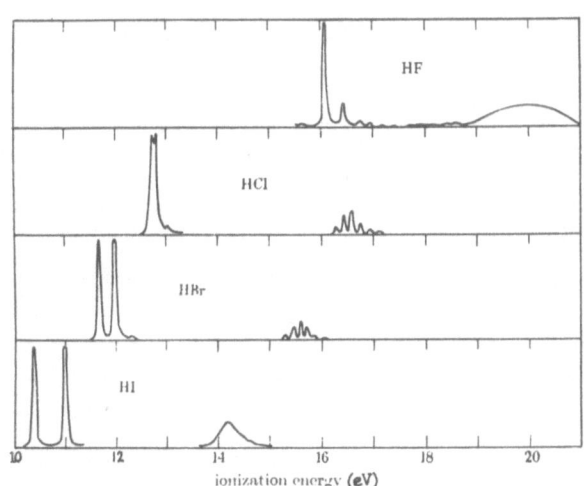

Figure 2.   He-I spectra of hydrogen halides (reference 3).
(From H.J. Lempka, T.R. Passmore and W.C. Price, Proc. Roy. Soc.
A 304, 53 (1968.)

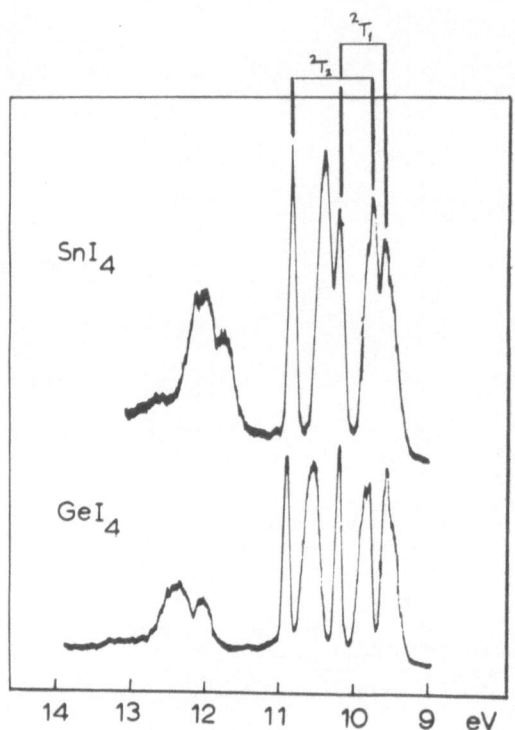

Figure 3. He-I spectra of SnI_4 and GeI_4.

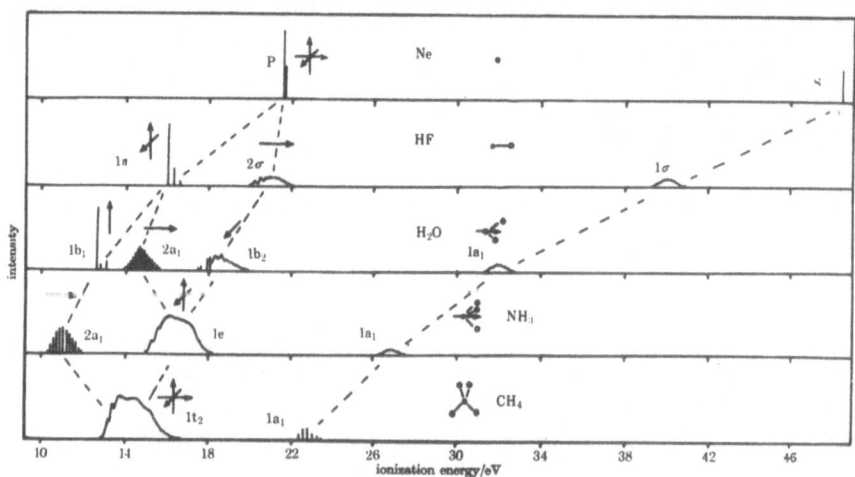

Figure 4. Diagrammatic photoelectron spectra of hydrides 'formed' by proton-withdrawal from neon. (From reference 3.)

of P_4 shows substantial band splittings which are assigned to the operation of the Jahn-Teller effect.[10]

A further source of band splitting arises in open shell molecules, from interaction of the unpaired spin(s) in the outer-most orbitals with the spin of the electron hole resulting from ionisation; the same phenomenon is observed in X-PES (see AFO I, Figure 24). An example is shown in Figure 5 where the $b^3\Pi$ and $A^1\Pi$ states both arise from ionisation of the same π orbital.

(b) Vibrational states

We also expect to observe structure on bands from vibrational excitation since in small molecules vibrational quanta are considerably greater than the instrumental resolution (at least for He-I work); 800 cm^{-1}, a fairly low vibrational frequency, corresponds to 0.1 eV. At room temperature most diatomic molecules have a high probability of being in their ground vibrational states; only if $h\nu$ (ν is the molecular vibration frequency) becomes comparable with kT will there be appreciable population of the higher vibrational states. Therefore in the ionisation process we can concentrate on the vibrations of the ion, and we can consider the probabilities of exciting particular vibrational states according to the Franck-Condon principle, exactly as for electronic transitions of neutral molecules (Figure 6). The extent to which the various vibrational states of the ion are excited depends on the displacement of the potential curves: if there is no displacement there is a high probability of exciting the ground vibrational state of the ion and very low probability for the others; if there is a substantial displace-ment then there will be an increasing probability up to some point (v = 3 on the second ionic state of Figure 6) and then a decreasing probability for the higher vibrational states.

The shapes of the PE bands resulting from the transitions to vibrational states of the ion are shown at the right of Figure 6. Removal of an electron which leaves the molecular geometry un-changed, i.e. a non-bonding electron, gives a sharp PE band, whose energy corresponds to that required to produce the ion in the ground vibrational state, the so-called adiabatic IE. Removal of a bonding (or antibonding) electron produces a band with a broad envelope. The first vibrational component is still the adiabatic IE, but we can also define the vertical IE as the most probable transition. This vertical IE is the appropriate quantity to compare with calculated orbital energies; if the vibrational structure is not resolved a vertical IE is measured as the band maximum, and a measure of bonding character of the orbital can be inferred from the width of the band. Where structure is resolved, direct information on bonding character is available from the change in vibrational frequency from that in the molecule.

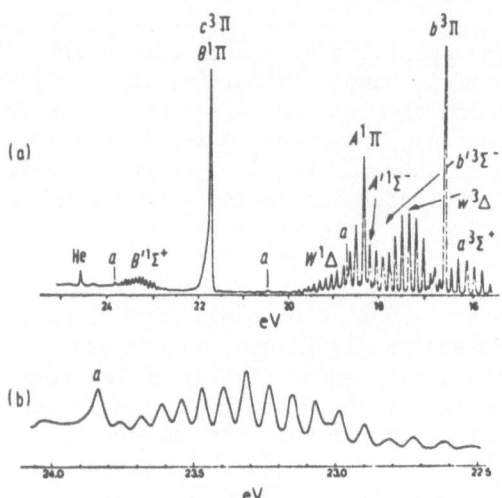

Figure 5.   He-II spectrum of NO: (a) survey;   (b) B' $^1\Sigma^+$
region.   (From O. Edqvist et al., Z. Naturforsch. 26a, 1407
(1971).)

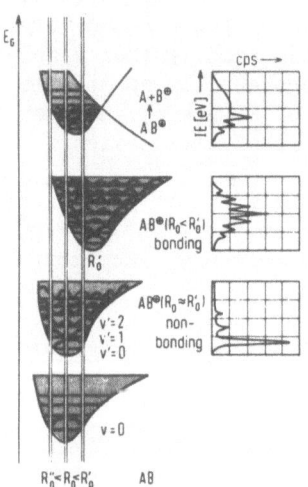

Figure 6.   Vertical transitions between the vibrational ground
state, v=0, of a diatomic neutral molecule AB and vibrational
levels in different types of radical cation states AB$^+$ (from H.
Bock and B.G. Ramsey, Angew. Chem. Internat. Edit. 1973, 12, 734).

However, some reservations about the idea of bonding character
will be mentioned below.

For high temperature studies kT is quite likely to approach
hν and the PE bands will be complicated by ionisations from
vibrationally excited molecules, so-called 'hot bands'.   Some
idea of the problem can be obtained from a room temperature study
of I_2, where ν is only 213 cm^{-1}, so that more than 30% of the
molecules are in the first vibrationally excited state, with
smaller but significant amounts in higher levels.   Figure 7 shows
the first band.   From temperature variation it can be shown that
the adiabatic IE corresponds to the peak just above 9.3 eV and
that all the lower IE bands are hot bands.   The vibrational
frequency changes only slightly in the ion (to 230 cm^{-1}) so the
various sets of transitions are almost coincident above 9.3 eV,
but even so there is some loss of resolution from this change.[4]
If the shift in frequency were much larger then the structure
would be hardly resolvable, and situations of this type can be
expected at high temperatures with many molecules.

In polyatomic molecules vibrational structure is much less
commonly observed than for diatomics.   There are several possible
reasons for this: more complex molecules have many possible
vibrations and if several are excited the overlapping of progressions would prevent resolution of the separate components.   In
polyatomic molecules there are often several low-frequency bending
vibrations and hot bands due to population of these levels could
also complicate the spectrum.   Even in diatomic molecules,
vibrational structure is not always observed;   if the ionic state
interacts with another state which is dissociative (see Figure 6)
then structure can be lost.

(c) Rotational

Rotational fine structure has been resolved in only one
special case, that of the H_2 spectrum.[11]   Here the rotational
spacings, of a few meV, cannot be observed with He radiation
because of the effects of thermal broadening, but by reducing the
electron kinetic energy by using Ne radiation the individual
rotational lines have been observed (AFO I, Figure 10b), but it
seems very unlikely that this will be repeated for other molecules.
In most molecules the only observable effect of rotational
transitions is a broadening of the vibrational lines, since many
rotational states of the molecule are populated at room temperature.   At high temperatures it is quite possible that the
extension of the rotational populations will mean complete overlapping between one vibrational transition and the next, and this
may be the most serious limitation on resolution at high
temperatures.[12]

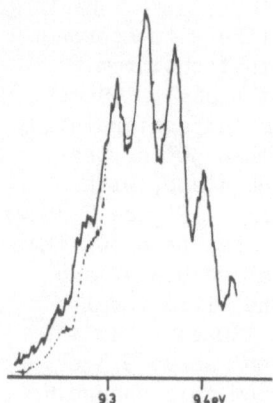

Figure 7.   The first band in the Ne I photoelectron spectrum of
I_2, $^2\Pi\frac{3}{2}g$.   The solid line is the spectrum at $30^{\circ}C$, and the dotted
line is that obtained at $-10^{\circ}C$.

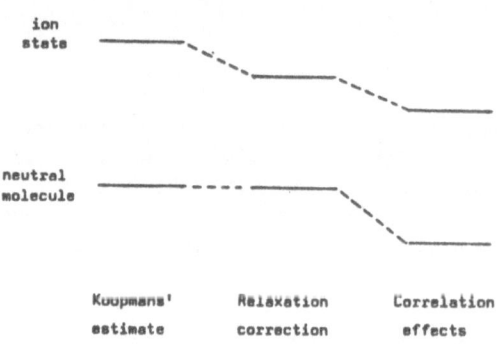

Figure 8.   The calculation of molecular ionisation energies.
Koopmans' theorem: $I_k = - E_k^{SCF}$.   Strictly, PES yields the
relative energies, $I_k - I_1$, of the various states of the molecular
ion that happen to be accessible.

Koopmans' theorem

When we have measured a PE spectrum we are concerned to compare it with theory, and it is commonplace to assume Koopmans' theorem, i.e. to equate orbital energies with (-) the ionisation energies. However, this assumes that the remainder of the electrons are unaffected by the loss of the ionized electron, and this is clearly not correct. As an electron is removed the other electrons will 'relax' to a new charge distribution: this leads to a reduction of the IE. It is possible to calculate ionisation energies corrected for this effect by performing separate SCF calculations on the ion states: the difference between these calculated IE and those from Koopmans' theorem is often called the relaxation energy. Relativistic effects are generally important only for core electrons, but correlation of the electron motions also needs to be considered, and the effect of this is usually greater on the molecule than on the ion, so in many cases there is an approximate cancellation of the two main corrections to the theorem (Figure 8). A calculation of these effects is shown for the Ne atom in Table 1. In many cases agreement between (-) the experimental IEs and the SCF eigenvalues is quite good, and in the majority of molecules it seems that at least the sequence of IEs is the same as the sequence of orbital energies. There are some cases however in which these sequences are not the same. The molecule N_2 is discussed below, and recent work on pyridine[13] shows that the N lone pair orbital lies nearly 1 eV below the highest filled π orbital but that the lone pair has the lowest IE. More serious from the point of view of inorganic chemistry is the annoying behaviour of d electrons in metal

Table 1. Calculation of the ionisation energies of the neon atom by SCF methods (Verhaegen et al., Chem. Phys. Letters, 9, 479.

Method of calculation	IE/eV	
	1s	2s
Koopmans' theorem (HF theory)	891.7	52.5
Direct method:		
HF theory	868.6	49.3
Relativistic HF theory	869.4	49.4
Relativistic HF method + correlation effects	870.8	48.3
Experimental values	870.2	48.4

complexes; according to the calculations by the groups of
Veillard[14],[15] and of Hillier[16],[17] the relaxation energy for these
electrons can be as high as 5 eV while for ligand electrons the
relaxation energy is only about 1 eV.   This leads to substantial
differences between the sequence of orbitals and the sequence of
IEs for these compounds.   On the other hand, for the majority of
main-group compounds and for the ligand levels of the metal
complexes, the agreement between - IE and calculated SCF eigenvalue is generally good.   However, the theorem only applies to
closed-shell molecules, at least in its simple form.

II.   SPECTRA OF MAIN GROUP MOLECULES

Diatomic molecules

As a very simple case we can take the He-I spectrum of H_2
shown in Figure 9. · There are only two electrons, paired in a
strongly bonding orbital, and ionisation of one of them gives a
substantial change in the internuclear distance so that a long
vibrational progression is observed which can be followed almost
to the dissociation limit of H_2^+.   Substantial contraction of the
spacing of adjacent vibrational elements occurs through the
progression, and a $\Delta G_{1/2}$ plot is linear (Figure 10).   Such
deviation from harmonic behaviour is frequently within experimental error for larger molecules, and vibrational spacings are
often measured as averages over the band.

The hydrogen halides are only slightly more complicated;   in
addition to a σ bonding orbital there are electrons in π non-
bonding orbitals, essentially unchanged halogen p orbitals.   The
spectra (Figure 2) show sharp bands at low IE due to the ionisations of the π orbitals and broad bands at higher IE from the σ
bonding ionisation.   The doubly degenerate $^2\Pi$ ionic state is
split by spin-orbit coupling, and as the nuclear charge increases
down the Group the energy separation of the multiplet components
$^2\Pi_{1/2}$ and $^2\Pi_{3/2}$ increases;   that in HF is only $\sim$ 30 meV[18] and is
not resolved in Figure 2.   For comparison later with the thallium
halides, it is to be emphasised that these   bands are sharp, i.e.
'typically bon-bonding'.   The IE of each band decreases down the
group, reflecting the falling electronegativity of the halogen
atom (see below).

The He-I spectrum of N_2 is shown in Figure 9 (AFO,I) and a
complete He-II spectrum using filtered radiation appears here in
Figure 11.   From a comparison with optical spectroscopy data it
is known that the first three bands, in order of increasing IE,
correspond to ionisation from the highest lying $\sigma_g(2p)$, π_u and
$\sigma_u(2s)$ levels.   From the band shapes it is clear that only π_u has
appreciable bonding character(and this is confirmed by examination

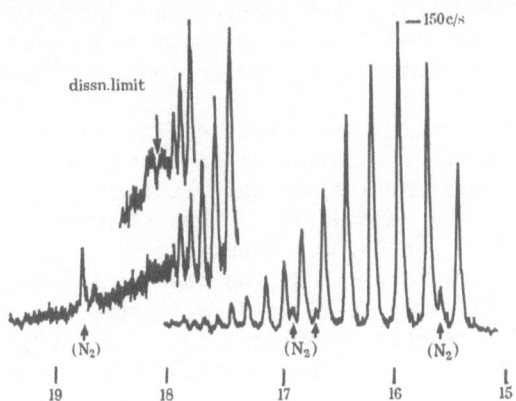

Figure 9.   He-I spectrum of H_2.   (From reference 6.)

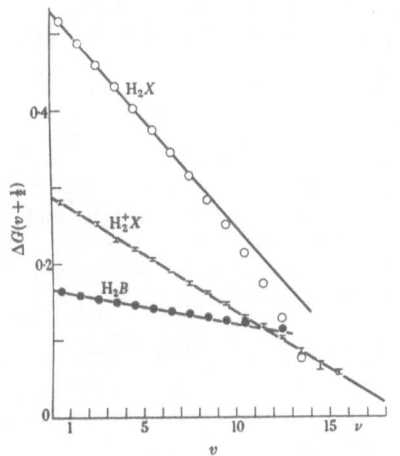

Figure 10. $\Delta G\ (v+\tfrac{1}{2})$ curves for H_2^+, H_2, $X^1\Sigma_g^+$ and H_2, $B^1\Sigma_u^+$. (Herzberg 1959.)

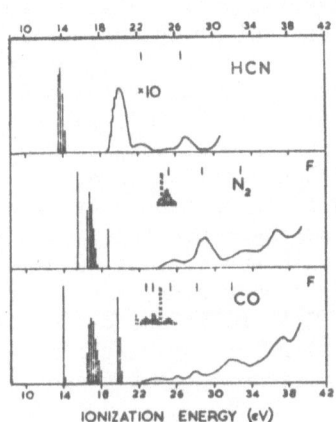

Figure 11.   He-II spectra of N_2, CO and HCN (from reference 6).

of the vibrational intervals). The non-bonding character of
σ_u and σ_g is not that expected from the elementary MO picture,
which assigns bonding character to σ_g and antibonding character
to σ_u, if σ_g is assumed to have pure 2p and σ_u pure 2s composition. However, if s-p mixing is introduced the orbital energy
of σ_g is raised and the bonding character is decreased. Although
these conclusions are substantiated by calculations, the highest
lying occupied orbital is π_u and not σ_g; thus Koopmans'
approximation breaks down for N_2.[19] When we come to look for
the only other valence orbital, $\sigma_g(2s)$, we need either X-PE
(Figure 11, AFO,I) or filtered He-II spectra, Figure 11, and in
both cases the spectra are quite complicated. The $\sigma_g(2s)$ level
lies very deep at an IE of 37 eV, substantially lower than the
2s level of the N atom. The stabilisation is presumably the
result both of s-p mixing and of a strongly N-N bonding character,
but resolution in the spectra is insufficient to observe any
vibrational structure. In addition to the $\sigma_g(2s)$ level ionisation, there are a number of other ionic states which must be
assigned as 'shake-up' states. The true $\sigma_g(2s)$ ionisation can
be picked out by the very substantial increase in intensity on
changing from the He-II to the X-PE spectrum. Comparison of
UV-PES with valence X-PES can often be illuminating - in
particular ionisations from orbitals of mainly 2s character are
usually weak in UV-PE spectra but very strong in X-PE spectra.
Other examples of this can be seen in Figures 16 and 17 (AFO,I)
and 21 (AFO,I). The correlation between N_2 and CO is straightforward: $\sigma_g(2p)$ becomes 5σ of CO, mainly localised on carbon,
and $\sigma_u(2s)$ becomes 4σ, mainly localised on oxygen. The PE
spectra show that the main bonding character in CO is still in
the π orbital, though 4σ has some bonding character. The IE
has decreased from 15.7 eV in N_2 to 14.0 eV in CO, reflecting the
smaller nuclear charge, or lower electronegativity, of C compared
to N. The high IE region of the CO UV-PE spectrum is similar to
that of N_2, with several shake-up processes evident. The 3σ
level, related to $\sigma_g(2s)$ of N_2, again has a very high IE. The
He-I spectrum of CS is shown in Figure 21 (AFO,I), and the first
three bands relate directly to the first three of CO, though they
are all at substantially lower IE. Orbitals 1π and 4σ are now
both strongly bonding. The band at 18 eV can only be assigned
as a shake-up process, since the 3σ orbital, with mainly C 2s
and S 3s character, is not likely to have an IE of less than 21
eV. Some relations between the PE spectra of N_2, CO and CS,
and their bonding capabilities toward transition metals, have
been discussed briefly elsewhere.[20]

The molecule NO has one more electron than CO, and this
occupies the high-lying antibonding 2π orbital, so the IE of NO
is quite low, 9.3 eV in fact. The first band in the PE spectrum
shows a simple vibrational progression with an increase in the
spacing as compared with that in the molecule. The spectrum in

Figure 5 lacks this first band, while the remaining spectrum is clearly more complex than that of N_2 or CO.   When an electron from 5δ is ionised, coupling of the two unpaired electrons in the ion, one in 5δ and the other in 2π, gives two electronic states $b^3\Pi$ and $A^1\Pi$, and two sharp bands are observed in the spectrum with intensity ratio 3:1.   The sharpness indicates that 5δ is non-bonding in NO just as in CO.   Similarly ionisation from 4δ gives states $c^3\Pi$ and $B^1\Pi$, and again the corresponding bands are sharp though the energy separation is very small.   Ionisation from the 1π orbital gives singlet and triplet spin states, but also there are three possibilities for the coupling of the orbital angular momenta, so that in all there are six electronic states from this ionisation.   Very careful high resolution work has identified all these states.[21]

    The diatomic halogen molecules are related to N_2, with four more electrons filling the antibonding π_g orbitals.   All halogen s ionisations are expected beyond 21 eV, and the He-I spectra should show bands from ionisation of π_g, π_u and $\delta_g(np)$.   Since s-p mixing is expected to be less in the halogens than in N_2 it is reasonable that σ_g should appear at higher IE than π_u.   The spectra are shown in Figure 12.   As in the halogen acids, the effect of spin-orbit coupling increases substantially down the group: this is seen most directly in the structure of the first ionisations in the four molecules (right hand side of Figure 12); in F_2 the spin-orbit splitting is less than the vibrational quantum so each vibrational component appears to be split into two, while in Br_2 and I_2 the spin-orbit splitting is large enough that two distinct Franck-Condon envelopes are seen for the split states $^2\Pi_{3/2g}$, $^2\Pi_{1/2g}$.   In Cl_2 there is an accidental coincidence of the vibrational interval and the spin-orbit splitting so that only a single envelope can be seen.   Hot bands can be detected in Cl_2 and Br_2, while in I_2 they are quite strong (see above and Figure 7).   The remaining bands in the He-I spectra are associated with ionisation of the π_u and δ_g (np) subshells.   The π_u ionisations, which occur at the lower energy, may be distinguished by virtue of the multiplet splittings observed in the Br_2 and I_2 spectra and also, albeit less reliably, by the relative intensities of the bands.   No vibrational structure has been resolved for either the π_u or δ_g bands, probably because the life-times of the ion states concerned are too short, but the breadth of the bands supports the idea that both π_u and δ_g (np) are important bonding orbitals.   Unfortunately, there have as yet been no He-II or X-PES studies to determine the position of the δ_g (ns) and δ_u (ns) levels.

    Figure 13 shows the spectra of two diatomic interhalogens, ICl and IBr.   A point of considerable interest is the extent of the spin-orbit splitting.   In both molecules the splitting of 2π is close to that of iodine, and the splitting of 1π is close to

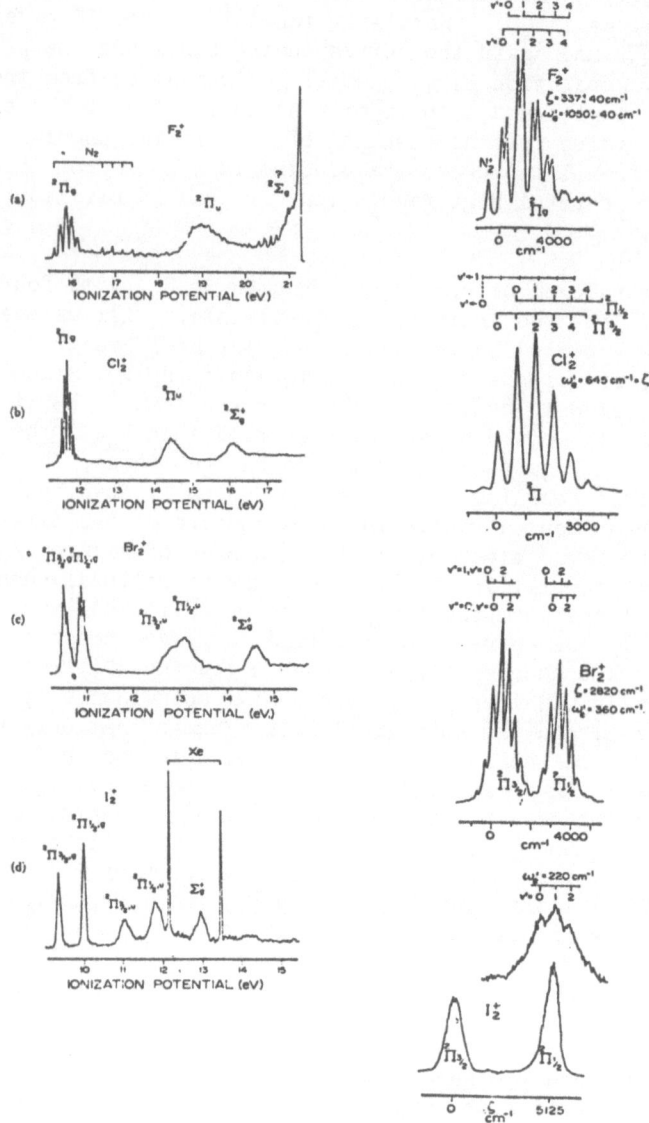

Figure 12.   He–I spectra of (a) F_2, (b) Cl_2, (c) Br_2, (d) I_2.

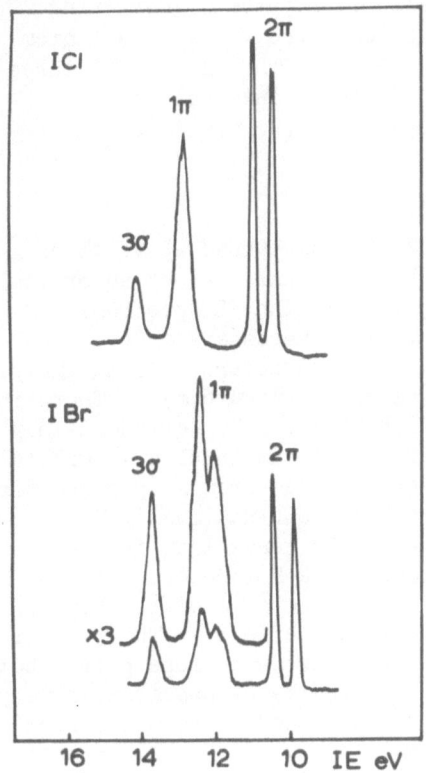

Figure 13.   He–I spectrum of ICl and IBr.   (S. Evans and A.F. Orchard, Inorg.Chim Acta 5, 81 (1971).)

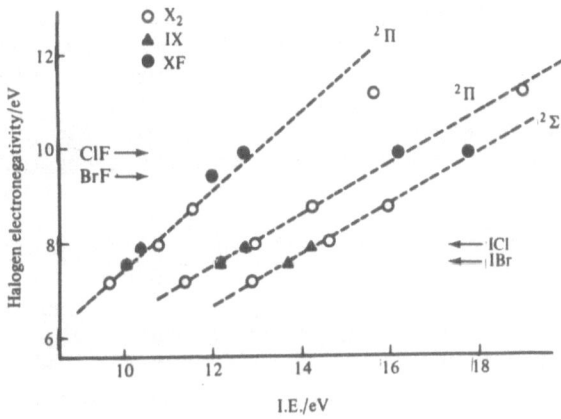

Figure 14.   Halogen electronegativity and ionization energies of interhalogens.

that of the other halogen (not detected in ICl). This gives an
indication of the orbital localisation: 2π, the higher lying
orbital, is mainly localised on the less electronegative I atom,
as expected from elementary MO arguments. The IEs of all three
outer orbitals in the halogens and diatomic interhalogens
correlate very well with electronegativity (Figure 14).

Polyatomic molecules: hydrides

For the covalent diatomic molecules discussed above there is
an immediate relationship between bonding character of an orbital
and the shape of the corresponding band in the PE spectrum.
With polyatomic molecules, however, it is necessary to be more
careful about the definition of bonding character. As an example
we can consider the PE spectrum of ammonia (Figure 4). There is
a strongly bonding orbital* $1a_1$, with IE ~ 27 eV, derived mainly
from the N 2s and a totally symmetric combination of H 1s orbitals,
and a degenerate pair 1e, IE ~ 17 eV, derived from N 2p perpendicular to the C_3 axis and appropriate H 1s combinations. These
three orbitals are strongly bonding in the sense that the N-H
distance is expected to increase, and the N-H stretching force
constant to decrease, upon ionisation. The asymmetric shape of
the 1e band is ascribed to the Jahn-Teller effect. The remaining
orbital in NH_3, $2a_1$, is usually referred to as a 'lone pair', but
the envelope of the corresponding band in the PE spectrum is very
broad, apparently indicating bonding character. However, from
the vibrational interval of the fine structure it is clear that
the vibration excited is ν_2, the 'umbrella' bending mode. In
terms of current ideas about determination of bond angles by lone
pairs this is of course quite reasonable; in fact removal of one
electron from the lone pair $2a_1$ produces a planar ion. Even in
the heavier Group V hydrides which have bond angles close to 90°
the ground ionic state is almost planar.[3] This clearly introduces a complication into ideas about relating bonding character
to PE band width; it is always possible in polyatomic molecules
that removal of an electron may set up bending rather than
stretching frequencies, and where structure is not resolved it is
not certain that a broad band can be used to deduce a change in
bond length on ionisation.

The PE spectrum of water (Figures 4,15) shows some similarities to that of ammonia. There are two strongly O-H bonding
orbitals, $1a_1$ with IE ~ 32 eV, derived mainly from O 2s, and $1b_2$,
IE ~ 18 eV, derived from in-plane O 2p, perpendicular to the
angle bisector. Ionisation from $1b_2$ excites mainly the O-H
stretch, though there are some complications.[3] The remaining
orbitals are 'lone pairs'. It should be noted that in terms of

---

* Orbital numberings ignore the core electrons.

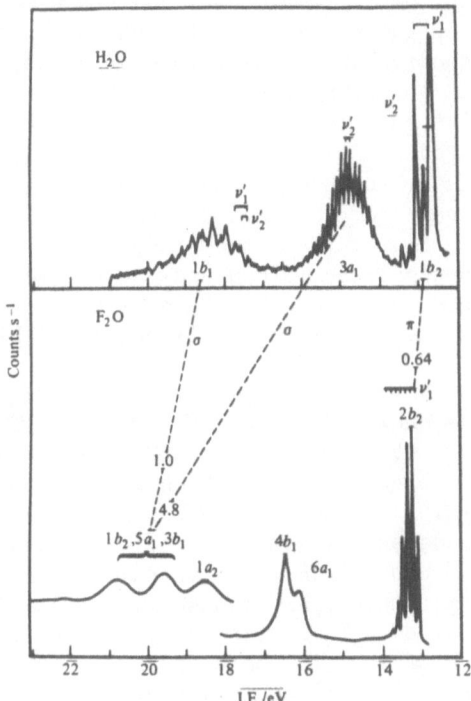

Figure 15.   He-I spectra of H_2O and F_2O.   (C.R. Brundle et al., J. Amer.Chem. Soc. 94, 1451 (1972).)   (The labels b_1 and b_2 are interchanged.)

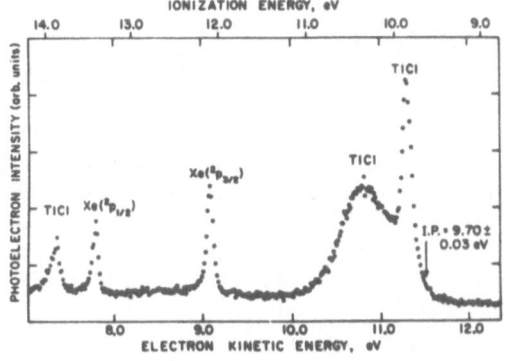

Figure 16.   He-I spectrum of TlCl (reference 12(a)).

delocalised molecular orbitals the two lone pairs in H_2O are not
equivalent.   One lies along the angle bisector and has some O
2s as well as O 2p character, and has a_1 symmetry,while the other
is a pure O 2p orbital perpendicular to the molecular plane, of
b_1 symmetry.   Ionisation of $1b_1$ gives a very sharp band with
almost no vibrational excitation;   thus it leaves the molecular
geometry essentially unchanged.   However, ionisation of $2a_1$
gives a long progression in ν_2, the H-O-H bending frequency.   In
ammonia removal of an electron from $2a_1$ gave a planar ion;   in
water removal of an electron from $2a_1$ gives a linear ion.[3]
However the corresponding ionisations in H_2S and the heavier
hydrides in Group VI give bent vibrational ground states, and the
PE bands are complicated by the transition to linearity at high
vibrational excitation.[3]

The spectrum of methane shows two bands (Figure 4).   The
weaker one at an IE of ca. 23 eV shows a simple progression and
is assigned to $1a_1$, strongly bonding between C 2s and a symmetric
combination of H 1s orbitals.   The more intense band at 14 eV is
assigned to $1t_2$, the triply degenerate set of bonding orbitals
derived from C 2p orbitals.   Unlike $1a_1$, the vibrational structure of this PE band is very complex, and two subsidiary maxima
can be seen on the high IE side.   This complexity is due to the
very substantial distortions of the $^2T_2$ state of the ion by Jahn-
Teller forces, and the vibronic detail has not been fully
analysed.[3]

Several points of interest arise from the correlation $Ne-CH_4$
shown in Figure 4.   The s-type orbital rises in energy very
steeply through this series as a result of changing nuclear
charge, but the effect on the p-type orbitals is less.   The lone
pair orbitals successively join the bonding orbitals through the
series, and the stabilisations from this change to bonding from
non-bonding are more than enough to counteract the destabilising
effect of changing the nuclear charge on the central atom.   A
point of direct relevance to other studies is the comparative
weakness of PE bands associated with the 2s-derived orbitals;
generally in UV-PES the intensities of orbitals with substantial
s character are low compared to those with p character, while in
X-PES the situation is reversed (see above).   Finally, although
the orbital discussion above has described the molecular orbitals
in terms of their constituent atomic orbitals, an equivalent
description is possible in terms of combinations of bond orbitals.
The four bonding MOs of CH_4 ($1a_1 + 1t_2$) can be built up from
combinations of the four bonds, the three MOs ($1a_1 + 1e$) of NH_3
from the three bonds, and $1a_1$ and $1b_2$ of H_2O from two bond
orbitals.   Although there is no particular advantage in using
such a description for these very simple molecules, this approach
can be a useful simplification for larger molecules.

Ionic systems

Because of volatility problems very few systems with
substantial ionic character have been examined so far, but
Berkowitz and co-workers have reported the spectra of the
thallium(I) halides.[12]   An example, TlCl, is shown in Figure 16.
Three bands are observed.   By analogy with the hydrogen halides
a π non-bonding pair of orbitals and a δ bonding orbital are
expected, and in TlX compounds there will also be a δ ionisation
from the Tl $6s^2$ 'inert pair' of electrons.   The weak band at an
IE of 13.8 eV corresponds to the δ bonding orbital, though the
band is not very broad, suggesting that there is not very much
bonding character.   Of the two bands around 10 eV, one is broad
and the other sharp.   By analogy with the covalent hydrogen
halides, ionisation from the π orbital would be expected to give
a sharp band.   However, the interpretation proposed by Berkowitz
assumes that the TlX molecules are predominantly ionic.   Removal
of an electron from an orbital localised on X^- in Tl^+X^- produces
a molecular ion Tl^+X in which the ionic bond has been destroyed,
so a substantial increase in internuclear distance, and a correspondingly broad PE band, are expected.   Calculations on Al^+Cl^-
indicate that the Al^+Cl state in which an electron is removed
from the Cl π orbitals has a potential curve with no minimum, i.e.
which is dissociative.   Thus, although the ionisation of Cl π
levels in HCl produces a sharp band, ionisation of Cl π levels in
TlCl gives a broad band because of the ionic character of TlCl.
In contrast ionisation from a Tl localised orbital gives a
molecular ion $Tl^{2+}Cl^-$ in which the electrostatic binding is
increased.   If the repulsive potential is rising steeply at the
equilibrium distance in Tl^+Cl^- then production of $Tl^{2+}Cl^-$ will
only shorten the distance slightly, and it is quite reasonable
therefore that ionisation of the Tl 'inert pair' should give a
fairly sharp PE band.   Support for this interpretation of the
TlCl spectrum rather than one in which the sharp band is assigned
to halogen π levels is provided by the alkali halide monomer
spectra in which ionisations from the halogen levels, the only
ones observed in He-I spectra, are all broad.*   Thus, in molecules
with a considerable ionic character, broad PE bands can be
observed on ionisation of electrons in orbitals which from a
covalent viewpoint are completely non-bonding.

Polyatomic molecules: symmetric halides

The halides of Groups II, III, IV, V and VI provide a number
of highly symmetric molecules of reasonable volatility for

---

*These spectra were presented by Professor W.C. Price during the
symposium after the Summer School.

investigation.   In several cases the bonding orbitals in these
molecules can be directly related to those in the corresponding
hydrides or alkyls.   These bonding orbitals can be directly
related to more conventional localised bond orbital descriptions.
In the halides, there are substantial interactions between the
orbitals usually described as halogen lone pairs, and the interactions are sufficient to give easily resolvable detail in the
PE spectra.

AB$_2$ molecules

In a linear AB$_2$ molecule there are two bonding orbitals,
δ_g with the symmetry of the s orbital of A, and δ_u with the
symmetry of the axial p orbital of A.   This corresponds approximately to the 'sp hybrid' description.   In the He-I PE spectrum
of dimethyl mercury (Figure 17) the ionisations from these two
orbitals are clearly seen as the two bands at lowest IE.   The
ionisation region 13 eV-15 eV is due to orbitals localised mainly
on the methyl groups, and the bands to higher IE are due to the
ionisation of the Hg inner 5d^{10} shell.   Since 6s of Hg lies
several eV below 6p it is reasonable to assign the 9 eV band as
δ_u and the 11.5 eV band as δ_g.   In support of this, the 11.5 eV
band has appreciably lower intensity, as expected from the greater
s character.   In HgCl$_2$ the corresponding ionisations may be
picked out at about 12.5 eV and 13.7 eV (Figure 17), and again
the δ_g ionisation is the weaker.   The shift in IE from the
methyl to the chloride reflects the increase of atomic IE between
C 2p and Cl 3p.   The Hg 5d^{10} ionisations occur beyond 16 eV, but
at low IE there are two bands of approximately twice the intensity
of the δ_u ionisation.   These result from ionisation of the Cl
lone pair orbitals, π_u and π_g.   These orbitals correspond to π_u
and π_g of the Cl$_2$ molecule, except that the energy separation is
too large for Cl-Cl interaction.   The π_u orbital has the correct
symmetry for interaction with the Hg 6p(π) orbitals and the
energy separation of π_u and π_g suggests that there is an appreciable π contribution to the bonding in addition to the δ bonds.
Interestingly, π_g shows a spin-orbit splitting which is greater
than that in Cl$_2$.

The above description of the bonding in HgCl$_2$ may also be
used for a variety of other 16 valence electron molecules with
D$_{\infty h}$ symmetry, such as CO$_2$, CS$_2$, CSe$_2$, and, with appropriate
modifications, to C$_{\infty v}$ molecules such as N$_2$O and COS.[22]   However,
some of the valence electron IEs lie in the He-II region and
studies with this source have not yet been reported.   Triatomic
molecules with 17, 18, 19 and 20 electrons are bent;   useful
correlations can be made between orbitals in these molecules and
those in the linear systems.[23]   At 22 valence electrons, triatomic molecules revert to a linear form;   examples are XeF$_2$ and

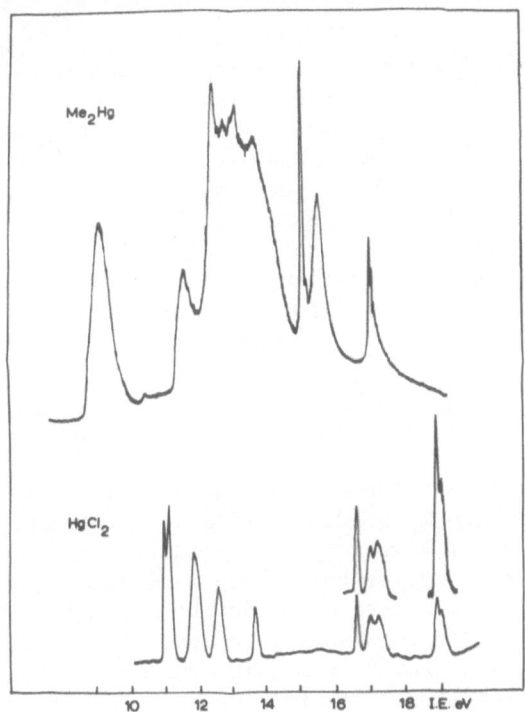

Figure 17.   He–I spectra of Me₂Hg and HgCl₂.   (S. Evans, A.F.
Orchard and N.V. Richardson, unpublished work.)

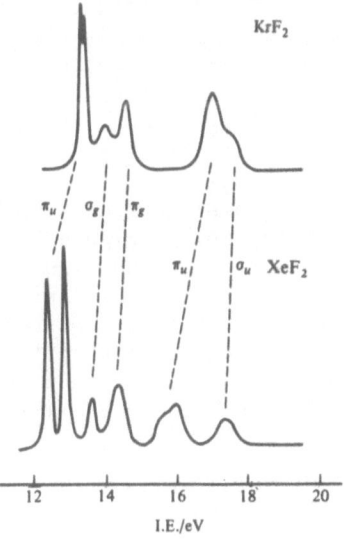

Figure 18.   He–I spectra of KrF₂ and XeF₂ (from reference 24).

KrF_2 whose PE spectra are shown in Figure 18. The lowest lying bonding orbital $1\delta_g$ has an IE beyond 21 eV, but δ_u, $1\pi_u$ and π_g, analogous to the orbitals of $HgCl_2$, are all accessible to He-I radiation. $1\pi_u$ has an appreciable heavy atom character, as is suggested by the shift to higher IE on changing from Xe to Kr, but π_g which can only have heavy atom character via the use of $\underline{d}$ orbitals, has an essentially constant IE. The additional six electrons are accommodated in $2\pi_u$ and $2\delta_g$. $2\pi_u$ is derived substantially from valence p orbitals of the heavy atom, as is shown by the spin-orbit splittings and the shift in IE between KrF_2 and XeF_2; the magnitudes of the splittings have been used to deduce percentage heavy atom characters for these orbitals in the two compounds.[24]

AB_4 molecules

The bonding orbitals of the tetrahedral molecules in Group IV are directly related to the a_1 and t_2 bonding orbitals in methane. The PE spectra of the chlorides in Figure 19 are typical. In CCl_4 the a_1 ionisation lies beyond 21 eV but the t_2 ionisation appears at about 17 eV; in the remaining chlorides both a_1 and t_2 are observed. The t_2 IE decreases steadily on descending Group IV, but the a_1 IE decreases from C to Si, increases slightly from Si to Ge, and decreases again from Ge to Sn. The same behaviour is found in the hydrides (Figure 20) and fluorides,[25] and in the atomic valence s IEs; it is probably related to the imperfect shielding of the nuclear charge by the $3d^{10}$ shell at Ge. In T_d symmetry the eight 'halogen lone pair' orbitals give rise to the orbital sets t_1, t_2 and e. The interaction between the halogen atoms is clearly substantial, since the t_1-e separation arises largely from 'ligand-ligand' interaction, and this separation amounts to about 2 eV in CCl_4. As the size of the halogen tetrahedron increases with increasing size of the central atom, so the t_1-e separation decreases. If there is any d-orbital participation in the bonding this will stabilise the e orbitals, but this effect is difficult to disentangle from the ligand-ligand interaction. In the tetramethyls (Figure 21) the t_2 M-C bonding orbitals have the lowest IE. The substantial M character of these orbitals is shown by the spin-orbit splitting in $Pb(CH_3)_4$. The corresponding a_1 ionisation can be picked out at high IE for the C, Si and Ge compounds, again with an irregularity at Ge, but for $Sn(CH_3)_4$ and $Pb(CH_3)_4$ this band is probably hidden by the ionisations from the methyl groups in the 13-16 eV region.

AB_6 molecules

The symmetry of the octahedron is well known in inorganic

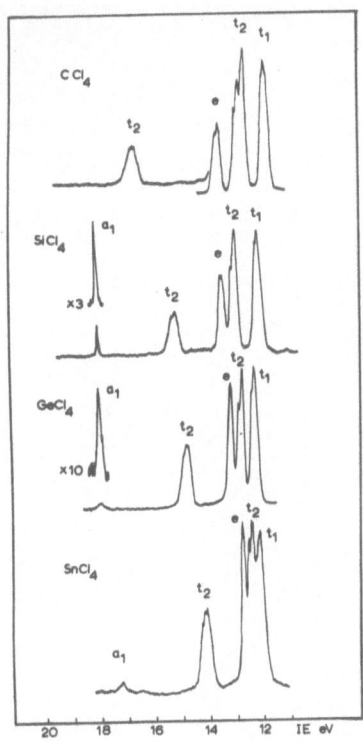

Figure 19. He-I spectra of Group IVB tetrachlorides. (From J.C. Green, M.L.H. Green, P.J. Joachim, A.F. Orchard and D.W. Turner, Phil. Trans. Roy. Soc. A268, 111 (1970).)

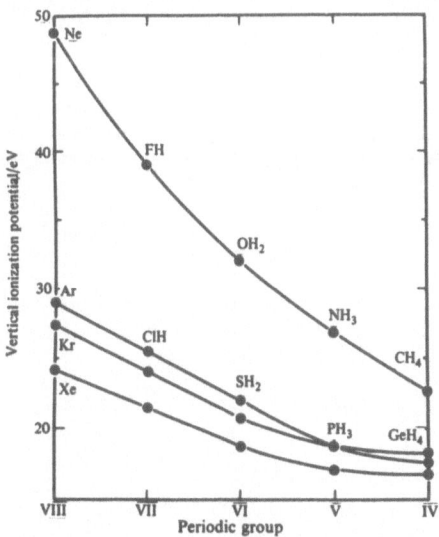

Figure 20. Ionisation energies of hydrides (reference 3).

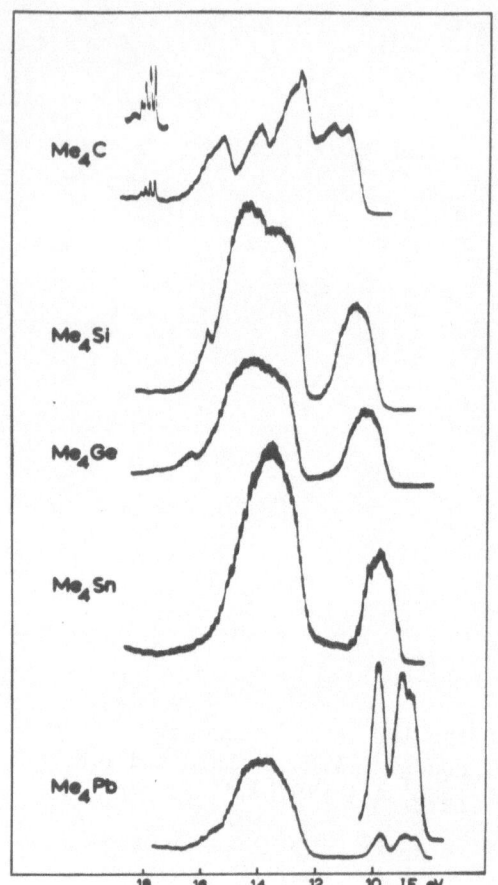

Figure 21.   He-I spectra of Group IVB tetramethyls.   (From
S. Evans, J.C. Green, P.J. Joachim, A.F. Orchard, D.W. Turner
and J.P. Maier, J. Chem. Soc., Faraday Trans. II, 68, 905 (1972).)

chemistry, but the only O_h molecules whose UV-PE spectra have been reported so far are the hexafluorides of Group VI shown in Figure 22. The bonding orbitals in an octahedron are the a_{1g} (s on central atom), $1t_{1u}$ (p on central atom) and e_g (d on central atom), corresponding to the hybrid description 'sp^{3}d^2'. The extent to which d orbitals are involved is by no means clear, but it is unlikely that e_g will be stabilised as much as a_{1g} and t_{1u}. The F lone pair orbitals are grouped in the four 'π' sets t_{2g}, t_{2u}, t_{1g} and t_{1u}. Although the expected seven bands are seen in the SF_6 spectrum, not all the details of the assignment are clear. However, it seems reasonable on grounds of energy and intensity that the 27 eV band corresponds to a_{1g} and the 23 eV band to $1t_{1u}$. In the X-PE spectrum the 27 eV band is much more intense, confirming the s character. Of the five bands at low IE the weakest is that at about 18 eV, and this is assigned to e_g. Of the remaining bands, that at about 20 eV is usually assigned to t_{2g} since this orbital is the most stable of four lone pair sets merely from F-F interaction, and any $d(\pi)$ orbital participation will give further stabilisation. The assignments of the remaining bands are disputed.[26] However, it is clear from Figure 22 that the pattern on descending Group VI is similar to that discussed above in the Group IV chlorides; the main bonding orbitals are progressively less stabilised, and as the size of the octahedron increases the spread of the π lone pair orbitals decreases. The IP increases from SF_6 to TeF_6, which is the opposite direction expected from electronegativity considerations; this is probably because of the loss in F-F antibonding interaction in the highest occupied level. A similar effect occurs with BrF_5 and IF_5.[27]

AB$_3$ molecules

The boron trihalide molecules are planar, while molecules such as the Group V trihalides with two more valence electrons are pyramidal; the effect is similar to the bending of the 18-electron triatomic molecules. The arguments for assignment follow those above and will only be outlined here: more detail can be found in references 26a and 28. The main bonding orbitals are a_1' (D_{3h}) or a_1 (C_{3v}) and e' or e. The halogen lone pairs form the set e' + e" + a_2' + a_2'' in D_{3h}, or 2e + a_1 + a_2 in C_{3v}. In the boron trihalides a_2'' is stabilised almost as much as the bonding orbitals, indicating substantial π bonding character.[26] The additional pair of electrons in the pyramidal molecules has the lowest IE and is broad for the same reasons as the lone pair in NH_3 is broad. A variety of correlations is shown in Figure 23.

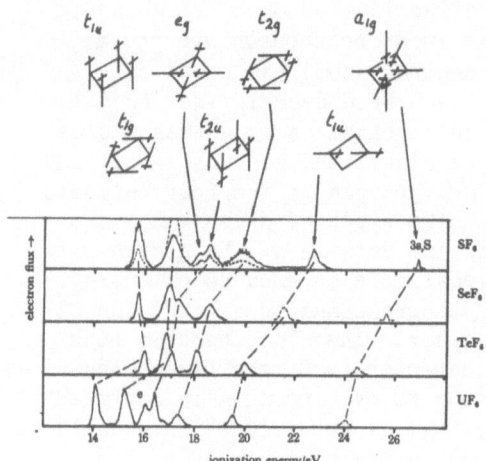

Figure 22.   Photoelectron spectra of the hexafluorides of S, Se,
Te and U (from reference 26(a)).

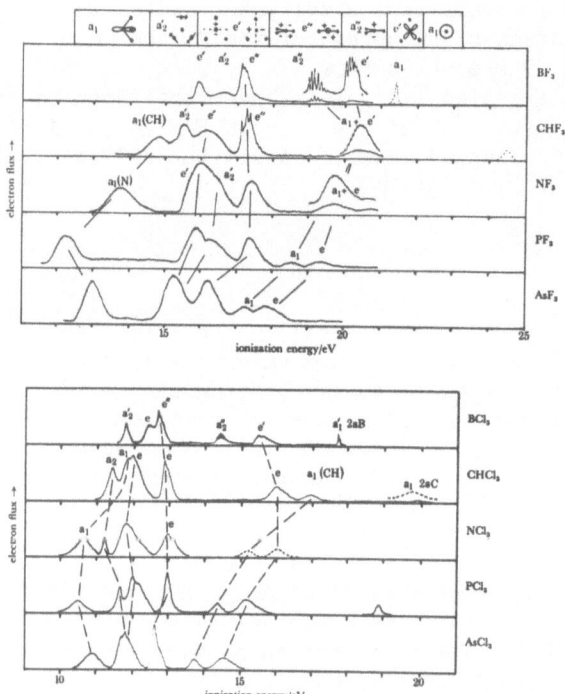

Figure 23.   (a) Photoelectron spectra of the trifluorides of B,
CH, N, P and As;   (b) photoelectron spectra of the trichlorides
of B, CH, N, P and As (from reference 26(a)).

Donor-acceptor interactions

Lone pair orbitals on all but the most electronegative atoms are capable of acting as donors to bind Lewis acids. Studies of the relationship between the PE spectra of free donor molecules and their complexes have been reported;[29,30] Figure 24 shows a typical result for trimethylamine and its complex with BH_3. The only band in the PE spectrum of the complex which can be attributed to BH_3 is band I, and the remainder of the spectrum can be correlated with that of the free base as shown. The major effect of complex formation is the strong stabilisation of band 1, the IE increasing by about 3 eV. The remaining bands shift by only about 1 eV, with very little change in their appearance. Similar results are obtained for the other methylamines both with BH_3 (Figure 25) and with BF_3;[30] the interpretation proposed is that the 1 eV stabilisation is effectively an electrostatic effect from the charge transfer involved in complex formation, while the lone pair stabilisation of 3 eV is mainly a covalent effect associated with formation of the 'donor bond', though there will also be an electrostatic influence on this. For NH_3 the separation of these two effects is not so clear cut, but agreement with ab initio calculations (Figure 26a) is good.

The interpretation of the spectra of $COBH_3$ and PF_3BH_3 (Figures 27,26) is less clear. The spectra of the complexes can be correlated with those of the free bases, and it is particularly noticeable that the second band in $COBH_3$ is broad whereas the corresponding first band in CO (Figure 11) is very sharp. However, there is much less of an increase in IE for the lone pair orbital of CO on complex formation. This may be partly due to 'back donation' to the Lewis base,[29] but an alternative interpretation is that the relaxation energy on ionisation is different for free and complexed CO:[31] as can be seen from Figure 26b the agreement between IEs and SCF eigenvalues is poor. Agreement for PF_3 and PF_3BH_3 is better (Figure 26c). These complexes are related to the transition metal complexes discussed in the next lecture (AFO,II).

The 'perfluoro-effect'

Brundle and co-workers[32] have reported the spectra of a large number of planar hydride molecules and of their fully fluorinated derivatives. As would be expected, the effect of fluorination is to stabilise the σ levels in the molecules very substantially. However the π levels in these planar molecules are much less stabilised. Thus, comparing H_2O and F_2O (Figure 15) the first IE increases by only about 0.5 eV. The first ionisation of H_2O is from the 'lone pair' orbital which is perpendicular to the molecular plane, i.e. from a π orbital.

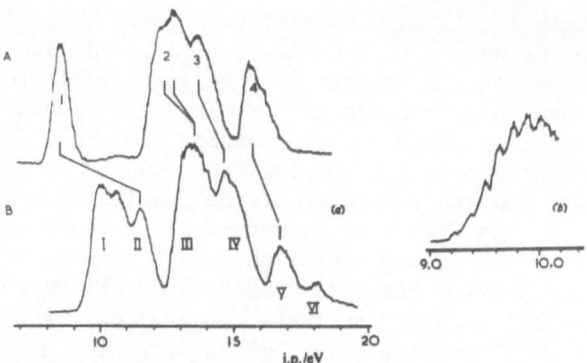

Figure 24.   (a) Photoelectron spectra of trimethylamine- and
trimethylamine-borane.   A, $(CH_3)_3N$;   B, $(CH_3)_3N,BH_3$;   (b) the
fine structure on the first band of the spectrum of $(CH_3)_3N,BH_3$
(reference 29).

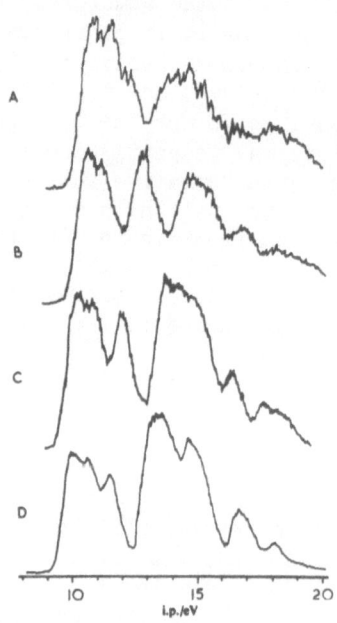

Figure 25.   Photoelectron spectra of ammonia-borane and the
methylamine-boranes: A, NH_3BH_3;   B, $CH_3NH_2BH_3$;   C, $(CH_3)_2NH,BH_3$;
D, $(CH_3)_3N,BH_3$ (reference 29).

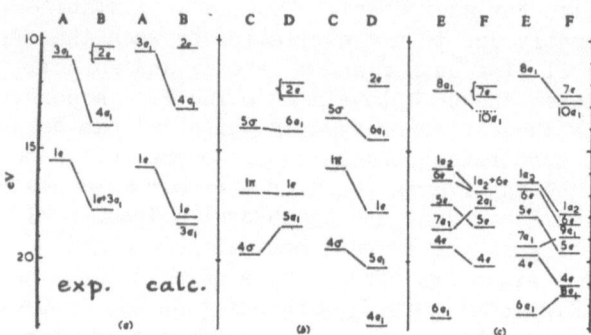

Figure 26. Comparison of experimental i.p. (exp) and the eigenvalues from ab initio calculations (calc) for borane complexes and for the free bases. The orbital numberings include the core electrons. A, NH$_3$; B, NH$_3$BH$_3$; C, CO; D, COBH$_3$; E, PF$_3$; F, PF$_3$BH$_3$ (reference 29).

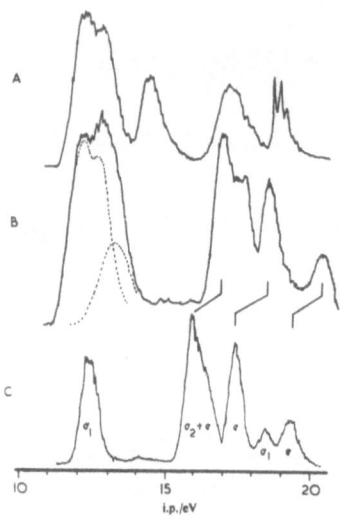

Figure 27. Photoelectron spectra of carbonyl borane, phosphorus trifluoride-borane and phosphorus trifluoride. A, COBH$_3$; B, PF$_3$BH$_3$; C, PF$_3$.

The effect seems to be quite general for planar molecules and can be a very useful criterion for assignment. The lack of stabilisation in F_2O is presumably due to a cancellation between the 'inductive effect', stabilising both σ and π levels, and an antibonding interaction between the b_2 π levels on O and F. A point to be emphasised here is the differing ways in which 'π' can be used by chemists. The spectroscopic meaning of course refers to angular momentum about the major axis (C_∞) of a linear molecule. For the highly symmetric tetrahedral and octahedral molecules 'σ' and 'π' are ways of distinguishing between orbitals along and perpendicular to the bond axis; in effect it is assumed that the local symmetry for the ligand atom is $C_{\infty v}$ rather than C_{3v} or C_{4v}. However, in planar molecules the usage of π is different and means antisymmetric with respect to reflection in the molecular plane, and the perfluoro effect is observed only with orbitals which are π in this sense; there is no comparable effect in non-planar molecules such as acetone.[32]

REFERENCES

1. J.H.D. Eland, 'Photoelectron Spectroscopy', Butterworths, London, 1974.

2. J.A.R. Samson, Rev. Sci. Instrum. 40, 1174 (1969).

3. A.W. Potts and W.C. Price, Proc. Roy. Soc. A, 326, 181 (1972).

4. B.R. Higginson, D.R. Lloyd and P.J. Roberts, Chem. Phys. Letters, 19, 480 (1973).

5. R.B. Cairns, H. Harrison and R.J. Schoen, Phil. Trans. A, 268, 163 (1970).

6. A.W. Potts and T.A. Williams, J. Electron Spectroscopy, 3, 3 (1974).

7. N. Jonathan, A. Morris, M. Okuda, K.J. Ross and D.J. Smith, Faraday Disc. Chem. Soc. 54, 48 (1972).

8. S. Cradock and W. Duncan, Mol. Phys. 27, 837 (1974).

9. F. Brogli, E. Heilbronner, E. Kloster-Jensen, A. Schmelzer, A.S. Manocha, J.A. Pople and L. Radom, Chem. Phys. 4, 107 (1974).

10. C.R. Brundle, N.A. Kuebler, M.B. Robin and H. Basch, Inorg. Chem. 11, 20 (1972); S. Evans, P.J. Joachim, A.F. Orchard and D.W. Turner, Internat. J. Mass Spectrom. Ion Phys. 9, 41 (1972).

11.   L. Åsbrink, Chem. Phys. Letters, <u>12</u>, 137 (1971).

12.   (a) J. Berkowitz, J. Chem. Phys. <u>56</u>, 2766 (1972); (b)
      J. Berkowitz and J.L. Dehmer, J. Chem. Phys. <u>57</u>, 3194
      (1972).

13.   D.T. Clark and I.W. Scanlan, J. Chem. Soc. Faraday II, <u>70</u>,
      1222 (1974) and references therein.

14.   M.M. Coutière, J. Demuynck and A. Veillard, Theoret. Chim.
      Acta, <u>27</u>, 281 (1972).

15.   M.M. Rohmer and A. Veillard, J. Chem. Soc. D. 250 (1973).

16.   I.H. Hillier, M.F. Guest, B.R. Higginson and D.R. Lloyd,
      Mol. Phys. <u>27</u>, 215 (1974).

17.   S. Evans, M.F. Guest, I.H. Hillier and A.F. Orchard, J.
      Chem. Soc. Faraday II, <u>70</u>, 417 (1974).

18.   J. Berkowitz, Chem. Phys. Letters, <u>11</u>, 21 (1971).

19.   W.G. Richards, Trans. Faraday Soc. <u>63</u>, 257 (1967).

20.   R.L. DeKock and D.R. Lloyd, Adv. Inorg. Chem. Radiochem.
      <u>16</u>, 66 (1974).

21.   O. Edqvist, L. Åsbrink and E. Lindholm, Z. Naturforsch. A,
      <u>26</u>, 1407 (1971).

22.   A.D. Baker, C. Baker, C.R. Brundle and D.W. Turner,
      'Molecular Photoelectron Spectroscopy', Wiley-Interscience,
      London, 1970, p. 61.

23.   C.R. Brundle, D. Neumann, W.C. Price, D. Evans, A.W. Potts
      and D.G. Streets, J. Chem. Phys. <u>53</u>, 705 (1970).

24.   C.R. Brundle and G.R. Jones, J. Chem. Soc. Faraday II, <u>68</u>,
      959 (1972).

25.   M.B. Hall, M.F. Guest, I.H. Hillier, D.R. Lloyd, A.F.
      Orchard and A.W. Potts, J. Electron. Spec. <u>1</u>, 497 (1973).
      and references therein.

26.   (a) A.W. Potts, H.J. Lempka, D.G. Streets and W.C. Price,
      Phil. Trans. Roy. Soc. A <u>268</u>, 59 (1970); (b) R.L. DeKock,
      B.R. Higginson and D.R. Lloyd, Faraday Disc. Chem. Soc.
      <u>54</u>, 84 (1972) and references cited therein; (c) N.V.
      Richardson, <u>ibid</u>., 96.

27.   R.L. DeKock, B.R. Higginson and D.R. Lloyd, Faraday
        Discussions, 54, 84 (1972).

28. (a) P.J. Bassett and D.R. Lloyd, J. Chem. Soc. A, 1551
        (1971); (b) G.H. King, S.S. Krishnamurthy, M.F. Lappert
        and J.P. Pedley, Faraday Disc. Chem. Soc. 54, 70 (1972);
        (c) ibid. 94, 96.

29.   D.R. Lloyd and N. Lynaugh, J. Chem. Soc. Faraday Trans. II,
        68, 947 (1972).

30.   R.F. Lake, Spectrochimica Acta 27A, 1220 (1971).

31.   W. Fuss, Ph.D. Thesis, University of Frankfurt am Main,
        1971.

32.   C.R. Brundle, M.B. Robin, N.A. Kuebler and H. Basch,
        J. Amer. Chem. Soc. 94, 1451, 1466 (1972).

UV PHOTOELECTRON SPECTROSCOPY OF TRANSITION METAL COMPOUNDS

A.F. Orchard

Inorganic Chemistry Laboratory, University of Oxford

UV photoelectron spectroscopy has already been extensively employed in the study of transition metal compounds in the gas phase but doubtless there will be many further applications of the technique in this particular context. The aim of this review is merely to outline the present status of the subject without pretensions to a comprehensive survey.

$d^{\,0}$ complexes

Let us begin with closed-shell molecular species formally lacking d electrons which, so far as the earlier transition metal groups ($\overline{\text{IVA}}$-VIIA) are concerned, are quasi-isoelectronic with congeneric main group species. The He-I spectra of the group IVA molecules, $TiCl_4$ and $TiBr_4$ (Figure 1)[1-3] are interestingly different from the spectra of the corresponding IVB species (cf. Lloyd, Figure 19)[1] and confirm the essential details of the established one-electron picture of tetrahedral d^0 species (Figure 2).[4,5] Somewhat surprisingly, however, the energy sequence of orbitals accessible, as it were, to He-I radiation is not very different from that anticipated on the basis of orbital interactions within the halogen tetrahedron (Figure 2): there would appear to be no strong mixing of the metal valence orbitals (3d, 4s and 4p) with the ligand p orbitals. This indicates that the titanium tetrahalides, despite being relatively covalent materials in the everyday sense, are actually quite ionic.

The relationship between the spectra of $TiCl_4$ and $TiBr_4$ (Figure 1) is clouded by the effects of spin-orbit coupling (centred about the bromine nucleus) in the latter compound.

P. Day (ed.), Electronic States of Inorganic Compounds. 339–360. *All Rights Reserved.*
Copyright © 1975 *by D. Reidel Publishing Company, Dordrecht-Holland.*

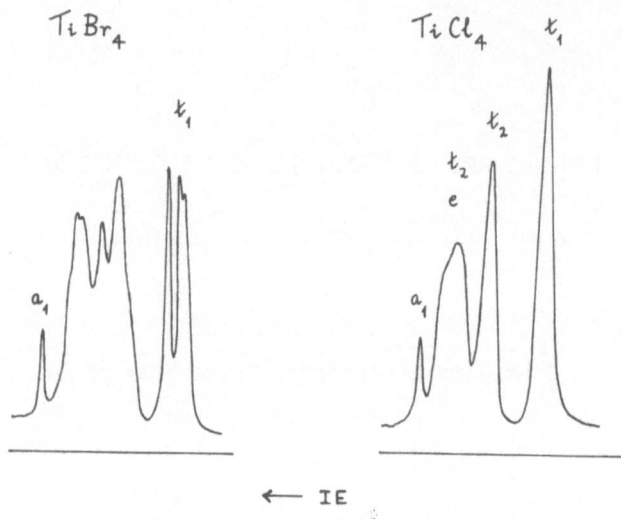

Figure 1.   A comparison of the He I photoelectron spectra of
$TiCl_4$ and $TiBr_4$.   (References 1-3.)

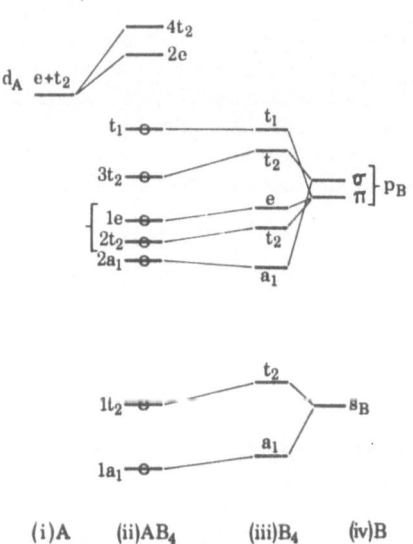

Figure 2.   LCAO-MO energy diagram for tetrahedral transition
metal compounds AB_4.

It is clear, however, that the first two bands in the $TiBr_4$ spectrum represent the multiplet components of a ^2T state of the molecular ion – almost certainly $^2T_1(1t_1^{-1})$, where the $1t_1$ MO is M–X non-bonding. The energy sequence of multiplet states inferred from the relative intensities of the two bands, namely $E' < U'$, is that expected for a hole in a $\underline{t}$ subshell primarily localised on the halogen atoms. The magnitude of the multiplet splitting (0.35 eV) is consistent with the assignment in terms of t_1 orbitals wholly localised on the bromine tetrahedron ($\mathcal{J}_{4p}$ = 0.31 eV for a free Br atom).

The He–I spectra of the d^0 tetroxo species RuO_4 and OsO_4 (Figure 3)[3,6,7] are on the other hand indicative of strong covalency. One of the t_2 levels, that represented by the broad PE band at high ionization energy, has dramatically shifted by comparison with the TiX_4 species. It would appear that the $t_2(d) - t_2(\delta)$ interaction dominates the covalent bonding: in other words, the strength of the metal-oxygen bonds apparently lies in the σ– rather than the π–component of the classical double bond M=O.[3] While the sequence of the p_π symmetry orbital energies ($t_1 > t_2 \sim e$, cf. Figure 2) is essentially preserved, the stabilisation of the $2t_2$ level is so marked that the relative energies of $2t_2$ and $2a_1$ are the reverse of the $t_2(\sigma)$ and $a_1(\sigma)$ symmetry orbitals. The large widths of the $2t_2^{-1}$ bands are indeed highly indicative of strongly bonding $2t_2$MOs. In the case of OsO_4 the band shows structure suggestive of an unusually strong Jahn-Teller interaction in the molecular ion.

The oxychloride species $VOCl_3$, CrO_2Cl_2 and MoO_2Cl_2 have also been investigated by UV-PE spectroscopy and, while the He-I spectra are difficult to interpret in detail (mainly because of the lower molecular symmetries), they do indicate a situation intermediate between the tetrachloride and tetroxide cases.[3] Thus, in the isoelectronic series $TiCl_4 - VOCl_3 - CrO_2Cl_2$, there appears to be a progressive stabilisation of the MOs correlating with $2t_2$ of the tetrahedral species.

But it should be said that all these conclusions are based on a naive application of Koopmans' approximation. There may be significant $\underline{differential}$ relaxation effects (specially for orbitals of very different localisation tendencies, $\underline{vide\ infra}$) the implications of which are difficult to assess. The differences in metal-ligand bonding between the TiX_4 and MO_4 species may be less extreme than we have suggested.

Ionic d^0 complexes, in particular the tetroxo-anions MnO_4^-, CrO_4^{2-} and VO_4^{3-}, which have been so extensively studied in crystal lattices by optical absorption spectroscopy,[5] cannot of course be investigated by gas-phase UV-PES. But they have instead been studied by X-ray photoelectron spectroscopy (X-PES).[8-10] The

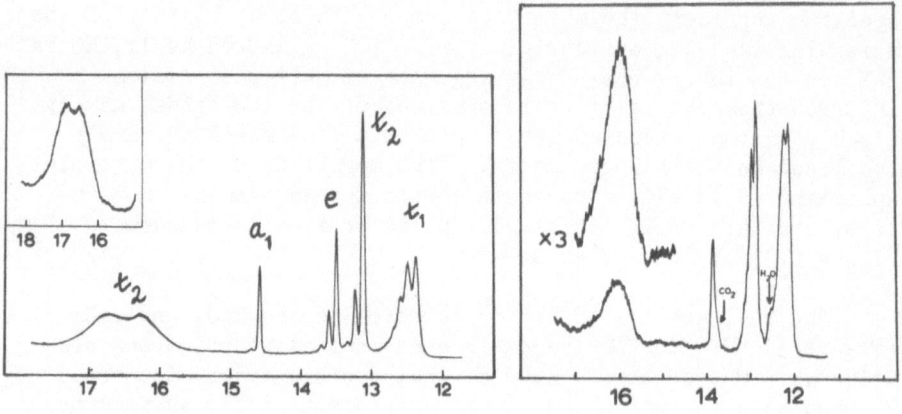

Figure 3.   He I PE spectra of OsO_4 and RuO_4.   (Reference 3.)

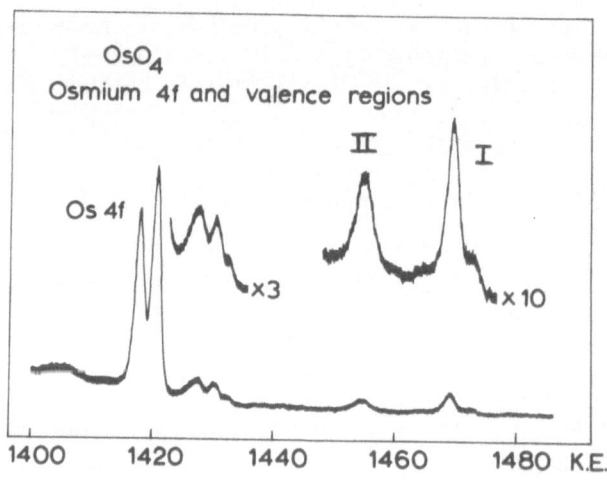

Figure 4.   High kinetic energy region of the X-ray (Al Kα) PE
spectrum of OsO_4.   (Reference 3.)

valence region X-PE spectra are however rather uninformative,
since the technique (at least with achromatic exciting X-rays)
does not permit the resolution of band structure comparable with
that observed in the UV-PE spectra of the neutral d^0 complexes.
Thus in the OsO_4 X-PE spectrum reproduced in Figure 4 the band
system labelled I corresponds to the entirety of the He-I
structure in Figure 3. The dominant peak of band I apparently
corresponds to the $2t_2$ ionisations while the shoulder to low IE
represents unresolved ionisations of $2a_1$, $3t_2$, $1e$ and $1t_1$
electrons.[3] X-PE spectroscopy does however enable us to locate
the $1t_2$ and $1a_1$ energy levels (which of course correlate strongly
with oxygen 2s), both of which are represented by band II.

The ionisation of d electrons

The outer d electrons of transition metal compounds have a
reasonable ionisation cross-section for incident UV photons and
normally give signals occurring in the low energy region (between
6 and 12 eV ionisation energy, say) of He-I or He-II spectra.
If the spectrum of the isostructural d^0 species, or that of an
isostructural main group analogue, is available for comparison
purposes, there is usually no difficulty in identifying the d
ionisation structure in a UV-PE spectrum. Two simple examples
are shown in Figures 5 and 6. Vanadium tetrachloride[2] illustrates
the particularly simple situation where we have the d^0 analogue
($TiCl_4$) to compare it with, and there is no ambiguity in assigning
the additional, low energy band to ionisation of the solitary d
electron occupying the 2e molecular orbital (cf. Figure 2).
Notice in the case of VCl_4 that the remaining part of the He-I
spectrum is distinctly more complex than the spectrum of $TiCl_4$.
This is the result of having an open shell ground state (viz.
^2E $(2e)^1$). Ionisation of a filled inner subshell γ can lead to
each of the ion states obtained by coupling $^2\Gamma$ with ^2E (e.g.,
$^{3,1}T_1 + ^{3,1}T_2$ in the case of t_1 or t_2 subshells). In all He-I
excitation generates twenty distinct ion states as compared with
just five in the case of $TiCl_4$. The technique is incapable of
resolving all these processes so the VCl_4 spectrum merely has a
somewhat amorphous appearance compared with that of $TiCl_4$.

In the case of ferrocene, first studied by Turner,[11,12] we
may compare the UV-PE spectrum with that of a main group bis
(cyclopentadienyl) such as $Mg(C_5H_5)_2$,[13] also known to have the
sandwich structure. This is rather less satisfactory but, given
the simplicity of the spectra in Figure 6, admits no serious uncertainty. The 8.5-10 eV band system of $Fe(C_5H_5)_2$ appears to
correlate with the two well-separated bands at ca. 8 eV and 9 eV,
respectively, in the $Mg(C_5H_5)_2$ spectrum - the latter corresponding
to electrons primarily localised on the ligand π system. The
lower energy structure, a doublet centred around 7 eV ionisation
energy in the ferrocene spectrum must accordingly be assigned to
the metal d electrons.

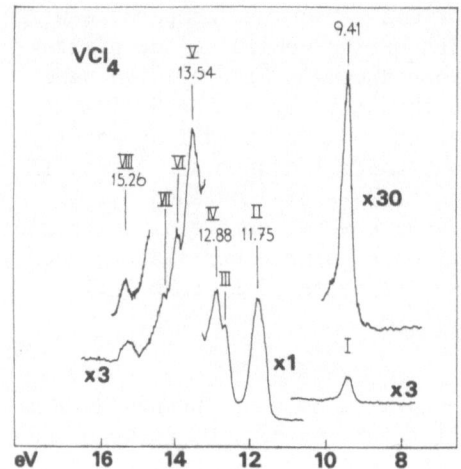

Figure 5.   He I PE spectrum of VCl$_4$.   (Reference 2.)

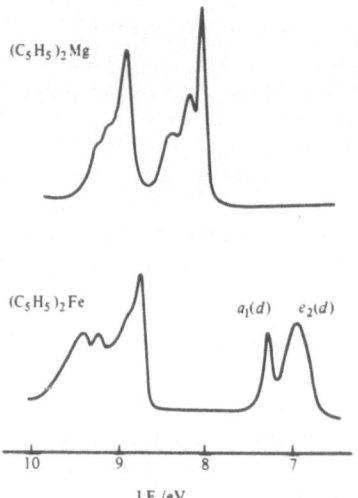

Figure 6.   High kinetic energy regions of the He I PE spectra of ferrocene and bis(cyclopentadienyl)-magnesium.   (Reference 13.)

When we can confidently identify the $\underline{d}$ structure in a PE spectrum then considerable interest may attach to the determination of the trends in $\underline{d}$ electron IEs with variation of metal atom within a series of isostructural compounds. The tris(hexafluoroacetylacetonato) complexes $M(hfa)_3$ of the transition elements provide a simple illustration of such a study.[14] Here we can follow the trends in $\underline{d}$ electron IEs, not only for an isostructural series but also (albeit to a limited extent) for constant configurational type, namely $(t_{2g})^n$. As shown in Figure 7, the $t_{2g}(d)$ ionisation energies increase steadily as we traverse the first transition series showing a trend almost parallel with that established by atomic emission spectroscopy for the average of configuration IEs of the 3d subshells of the free ions. It would appear, then, that the $t_{2g}(d)$ orbitals are progressively stabilised with increasing atomic number across the transition series. Of course, this had for some time been assumed to be the case, but UV-PES provided the first reasonably direct evidence on the point. The effect seems to be an essentially atomic one: thus, even in the case of the mononuclear carbonyls $M(CO)_n$, where a change of coordination number occurs, the average $\underline{d}$ electron IE is found to increase in the sequence $V(CO)_6 < Cr(CO)_6 < Fe(CO)_5 \sim Ni(CO)_4$. (The figures (in eV) are: $V(CO)_6$ ca. 7.7,[16] $Cr(CO)_6$ 8.40,[12,17] $Fe(CO)_5$ ca. 9.2,[15] $Ni(CO)_4$ ca. 9.2,[15,18] With the sole exception of $Cr(CO)_6$ the data relate to degeneracy-weighted averages over the components of split $\underline{d}$ signals (vide infra.)

However, the situation will not always be so straightforward, even for an isostructural series, since changing ligand field configurations, not to mention open-shell effects (vide infra), may frequently confuse the issue. The pattern of $\underline{d}$ electron IEs encountered for the first row metallocenes, discussed below, illustrates the problems that arise.

The variation in $\underline{d}$ electron IEs for constant configurational type as one descends a transition metal subgroup have also been investigated. The general trend observed parallels that of the average of configuration IEs of the free metal atoms or uni-positive ions. The $\underline{d}$ electron IEs tend to increase on passing from the first to the second transition series, but there is little difference between analogous compounds of the second and third series. Frequently, there is of course a change in spin state (the spin-pairing effect) on going from a complex of a first row transition element to its congeners of the second and third series (Table 1).

Before we move on, a word about the relative energies of metal $\underline{d}$ and ligand orbitals is perhaps called for. It had always been assumed that the metal $\underline{d}$ levels in transition metal complexes lay above the ligand electron levels (i.e. that the

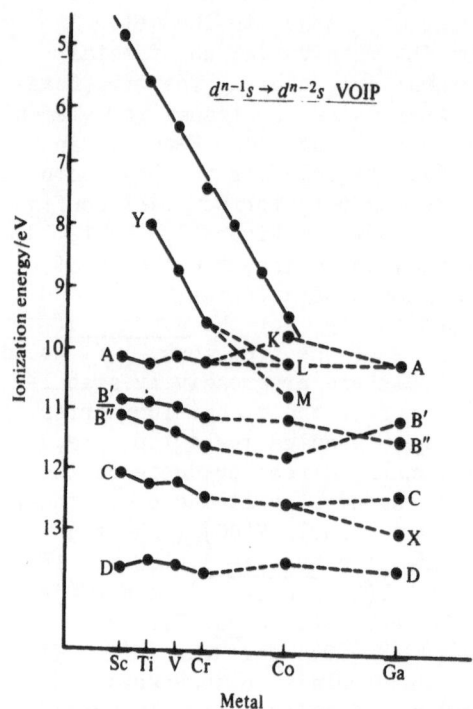

Figure 7.   Ionisation energy trends (from UV-PES) for the
tris(hexafluoroacetyl-acetonato) compounds of the first row
transition elements.   (Reference 14.)

metal $\underline{d}$ electrons had less negative orbital energies). However, UV-PES studies have shown that this is not always the case. Take, for example, the $M(hfa)_3$ series (Figure 7). By the time one reaches $Co(hfa)_3$ (t_{2g}^6 configuration) it would appear that the occupied metal $\underline{d}$ levels have stabilised to the extent that they actually lie below the highest occupied ligand levels. A similar reversal of the previously assumed sequence of metal $\underline{d}$ and ligand orbital energies apparently occurs in a number of other cases, including some open-shell complexes (e.g. $Mn(C_5H_5)_2$, vide infra). The significance of such observations has been discussed by Jørgensen.[20]

Table 1. Some data illustrating the variation of metal $\underline{d}$ ionisation energies (eV) within the transition metal subgroups

System	Metal	$\underline{d}$ IE/$_{eV}$	Ref.
$M(C_5H_4Me)_2$ [a,b]	Fe	6.2	13
	Ru	7.2_5	
	Os	7.2_5	
$M(CO)_6$	Cr	8.40	17
	Mo	8.50	
	W	8.56	
$M(PF_3)_4$ [a,c]	Ni	10.1	19
	Pd	10.8	
	Pt	10.9	

[a] Average values, weighted according to degeneracy, for the separate $\underline{d}$ signals.

[b] cf. Figure 10

[c] cf. Table 2 and Figure 9

Table 2.   A comparison of metal $\underline{d}$ ionisation energies (eV) for analogous carbonyl and trifluorophosphine complexes

System	Metal	CO		PF_3	
		t_2	e	t_2	e
ML_4[a]	Ni	8.90	9.77	9.69	10.74
	Pd	–	–	9.9	12.2
	Pt	–	–	9.83	12.45
		e'	e''	e'	e''
ML_5[b,c]	Fe	8.60	9.86	8.9	10.2
		t_{2g}		t_{2g}	
ML_6[d,e]	Cr	8.40		9.0	
	Mo	8.50		9.3	

[a] Bassett et al., ref. 19
[b] Lloyd and Schlag, ref. 15
[c] Nixon, ref. 28
[d] Higginson et al., ref. 17
[e] Eland, ref. 27

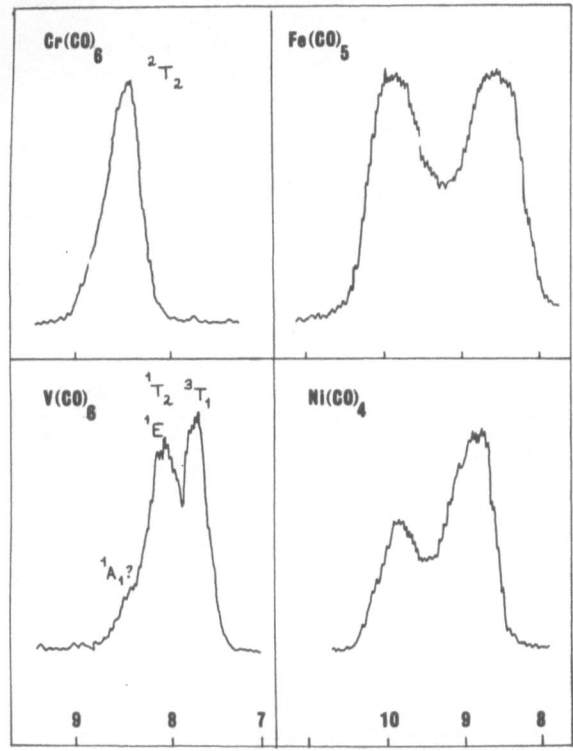

Figure 8.   Metal $\underline{d}$ ionisation structure in the He I PE spectra of the mononuclear carbonyls $M(CO)_n$.   (References 15–18.)

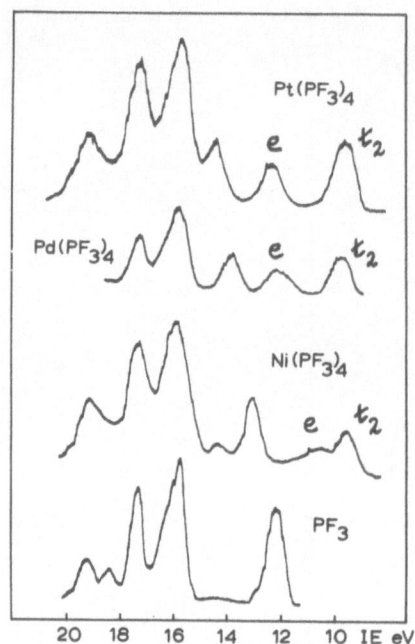

Figure 9.   A comparison of the He I PE spectra of the $M(PF_3)_4$ species (M = Ni, Pd, Pt) and the free ligand.   (Reference 19.)

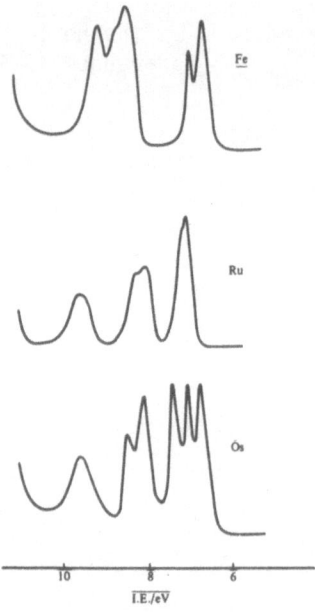

Figure 10.   High kinetic energy regions of the He I PE spectra of 1,1'-dimethyl-ferrocene and its ruthenium and osmium analogues – all formally d^6 species.   (Reference 13.)

Ligand field splittings

UV-PES has proved a fertile source of information concerning symmetry dependent splittings of the outer $\underline{d}$ subshells of transition metal compounds – a subject which is, of course, traditionally the preserve of ligand field theory.   Figure 8 shows the results obtained for the simple binary carbonyls of the first row transition elements.   In the case of octahedral $Cr(CO)_6$, where six $\underline{d}$ electrons are accommodated in the outer t_{2g} subshell, a single symmetrical band is observed in the low energy region of the PE spectrum.[12,17]   The spectrum of $Fe(CO)_5$, a trigonal bipyramidal molecule, shows two $\underline{d}$ bands of roughly equal intensity, consistent with the configuration $(e'')^4(e')^4$ expected for a formally d^8, diamagnetic species, while that of $Ni(CO)_4$, formally a d^{10} system, has also two low energy bands whose relative intensity ($\sim$ 3:2, the more intense band occurring at lower IE) is close to that expected for a $(e)^4(t_2)^6$ $\underline{d}$ electron configuration.[15]   The basic predictions of ligand field theory are thus elegantly confirmed.   The rather complex form of the $\underline{d}$ band observed in the case of $V(CO)_6$, outer electron configuration $(t_{2g})^5$, may also be understood in terms of conventional ligand field theory.[16]   The $\underline{d}$ electron configuration in the molecular ion, $(t_{2g})^4$, gives rise to various terms which, in order of increasing energy, are $^3T_{1g} < {}^1T_{2g} \sim {}^1E_g < {}^1A_{1g}$ and which, in the simplest scheme of things, are accessible with relative transition probabilities proportional to their total degeneracies (namely 9, 3, 2 and 1).   The $\underline{d}$ structure in the $V(CO)_6$ photoelectron spectrum (Figure 8) is consistent with this simple theory.

The validity of ligand field theory may be less obvious in more complicated situations, but a close examination of the data usually confirms its authority.   Take the case of ferrocene, for example.[13]   Formally a d^6 species, of molecular symmetry D_{5d}, ferrocene is known to be diamagnetic and must therefore have the outer electronic structure $(a_{1g})^2(e_{2g})^4$.   The relative intensities of the two $\underline{d}$ bands (Figure 6) argue clearly for assignment of the first band to ionisation of the $e_{2g}(d)$ electrons and the second to the $a_{1g}(d)$ electrons.   The sequence of orbital energies deduced via Koopmans' theorem, namely $e_{2g} > a_{1g}$, is the opposite of the order of crystal field energy levels, $a_{1g} > e_{2g}$, established by the analysis of $\underline{d}$-$\underline{d}$ optical absorption spectra.   There is no inconsistency here, however, once the effects of electron repulsion within the perturbed $\underline{d}$ sub-shell are allowed for.   In crystal field theory the energy difference of the $^2A_{1g}$ $(a_{1g}e_{2g}^4)$ and $^2E_{2g}$ $(a_{1g}^2 e_{2g}^3)$ states of the molecular ion $Fe(C_5H_5)_2^+$ is just

$$E(^2A_1) - E(^2E_2) = 20B - \Delta_2$$

where B is the usual Racah inter-electronic repulsion parameter and Δ_2 the a_{1g}-e_{2g} crystal field splitting parameter.   Thus the

$e_{2g}(d)$ ionisations occur at lower energy than the $a_{1g}(d)$
ionisations (i.e. the ground state of the ion is $^2E_{2g}$) simply
because 20B happens to be greater than Δ_2.   The relative
magnitudes of the parameters B and Δ_2 determined for the neutral
molecule by electronic absorption spectroscopy tend to substantiate the latter conclusion.

It has been well established by optical absorption studies
on classical (intermediate oxidation state) coordination compounds
that the principal crystal field splitting parameters tend to
increase, for isostructural complexes, down the transition metal
subgroups.   A number of applications of UV-PES have demonstrated
that this generalisation probably applies also to low oxidation
state systems.   Thus, the He-I spectra of the tetrakis(trifluoro-
phosphines) $M(PF_3)_4$ [19] (Figure 9) which, like $Ni(CO)_4$, are formally
d^{10} complexes, indicate that the $t_2(d)-e(d)$ orbital energy
difference increases progressively on going from $Ni(PF_3)_4$, through
$Pd(PF_3)_4$, to $Pt(PF_3)_4$ (cf. Table 2).[19]   In the case of d^{10} species
(d^9 molecular ion configurations) the sequence of one electron MO
energies is necessarily identical with the ligand field orbital
energy sequence.

Useful information has also been obtained about ligand field
effects in organometallic compounds.   For example, a comparative
study of $Fe(C_5H_5)_2$ and $Ru(C_5H_5)_2$ reveals that the $a_{1g}(d)$ and
$e_{2g}(d)$ signals have coalesced in the spectrum of the ruthenium
compound (cf. Figure 10).[13]   This can be understood in terms of
an increase in the crystal field parameter Δ_2 but is undoubtedly
due in part to the expected decrease in the inter-electronic
repulsion parameter B.   (An extended comparison embracing the
osmium analogue is unfortunately precluded by the confusing
effects of spin-orbit coupling, which induces a symmetrical
splitting of the $^2E_2$ ion state – Figure 10).   Comparable changes
are evident on passing from the PE spectrum of $Cr(C_6H_6)_2$ to that
of $Mo(C_6H_6)_2$ [21,22] (these being also d^6 systems, of course).
However, in the case of the bis(benzene) complexes the sequence
of $a_1(d)$ and $e_2(d)$ ionisations is inverted relative to the d^6
bis(cyclopentadienyl) complexes, the ion ground state now being
$^2A_1$ rather than $^2E_2$.   This indicates a very substantial increase
in the parameter Δ_2:   in other words, we might say that C_6H_6
occupies a higher position in the <u>spectrochemical series</u> than
$C_5H_5^-$.   (It should be said, however, that the sequence of one-
electron energy levels deduced from <u>d-d</u> spectra, viz.
$e_1 > a_1 > e_2$, is not altogether consistent with the theory of
axial crystal fields.)   Consistent with this supposition is the
observation that, in the mixed d^6 sandwich complex $(C_5H_5)Mn(C_6H_6)$,
the $^2E_2 - {}^2A_1$ energy separation is considerably reduced by
comparison with $Cr(C_6H_6)_2$ [21] (Figure 11).   On the other hand, the
PE spectrum of $(C_5H_5)Cr(C_7H_7)$[22] shows a substantial <u>increase</u> in
the separation of the $^2E_2$ and $^2A_1$ ion states.   (The $^2E_2 - {}^2A_1$

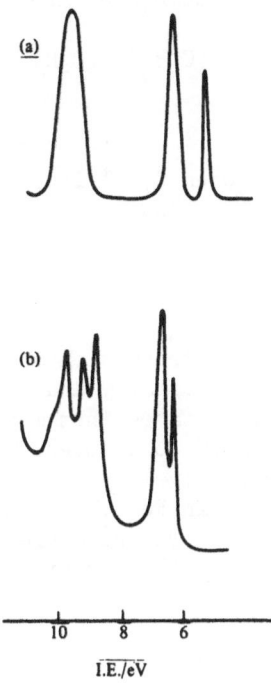

Figure 11.   The high kinetic energy structure in the He I PE spectra of the d^6 compounds: (a) $Cr(C_6H_6)_2$ and (b) (C_5H_5) $Mn(C_6H_6)$.   (Reference 21.)

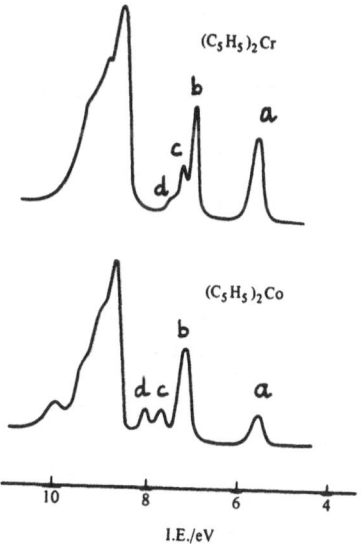

Figure 12.   High kinetic energy parts of the He I PE spectra of $(C_5H_5)_2Co$ and $(C_5H_5)_2Cr$.   (Reference 23.)

energy separations for the various d^6 sandwich complexes are:

$$(C_5H_5)Cr(C_7H_7) \ 1.60 \ eV, \ Cr(C_6H_6)_2 \ 1.0 \ eV, \ (C_5H_5)Mn(C_6H_6)$$

$$0.36 \ eV, \ Fe(C_5H_5)_2 \ 0.35 \ eV.)$$

This observation suggests that the ligand species $C_7H_7^+$ lies higher still in the spectrochemical series of ligands.

Ligand field effects in open shell complexes

The interpretation of the UV–PE spectra of transition metal complexes having open shell ground states is inevitably a difficult matter given the large number of molecular ion states that may be accessible on photoionisation. The spectrum of cobaltocene (Figure 12) illustrates the problem quite effectively. The $Co(C_5H_5)_2$ molecule, with one more d electron than $Fe(C_5H_5)_2$, has an electronic ground state $^2E_{1g}$ ($a_{1g}^2 \ e_{2g}^4 \ e_{1g}^1$). The possible one-electron ionisation processes can generate the following states of the molecular ion:[23,24]

Ion configuration	Ion states
$a_{1g}^2 \ e_{2g}^4$	$^1A_{1g}$
$a_{1g}^1 \ e_{2g}^4 \ e_{1g}^1$	$^{3,1}E_{1g}$
$a_{1g}^2 \ e_{2g}^3 \ e_{1g}^1$	$^{3,1}E_{1g} + ^{3,1}E_{2g}$

Thus seven distinct ion states are expected but only four d PE bands (labelled a – d in Figure 12) may confidently be identified. The first band a (vertical IE = 5.56 eV) presumably corresponds to ionisation of the lone $e_{1g}(d)$ electron, i.e. production of the $^1A_{1g}(a_{1g}^2 \ e_{2g}^4)$ state of $Co(C_5H_5)_2^+$, but the interpretation of the remaining d structure is problematical. It represents an obvious challenge to conventional ligand field theory as applied to the states of the molecular ion. Useful guidance regarding the assignment problem may also be obtained by reference to a theory of relative photoionisation cross sections developed by Cox et al.[25]

It seems quite clear from the ligand field analysis (due regard being given to the question of band intensities) that band b represents transitions both to the $^3E_{2g}$ state and the lower energy $^3E_{1g}$ state. However, the assignment of the higher IE structure in the $Co(C_5H_5)_2$ spectrum (bands c and d in particular) is not altogether clear.

The complex $\underline{d}$ ionisation structure commonly observed for open-shell compounds makes it difficult to divine the trends in $\underline{d}$ orbital energies when species having different electronic configurations are compared.   Consider Figure 13, for example, which shows the pattern of IE data for the first row metallocenes. The situation is evidently a great deal more complicated than in the case of the $M(hfa)_3$ species (Figure 7).   However, similar problems arise in comparing the PE spectra of $Fe(hfa)_3$ and $Mn(hfa)_3$, both of which are high-spin complexes (ground configurations $t_{2g}^3 e_g^2$ and $t_{2g}^3 e_g^1$, respectively), with the other $M(hfa)_3$ spectra (t_{2g}^n configurations).   The study of $\underline{d}$ orbital energy trends in, for example, the iron group series $\overline{Fe}(hfa)_3$, $Ru(hfa)_3$, $Os(hfa)_3$ is also greatly confused by a change in electronic configurational type, the Ru and Os compounds being low-spin t_{2g}^5 species.   However, the $Mn(C_5H_5)_2$ data excepted, the $\underline{average}$ of the observed $\underline{d}$ ionisation energies (not weighted in any way) does show a reasonably steady decrease on traversing the transition series, suggesting that the irregular pattern of $\underline{d}$ ionisation data points may be due simply to the subtle effects of electron repulsion in the various open-shell configurations involved.   A simple analysis of the open-shell effects does indeed suggest that the $\underline{\text{average of configuration}}$ $e_{2g}(d)$ ionisation energy increases steadily on going from $V(C_5H_5)_2$ to $Ni(C_5H_5)_2$ (apart from a discontinuity for the Mn compound).   The anomalous data for manganocene, the PE spectrum of which is consistent with the supposition of a high-spin $^6A_{1g}$ ($a_{1g}^1 e_{2g}^2 e_{1g}^2$) ground state, indicates dramatically higher ionic character than in the other metallocenes.

Two further aspects of the UV-PES work on the metallocenes are worth mention, the first concerning 1,1'dimethyl-manganocene, $Mn(C_5H_4Me)_2$.   The He-I spectrum of this compound (Figure 14) shows additional low IE bands ($\underline{a}'$, $\underline{b}'$ and $\underline{c}'$) presumably associated with ionisation of the $\underline{d}$ electrons.   This most unusual result was interpreted in terms of a high-spin/low-spin equilibrium, the additional $\underline{d}$ bands being ascribed to a low-spin isomer of Mn $(C_5H_4Me)_2$.   Subsequent magnetic measurements (EPR and bulk susceptibility studies) on $Mn(C_5H_4Me)_2$ in hydrocarbon solvents have substantiated this hypothesis (Professor M.F. Rettig, personal communication).   The paramagnetic susceptibility data fit a two-state equilibrium, $^2E \leftrightarrow {}^6A$, and a characteristic ^2E EPR spectrum is observed at 4.2°K.   Related work has been carried out by Dr. J. Ammeter at Zurich (personal communication), who also concludes that the low-spin form is ^2E ($a_2^2 e_3^3$), like the $Fe(C_5H_5)_2$ molecular ion, rather than $^2A_1$ ($a_1^4 e_2^2$).   This is apparently the first evidence of a spin-state equilibrium for a TM compound in the vapour phase.   Manganocene itself is clearly present almost entirely in the high-spin state, though careful examination of the low IE region of its PE spectrum reveals evidence in the form of a very weak band $\underline{a}'$ (Figure 14) for a

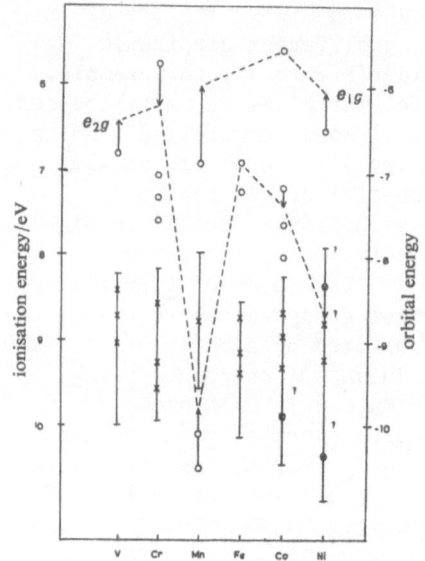

Figure 13.   Ionisation energy trends observed for the metallocenes $(C_5H_5)_2M$ by UV-PES.   (Reference 23.)

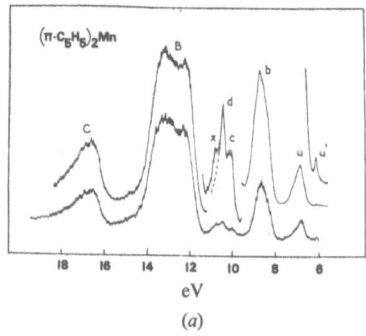

(a)

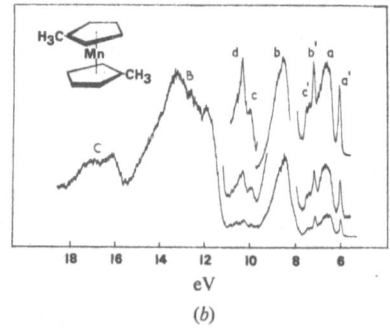

(b)

Figure 14.   A comparison of the He I PE spectra of manganocene
and its 1,1'-dimethyl analogue.   (Reference 23.)

small amount of the spin isomer. It would be quite interesting to study the PE spectra of both $Mn(C_5H_5)_2$ and $Mn(C_5H_4Me)_2$ as a function of temperature.

The second observation concerns chromocene, $Cr(C_5H_5)_2$, where there was some doubt regarding the ground state - whether it is $^3A_{2g}$ ($a_{1g}^2 e_{2g}^2$) or $^3E_{2g}$ ($a_{1g}^1 e_{2g}^3$). Interestingly, the UV-PE spectrum establishes that it is in fact $^3E_{2g}$. If it were $^3A_{2g}$ just three $\underline{d}$ bands, corresponding to the ion states, $^{4,2}A_{2g}$ ($a_{1g}^1 e_{2g}^2$) and $^2E_{2g}$ ($a_{1g}^2 e_{2g}^1$), would be expected. The PE spectrum (Figure 12), like that of its 1,1-dimethyl derivative, shows in fact $\underline{four}$ bands attributable to ionisation of the metal $\underline{d}$ electrons. This is consistent with a $^3E_{2g}$ ground state, from which transitions to each of the five ion states $^2E_{2g}(e_{2g}^3)$, $^{4,2}A_{2g}$ ($a_{1g}^1 e_{2g}^2$), $^2A_{1g}$ ($a_{1g}^1 e_{2g}^2$) and $^2E_{1g}$ ($a_{1g}^1 e_{2g}^2$), are permitted.

Metal-ligand bonding

As implied at several points in the above discussion, UV-PES can often furnish useful information concerning the nature of the metal-ligand bonding in transition-metal compounds. Thus, analysis of the PE spectra of sandwich complexes[13,21-24] confirms the established view that the main source of bonding resides in the $e_1(\pi)$ - $e_1(d)$ ring-metal interactions: moreover, in the case of the metallocenes $M(C_5H_5)_2$, the PE data lends strong support to the contention that the high-spin manganese compound is substantially more ionic than its congeners.[23] Comparative studies of notionally isoelectronic ligand species prove particularly informative. For example, a comparison of IE data for analogous carbonyl $M(CO)_n$ and trifluorophosphine $M(PF_3)_4$ complexes (Table 2) reveals that the average metal $\underline{d}$ IEs are systematically greater for the PF_3 complexes. The $M-PF_3$ bonds are thus apparently the more polar despite the fact that PF_3 (IP = 12.3 eV) is expected to be a better δ-donor than CO (IP = 14.0 eV). One concludes that PF_3 is an altogether more electronegative ligand than CO, the extent of $d \rightarrow \pi^*$ back-bonding in the $M(PF_3)_n$ complexes being sufficiently large to produce a greater net transfer of electron density from the metal atom than occurs in the carbonyl complexes: in other words, the $M-PF_3$ bond has the more pronounced synergic character. The relative magnitudes of the t_2 - e ligand field splittings observed for $Ni(PF_3)_4$ (1.05 eV) and $Ni(CO)_4$ (0.87 eV) tend to support this conclusion.

The 'ligand structure' observed in the UV-PE spectra of transition metal compounds is usually rather diffuse and therefore less informative than that associated with ionisation of the metal $\underline{d}$ electrons.   Only in relatively simple cases (for example, transition metal halides, vide supra) do the electrons largely localised on the ligands yield well-resolved and assignable PE bands.   However, the $M(PF_3)_4$ species afford an example of a relatively complex system for which a complete assignment of the UV-PE spectra (Figure 9) is possible.[19]   The structure that may loosely be associated with the fluorine lone-pair electrons of PF_3 is virtually unchanged in the $M(PF_3)_4$ spectra but that associated with the phosphorus lone pair and the P-F bonding electrons is observed to shift to higher IE on complexing.   Separate bands representing the t_2 and a_1 bonding MOs derived from the four equivalent PF_3 lone pairs are clearly discernible.   In the same region of the UV-PE spectrum of $Mo(PF_3)_6$ three bands, corresponding to t_{1u}, e_g and a_{1g} lone-pair combinations, are observed.[27]

REFERENCES

1.   J.C. Green, M.L.H. Green, P.J. Joachim, A.F. Orchard and
       D.W. Turner, Phil. Trans. Roy. Soc. $\underline{A}$ $\underline{268}$, 111 (1970).

2.   P.A. Cox, S. Evans, A. Hamnett and A.F. Orchard, Chem.Phys.
       Letters $\underline{7}$, 414 (1970).

3.   P. Burroughs, S. Evans, A. Hamnett, A.F. Orchard and N.V.
       Richardson, J. Chem. Soc., Faraday Trans. II, $\underline{70}$, 1895
       (1974).

4.   C.J. Ballhausen and H.B. Gray, Molecular Orbital Theory,
       Benjamin, New York (1964).

5.   A. Müller, E. Diemann and C.K. Jørgensen, Structure and
       Bonding, $\underline{14}$, 23 (1973) and refs. therein.

6.   E. Diemann and A. Müller, Chem. Phys. Letters $\underline{19}$, 538 (1973).

7.   S. Foster, S. Felps, L.C. Cusachs and S.P. McGlynn, J. Amer.
       Chem. Soc. $\underline{95}$, 5521 (1973).

8.   A. Calabrese and R.G. Hayes, J. Amer. Chem. Soc. $\underline{95}$, 2819
       (1973).

9.   R. Prins and T. Novakov, Chem.Phys. Letters $\underline{16}$, 86 (1972).

10. J.A. Connor, I.H. Hillier, V.R. Saunders, M.H. Wood and
       M. Barber, Mol. Phys. $\underline{24}$, 497 (1972).

11. D.W. Turner, in 'Physical Methods in Advanced Inorganic
     Chemistry, ed. H.A.O. Hill and P. Day, Interscience,
     London (1968).

12. D.W. Turner, A.D. Baker, C. Baker and C.R. Brundle,
     Molecular Photoelectron Spectroscopy, Wiley, New York
     (1970).

13. S. Evans, M.L.H. Green, B. Jewitt, A.F. Orchard and
     C.F. Pygall, J. Chem. Soc., Faraday Trans. II, 68, 1847
     (1972).

14. S. Evans, A. Hamnett, A.F. Orchard and D.R. Lloyd, Faraday
     Disc. Chem. Soc. 54, 227 (1972).

15. D.R. Lloyd and E.W. Schlag, Inorg. Chem. 8, 2544 (1969).

16. S. Evans, J.C. Green, A.F. Orchard, T. Saito and D.W. Turner,
     Chem. Phys. Letters 4, 361 (1969).

17. B.R. Higginson, D.R. Lloyd, P. Burroughs, D.M. Gibson and
     A.F. Orchard, J. Chem. Soc., Faraday Trans. II, 69, 1659
     (1973).

18. I.H. Hillier, M.F. Guest, B.R. Higginson and D.R. Lloyd,
     Mol. Phys. 27, 215 (1974).

19. P.J. Bassett, B.R. Higginson, D.R. Lloyd, N. Lynaugh and
     P.J. Roberts, J. Chem. Soc., Dalton Trans., to be published.

20. C.K. Jørgensen, Chimia 27, 203 (1973).

21. S. Evans, J.C. Green and S.E. Jackson, J. Chem. Soc.,
     Faraday Trans. II, 68, 249 (1972).

22. S. Evans, J.C. Green, S.E. Jackson and B.R. Higginson,
     J. Chem. Soc., Dalton Trans. 304 (1974).

23. S. Evans, M.L.H. Green, B. Jewitt, G.H. King and A.F. Orchard,
     J. Chem. Soc., Faraday Trans. II, 70, 356 (1974).

24. J.W. Rabelais, L.O. Werme, T. Bergmark, L. Karlsson, M. Husain
     and K. Siegbahn, J. Chem. Phys., 57, 1185 (1972);
     erratum, 57, 4508 (1972).

25. P.A. Cox, S. Evans and A.F. Orchard, Chem. Phys. Letters 13,
     386 (1972).

26. P.A. Cox and A.F. Orchard, Chem. Phys. Letters, 7, 273 (1970).

27.   J.H.D. Eland, 'Photoelectron Spectroscopy', Butterworths,
      London 1974.

28.   J.F. Nixon, J. Chem. Soc., Dalton Trans. 2226 (1973).

VALENCE LEVEL PHOTOELECTRON SPECTRA (XPS AND UPS) OF SOLIDS AND
INTERFACES

C.R. Brundle

School of Chemistry, University of Bradford, Bradford,
Yorkshire, BD7 1DP

1.   INTRODUCTION

In the introductory chapter on the band structure of solids,[1]
Professor Thomas explains how in crystalline materials a simple
one-dimensional M.O. energy level diagram, such as that for CO
(Figure 1a) is no longer appropriate or sufficient to describe
the photoemission process because of the introduction of a
further quantum restriction, conservation of crystal momentum, k.
The allowed levels undergo variations in energy according to
position in k- space as in Figure 2(a) for GaS.[2]   To arrive at
a density of states (DOS) picture of the entire crystal the
individual atomic orbitals of the atoms are combined into energy
bands (e.g., Figure 2(b)) made up of all the allowed states summed
through k-space.   As a guide, the relative intensity of states
can be judged from the E versus k diagram: the flatter the band
(i.e. small changes in E across k-space) the greater the DOS.
Thus the histogram of Figure 2b represents the one-electron DOS
for GaS.   For a metal the allowed states are filled up to the
Fermi level; for example, the DOS of Au is made up from the $5d^{10}$
$6s^1$ valence electrons of the individual atoms (Figures 1b and 1c).
The d and s parentage of the resultant DOS varies across its
width and in other examples, Zn for instance, $3d^{10} 4s^2$, the d and
s components are very well separated out into separate bands.
Since they are not involved in the bonding, the core-levels retain
their atomic character so that, for example, their characteristic
energies can be used for atomic analysis.

In the remainder of this Chapter, we shall primarily be
interested in obtaining the DOS since the nature of bonding in
the solid is related to the energies and distribution of electron

P. Day (ed.), Electronic States of Inorganic Compounds. 361–392. All Rights Reserved.
Copyright © 1975 by D. Reidel Publishing Company, Dordrecht-Holland.

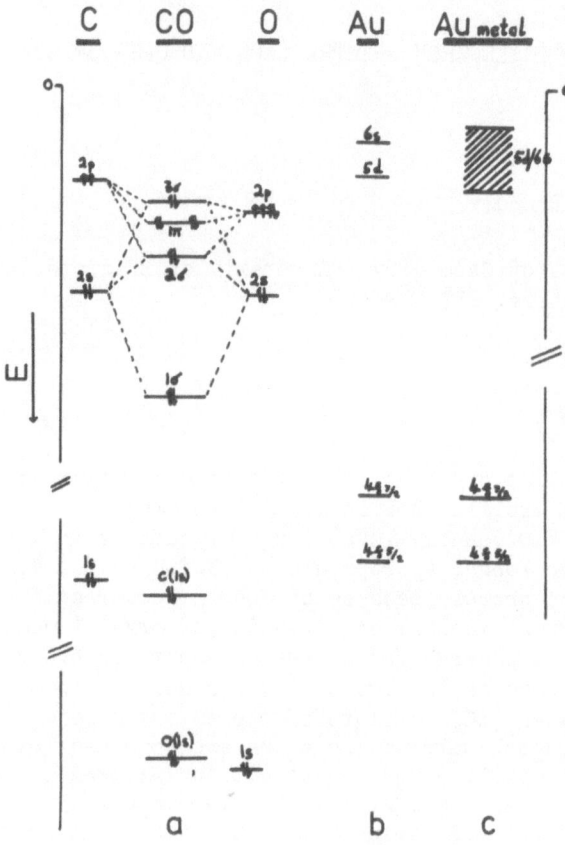

Figure 1.   Schematic molecular orbital energy level diagram for
(a) the CO molecule and (b) a gold atom and (c) solid gold.

states in a manner similar to that of the filled MO's of an
individual molecule.

     XPS and UPS are two of the more direct ways of obtaining
DOS information, and it is their use for this purpose that will
be described here.   Solid elements will be considered first,
followed by alloys, simple inorganic compounds and ionic salts.
The second part of the chapter deals with the study of electronic

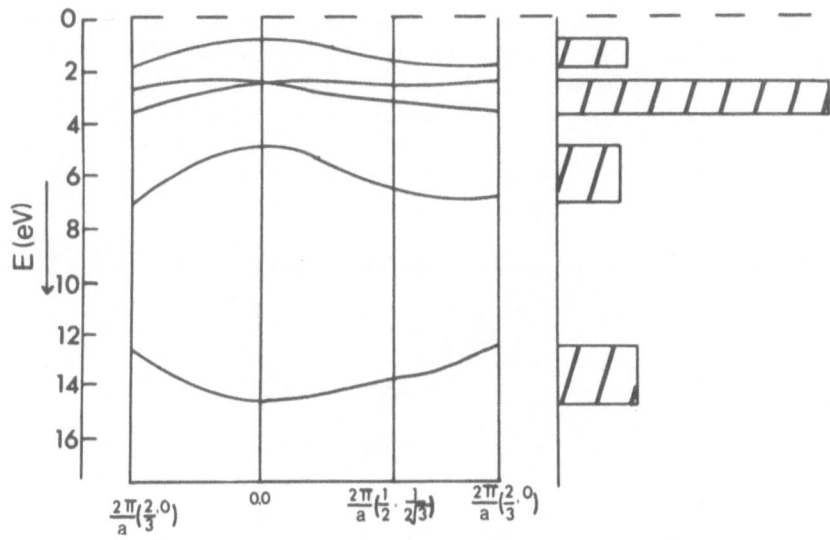

Figure 2.   (Left) part of the (E,k) band structure of GaS;
(right) approximate one-electron DOS for GaS (reference 2).

structure at the surfaces of solids, in particular bonding
between an adsorbate and a substrate.   This type of study is
feasible in XPS and UPS because the short escape length of the
ejected photoelectrons leads to a spectrum which is representative
of only the top 5Å-100Å of material, depending on the parameters
of the measurement.[3-5]

2.   ELEMENTS

2.1 Metals

      Figures 3(a) and (b) show the UPS spectra of polycrystalline
gold using HeI (21.2 eV) and HeII (40.8 eV) photons.[6]   Unlike
the situation in UPS spectra of gaseous molecules for which, at
the most, relative intensity changes owing to variations in
cross-section would be expected, the two spectra are quite
dissimilar.   The dissimilarity of the spectra reflects an

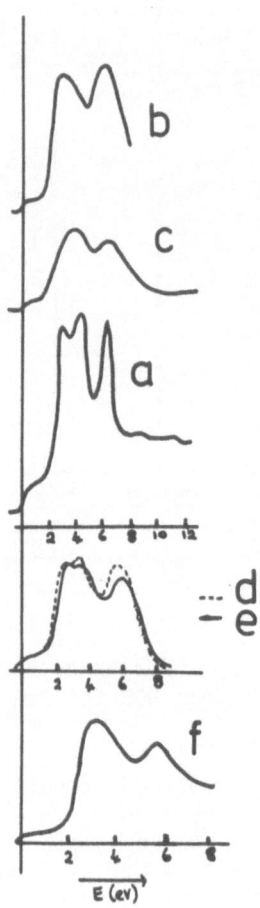

Figure 3. Experimental photoemission DOS of Au and calculated
DOS of Au.   (a) HeI spectrum of polycrystalline Au (reference 6);
(b) HeII spectrum of polycrystalline Au (reference 6);   (c) AlKα
spectrum of polycrystalline Au (reference 6);   (d) calculated DOS
of Au (reference 10);   (e) AlKα (monochromatized) spectrum of a
single crystal of Au (reference 10), and (f) HeI spectrum of
molten Au (reference 12).

important modifying effect of <u>final-state</u> structure for solids,
which is absent for free molecules.   If we consider the hypothetical E,k diagram of Figure 4(a), the photoionization process
may be considered as raising an electron in a filled allowed E
value (at a particular k value) by hν (21.2 eV if HeI is being
used).   However, unlike the situation in a free molecule, the
electron is still within the crystal at this point and must end
up occupying an unfilled allowed E state which is part of the E,k
diagram of the crystal (cf. the continimum of free-electron states
available to the photoelectron from a free gaseous molecule).
If the quantum restriction of k-conservation is maintained a
transition can only occur at those values of k for which there is
a gap of exactly hν between occupied and unoccupied E states
(represented by vertical transition arrows only in Figure 4a).
Having reached the final state (the first step of the three-step
model)[7] the electron is then transported through the crystal
(second step), and ejected across the surface (third step) as a
free electron.   Thus, changing hν will, in general, change the

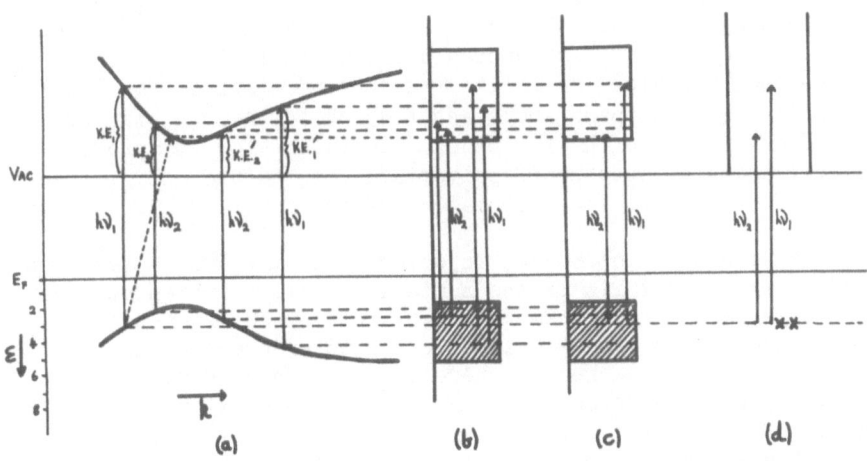

Figure 4.   (a) and (b) hypothetical (E,k) diagram and DOS
illustrating the effect of k-conservation and final states on the
photoelectron spectra at different incident photon energies, hν;
(c) effect of removing k-conservation; (d) equivalent situation
for a free molecule.

transitions which are allowable (e.g., Figure 4a in which $h\nu_1$ is changed to $h\nu_2$) and the resultant UPS spectra will be different for different $h\nu$, unlike gas phase spectra where a continimum of levels is always available to connect with the initial state no matter what the value of $h\nu$ (Figure 4d).

The nearest analogy in free molecules to the solid state situation described above is probably the phenomenon of auto-ionization where an electron is initially excited into a bound state above the ionization limit, subsequently decaying into the ionization continimum.   The resultant spectrum is governed by the transition probability to the bound state (autoionizing state) as explained by A.F. Orchard.[8]

From what has been said above the UPS spectrum of crystalline Au would be expected to vary with $h\nu$, and not to be representative of the initial DOS.   This is in fact so for $h\nu$ up to about 35 eV.[9]

There are practical difficulties in obtaining a continuous range of $h\nu$ sources, which will not be dealt with here.   The most difficult range, 20 eV up to soft X-ray values can now be covered by synchrotron radiation,[9] when available.   Above about 35 eV, however, the variation with h becomes much less drastic.   For example, the HeII spectrum of Figure 3b is quite similar to the XPS spectrum[6] ($h\nu$ = 1486.6 eV) of Figure 3c, and both are quite close representations of the calculated DOS (Figure 3d).[10]   This comes about because at high enough h , the final states reached are essentially featureless, like the continuum states for free molecules.   This happens above about 35 eV for Au, but may be at different values of h for other materials: for example the HeI, HeII and AlK spectra of polycrystalline Hg are all similar and therefore probably representative of the initial DOS.[11]

Thus if one is interested in determining the DOS, as opposed to examining the details of the E,k diagram, it is often necessary to use high-energy $h\nu$ to remove final-state effects, and it is best to check using several values of $h\nu$ to ensure that such a situation has been reached.   Therefore one might think that there is no obvious reason why UPS should be used at all for obtaining DOS information, when XPS can always guarantee that h is sufficiently high for final-state modulation effects to be removed.   In fact, there are three reasons for using UPS, two of them practical.   First, sensitivity is considerably higher for valence levels using UPS photon energies because ionization cross-sections are higher, and second, resolution is higher (a few meV compared to about 1 eV for non-monochromatic X-radiation). The use of an X-ray monochromator in XPS substantially improves resolution (to ca. 0.5 eV) allowing more detail to be observed in the experimental curves (e.g., Figure 3e for a single crystal of gold).[10]   The third point is related to the first, namely that even if one has reached sufficiently high hν to eradicate final-state modulation, there is still the remaining possibility that the experimental spectrum is a modified version of the initial DOS owing to variations in the ionisation cross-sections of electrons of different character (s,p,d). Since relative cross-sections as well as absolute cross-sections vary significantly with photon energy (e.g., C 2s/2p increases by more than an order of magnitude on going from HeII to MgKα), the relative intensities of different parts of the experimental DOS will alter with hν if the s,p,d parentage varies across the spectrum. Examples of cross-section effects are given in sections 2b and 3.

Up to now we have been considering only crystalline material. In amorphous materials or alloys, in which there might be insufficient order for k to be of any importance, the photoionisation may be represented by a simple band to band transition (Figure 4b), which makes it more likely that the UPS spectrum at low hν energies will (a) not alter with small changes in hν, (b) that it may represent the initial DOS. An example of this effect is observed with amorphous Au,[12] obtained by melting, where the HeI spectrum of the molten sample no longer shows the final-state modulations, and looks quite similar to the HeII spectrum (Figure 3f). It is not well established that k-conservation is an inviolate rule in crystalline samples and if the restriction is relaxed, many more transitions (represented by non-vertical arrows in Figure 4a) become allowed at each hν value, so it is more likely that small changes in hν will not change the experimental UPS.

Over the last two years XPS instruments capable of working at the UHV conditions required (because of the surface sensitivity of the technique - see section 4) and fitted with X-ray monochromators have begun to be used to study band structures. The paper by Hüfner et al.[13] reviews the situation to date for the metals Cu, Ni and Ag (and some alloys) and should be referred to by those interested in a more detailed discussion of this field.

2.2 Semiconductors

Figure 5 shows the XPS spectra of crystalline and amorphous Ge and Si, together with available calculations giving the one-electron DOS.[14] One can see that all the major features in the DOS calculated for the crystalline samples are in fact observed, although there are discrepancies in their relative intensities. In particular, whereas the calculated relative intensity of the band nearest E_F relative to the other two is almost the same for Ge and Si, the experimental ratio decreases considerably on going from Ge to Si. The atomic parentages of the bands estimated from the calculations indicate that the first band is mainly p-like in character (3p Si; 4p Ge), whereas the two higher

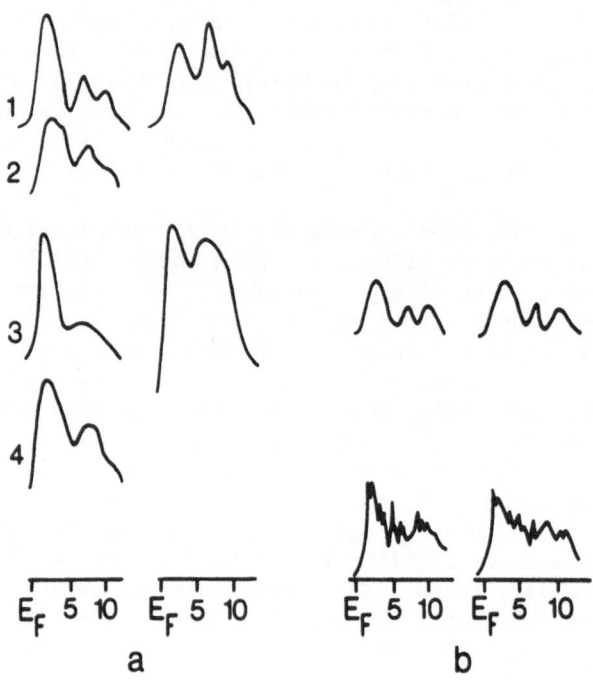

Figure 5.   (a) Photoelectron spectra of crystalline and amorphous
Ge and Si.   Trace (1) XPS of crystalline Ge and Si;   (2) UPS of
crystalline Ge (hν = 25 eV);   (3) XPS of amorphous Ge and Si;
(4) UPS of amorphous Ge (hν = 24 eV).   (References 14 and 15.)
(b) Calculated DOS of crystalline Ge and Si (upper trace) and
amorphous Ge and Si (lower trace).

peaks are s-like (3s Si;   4s Ge).   It might therefore be
suspected that the discrepancies in the relative intensities are
a result of differing cross-section for s and p states.   In fact
an extrapolation of relative intensities can be made from available
data on 3s, 3p, 4s and 4p <u>core-levels</u>, from which it can be shown
that:

$$3s/3p : 4s/4p \simeq 2.3,$$

i.e., in Si the ionization cross-section of the second two bands
(3s-like) relative to the first (3p-like) is expected to be about
2.3 times greater than that of Ge, which is found to be the case
experimentally, assuming the DOS to be the same in both cases
(Figure 5).

Turning to the amorphous results one can still see the effect of the cross-section variations, but in addition the band structure is significantly different to that of the crystalline case (cf. Au where the DOS are the same). The s bands have broadened and collapsed into one band, and the p band has moved about 0.5 eV nearer to E_F. The experimental results, allowing for the cross-section modulation are broadly in agreement with a theoretical scheme which is apparently based on the presence of five- and seven-membered rings in the amorphous state, for which the calculated DOS is shown in Figure 5b.

The band structures of Ge and Si have been studied many times by UPS, and a recent review has been given by Spicer.[15] There are some ambiguities which still remain. For example the UPS spectra of crystalline and amorphous Ge, taken using 25 eV and 24 eV radiation, are also shown in Figure 5.[15] Though hν is apparently sufficiently high to remove k-conservation effects the spectra show some unexplained differences when compared to the XPS results. The most significant is that the p band appears to be broader in the UPS even though the resolution is better. This effect has been observed for several other elements and may possibly be associated with the shorter escape depth in the UPS case, resulting in a spectrum which is more characteristic of a surface DOS. This point is referred to again in section 3.

Other semiconductor elements have also been well covered by UPS, notably Se and Te, and the interested reader is referred to the original publications for discussions of the results.[16,17]

3.   COMPOUNDS

3.1 Alloys

Originally it was believed that the DOS of a binary alloy was best represented theoretically by the rigid-band model,[18] (sometimes called the common band model) which provides a DOS invariant with percentage component composition but partially filled to different E_F values as the composition changes. There is little experimental evidence to support such a model and one could be forgiven for thinking that the XPS and UPS data for many alloy systems in which the pure component d levels are well separated (split band systems) suggested an alloy DOS which was just a weighted construction of the DOS of the pure components. This would indicate that the d-electrons within the valence region behave in a predominantly atomic fashion and that the influence of neighbouring atom potentials is not dramatic, a premise which is the basis of the tight-binding model.

Figure 6 shows the XPS valence region spectrum for a range of AgPd alloys, obtained using monochromatised X-rays.[13]   One can see that it is an oversimplification to describe the spectra just as a weighted superimposition of pure Ag and Pd since, as the concentration of one of the components becomes diluted, features resulting from it in the composite DOS become narrower than for the major component.   This may be taken to indicate a loss of band character by the dilute component owing to the separation of the individual atoms by the host metal.   The width of the d level of the dilute component is not entirely an atomic property, however, since interaction between this level and the free-electron component of the host metal remains.   This is accounted for by the virtual bound state model[19] for very dilute compositions, and by the coherent potential model[20] at all concentrations.   The spectrum at more equal concentrations of

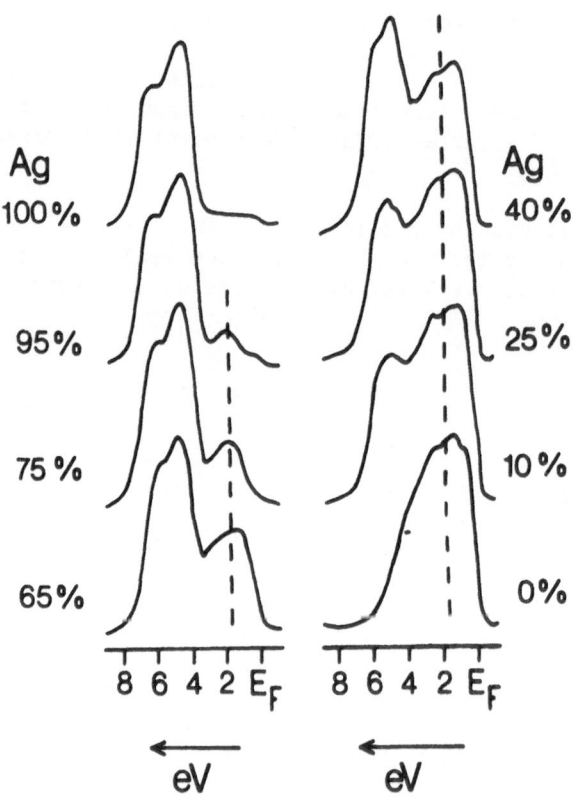

Figure 6.   XPS spectra for Ag/Pd alloys (reference 13).   The percentage compositions are as shown on the spectra.

components is closer to a superimposition of the pure components, but there is still some narrowing present. Throughout the whole range of composition, however, the positions of the bands remain unchanged with respect to those of the pure components. Similar results have been obtained for the Cu/Ni alloy system, as shown in Figure 7a.

If instead of using two similar transition metals, one looks at alloys between a free-electron-like element and a transition metal, the diluting power of the s,p electrons from the free-electron-like component becomes very obvious. Fuggle et al.[21] have examined the series Au,Au$_2$Al, AuAl and AuAl$_2$ and observe a narrowing of the Au d-band width from 5.3 eV (Au) to 3.2 eV (AuAl$_2$), but this time the centre of the band also moves (from about 5 eV to 7 eV below Ef). The Ag/Al series shows similar effects on the Ag d-band.

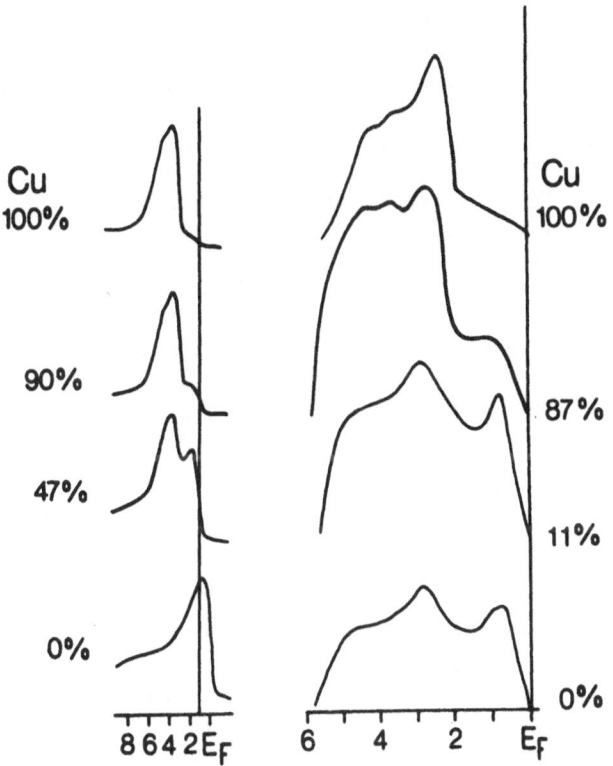

Figure 7.   XPS and UPS (hν = 10.2 eV) spectra for Cu/Ni alloys (references 13 and 23).   Compositions are as shown on the spectra.   XPS, left-hand column; UPS, right-hand column.

In the above discussion it has been supposed that the
concentration of the alloy components at the surface is the same
as that of the bulk for which the concentration analyses apply.
This may not in fact be so since it is well known that quite
dramatic differences can occur in the surface region (often
explainable by simple thermodynamics arguments) and that composition is sometimes dependent on the nature of the gas phase at
the interface.[22]   This complicates interpretation, particularly
for the more surface-sensitive UPS, but indicates the potential
of these techniques for studying such surface phenomena, which
may be of extreme importance in the behaviour of alloys as
catalysts.

     UPS spectra for a range of Cu/Ni alloys are shown in Figure
7b.[23]   The interpretation of the results is in general agreement
with that from the XPS data;   namely that the d electrons of the
two components form largely independent levels, because they are
well separated in energy.   There is no transfer of electrons
from Cu to Ni to fill the Ni 3d shell, as would be suggested by
the rigid-band model.   The broadening of the Ni 3d level on
going from very dilute Ni concentrations through to higher concentrations is ascribed to the building up of interaction between
the d-states on the individual Ni atoms, plus some interaction
between the d states and the free-electron-like component from
Cu.   In Ni-rich Ni/Cu alloys, coherent potential approximation
theory suggests some d overlap between Cu and Ni sites,[20] but
this effect, if present, is not obvious in the UPS spectra where
structure does not alter in position with change in concentration.

     An interesting result of the lack of common bands for the Cu
and Ni d electrons is that k becomes less and less valid as a
quantum restriction as more and more impurity atoms are dissolved
into the host metal.   When comparing HeI and HeII spectra, this
should show up as a gradual change in HeI towards the form of
HeII (i.e., closer to the DOS), and therefore might provide an
experimental test of the non-interaction of the alloy component
states.   Such behaviour has indeed been observed when Hg is
allowed to diffuse into polycrystalline Au films after being
initially retained at the surface at 77 K.[24]   The Au and Hg 5d
levels are well separated in energy, but as the alloy is formed
the Au d band part of the alloy HeI spectrum gradually changes
as a function of diffusion time to look like the HeII spectrum.
Of course, the destruction of k as a quantum mumber is also
explicable by assuming that the alloy is amorphous rather than
crystalline.

3.2 Layer chalcogenides

     Some of the transition metal chalcogenides have excited a
lot of experimental and theoretical interest recently because of

their layer structure (chalcogen-metal-chalcogen; weak bonds
between chalcogen atoms of successive layer) and because their
electrical properties range from insulating (ZrS_2) to superconducting $(NbSe_2)$.   Figure 8 shows the UPS spectra of a number of
such compounds[25] to illustrate the general form of their DOS which
approximately conforms to the Wilson and Yoffe rigid-band model[26]
for the series.   The model predicts a filled d-bonding level,
largely of chalcogen p-character, a non-bonding metal d_z^2,
$d_{x^2-y^2}$, and d_{xy} group and an empty d^* antibonding level well above
E_F.   It is the position and occupancy of the non-bonding metal
d levels which determines the electrical properties.   In Figure
8 we observe the progressive filling of a d band near E_F from
Zr to Mo. ZrS_2 is an insulator because there are only sufficient
electrons to populate the d bonding band (p states) which is well
below E_F. TiS_2 should be the same but there is a weak band at
E_F which, it has been suggested, is due to electron donation from
excess Ti into the non-bonding d level.   Nb and Ta have an

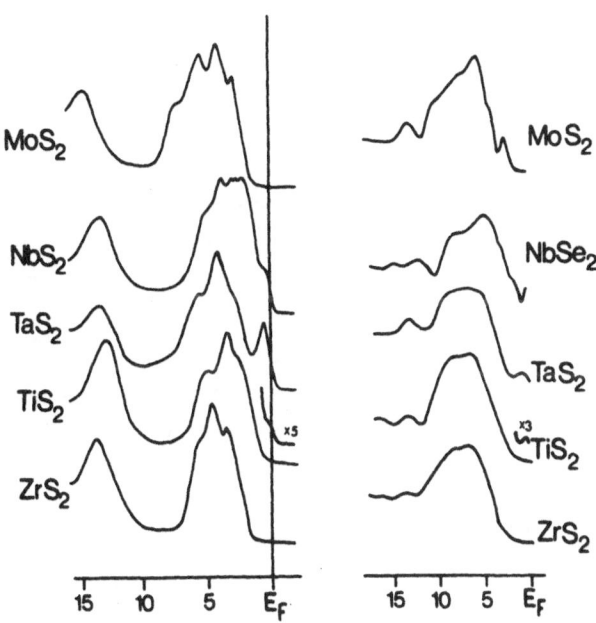

Figure 8.   (Left) XPS of some layer chalcogenides (reference 28);
(right) UPS $(h\nu = 40.8$ eV) of some layer type chalcogenides
(reference 25).

additional electron, and so their chalcogenides have a partly
occupied d band, and thus become semiconductors or metals
depending on its exact position.   Mo compounds have two electrons
available for the d band, but the band is lower lying and now
overlaps the p band.

The two-dimensional nature of the band structure of these
compounds suggests that the DOS of the first layers as given by
UPS, should also be characteristic of the bulk and that a simple
molecular orbital approach to the electronic structure might be
particularly appropriate.   Figure 9a, a UPS spectrum of $TaSe_2$[27]
shows the occupied bon-bonding d-band split off from the δ-bonding
p levels.   Pronounced changes in the relative intensities of the
p and d bands have been observed as a function of azimuthal angle
of photoelectron emission.   The change in intensity of the d
component is shown in Figure 9b, from which we see that the
favoured direction of photoemission avoids both the nearest
neighbour and second nearest neighbour Se atoms.   It would be
nice to be able to draw the conclusion from this result that the
d electron density points in the same directions, thus avoiding
the Se atoms, as expected for non-bonding orbitals, but at the

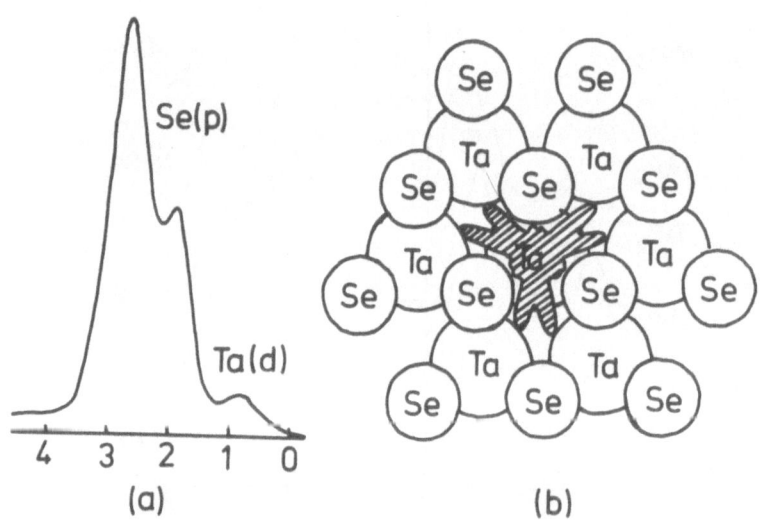

Figure 9.   (a) UPS (hν = 10.2 eV) of $TaSe_2$ (reference 27);
(b) radial plot of the azimuthal dependence of the d-emission
intensity from $TaSe_2$ (hν = 10.2 eV) superimposed on the geometry
of the crystal face (reference 27).

present time the connection between the <u>direction of photoemission</u>
and an electron density map of the original orbitals remains
tenuous.   What is clear is that the angular effects can be very
large, so that it is necessary when determining a DOS to beware
of cross-sectional modifications due both to different transition
moment matrices for different electron types and to different
angular properties.

The XPS spectra of the transition metal chalcogenides[28]
confirm most of the features of the UPS studies, though the
resolution is lower.   However, they reveal clearly the where-
abouts of the chalcogen s levels, which is not obvious from the
UPS data, and also indicate that part of the structure ascribed
to chalcogen p states in the UPS data must be due to scattered
electron density.   The monochromatised AlKα XPS results for
NbSe$_2$ and MoS$_2$ are included in Figure 8 for comparison with UPS.
Note that, probably as a result of cross-section and (for UPS)
scattered background effects, the relative intensities of the
different sets of fine structure differ in the two spectra.

3.3   Transition metal oxides

The XPS measurements on the valence levels of transition
metal oxides all exhibit the same general pattern: if there are
d-electrons in the compound one finds the d-electron density of
the metal component near E_F, and $O(2p)$ and $O(2s)$ bands lying
deeper (cf. the metal d, p and s levels of the layer chalcogenides).   The separation between $O(2p)$ and $(2s)$ levels is always
about 16 eV,[29] close to the free atom value.   This is what would
be expected in a fully ionic model, but in fact cannot be considered as good evidence for the validity of such a model since
various properties of the transition metal oxides indicate that
hybridization between metal d and oxygen 2p levels (covalency) is
important in some cases.   The XPS spectra of some monoxides and
dioxides are shown in Figure 10.[29]   One can see that for MnO the
levels derived from 3d and 2p overlap strongly, suggesting that
hybridization may be significant.   NiO presents an instructive
example since although it is the oxide for which most experimental
work (XPS and many other techniques) and theoretical work has been
done, yet it is the least understood.   In the XPS spectrum it is
not at all obvious where the $O(2p)$ levels lie, though $O(2s)$ is
clear enough.   There are four possible explanations for the
structure between 0 and 14 eV.   The first is that only the
feature at 2 eV represents the Ni3d levels, the peaks at 3.8 and
9 eV being satellites due to multi-electron excitation and the
$O(2p)$ level being weak and buried in the 6 eV region.   Support
for this view comes from the Ni 2p core-level spectrum which
shows similar satellite structure.   However in other Ni compounds
where satellites are observed in the (2p) level, similar structure
is <u>not</u> observed in the valence region, which suggests that the

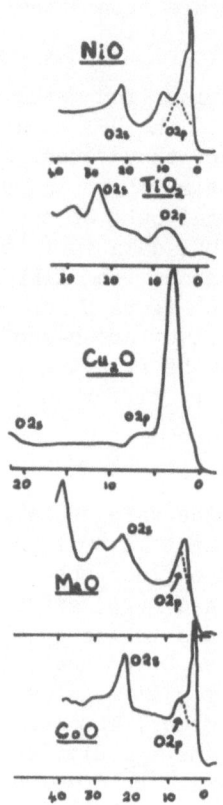

Figure 10.   XPS of some mono- and dioxides of transition metals
(reference 29).

structure in the NiO valence region may be due to something else.
The second possibility is that the entire structure from 0-14 eV
is Ni3d, with O(2p) being weak and buried at 6 eV.   There are
theoretical models which fit either of these cases, i.e., narrow
3d level and separate O(2p) level;   or 3d and O(2p) strongly
hybridized to spread the 3d band considerably.   A third possibility is that the first two features are genuine Ni3d structure and
only the 9 eV peak a satellite, perhaps a surface plasmon.
Finally, a fourth and more unlikely possibility is that the 9 eV
feature represents O(2p) and that the O(2p)-(2s) separation is
not 16 eV in NiO.   If the latter explanation is discounted, why

is the $O(2p)$ level so weak so as not to be observed?   Here we
return to arguments about relative cross-sections.   The startling
effect that gross differences in cross-section can have on a
spectrum is illustrated by the case of ReO_3.[30]   Covalency
effects are very strong in ReO_3 so that the calculated DOS in the
0-10 eV region has a combination of Re 5d and $O(2p)$ character
throughout (Figure 11a).   It is clear from a comparison of
Figures 11a and b, representing the XPS valence spectrum of ReO_3,
that the strong features in the DOS at ca. 3 eV, almost entirely
$O(2p)$ in character, make little contribution to the XPS spectrum.
This suggests that the relative cross-section for $O(2p)$ is very
low (perhaps twenty times less than for Re(5d)) and that the
observed XPS spectrum really represents only the Re(5d) character
in the total DOS.   It has been suggested that the experimental

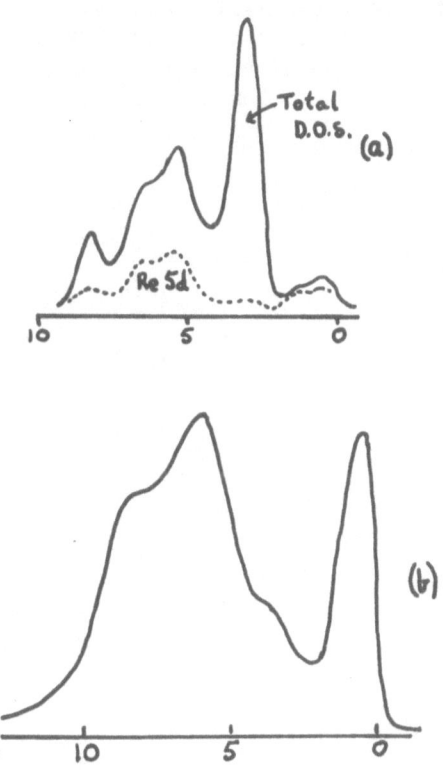

Figure 11.   (a) Calculated DOS for ReO_3;   (b) XPS of ReO_3
(reference 30).

XPS spectra of other oxides (particularly NiO) should be re-
considered in the light of the ReO_3 results.

Perhaps because of instrumental difficulties (e.g. charging
effects) or from difficulty in obtaining samples which are
stoichiometric in the surface region, UPS studies of bulk metal
oxides have not as yet been very common, particularly in the HeI,
HeII photon energy range.   More common are experiments which
trace the effect of adsorption of oxygen and the later stages of
oxidation of metal surfaces (see section 4).   Figure 12 shows
the HeII and XPS spectrum of Cu_2O grown from the metal substrate,[31] and the HeI[32] and XPS[29] spectrum of NiO produced in a
similar manner.   It must be emphasised that there is no guarantee
that these are stoichiometric samples, and in fact for NiO there
is strong evidence for two oxygen species, one of which is
probably adsorbed atoms.[33]   Apart from the improved resolution
for the HeII case and a difference in the relative intensities of
the $Cu(3d)/O(2p)$ features reflecting the lower relative cross-
section of the 2p level at XPS photon energies, the HeII and XPS
spectrum of Cu_2O are quite similar.   Just as for the XPS results
discussed above, the interpretation of the UPS of NiO is still
not definitely settled.   Relative $O(2p)$ cross-sections at 21.2 eV
photon energy are even higher than at 40.8 eV and so the $O(2p)$

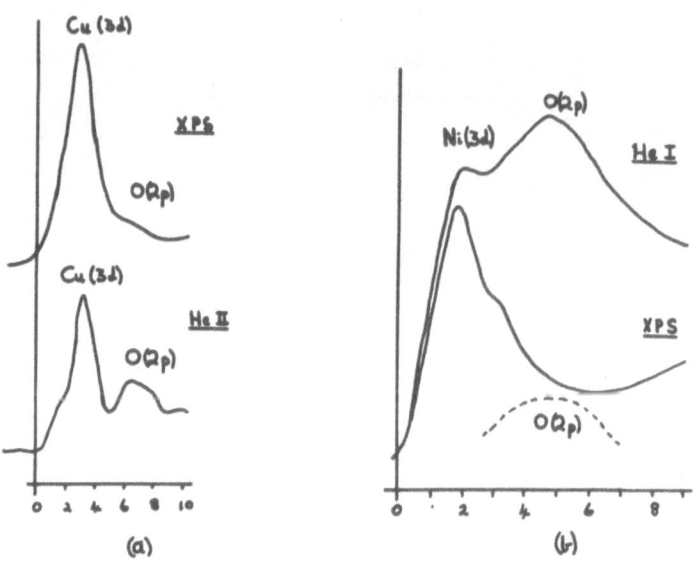

Figure 12.   (a) HeII and XPS spectra of Cu_2O (reference 31);
(b) HeI and XPS spectra of NiO (references 32 and 30).

level centred at about 5 eV below E_F, shows up strongly.   This adds support to the idea that in the XPS, O(2p) is present at about 5 eV but is not observed because of its low cross-section and the superimposition of satellite structure (see above). Structure corresponding to these satellites (3.8 and 9 eV) is not obvious in the UPS spectra, but the signal intensity is high over the whole of this region with the superimposition of the O(2p) level at 5 eV, so they could be present though not observable.

3.4 Polyatomic ionic salts

It is normally assumed that the interactions between ions in a lattice containing polyatomic species is very small and that, to a first approximation, the effects of crystal environment are negligible.   Thus the valence region spectra of $M^+ClO_4^-$ and $M^{2+}SO_4^{2-}$ are nearly independent of the cation and characteristic of the valence region of the anion.   The valence shell XPS spectrum of $LiClO_4$ is shown in Figure 13 together with an assignment based on a molecular orbital scheme for the ClO_4^- anion.[34] The m.o. calculation is in fair agreement with the experimental results, and it can be seen that a_1 and t_2 are O(2s) derived, a_1 and t_2 are a combination of Cl(3s) and (3p), while the peaks labelled O(2s) and (2p), and e t_2 and t_1 are largely O(2p).   One cannot be sure of the assignment of the individual orbitals from the XPS spectrum alone, but X-ray fluorescence data helps in defining the symmetries of the orbitals, as described by D. Urch[35] in Chapter   .   In fact, the intensities of the peaks in the XPS spectrum are not even a direct measure of the orbital degeneracies because of the different photoionization cross-sections of orbitals with predominantly Cl(3p), (3s) and O(2s) and (2p) character.   If the relative cross-sections could be established in some way, a further check on the validity of the calculation would be to compare the experimental valence level intensities with those calculated from an orbital population analysis. Actually, the required information can be obtained directly from the XPS spectra of MgO and LiCl because to a good first approximation the valence levels in these ionic lattices have essentially the characteristics of atomic orbitals.   Thus the peaks at 8 eV and 24 eV in MgO represent O(2p) and O(2s) orbitals respectively, and those at 6 eV and 17 eV for LiCl represent Cl(3s) and (3p) respectively.[34]   The relative cross-sections for these four orbitals can then be worked out from the peak intensities, provided that one spectrum is normalised with respect to the other.   This can be done by measuring the ratio between intensities of O(1s) and Cl(2p) in $LiClO_4$, or in any other molecule containing both O and Cl.   The experimental cross-sections determined in this way are:

O(1s) 1.000;   O(2s) 0.048;   O(2p) 0.005
                    Cl(3s) 0.058;   Cl(3p) 0.024.

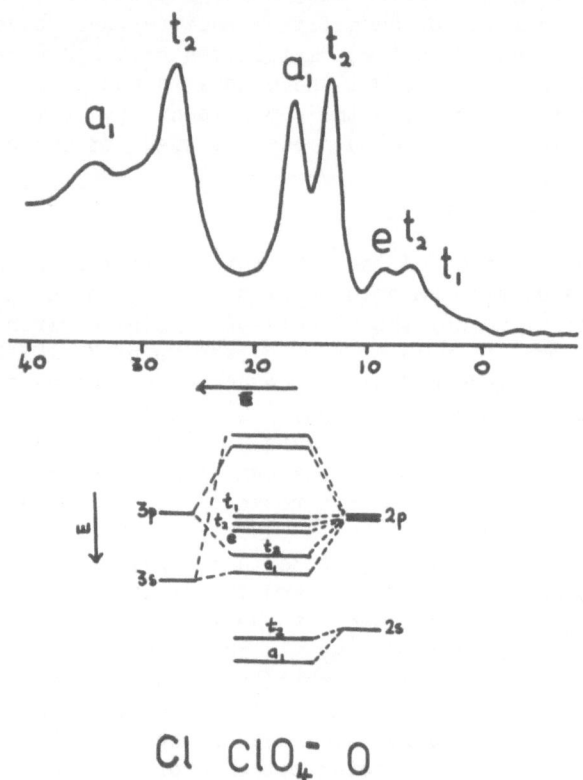

Figure 13.   XPS spectrum of LiClO₄ with MO assignments
(reference 34).

These cross-sections may then be used to test any calculation
which derives atomic populations of the valence level orbitals of
LiClO4 by comparing the relative intensities of the orbitals
determined from the calculated atomic populations and their known
cross-sections to the experimentally measured relative intensities.

     In general one will not be able to obtain cross-section data
pertaining to valence level MO's directly from experiment since
systems in which the valence levels are essentially atomic (such
as O^{2-} and Cl^- in MgO and LiCl) are not generally available.
One then has to rely on the accuracy of calculated atomic populations.   For example, the relative valence level cross-section
for $C(2s)$ and $C(2p)$ may be estimated from an <u>ab initio</u> population

analysis of the t_2 and a_1 MO's of CH_4 (made up of $C(2s)$; $C(2p)$, and $H(1s)$) and the experimental relative intensities of these orbitals (assuming the $H(1s)$ cross-section to be negligible for AlKα radiation). Similarly $\delta O(2s)/\delta C(2p)$ can be found from calculation plus experiment on H_2O. The relative cross-sections of the atoms can then be obtained in the same way as for O and Cl in $LiClO_4$, i.e. by normalising them through core level intensities. For example, $\delta O(2s)/\delta C(2s)$ can be found from the relative intensities of $O(1s)$ and 1a, in H_2O, the relative intensities of $C(1s)$ and 1a, of CH_4, and the relative intensities of $C(1s)$ to $O(1s)$ in some molecule containing both atoms.

4.   ELECTRONIC STRUCTURE AT SURFACES

The way in which the electronic structures of adsorbates and substrates are modified during an adsorption process is of extreme theoretical and technological importance because of the role of such phenomena in catalysis, biology and solid-state devices. However, the present level of information and understanding of this topic might be compared to that prevailing in classical structural chemistry at the beginnings of X-ray crystallography. It seems likely though that after several decades of relatively slow progress a rapid expansion of knowledge of surface states, and application of this knowledge, will take place over the next decade because of the invention of physical techniques which will play the same role for surfaces as X-ray crystallography did for classical structure determination. The author believes that the electron spectroscopic techniques will play a major role, having already provided considerable information over the last four years.

Several reviews of the application of electron spectroscopy to surface studies have appeared recently,[3-5] so only a brief account will be presented here with a few specific examples of the work pertaining to valence level studies. Instrumental aspects will not be mentioned at all, since they are adequately dealt with elsewhere,[36] except to say that all the normal conditions and facilities for studying clean surfaces must be met. Thus the electron spectrometer must have a clean ultra-high vacuum, i.e. no hydrocarbons and a base pressure of less than 1×10^{-10} torr, since at 1×10^{-6} torr a reactive surface can be completely contaminated in a few seconds. Facilities for heating, cooling and cleaning samples within the instrument at these pressures must also be provided. Such systems are very expensive so it makes economic sense to combine as many techniques as possible within one vacuum system. It is also sound practical sense because comparisons between the results of different techniques applied to the same surface become more meaningful.

The rest of this section is divided into two parts: (1) a
discussion of the surface sensitivity of the techniques and (2)
examples of electronic structure studies.

4.1 Surface sensitivity

There are several ways of defining the sensitivity of a
technique: speed of detection, detection limit and sensitivity
to chemical information. Electron spectroscopy can be used for
the study of surface composition and electronic structure because
of the short inelastic scattering mean free path lengths, l_e, of
the electrons concerned. The impacting photons (vacuum UV or
X-ray) always penetrate a sample to a considerable depth, but the
l_e value of the ejected photoelectron is a function of its kinetic
energy, the relationship and approximate absolute (for metals)
values being shown in Figure 14a.[5] Thus it can be seen that a
HeII photoelectron spectrum might consist of electrons with l_e
values as low as 5Å, whereas an AlKα XPS spectrum of the same
valence levels will represent l_e values of 25-30Å. Only those
electrons which do not undergo inelastic scattering will contribute to the primary photoelectron spectrum, the rest appearing in
the scattered electron background at lower kinetic energies. In
practice this results in the less surface-sensitive XPS spectrum

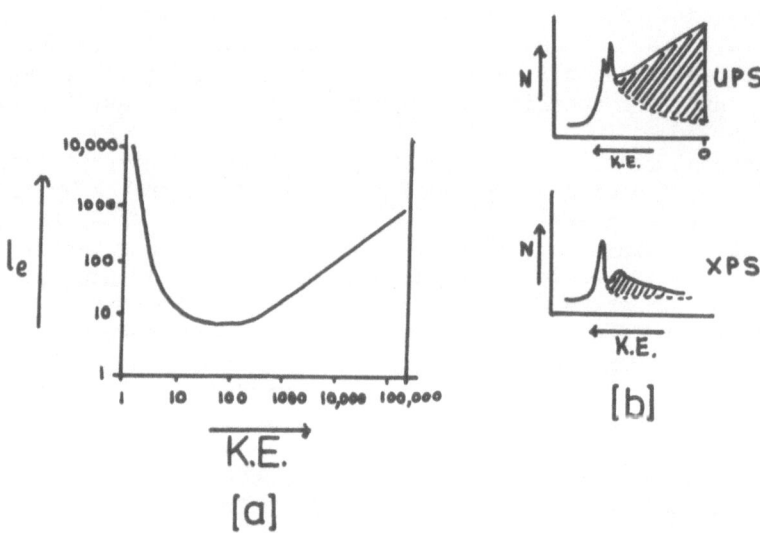

Figure 14. (a) Relationship between electron inelastic scattering mean free path length, L_e, and electron K.E. (reference 5);
(b) schematic XPS and UPS spectra showing likely scattered
electron contribution.

of the valence levels being 'cleaner' than the UPS because
multiple scattering results in the majority of scattered electrons
having energies near zero, thus leading to a high electron background for the UPS spectrum but a lower one for XPS (Figure 14b).

Hence, comparing the two techniques, UPS may be up to five
times as surface-sensitive in terms of the depth contributing to
the primary spectrum, but that spectrum is likely to be superimposed on a scattered background.   This will be less true for
a HeII spectrum than for UPS results at lower photon energy.   A
further important practical difference is that because both photon
flux and adsorbate cross-section (for valence levels) are lower
in XPS, the total signal strengths are weaker.   In addition it
is often the case that one is looking at adsorbed materials whose
valence levels have relatively lower cross-section, compared to
the substrate levels, in the XPS than in UPS (e.g., $O(2p)$: Metal
$(3d)$).   Taking all these considerations together, it is not
surprising that whereas UPS is used fairly routinely to study
valence levels of adsorbates at as little as a few percent monolayer coverage, there have hardly been any reported cases of
adsorbate valence levels being studied by XPS.   This section will
therefore only deal with the former studies.   XPS still has a
crucial role to play, however, in giving core-level spectra which
generally consist of electrons with much shorter escape depths
(lower K.E.), and high cross-section.   In addition to providing
an elemental analysis (which UPS on its own cannot supply) the
chemical shifts in the XPS results are a valuable source of
information on the surface bonding.[4,5]

4.2 Surface electronic structure studies

Even for a material which has its surface completely free of
adsorbates it is to be expected that the DOS of the surface will
be different from that of the bulk because of the termination of
the bulk structure.   Exactly how different it might be.is often
not clear.   Experimental evidence from UPS is generally incon-
clusive, both because of lack of a known experimental bulk UPS
spectrum to compare the surface-sensitive results with and the
difficulty in knowing whether the surface region has exactly the
same stoichiometry as the bulk.   In special cases, where well-
defined strong surface states are present which are destroyed on
adsorption, UPS has been able to follow the changes and therefore
confirm the importance of the surface states.   For example, the
(100) surface of W exhibits a strong surface state which is
destroyed on adsorption of a quarter of a monolayer of H_2.[37]

More subtle changes are observed in substrate band structure
on adsorption owing to electron transfer to or from the adsorbate,
(see subsequent Figures) but interpretations are usually lacking,
so interest at this point is concentrated on the striking and more

interpretable effects observed in the adsorbate valence levels.
Figure 15 is a hypothetical schematic representation of the UPS
spectra of an adsorbate for different possible types of adsorption.   Figure 15a shows a gas phase UPS spectrum with resolved
vibrational fine structure, and Figure 15b shows what might be
expected if the same molecule were condensed as an inert metal
substrate.   Apart from loss of fine structure the features shown
should be rather similar to those of the gas phase both in
relative positions and intensities if substrate-adsorbate interactions are negligible.   There is a problem in comparing the
absolute IP values in the two cases, as discussed later, and
Figures 15a and b are aligned arbitrarily to the same energy
scale.

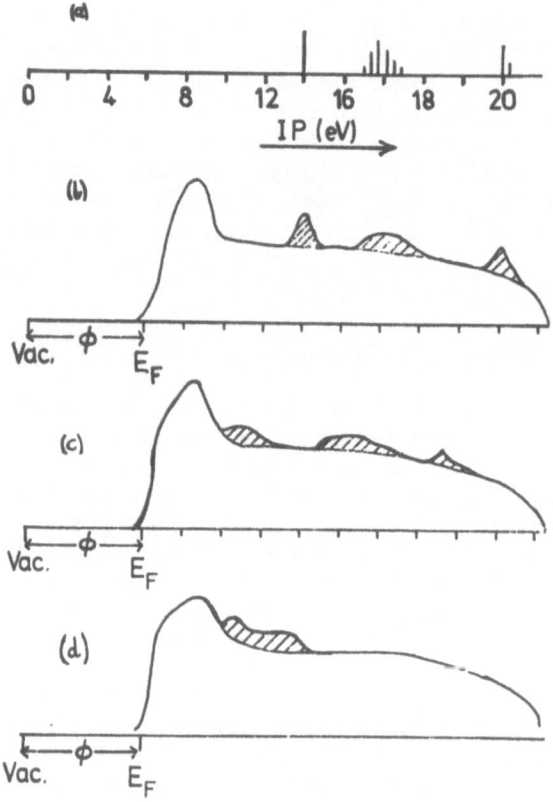

Figure 15.   Schematic representation of UPS spectra for different
types of adsorption of a hypothetical molecule on a metal (see
text).

Figure 15c represents a possible modification to the UPS spectrum when chemisorption occurs (i.e. a chemical bond is formed between adsorbate and substrate), but the adsorbate molecule still retains its molecular identity. All the orbitals are still observable but their relative positions and intensities have changed because some are now involved in the adsorption bonding. Figure 15d represents the case where the chemisorption process has been sufficiently strong to dissociate the adsorbed molecule in such a way that no orbital characteristics of the molecule remain and new features representative of bonding between the fragments and the substrate take their place. It could be difficult to distinguish between (c) and (d) if the movement of levels in (c) is large, or if the spectrum is very complex.

Condensation of CO_2 on Au at 77K[38] provides an example of case (a). The valence levels have relative I.P.'s close to those of the CO_2 molecule in the gas phase, although even in this simple case an interesting complication is observed in that condensed CO_2 molecules are distinguishable from physically adsorbed molecules (physically adsorbed molecules being defined as those where some influence of the substrate is felt) by a shifted set of valence levels. Thus, below the saturation vapour pressure (s.v.p.) at 77K only one set of orbitals is observed, but above s.v.p. two are observed, the second set at about 1.4 eV higher I.P., which grow in intensity as a function of time, representing the growth of multi-layers of condensed CO_2. When H_2O is adsorbed,[39] only one set of levels is observed, from which one may conclude that strong H_2O–H_2O interactions are always dominant. In agreement with this the separation between valence levels of adsorbed H_2O is not the same as in the gas phase, in other words the adsorbate level positions are modified by adsorbate–adsorbate interactions rather than adsorbate–substrate interactions.

Situations corresponding to those in Figures 15c and 15d may be found with various conditions of adsorption of CO on Mo.[40] Figure 16 shows the UPS spectra of CO adsorbed at 77K and at 300K. At 77K either two or three molecular orbitals are observed (the band nearer E_F may be two overlapping bands) which correspond either to the 5δ and 1π levels of CO, or the $5\delta + 1\pi$, and 4δ levels (there has been considerable discussion as to the correct assignment of the levels).[41] Thus there is a strong chemisorption bond which perturbs the levels considerably, but does not cause dissociation. Adsorption at 300K, or warming the surface from 77K to 300K, produces only features characteristic of carbon and oxygen atomic levels, indicating that a dissociation reaction occurs at the higher temperature. This interpretation of the data is fully supported by simultaneous XPS core-level results which show an O(1s) peak at 300K identical to that produced by O_2 adsorption, but a peak at 77K which is about 1 eV higher.[40]

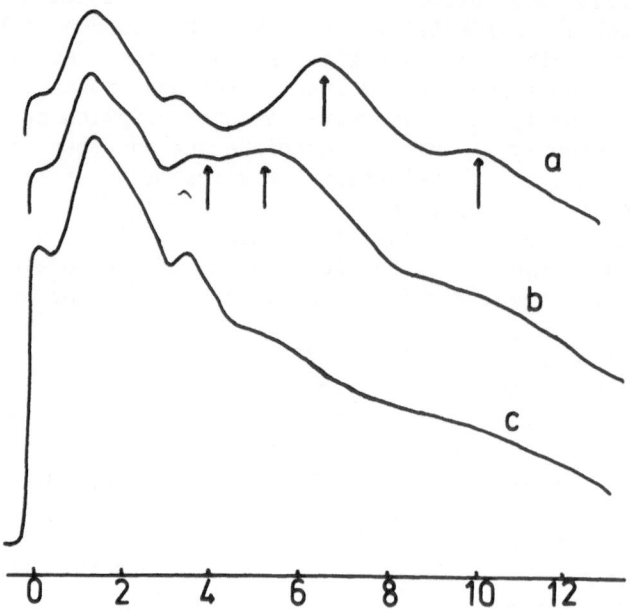

Figure 16.   HeI spectra for CO adsorbed on Mo (a) at 77K and
(b) at 300K;   (c) represents the clean Mo spectrum (reference
40).

The adsorption of ethylene on nickel[42,43] (Figure 17)
provides an example for which the whole range of interactions
may be observed.   Condensation at 77K of a few monolayers
results in the spectrum shown in Figure 17a.[44]   Warming to
200K desorbs most of the ethylene, and only a strongly chemisorbed monolayer remains.   The relative position and intensity
of the upper π levels are then found to have altered compared to
those of the other orbitals, an indication that the upper π level
is the one involved in the chemisorption bonding.[42]   Warming
the sample further to 300K (Figure 17d) completely changes the
spectrum, all structure resembling the C_2H_4 spectrum being lost,
but replaced by structure identical to that obtained from the
adsorption of acetylene, i.e. a dissociation reaction has
occurred.   An interesting feature of the acetylene spectrum is
the great difference of the outer $(\pi)^4$ level compared to that in
the gas phase or of condensed acetylene.   A sharp intense peak
has been replaced by a weak broad feature.   From the difference
spectrum $(Ni + C_2H_2 - clean\ Ni)$ it can be seen (Figure 17e) that

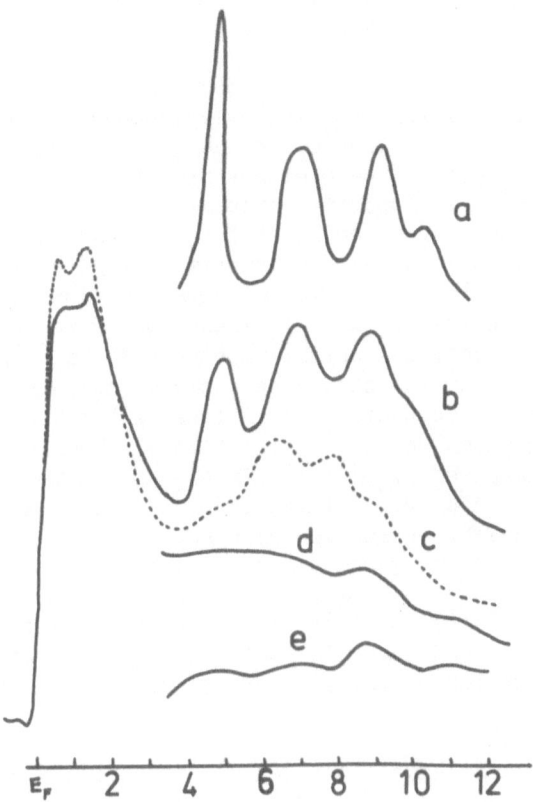

Figure 17.   HeI spectrum for the adsorption of ethylene on polycrystalline Ni (reference 43):   (a) gaseous ethylene (schematic);   (b) condensation of a few monolayers at 77K; (c) sample warmed to 200K;   (d) sample warmed to 300K;   (e) difference spectrum of (d) and clean Ni spectrum.

two features are present in the $(\pi)^4$ region, one displaced from the gas phase spectrum and one not.   This is what would be expected of an acetylene molecule using one of its   orbitals to bond to the surface, the other being non-interacting and parallel to the surface.   The removal of orbital degeneracies during adsorption is likely to be an area suitable for study by UPS in many other systems (e.g. rings, conjugated double bonds, etc.). When adsorption behaviour is complicated, or when it is not just confined to the surface, the UPS results are not nearly so helpful.

For instance XPS core-level measurements during oxidation of
nickel reveal three types of oxygen species[33] with at least one
corresponding to penetration below the surface, and with all three
types co-existing under some conditions, but in the UPS one cannot
even distinguish all three types.

Ideally one would like to interpret changes in the absolute
binding energies of orbitals on going from gaseous to adsorbed
states directly in terms of changes in the geometry of the
adsorbate and the nature of the chemisorption bond. This cannot
be done yet because of two unsolved problems. The first is to
find a suitable common reference level for gas phase and solid-
state measurements; the former are experimentally referenced to
the vacuum level, the latter to E_F. A work function term must
therefore be added to the solid state measurements to reference
them to the vacuum level, but unfortunately there is no general
agreement on what correction should be included to take account
of the fact that the adsorbate molecule lies within the surface
dipole which contributes to the work-function. The second point
is that equating measured I.P.'s with initial state valence level
energies ignores the electron relaxation that takes place into
the core-hole during the ionisation process. If the relaxation
energy is different in the gaseous and adsorbed states (relaxation energies are known to be larger in the solid-state than
for free molecules)[45] any absolute comparison between the I.P.'s
of a gaseous and adsorbed molecule includes this unknown
difference in relaxation energies, thereby negating a comparison
in purely initial state terms.

Demuth and Eastman have attempted to circumvent such
problems.[42] They use condensed phase values as 'free-molecule'
references and assume that in a spectrum of a chemisorbed species
all orbitals which are not involved in the bonding would remain
at the condensed phase values, except for the difference in
relaxation energies in the two cases. A relaxation energy
difference is thus obtained experimentally for the non-bonding
levels. Demuth and Eastman then assume that the same relaxation
energy shift is appropriate for the levels involved in the bonding,
a highly questionable assumption which nevertheless enables them
to separate the experimentally observed shift into its initial
state and relaxation effect components. In this fashion the
initial state shift for the $(\pi)^4$ level of ethylene when adsorbed
on nickel was estimated as 0.9 eV. Using the treatment of
Grimley[46] they then calculated a chemisorption bond energy from
the shift. The value found was compatible with the experimentally
observed dissociation of ethylene to acetylene on the surface at
300K, though it remains to be seen whether similar successes can
be achieved with other systems.

The above discussion indicates that it is relatively easy in UPS to pick out the adsorbate-induced structure and often to tell whether dissociation has occurred, and what fragments are produced. The orbitals involved in the chemisorption process are sometimes identifiable, particularly when degeneracies are lifted. Some progress has been made in treating the shifts observed in these levels, but a full analysis awaits a solution to the problem of the reference level and a better understanding of relaxation energies.

Note added in Proof

Since the preparation of this Chapter, the Proceedings of the International Conference on Electron Spectroscopy, held at Namur, Belgium, April 16-19, 1974, have been published (Journal of Electron Spectroscopy, Volume 5). A section of these proceedings (pages 531-642) is devoted to valence band studies using XPS and UPS (including synchrotron radiation). Papers include studies of pure metals (Au, Ag, Cu and the entire lanthanide series); alloys (Cu/Ag, Cd/Mg and Cu/Zn); and a number of compounds (GaAs, GaP, halides of Ag, Cu, Pb and Cd, and a series of fluoropolymers). The importance of improving information on photoionisation cross-sections and of taking measurements at several photon energies is clear from these studies. The same volume also contains a section on surface studies (pages 291-447), including a review by the present author, and a study by UPS of the decompositions of methanol and formaldehyde at a W(100) surface.

REFERENCES

1. J.M. Thomas, chapter 2 , this volume.

2. J.M. Thomas, I. Adams, R.H. Williams and M. Barber, J. Chem. Soc. Farad. II, 68, 755 (1972).

3. C.R. Brundle in 'Defect and Surface Properties of Solids', Vol. 1, ed. J.M. Thomas and M.W. Roberts, Chemical Society Specialist Periodical Report.

4. C.R. Brundle, J. Vac. Sci. Tech. 11, 212 (1974).

5. C.R. Brundle, Surface Science, to be published, March Issue, 1975.

6. C.R. Brundle and M.W. Roberts, Proc. Roy. Soc. (Lond.) A331, 383 (1972).

7.   C.N. Berglund and W.E. Spicer, Phys. Rev. A136, 1030; 1044
     (1964).

8.   A.F. Orchard, chapter   , this volume.

9.   J. Freeouf, M. Erbudak and D.E. Eastman, Solid State
     Communications, 13, 771 (1973).

10. D.A. Shirley, Phys. Rev. B5, 4709 (1972).

11. C.R. Brundle and A.F. Carley, unpublished work.

12. D.E. Eastman, Phys. Rev. Letters, 26, 1108 (1971).

13. S. Hüfner, G.K. Wertheim and J.H. Wernick, Phys. Rev. B8,
     4511 (1973).

14. L. Ley, S. Kowalczyk, R. Pollak and D.A. Shirley, Phys.
     Rev. Letters, 29, 1088 (1972).

15. W.E. Spicer 'Photoemission Spectroscopy and the Electronic
     Structure of Amorphous Materials - Studies of Ge and Si',
     to be published.

16. N.J. Shevchik, J. Tejeda, M. Cardona and D.W. Langer,
     Solid State Communications, 12, 1285 (1973).

17. N.J. Shevchik, M. Cardona and J. Tejeda, Phys. Rev. B8,
     2833 (1973).

18. N.F. Mott and J. Jones 'Theory of the Properties of Metals
     and Alloys', Dover, N.Y. (1958).

19. J. Friedel, Can. J. Phys. 34, 1190 (1956); P.W. Anderson,
     Phys. Rev. 124, 41 (1961).

20. G.M. Stocks, R.W. Williams and J.S. Faulkner, Phys. Rev.
     B4, 4390 (1971).

21. J.C. Fuggle, L.M. Watson, D.J. Fabian and P.R. Norris,
     Solid State Communications, 13, 507 (1973).

22. W.M.H. Sachtler, Le Vide, No. 164, p. 67 (1973).

23. D.H. Seib and W.E. Spicer, Phys. Rev. B2, 1676 (1970);
     2, 1694 (1970).

24. C.R. Brundle, unpublished work.

25. P.M. Williams and F.R. Shepherd, J. Phys. C6, L36 (1973).

26.   J.A. Wilson and A.D. Yoffe, Adv. Phys. 18, 193 (1969).

27.   M.M. Trawn, N.V. Smith and F.J. Di Salvo, Phys. Rev. Lett. 32, 1241 (1974).

28.   G.K. Wertheim, F.J. Di Salvo and D.N.E. Buchanan, Solid State Communications, 13, 1225 (1973).

29.   S. Hüfner and G.K. Wertheim, Phys. Rev. B8, 4857 (1973).

30.   G.K. Wertheim, L.F. Mattheiss, M. Campagna and T.P. Pearsall, Phys. Rev. Letters 32, 997 (1974).

31.   S. Evans, Farad. II, to be published.

32.   C.R. Brundle, unpublished data.

33.   C.R. Brundle and A.F. Carley, Chem. Phys. Letters, to be published.

34.   R. Prins and T. Novakov, Chem. Phys. Letters, 9, 593 (1971); R. Prins, Chem. Phys. Letters, 19, 355 (1973).

35.   D. Urch, chapter   , this volume.

36.   C.R. Brundle, M.W. Roberts, D. Latham and K. Yates, J. Electron Spectrosc. 3, 241 (1974).

37.   B.J. Waclawski and E.W. Plummer, Phys. Rev. Letters 29, 783 (1972).

38.   S.J. Atkinson, C.R. Brundle and M.W. Roberts, Disc. Farad. Soc. No. 58, to be published.

39.   C.R. Brundle and M.W. Roberts, Surface Science, 38, 234 (1973).

40.   S.J. Atkinson, C.R. Brundle and M.W. Roberts, Chem. Phys. Lett. 24, 175 (1974).

41.   See reference 5 and the General Discussion of Disc. Farad. Soc. No. 58, to be published.

42.   J.E. Demuth and D.E. Eastman, Phys. Rev. Lett.

43.   C.R. Brundle – General Discussion Farad. Soc. No. 58, to be published.   (Also in reference 5.)

44.   D.W. Turner, A.D. Baker, C. Baker and C.R. Brundle, 'Molecular Photoelectron Spectroscopy', Wiley, N.Y. (1970).

45.   D.A. Shirley, Chem. Phys. Lett. 16, 220 (1972); P.A.
      Citrin and T.D. Thomas, J. Chem. Phys. 57, 4446 (1972).

46.   T.B. Grimley in 'Molecular Processes on Solid Surfaces',
      ed. E. Drauglis, R.D. Gretz and R.I. Jaffee, McGraw-Hill,
      N.Y. (1969).

MULTIPLET SPLITTING IN THE X-RAY PHOTOEMISSION SPECTRA OF OPEN-SHELL IONS

G.K. Wertheim

Bell Laboratories, Murray Hill, New Jersey 07974, U.S.A.

I.   INITIAL AND FINAL STATES IN PHOTOEMISSION

The photoemission process is characterized by the initial
state, consisting of the unperturbed atom, molecule or solid and
photon, and the final state comprising the ionized atom and photoelectron.   A major distinction between UV photoemission and X-ray
photoemission in solids arises from the properties of the final
state photoelectron.   At UV energies the electron is in a region
of the unoccupied band structure where the density of states is
strongly modulated.   The photoelectric transition probability
therefore depends on both the occupied and empty density of states.
At X-ray energies the excited electron is in a region where the
density of states has no significant modulation and may be
considered to be plane-wave-like.   The transition probability is
then proportional to the occupied density of states and, of course,
to the transition matrix element.   This is an important simplification which accrues from the use of energetic photons.

In X-ray photoemission the information obtained thus relates
almost exclusively to the final state of the ion which is left
behind.   Core electron ionization of closed-shell ions provides
the simplest case.   The final state is described by the quantum
numbers S, L and J of the ionized shell, e.g., $^2S_{1/2}$ for s-shells,
$^2P_{1/2}$ and $^2P_{3/2}$ for p-shells, etc...   The separation between the
two P states, i.e., the spin-orbit splitting, represents the
difference in the energy of the final state in which S is parallel
or antiparallel to L.   The intensity ratio is calculated from
the multiplicity, 2J+1, and is 2:1 for p-shells, 3:2 for d-shells
and 4:3 for f-shells.   It requires more energy to remove the
electron with spin antiparallel to the orbital angular momentum.

P. Day (ed.), Electronic States of Inorganic Compounds. 393–408. *All Rights Reserved.*
Copyright © 1975 by D. Reidel Publishing Company, Dordrecht-Holland.

There are, however, a number of other significant considerations.   Removal of a core electron requires readjustment of the
outer electronic orbitals which now see a different screened
nuclear potential.   In the final state this relaxation has, in
general, taken place.   In solids a redistribution of valence
electrons, in order to screen the lattice site with unit excess
positive charge, will similarly take place.   These processes
will greatly modify the final state energy and have been discussed
elsewhere.[1,2]

The overall process can be described in the following way.
As a first step the photoelectric process suddenly imparts an
energy h to one electron, raising it to a highly excited state
in the band structure.   If the electron were now to leave the
site of the photoelectric process without any readjustment or
relaxation of the electronic structure or lattice, we would reach
the Koopmans state, which is not an eigenstate of the solid.   In
actuality the photoelectron continues to interact electrostatic-
ally with the other electrons, and emerges from its lattice site
with an energy which reflects the energy of the final state
modified by relaxation, screening, as well as electronic and
lattice[2] excitations.   The final states are eigenstates, and all
states with the symmetry allowed by the photoelectric dipole
selection rule must be considered.   The relative probability of
populating a particular final state is determined by the projection
of the Koopmans state on all final states.   As we shall see,
these final states may be modified by exchange and crystal field
interactions.

II. PHOTOIONIZATION OF OPEN SHELLS

(a) Localized electronic states

Open-shell ions, e.g., those of the transition metals, rare
earths and actinides, present a more complex picture because the
initial state may have both spin and orbital angular momentum.
We consider first the photoionization of such shells themselves.
Note that the final states are different from those reached by
optical absorption because in photoemission the final state has
one less electron than the initial state.   Nevertheless, a
correspondence with optical spectra can be made, but it will be
necessary to use data for an ion with the same number of electrons
as the final state which will often be one with next lower atomic
number.

This is perhaps best clarified by some concrete examples
taken from recent work on the rare earths in which the 4f-electrons
are well localized.[4-7]   Consider Gd^{3+}, $4f^7$, with a Hund's rule
coupled $^8S_{7/2}$ ground state.   Removal of one electron, in the

absence of other electronic transitions, clearly leads to a ^7F-
state.   From the optical spectra of Eu^{3+} we know that the F-
multiplet spans 5000 cm^{-1} 0.62 eV, indicating that X-ray photoemission from Gd^{3+} should result in a single line containing the
unresolved ^7F-multiplet.   However, when we consider Hund's rule
coupled ions with shells which are more than half-filled, the
picture changes completely.   For example Tb^{3+}, $4f^8$, with $^7F_6$
ground state has seven electrons coupled with spin parallel and
one antiparallel.   Photoionization can therefore lead both to an
^8S-state with 7 spins parallel or to ^6P, ^6I, ^6D ^6G, ^6H and ^6F-
states with 6 parallel and 1 antiparallel.   These are of course
the ground and excited states of Gd^{3+}.   The photoemission spectra
of GdSb, TbSb and DySb which illustrate these observations are
shown in Figure 1, taken from a forthcoming publication[7] in which
the whole rare earth antimonide series is discussed.

The next question concerns the intensities of the lines.
In the case of Tb^{3+} a crude estimate is obtained by recognizing
that photoionization of the single spin-down electron produces
the ^8S-state, while removal of any one of the seven spin-up
electrons produces the sextet states.   More detailed information
is obtained by the method of fractional parentage which expands
the n-electron initial state in terms of the (n-1)-electron final
state and an outgoing plane-wave-like photoelectron.[8]   The
amplitudes of the bars in Figure 1 are obtained by this technique,
and correspond quite well to the experimental observations.

In transition metal compounds, where the crystal field
splitting is non-negligible even with the resolution currently
achieved by X-ray photoemission, the final state must be described
in terms of the electronic levels of the final state ion in a
suitable crystal field.   Such splittings have been identified in
FeF_2 and NiO. [9]

(b) Band electrons

How does this essentially atomic picture change when we
examine delocalized electronic states by photoemission?   In the
case of the noble metals, Ag and Au, one can demonstrate a very
good correspondence between XPS data and one-electron APW band
structure calculations.[10,11]   Except for the case of Ni, similar
agreement also obtains for the transition metals with open d-
shells.[12]   One can understand this by thinking of the photo hole
as delocalized, so that an ionic description based on a single
perturbed atom is inappropriate.   The behaviour in Auger spectroscopy where one has a two correlated final state holes, appears
to be entirely different.   Auger spectra cannot be understood in
terms of the folding of normal density of states into itself, but
indicate the production of a localized ionic final state, with
two holes on one atom.[13]   In the case of XPS no evidence has so

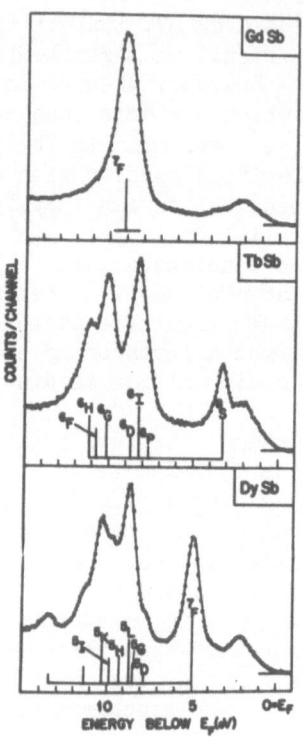

Figure 1.   Final state multiplet structure for 4f-photoionization of trivalent rare earth ions in GdSb, TbSb and DySb (from Ref.7).

far been presented to show that similar localization has taken place.

Even in the case of transition metal compounds, entirely band-like behaviour has been demonstrated, at least in systems with metallic conductivity.   For example, the XPS data for ReO_3 are readily interpreted in terms of an APW band structure calculation, and exhibit a d-electron conduction band with a clearly resolved Fermi edge.[14]   Similar d-electron conduction bands have also been seen in VO_2, V_2O_3 and in the sodium tungsten bronzes.[15]

The interpretation of data on compounds like NiO where the degree of d-electron delocalization is less, presents a considerable problem.   In fact the same data which have been cited to support an entirely localized molecular cluster calculation[16] have also been shown to be compatible with a band model.   In large part this dilemma can be traced to the lack of resolution currently available in XPS.

III. PHOTOIONIZATION OF CORE LEVELS IN OPEN-SHELL IONS

The extra complexity which appears when core levels of ions with incomplete outer shells are examined by XPS becomes apparent from the following simple consideration. Imagine an ion with a net spin S in an outer d or f shell, and remove one core electron. Clearly there will be two distinct final states depending on whether the core spin is parallel or antiparallel to the spin of the outer shell. The energy difference is given by an exchange integral. This description is of course not limited to core s-shells but applies to all core levels. The presence of both spin and orbital angular momentum in both the core and outer shell can clearly result in very complex spectra. This description in terms of exchange leads to the erroneous idea that a core s-electron spectrum will be split into two lines of equal intensity.

A proper description requires that we consider the final state in which the two incomplete shells are multiplet coupled.[17,18] A number of examples will serve to clarify these concepts.

1. <u>Core hole in s-shell; spin-only outer shell.</u> The spin of the core shell is coupled to that of outer shell (denoted by S), yielding two final states with spin $S \pm 1/2$. Mn^{2+}, $3d^5$, ^6S, according to this theory, should have a 3s-spectrum corresponding to ^7S and ^5S final states with intensity ratio 7/5, see Figure 2a. Gd^{3+}, $4f^7$, ^8S has ^9S and ^7S final states.

2. <u>Core hole in p-shell; spin-only outer shell.</u> All final states will be P-states. The net spin is again $S \pm 1/2$. The 3p-spectrum of Mn^{2+} will consist of ^7P and ^5P final states. Neglecting the spin-orbit interaction in the p-shell, the ^7P state can be formed only by coupling the core spin to the ^6S state of the $3d^5$ shell. On the other hand ^5P-states can be obtained by coupling $S = 1/2$ and $L = 1$ to the ^6S, ^4P, and ^4D-states of $3d^5$. As a result there are four energetically distinct final states ^7P, $^5P_3$, $^5P_2$ and $^5P_1$.

3. <u>Core hole in s-shell; outer shell with S and L.</u> All final states will have the L of the outer shell. Tb^{3+}, $4f^8$, $^7F_6$ will have ^6F and ^8F final states.

4. When there is orbital angular momentum in both shells, the number of final states tends to become large.

The most important new concept that emerges from the above examples is that the outer incomplete shell may change its net S and L during the photoemission process. The coupling responsible for these transitions is electrostatic and falls into the general concept of electron-electron correlations. This introduces the

multi-electron aspects of photoemission which must be more fully
explored to understand core electron multiplet splitting.   In
fact a critical comparison of the 3s-multiplet structure in
Figure 2a with the theoretical description given above clearly
shows that the 7/5 intensity ratio is not realized experimentally.
We shall see that this is also a result of electron-electron
correlations.

IV. MULTI-ELECTRON ASPECTS OF PHOTOEMISSION

Up to this point we have treated photoemission in terms of a
final state defined by the removal of one electron (by the photoelectric process), and the multiplet coupling of the resulting
spin and orbital angular momenta.   This is, however, overly
restrictive because the quantum states of other electrons can
change during photoemission, subject to the rule that the final
states have the same symmetry as the Koopmans state produced by
the removal of one electron.   It is not correct to artificially

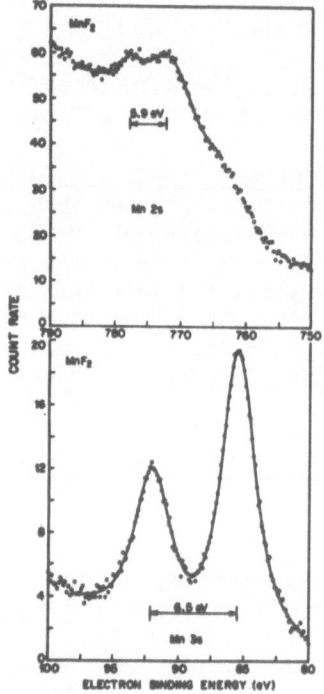

Figure 2a.   Core electron multiplet splitting for the 2s- and
3s-electrons in MnF_2.

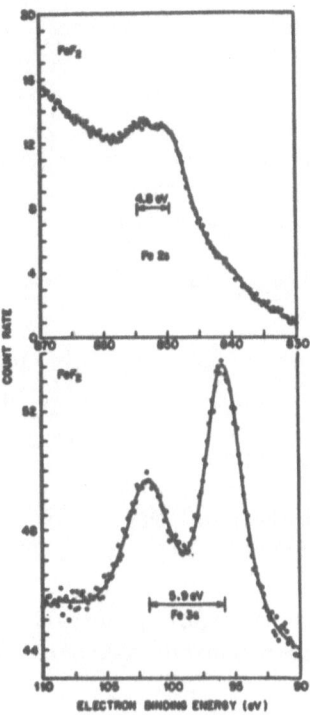

Figure 2b.   Core electron multiplet splitting for the 2s- and
3s-electrons in FeF_2.

separate the total process into a dipole photoemission process
and a monopole shake-up process as has long been customary.

This was initially demonstrated in a photoemission study of
the alkali halides[19] in which strong satellites appeared below
the photoemission line of the outermost core s-electron, e.g.,
the 3s of K^+, see Figure 3, and the 4s of Rb^+.   The unusual
intensity of the satellite, together with the fact that other core
lines did not have similar structure, clearly indicated that the
conventional shake-up treatment could not be applied.   The final
states which can contribute to the photoemission process must have
the same ^2S character as the final state due to the simple removal
of one core s-electron from the ^1S monovalent initial-state ion.
We can consequently think of the satellites as arising through
configuration mixing.

In the case of K^+, initially in the Ne $3s^2 3p^6$, ^1S configuration, the main line corresponds to Ne $3s^1 3p^6$, ^2S and the

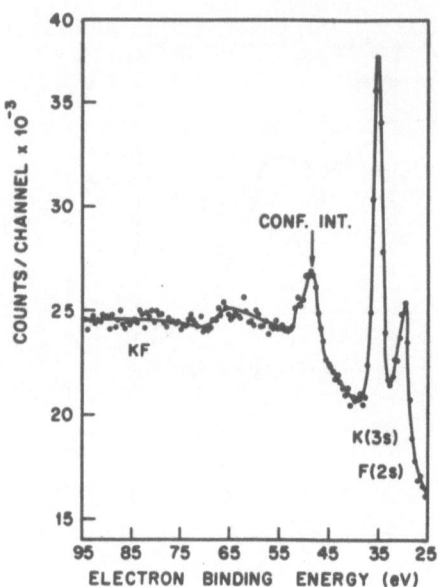

Figure 3. Configuration interaction satellite due to multi-electron excitation in 3s-photoionization of KF.

the satellite to Ne $3s^2 3p^4 3d$, ^2S with a much weaker contribution from Ne $3s^2 3p^4 4s$, ^2S. These cannot be described as monopole excitation satellites of either the 3s or 3p photoemission line. The energies of these states are known from atomic spectroscopy and are in good agreement with the measured satellite separation. One significant negative experimental result, the absence of a similar satellite on the 2s lines of Na^+, now makes very good sense simply because there are no 2d electronic states![20,21]

One may now wonder why such satellites are not observed more generally. In part the answer is that they had been overlooked, but mainly it reflects the fact that special conditions must be met to give them appreciable intensity, i.e., the other states with the same symmetry must be close in energy to the one electron state. This is usually possible only in photoionization of the outermost shell. The fact that electron correlations are strongest between electrons within a shell with the same principal quantum number is another way of stating the same limitation.

We now turn to apply these concepts to the photoionization of atoms with incomplete shells. During the photoionization of the incomplete shell itself this effect remains unimportant, but during core level ionization it adds greatly to the number of

available final states.   In the case of Mn^{2+} discussed above
satellite structure analogous to that of K^+ is formed, i.e. the
$3s3p^63d^5$ ^5S, final state is accompanied by $3s^23p^43d^6$, ^5S structure.[22]   A little thought will show that there is no analogous
satellite for the ^7S state, see Table I.   As a result the ^5S line
is weakened and shifted while the ^7S line remains unaffected.
This has the result of reducing the apparent multiplet splitting
and explains the discrepancy in the multiplet splitting between
theory and experiment discussed below.   The main components of
the 3s-multiplet structure of Co^{2+} in CoF^2 are shown in Figure 4.

TABLE I. ^7S- and ^5S-states produced by photoionization of 3s-
and 2s-shells.   The 3s-case is represented by considering ways
of distributing 12 electrons in the M-shell consistent with the
symmetry constraint.   The 2s-case by putting 7 into the L-shell
and 13 into the M-shell.   The initial state is $3s^23p^63d^5$, ^6S.

	$n^{(a)}$		$_vE_{ex}^{(b)}$
3S PHOTO-IONIZATION			
^7S	1	$3s^1(^2S)3p^6(^1S)3d^5(^6S)$	—
	2	None	
	3	$3s^1(^2S)3p^4(^3P)3d^7(^4P)$	116
^5S	1	$3s^1(^2S)3p^6(^1S)3d^5(^6S)$	—
	2	$3s^2(^1S)3p^4(^3P)3d^6(^3P_1)$	21
	2	$3s^2(^1S)3p^4(^3P)3d^6(^3P_2)$	21
	2	$3s^2(^1S)3p^4(^1D)3d^6(^5D)$	21
	3	$3s^1(^2S)3p^4(^3P)$ $(^4P)3d^7(^4P)$	116
	3	$3s^1(^2S)3p^4(^3P)$ $(^2P)3d^7(^4P)$	116
	3	$3s^1(^2S)3p^4(^3P)$ $(^4P)3d^7(^2P)$	116
2S PHOTO-IONIZATION			
$^7S,^5S$	1	$2s^12p^63s^23p^63d^5$	—
	2	None	
	3	$2s^12p^63s^23p^43d^7$	116
	3	$2s^22p^53s^23p^53d^6$	-68

(a) number of excited electrons;   (b) estimated energy relative to
n = 1 line in eV.

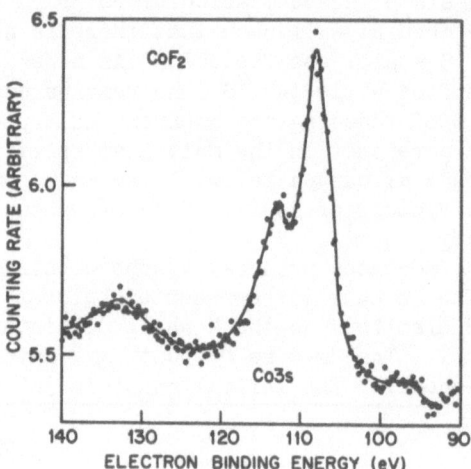

Figure 4.   The major components of the 3s-multiplet structure of
the 3s-electrons in CoF_2.

       Since the s-multiplet splitting is a result of the exchange
interaction between the residual core s-electron and the spin of
the unfilled shell, systematic behaviour is expected as one traverses the 3d or 4f shell.   To first approximation one might
expect the splitting to be simply proportional to the spin of the
d- or f-shell.[24-26]   (The data for FeF_2 in Figure 2b show that
orbital angular momentum in the 3d-shell does not cause any
additional resolved splitting in the 3s-shell.)   In Figure 5 we
consequently plot the measured 3s splitting as a function of
atomic number for a set of isostructural, ionic, 3d-group transition metal compounds, the rutile structure fluorides.   The data
do not fall on the dotted line corresponding to simple proportionality to d-shell spin.   (Even the more proper proportionality to
$S+1/2$ is not realized.)   The sense of the deviation is such as
to indicate increasing s-d overlap with increasing atomic number.
The theoretical splittings, which are shown[23] are consistently
largor by almost a factor of two, but show similar deviation from
proportionality with S.

       The most disturbing observation is the large discrepancy
between theory and experiment, especially considering the success
of the theory in accounting for core polarization hyperfine fields.
Stated most plainly, the difference is due to the fact that what
is measured is not the same as what is computed.   The calculations
do not include the effects of electron correlations, while the
experiment is greatly affected by them.   The splitting-off of

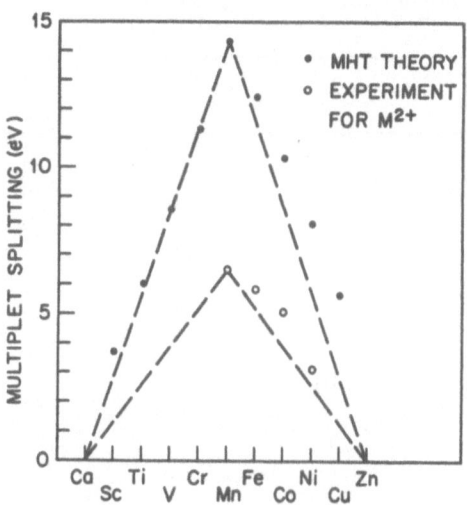

Figure 5.   Systematics of the multiplet splittings of the 3s-electrons in MF_2 compounds compared with multiplet hole theory taken from Ref. 23.

multi-electron satellites from the ^5S line of Mn^{2+} not only reduces its intensity below 5/7 of ^7S intensity but also shifts it towards the ^7S line reducing the splitting.   The data in Figure 5 show the surprising fact that the effect of electron correlations on the splitting results in proportionate reductions in both S-state Mn^{2+} and non-S-state Fe^{2+} and Co^{2+}.   A detailed theoretical explanation of this observation has not yet been given.

A more detailed quantitative comparison of both 2s- and 3s-splittings is made in Table II.   Note first of all that the 2s-splittings are in quite good agreement with theory.   The reason for this is found in the fact that multi-electron excitations are quite unimportant for the 2s-shell.   (This follows from the large excitation energy of the possible final states for 2s photoexcitation in Table I.)   Note also that when correlation effects are included in the calculation of the 3s splitting, the agreement with experiment is greatly improved.   Table II also shows that small but significant changes in multiplet splitting are found for a given ion, depending on its coordination, suggesting co-valency effects.

Data for rare earth compounds are more extensive, but show quite similar behaviour.[24]   Figure 6 exhibits the 4s-multiplet splitting in some rare earth trifluorides.   (For the rare earths covalency has negligible effect on the multiplet splitting.)

TABLE II.   Comparison of theoretical and experimental multiplet splittings for 2s and 3s electrons of 3d-group transition metal ions and compounds

Splitting	2s (eV)	3s (eV)
Mn^{2+}[a]	6.10	14.32[b]
MnF_2	5.9	6.5
MnO	5.6	6.1
Fe^{2+}[a]	5.40	12.40
FeF_2	4.8	5.9
Co^{2+}[a]	—	10.34
CoF_2	—	5.1

(a) calculated for multiplet hole theory using optimized orbital from Ref. 23;   (b) reduced to 8.2 eV by correlation effects (Sasaki and Bagus, quoted in Ref. 23).

In fact quite similar values have been obtained not only in Gd_2O_3 and GdF_3 but even in the metallic RSb and the corresponding insulating RF_3.   This is no surprise since the 4f-electrons are well shielded by the 5s- and 5p-electrons.

The systematics of the 4s- and 5s-splitting are shown in Figure 7, together with Hartree-Fock values.[27]   The agreement between theory and experiment in the case of the 5s-splittings is as striking as the constant factor representing the discrepancy between calculated and measured 4s-splittings.   The fact that the discrepancy in 4f-ions appears in the 4s-splitting (in 3d-ions it was in the 3s-splitting) points strongly towards electron correlations as the cause because these are most important within a shell of given principal quantum number.   The detailed analysis required to substantiate this conjecture for the rare earths has not yet been carried out.

V.   CORE LEVEL SPLITTING AND HYPERFINE INTERACTIONS

In spin-only ions with incomplete shells there is a clear

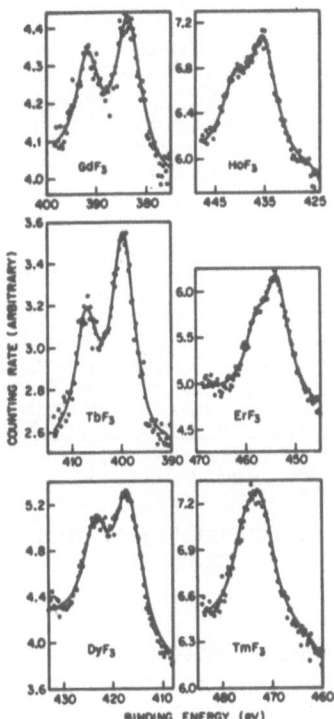

Figure 6.   4s-multiplet splitting in rare earth trifluorides
(from Ref. 24).

connection between the core s-shell multiplet splitting and the
nuclear hyperfine interaction.[28]   The latter is due to the interaction between the nuclear magnetic moment and the unbalanced s-
electron spin density at the nucleus.   The point is that it is
the core s-electrons which produce the net spin density at the
nucleus.   The radial wavefunction of the spin-up s-electron is
attracted towards the partially filled shell while the spin-down
electron is not affected.   This process is called core polarization and can lead to either an increase or a decrease in spin-up
density at the nucleus depending on whether the main lobe of the
s-wavefunction lies outside or inside the incomplete outer shell.

The same core polarization process is of course also responsible for the core s-electron multiplet splitting seen in X-ray
photoemission.   It would be rash, however, to assume that multiplet splittings and hyperfine interactions must be proportional,
because there are large cancellations between the contributions
of the various inner shells to the contact hyperfine interaction.

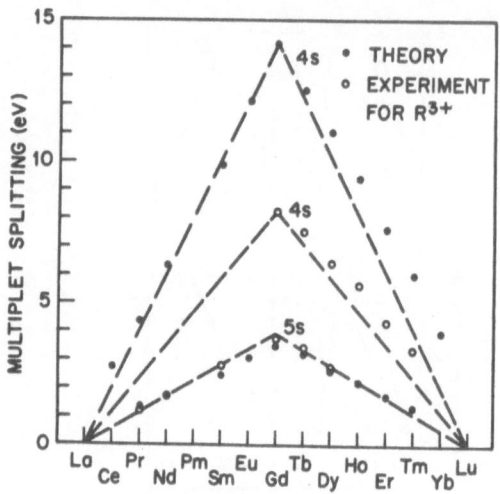

Figure 7.   Systematics of RF_3 multiplet splittings compared with
Hartree-Fock theory taken from Ref. 27.

Moreover the multi-electron aspects of photoemission strongly
affect the multiplet splitting but are not relevant to hyperfine
interactions.   The two types of measurement are therefore not
strictly comparable.

Nevertheless, it is worthwhile to make the empirical comparison suggested above.   As shown in Figure 8 there is in fact a
reasonably good correlation between these two parameters for Fe^{3+}
and Mn^{2+}, which are both ^6S ions.   The data suggest that covalency must change the effective number of d-electrons.   Note that
even metallic iron falls close to the straight line through the
origin, as it should if the significant parameter in the plot is
the d-shell moment.   This subject has been more fully discussed
in Ref. 28.

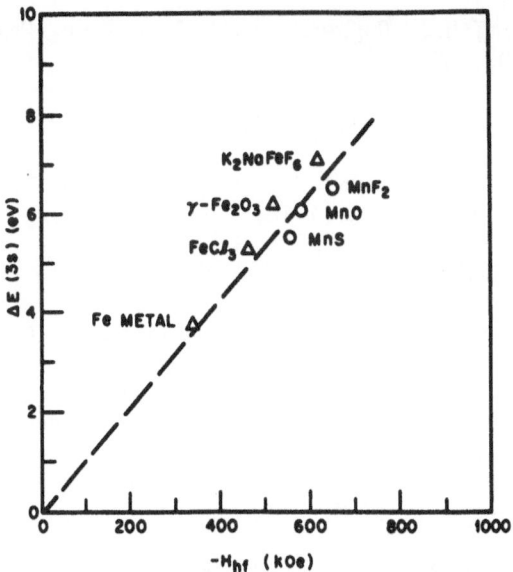

Figure 8.   Comparison of multiplet splitting and hyperfine fields in spin-only ions (from Ref. 28).

REFERENCES

1.   L. Hedin, Ark. Fyz. 30, 231 (1965); L. Hedin and G. Johannson, J. Phys. B2, 1336 (1969).

2.   P.H. Citrin and T.D. Thomas, J. Chem. Phys. 57, 4446 (1972); P.H. Citrin and D.R. Hamann (to be published); P.H. Citrin, P.M. Eisenberger and D.R. Hamann (to be published).

3.   D.A. Shirley, Chem. Phys. Letters 16, 220 (1972; L. Ley et al., Phys. Rev. B8, 2392 (1973).

4.   G.K. Wertheim, A. Rosencwaig, R.L. Cohen and H.J. Guggenheim, Phys. Rev. Letters 27, 505 (1971).

5.   F.R. McFeely, S.P. Mowalczyk, L. Ley and D.A. Shirley, Phys. Rev. Letters A45, 227 (1973).

6.   M. Campagna, E. Bucher, G.K. Wertheim, D.N.E. Buchanan and L.D. Longinotti, Phys. Rev. Letters 32, 885 (1974).

7.   M. Campagna et al., 11th Rare Earth Research Conference (to be published).

8.   P.A. Cox, Y. Baer and C.K. Jørgensen, Chem. Phys. Letters 22, 443 (1973).

9.   G.K. Wertheim, H.J. Guggenheim and S. Hufner, Phys. Rev. Letters $\underline{30}$, 1050 (1973).

10.   D.A. Shirley, Phys. Rev. $\underline{B5}$, 4709 (1972).

11.   S. Hufner, G.K. Wertheim, N.V. Smith and M.M. Traum, Solid State Commun. $\underline{11}$, 323 (1972).

12.   N.V. Smith, G.K. Wertheim and S. Hufner (to be published).

13.   P.J. Bassett, T.E. Gallon, J.A.D. Matthew and M. Button, Surf. Sci. $\underline{35}$, 63 (1973).

14.   G.K. Wertheim, L.F. Mattheiss, M. Campagna and T.P. Pearsall, Phys. Rev. Letters $\underline{32}$, 997 (1974).

15.   Unpublished results of M. Campagna $\underline{et\ al}$.

16.   K.H. Johnson, R.P. Messmer and J.W.D. Connolly, Solid State Commun. $\underline{12}$, 313 (1973).

17.   C.S. Fadley, D.A. Shirley, A.J. Freeman, P.S. Bagus and V.J. Mallow, Phys. Rev. Letters $\underline{23}$, 1397 (1969).

18.   C.S. Fadley and D.A. Shirley, Phys. Rev. $\underline{A2}$, 1109 (1970).

19.   G.K. Wertheim and A. Rosencwaig, Phys. Rev. Letters $\underline{26}$, 1179 (1971).

20.   J. Reader, Phys. Rev. $\underline{A7}$, 1431 (1973).

21.   Y. Yafet and R.E. Watson, Int. J. Quantum Chem., 93 (1973).

22.   S.P. Kowalczyk, L. Ley, R.A. Pollak, F.R. McFeely and D.A. Shirley, Phys. Rev. $\underline{B7}$, 4009 (1973).

23.   A.J. Freeman, P.S. Bagus and V.J. Mallow, Int. J. Magnetism $\underline{4}$, 35 (1973).

24.   R.L. Cohen, G.K. Wertheim, A. Rosencwaig and H.J. Guggenheim, Phys. Rev. $\underline{B5}$, 1037 (1972).

25.   J.C. Carver, G.K. Schweitzer and T.A. Carlson, J. Chem. Phys. $\underline{57}$, 973 (1972).

26.   G.K. Wertheim, S. Hufner and H.J. Guggenheim, Phys. Rev. $\underline{B7}$, 556 (1973).

27.   J.F. Herbst, D.N. Lowy and R.E. Watson, Phys. Rev. $\underline{B6}$, 1913 (1972).

28.   S. Hufner and G.K. Wertheim, Phys. Rev. $\underline{B7}$, 2333 (1973).

X-RAY PHOTOEMISSION FROM CORE ELECTRONS IN SOLIDS

C.E. Johnson

Oliver Lodge Laboratory, University of Liverpool

The energies of outer electrons of atoms in solids are broadened into bands, and our knowledge of the shape of these bands has been increased very considerably during the past few years by measurements of their photoelectron spectra. The spectra may be excited by uv or x-radiation, although it is generally recognized that excitation by x-rays, where the electron final state lies in the continuum, gives a more direct picture of the density of states versus energy curve than uv excitation, where the electron may finish up in an empty band state. It is found that conduction or valence bands are several eV in width and that their density of states curves may show considerable structure.

X-ray photoelectron spectroscopy (XPS) may be used to probe deeper with the atoms and so the core electrons to be studied. By contrast with the outer electrons they generally give narrow lines, their observed linewidth usually being of the order of that of the exciting X-rays, i.e. 1 eV or less. At first sight it might seem that the core lines do not give much interesting information: after all it is the outer electrons which take part in the chemical bond and which give cohesion in solids. However, the difference in the distribution of the outer electrons in different solids causes different amounts of screening of the core electrons and this appears as a shift of the core lines. This of course is the well known chemical or ESCA[1] shift and it has been measured for a large number of compounds and used as a method of chemical analysis, and as a probe of charge transfer in covalent bonds. The interpretation is necessarily empirical or semi-empirical, but nevertheless the XPS of core electrons has contributed greatly to our knowledge of chemical bonding.

The position and widths of the lines in the photoelectron
spectrum of core electrons provide valuable information, therefore,
though in general it is not as detailed as that obtained about
the band structure of the outer electrons.   The theory of the
chemical shift has been extensively discussed during this Summer
School so I shall not discuss it further here.   In several cases,
however, the spectrum of core electrons is more complicated than
expected and these will be the subject of this lecture.   Three
features commonly observed in core spectra are:-

(A)   chemical reduction of the specimen

(B)   the appearance of multi-electron ('shake-up') satellite
      lines and

(C)   core polarization splitting of lines in magnetic atoms.

These are listed in order of increasing fundamental importance.
Reduction of course is not at all fundamental, being more of an
experimental nuisance.   Shake-up is more fundamental, i.e. it
has a definite physical origin although it is not understood in
detail, and if we knew what it was it should enable information
about outer electron levels to be deduced.   The third is a
fundamental effect - it was predicted, looked for and eventually
observed and it is a valuable source of magnetic information in
atoms with unpaired electron spin.   I shall describe and give
examples of each of these in turn.

(A) REDUCTION

This is simply the chemical reduction of the specimen when
in the spectrometer and it shows up through the chemical shift
which allows lines of the reduced species to be distinguished
from those of the origin sample.   It is presumably caused by the
secondary electrons present when the X-ray source is on, rather
than a direct effect of the X-rays themselves.   Its effect varies
for different samples and fortunately most substances do not show
a very big effect during the course of a measurement.   If they
did, of course, the value of the ESCA technique would be severely
limitod.   It is also different for different spectrometers,
presumably owing to their differences in the fluxes of electrons
falling on the sample.   This phenomenon is not at all fundamental,
but it is important to be aware of its possibility when interpret-
ing unexpected lines in the XPS spectrum.

Figure 1 illustrates a case of particularly strong reduction
effect which was observed[2] in cupric fluoride CuF_2 in an AEI
ES200 spectrometer.   The spectra were taken after different
intervals of time and it is clearly seen that they were altering

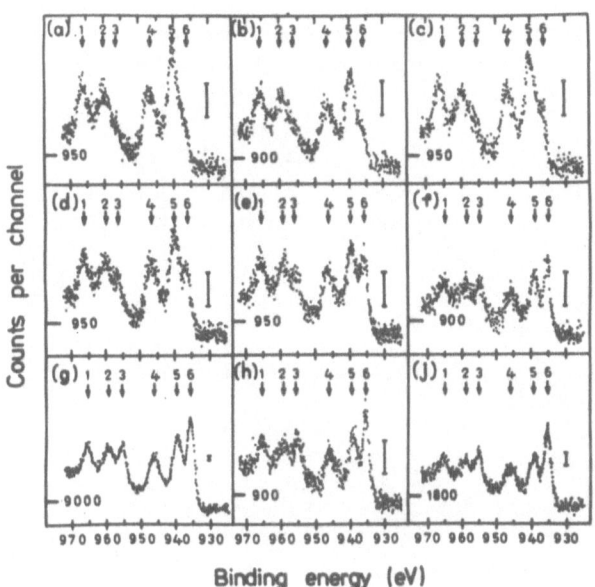

Figure 1. XPS of the Cu 2p region in cupric fluoride at different irradiation times increasing from (a) to (j) in an AEI ES200B spectrometer. Lines 1, 2, 4 and 5 are the 'original' lines due to CuF_2: lines 3 and 6 are the 'additional' lines due to the reduced species.

with time. The Cu 2p region of the spectrum is shown and the Cu^{2+} ion shows four lines ($2p_{1/2}$ and $2p_{3/2}$ each together with a 'shake-up' satellite — to be discussed later in B) marked 1, 2, 4 and 5. In the first spectrum (Figure 1(a)) these lines clearly stand out as the most prominent ones. As time goes on two more lines (marked 3 and 6) appeared at lower binding energies compared with the Cu^{2+} 2p lines. These were shifted in the direction corresponding to reduction, and they grew in intensity as time went on. No satellites were observed accompanying these lines. This is consistent with their arising from Cu^+ ions, since satellites are not observed in transition metal ions with a closed 3d shell. These extra lines were narrow compared with those of Cu^{2+}; also suggesting that they are due to Cu^+. Cuprous fluoride CuF does not exist as a stable compound in the laboratory, but apparently it can exist in an XPS spectrometer, produced by the action of electrons on CuF_2. In the last spectrum (Figure 1(j)) the additional lines gave higher peaks than those due to the original CuF_2. Figure 2 shows the peak height of the Cu $2p_{3/2}$ line from the Cu^+ relative to that from

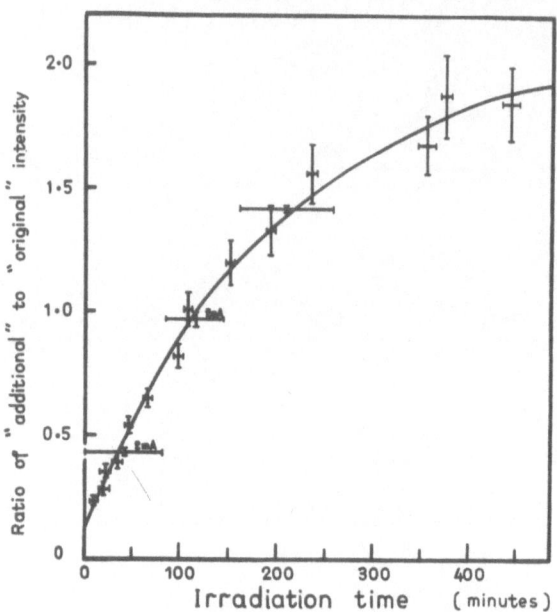

Figure 2.   Plot of the ratio of the 'additional' intensity and
the 'original' intensity against irradiation time for the
Cu 2p3/2 lines in cupric fluoride.   The points marked 2mA were
taken with one tenth of the X-ray emission current and have been
plotted at one tenth of their actual irradiation time.

the original Cu^{2+} plotted as a function of time.   It is seen
that the effects of reduction appeared almost at once and were
still increasing after seven hours.   The rate of reduction did
not depend upon the temperature of the sample, the pressure in
the vacuum or the X-ray flux, but was (for this particular
spectrometer) a function of the integrated X-ray dose.   When
the X-ray emission current was decreased by a factor of 10 from
20mA, the reduction proceeded 10 times as slowly.   While the
X-rays were off no reduction occurred.   The changes were
permanent as long as the sample was left in the spectrometer.
But they can be reversed.   When the sample was withdrawn from
the spectrometer, exposed to air and replaced in the spectrometer,
the spectrum initially obtained was similar to that of a freshly
prepared sample.

        Similar effects have been found in other cupric salts, e.g.
$CuCl_2$.   No such strong reduction effects have been observed in
the fluorides of other transition metals, although a slight effect
was detected in ferric fluoride FeF_3.

Strong reduction has been observed in the low-spin ferric
salt potassium ferricyanide $K_3Fe(CN)_6$. Spectra of the Fe 2p
lines are shown in Figure 3 (a) - (c) at various times. The
two lines ($2p_{1/2}$ and $2p_{3/2}$ - there are no satellites in low-spin
ions) of ferricyanide become shifted towards those of potassium
ferrocyanide. After a few hours there are equal amounts of the
two spectra, and after 20 hours the ferricyanide has been almost
completely reduced to ferrocyanide, only a trace of the original
lines remaining. Similar charges are observed on the nitrogen
1s lines, which shift by about 1.2 eV toward lower binding energy
Figure 3(d) to (f)), the lines being measured relative to the

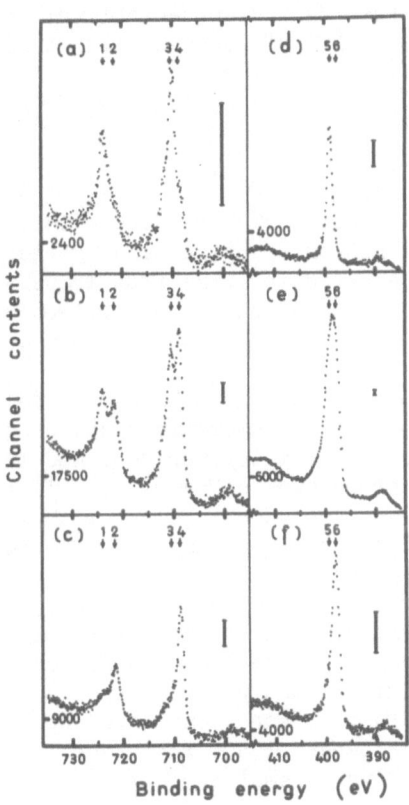

Figure 3.   XPS of potassium ferricyanide, showing the Fe 2p
region at times increasing from (a) to (c) and the N 1s region at
roughly similar times increasing from (d) to (f).   Lines 1, 3 and
5 are the 'original' lines; lines 2, 4 and 6 are the 'additional'
lines due to the reduced species.   The spectrometer was an AEI
ES200B instrument.

potassium 2s line.   That this is not due to specimen charging
is confirmed by Figure 3(e) where the line appears broadened
(actually two unresolved lines) when the reduction was about half
complete as judged by the iron lines (Figure 3(b)), and later
narrows (Figure 3(f)) when the reduction is total.   In low spin
cobalt(III), e.g. potassium cobalticyanide $K_3Co(CN)_6$ there appears
to be no appreciable sign of reduction either on the Co 2p or
N 1s lines.   This is probably because a more energetic e_g
electron must be transferred in the reduction of cobalt(III)
since the t_{2g} shell is full, whereas to reduce low-spin iron(III)
which has five t_{2g} electrons involves transfer of another t_{2g}
electron which requires a much smaller energy.

Chemical shifts in the metal 3s and 3p lines also of course
accompany reduction, but they are smaller and tend to get lost
amongst the multiplet (see C) and spin-orbit splittings.   If
reduction is suspected in XPS of these levels it is most easily
confirmed by monitoring the 2p lines.

(B)   'SHAKE-UP'

The photoelectron spectra of core levels in transition metal
salts show satellite lines on the high binding energy side of the
main peaks which are believed to arise from transitions where
there is a simultaneous excitation of an electron in an outer
shell.   Such multi-electron excitation is known as 'shake-up'.

The first examples of these in solids were found by
Rosencwaig et al.[3] who made a systematic study of the metal 2p
electrons in the transition metal fluorides, and their spectra for
the ions Mn^{2+} to Zn^{2+} are shown in Figure 4.   In each spectrum
the $2p_{1/2}$ and $2p_{3/2}$ lines are seen together with satellite lines
lying at about 6 eV higher binding energy.   The intensity of
these satellites in the second half of the 3d shell is seen to
increase as the number of 3d electrons increases, being small for
Mn^{2+} and almost as large as that of the parent lines for Cu^{2+}.
This shows that the origin of the lines is not connected with
magnetic effects.   Similar spectra[4] for the fluorides of Sc^{3+} to
Cr^{3+} are shown in Figure 5.   For these ions the separation of
the satellites from the parent lines is larger, being about 12 eV.
As the spin-orbit coupling is smaller in the first half of the 3d
shell the satellite lines lie together and clear of their parents.
Satellite lines with a similar large separation were observed[4] to
accompany the 2s lines in Sc^{3+}, Ti^{3+} and V^{3+} ions (see Figure 6).
They presumably also occur in ions with a more than half-filled
3d shell, but they are more difficult to observe as their
separation from the parent line would be smaller and so they
would be mixed up with the broad 2s lines.

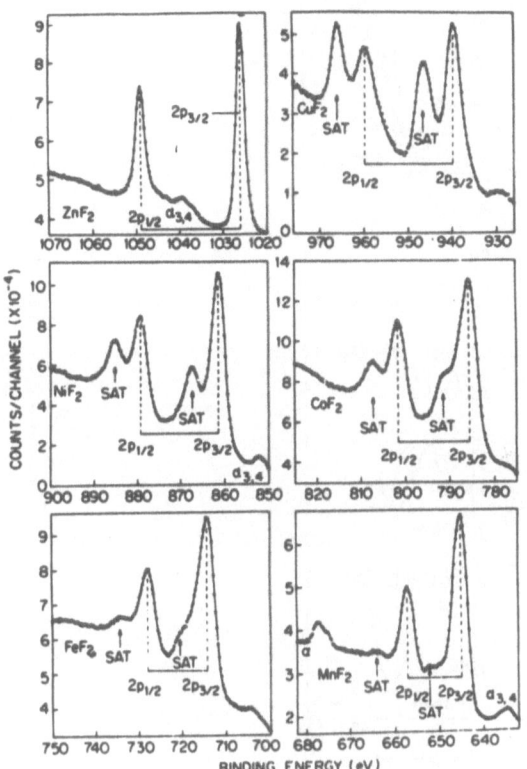

Figure 4. XPS of 2p electrons of transition metal difluorides of ions with 0-3 3d electrons (Rosencwaig et al.). Note that there are no satellites for Zn^{2+} which has filled d shell. Note also that no noticeable reduction of CuF_2 was observed in these measurements which are made with a Varian IEE 15 spectrometer.

The processes by which the main lines and the satellites are believed to be generated are illustrated in Figure 7. In Figure 7(a) a normal photoelectron is emitted. In Figure 7(b) an electron in an outer shell is simultaneously excited (i.e. shaken-up) and the emitted photoelectron has correspondingly less kinetic energy or a greater apparent binding energy. The energy required to drive the shake-up is derived from the relaxation energy due to the increase in the screened ionic potential seen by the outer electrons when the core electron is removed. The process has $\Delta J = 0$ selection rules.

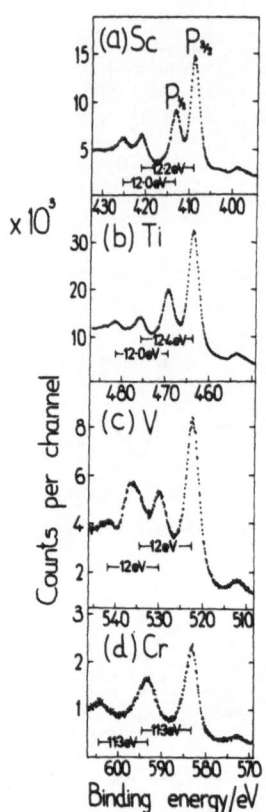

Figure 5.   XPS 2p electrons of transition metal trifluorides of
ions with 0-3 3d electrons.   The V 2p satellite is concealed by
the 1s photoemission peak from oxygen absorbed on the sample.

        If the initial and final outer electron states involved in
the shake-up could be established, measurements of the satellites
could be used to provide valuable information about them.   Let
us then list some of the relevant facts about the satellites
which might have a bearing on their origin.   These are:-

(1)   no satellites are observed when the d-shell is full

        e.g. for the ions Cu^+ and Zn^{2+} which have the d^{10} configur-
        ation.   This suggests that the final state in the shake-up
        is the d-shell, so that when it is full transitions to it
        are not allowed, as observed.

(2)   satellites are observed for the d^o ions

        e.g. in Sc^{3+} and Ti^{4+}.   This shows that the d-state cannot

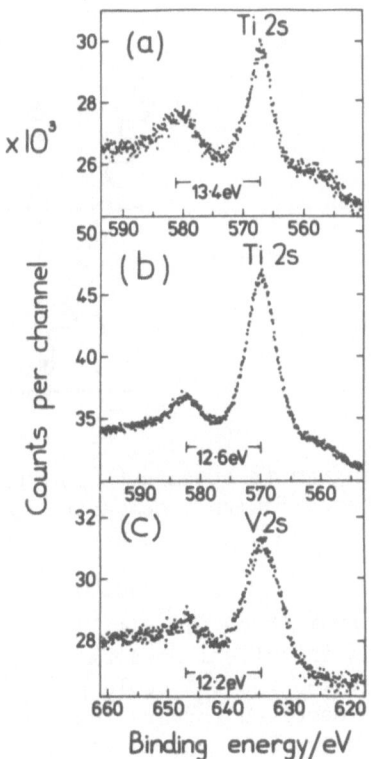

Figure 6. XPS of 2s electrons for (a) TiO_2 (b) TiF_3 and (c) VF_3, which have metal ions with 0, 1 and 2 3d electrons respectively.

be the initial state, and is consistent with the observation (1) for the full d-shell that it is the final state.

(3) the satellites are ligand dependent

e.g. Figure 8 shows the 2p spectrum for $MnCl_2$ and $CoCl_2$.[4] The energy separation of the satellite lines from their parents and the intensities are different in the chlorides from those in the corresponding fluorides.

These observations are consistent with the shake-up being frmm a ligand (e.g. F 2p) to a metal 3d orbital, though this is by no means proved. This is a charge transfer transition, though the selection rules ($\Delta J = 0$, connecting states of the same

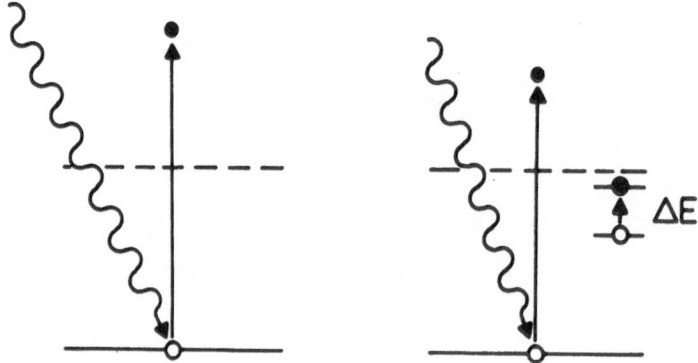

Figure 7.   Schematic diagram to compare (a) normal core electron
photoemission and (b) photoemission accompanied by a shake-up
process.

symmetry) are different from those $(\Delta J = \pm 1)$ observed for charge
transfer absorption bands in optical spectra.   Transitions of
the kind $\pi t_{2g} \to 3d\ t_{2g}^*$ or $6e_g \to 3d\ e_g^*$ could possibly be involved.
Satellites have not been observed for any low-spin ions,
presumably because the outer electrons do not have the right sort
of states.   Satellite structure found in rare-earth trifluorides[5]
is satisfactorily explained by ligand metal 4f transitions, and
the ligand (valence) and metal 4f photoelectrons were measured
and their energy difference correlated with the satellite
separations.   In the 3d compounds the 3d electrons lie in the
valence band and are broadened by crystal-field effects, and so
far no comparison of the ligand-3d energy difference with the
satellite separation has been possible.

Many other facts concerning the occurrence of shake-up
satellites need to be explained, and their study may lead to an
understanding of their origin.   Larger satellite separations
(12 eV) are found in the first half of the 3d-shell compared
with those (6 eV) in the second half as already mentioned and
this is illustrated[4] in Figure 9.   Assuming the shake-up is a
ligand 3d transition, this could be associated with ligand-metal
exchange interaction in the final state.   For less than a half-
filled shell the net spin on the metal after shake-up is antiparallel to the unpaired spin left on the ligand, while for ions
with five or more d-electrons these spins are parallel.   Generally
the satellite of the $2p_{1/2}$ line is broader and slightly further
separated from its parent than that of the $2p_{3/2}$ line.

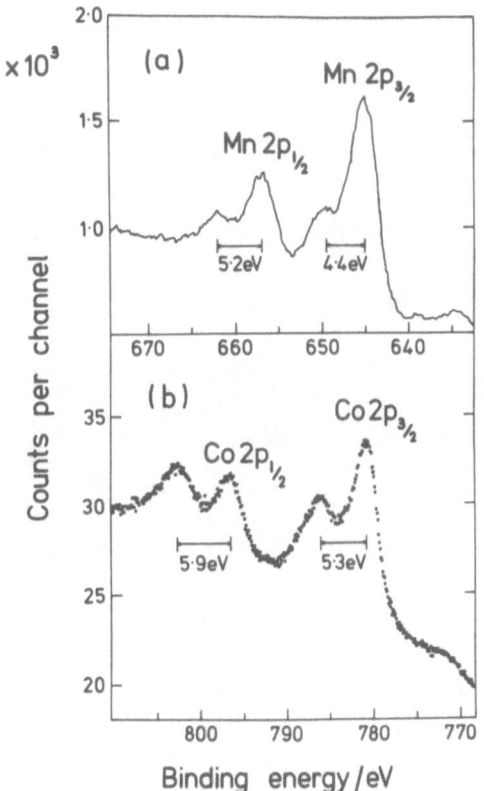

Figure 8.   XPS in the metal 2p region for (a) $MnCl_2$ (b) $CoCl_2$.

Clearly quite a lot of systematic work needs to be done before shake-up transitions are explained.

(C) CORE POLARIZATION SPLITTING

In magnetic atoms, the energies of the core levels are split by exchange interaction with the spins of the electrons in the unfilled shell.   This is most easily demonstrated for the 3s electrons in a transition metal ion, and arises because the interaction which the unpaired electron spin of the d-electrons is different for the 3s and the 3s electrons.   The motion of electrons with parallel spins is correlated so that they keep out of each others' way according to the Pauli exclusion principle. This lowers the energy of the electrostatic interaction between them, so that the 3s level lies lower than the 3s , i.e. the

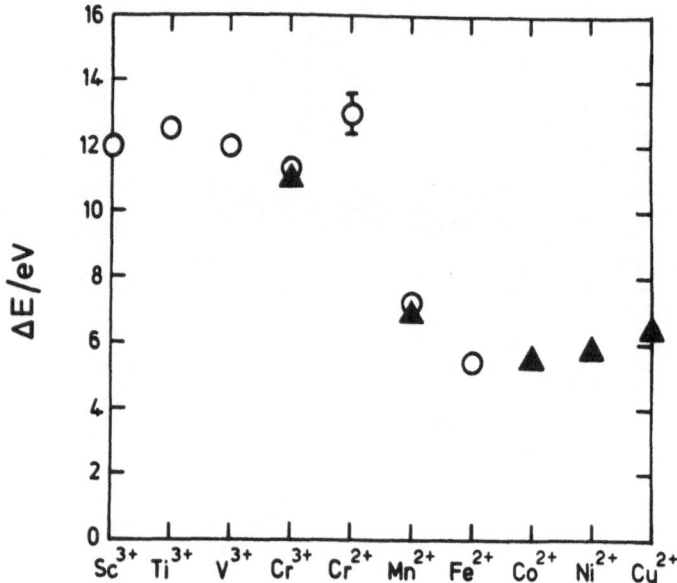

Figure 9.   Shake-up separation energy E (averaged for $2p_{1/2}$ and $2p_{3/2}$ satellites) plotted as a function of n the number of 3d electrons in the ion.   The value of n runs from 0 at Sc^{3+} to 9 at Cu^{2+}.   Closed triangles are from Rosencwaig et al., open circles from Wallbank et al.

core is polarized.   (This polarization of the core produces a large contribution to the magnetic hyperfine field since the s electrons have a finite density at the nucleus.)   In photo-emission from an s-shell of a magnetic atom therefore, two lines should appear instead of one.

    Magnetic splittings in solids were first seen by Fadley et al.[6] and their data for iron and two of its salts are shown in Figure 10.   Note that in $K_4Fe(CN)_6$, where the iron is non-magnetic, i.e. low-spin, the 3s level shows one line only, whereas in FeF_3, where the iron is high-spin with a magnetic moment of 5 Bohr magnetons two 3s lines are clearly seen separated by about 6 eV.   The Fe^{3+} ion has a ^6S ground state and the multiplet

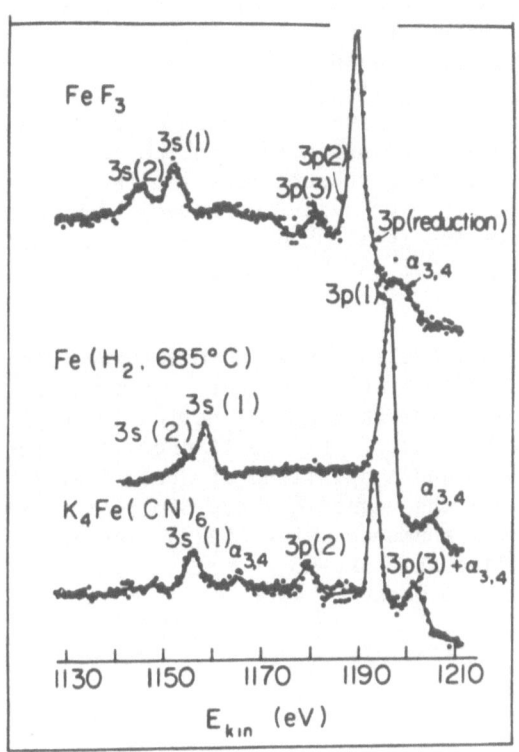

Figure 10.   XPS in the 3s and 3p region of iron in various solids, using Mg Kα radiation (Fadley et al.).

coupling of the d-electrons with the spin of the photohole gives the two final states as ^7S and ^5S, the latter lying lower in energy. Hence the two lines should have an intensity ratio 7:5.   The observed intensities are more like 2:1.   In general for 3d ions with spin S the splitting of the 3s lines is theoretically given by $\frac{2S+1}{5}$ $G^2(3s,3d)$ and the intensity ratio is S: (S + 1), where G^2 (3s,3d) is a Slater overlap integral.   Inclusion of correction effects brings the theory[7,8] into closer agreement with the experimentally observed values of both the splitting and the intensities of the lines.   Of course p-electron levels are also split by exchange interaction although the situation here is more complex and more than two lines result, which can also be seen for the 3p lines in the FeF_3 spectrum.   The magnetic splittings can also be seen for iron metal, though they are smaller because the atomic magnetic moment is smaller, i.e. 2.2 Bohr magnetons.

Observation of the core polarization splitting provides, therefore, a probe of magnetism in solids. This probe is highly localized, both in space and in time. The splitting is not dependent upon spin relaxation rates (as is for example the observation of magnetic resonance) nor is it affected by ferromagnetic or antiferromagnetic coupling. It could, for example, provide a method for studying the magnetic properties of dimeric (or polymeric) molecules where the spins of two (or more) atoms are coupled together to give zero resultant spin. A conventional magnetic susceptibility or hyperfine interaction measurement in this case measures the total magnetic moment which is zero, whereas the XPS spectrum which is a local measurement shows magnetic splitting.

An example[9] which illustrates this oxidized Clostridium ferredoxin, a biological molecule where clusters of four iron atoms are believed to be antiferromagnetically coupled to give a non-magnetic ground state. Figure 11 shows the iron 3s region of the XPS spectrum and a splitting of about 5 eV is clearly seen, which confirms that the magnetic state of the iron atoms comes from observation of shake-up satellites on the Fe 2p lines (Figure 12). Although shake-up is not a magnetic effect it has only been observed in magnetic atoms.

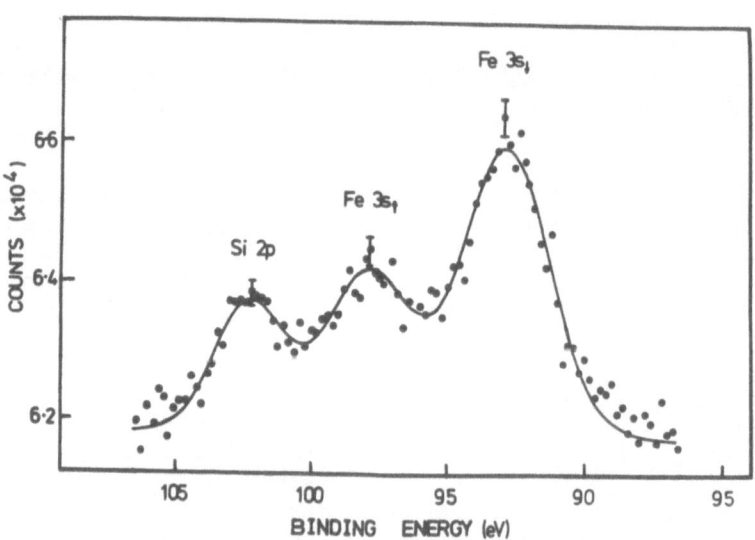

Figure 11.   XPS of Fe 3s in oxidized Clostridium ferredoxin. The line on the left of the spectrum is due to a silicon impurity.

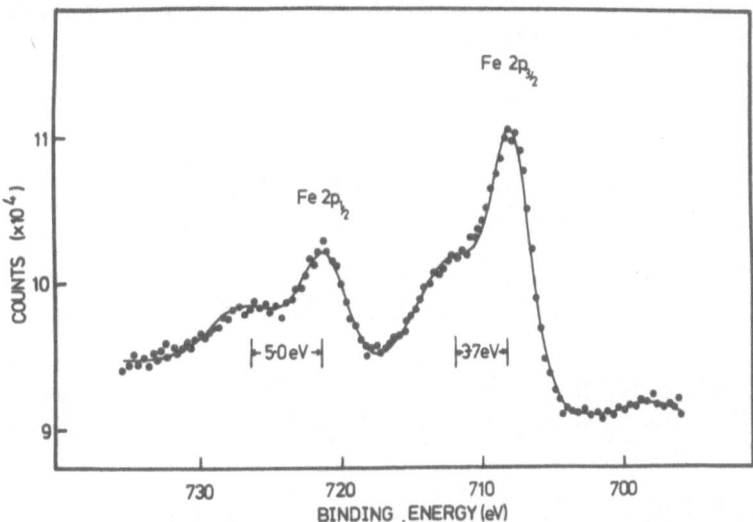

Figure 12.   XPS of Fe 2p in oxidized <u>Clostridium</u> ferredoxin.
The energy shifts of the shake-up satellites from the parent
peaks are shown.

CONCLUSIONS

     Observation of core electron lines in photoemission spectra
from solids thus provides a check on sample impurities, including
impurities produced by chemical reduction within the spectrometer
Measurements of the chemical (ESCA) shifts provides information
about the chemical state of the atoms in the particular solid
which is being studied.   The measurement of magnetic splittings
gives further valuable and detailed knowledge on the electronic
state of magnetic atoms on ions, often in cases where conventional
magnetic techniques are limited.   Data on shake-up satellites may
be used to identify magnetic atoms, and their shape may ultimately
yield information on valence band structure.

     In addition to the features discussed, photoelectron spectra
of core states may show many other effects such as line broadening
due to thermal vibrations, asymmetry of the lines due to hole
interactions, etc., so that the technique looks rich in future
powerful applications.

REFERENCES

1. K. Siegbahn et al., Nova Acta Regiae Soc. Sci. Uppsal.,
    Ser. IV, 20, 1 (1967).

2. B. Wallbank, C.E. Johnson and I.G. Main, J. Elec. Spect.
    4, 263 (1974).

3. A. Rosencwaig, G.K. Wertheim and H.J. Guggenheim, Phys. Rev.
    Lett. 27, 479 (1971).

4. B. Wallbank, C.E. Johnson and I.G. Main, J. Phys. C 6, L340
    (1973) and 6, L493 (1973) and J. Elec. Spect., to be
    published.

5. G.K. Wertheim, R.L. Cohen, A. Rosencwaig and H.J. Guggenheim,
    'Electron Spectroscopy' (ed. D.A. Shirley: North Holland
    1972) p. 813.

6. C.S. Fadley, D.A. Shirley, A.J. Freeman, P.S. Bagus and
    J.V. Mallow, Phys. Rev. Lett. 23, 1397 (1969).

7. A.J. Freeman, P.S. Bagus and J.V. Mallow, Int. J. Magnetism
    4, 35 (1973).

8. P.S. Bagus, A.J. Freeman and F. Sasaki, Phys. Rev. Lett. 30,
    850 (1973).

9. P.T. Andrews, C.E. Johnson, B. Wallbank, R. Cammack, D.O.
    Hall and K.K. Rao, to be published.

TECHNIQUES RELATED TO PHOTOELECTRON SPECTROSCOPY

C.R. Brundle

School of Chemistry, University of Bradford, Bradford, England

and   A.D. Baker

Department of Chemistry, Queens College, City University of New York, U.S.A.

1.   INTRODUCTION

Electron spectroscopy is the name given to a collection of individual techniques based upon the analysis of electron energies, following the interaction between an impacting particle or photon with an atom, molecule, or solid. Photoelectron spectroscopy, where the impacting species are vacuum ultraviolet or X-ray photons, is merely one of these individual techniques. The remainder are summarised in Table 1. Previous reviews or leading references are also listed. The individual techniques were initially developed more or less independently by groups of workers in diverse areas. The rapid advances of recent years in most of these techniques can at least partly be attributed to a belated interaction between workers in the different fields.

This chapter briefly sets out the principles behind four of the techniques listed in Table 1 - Electron Impact Energy Loss Spectroscopy (ELS); Penning Ionization (PI); Auger Electron Spectroscopy (AES), and Ion Neutralization Spectroscopy (INS) - particularly emphasising their relationship to photoelectron spectroscopy. Some examples of the type of applications for which these techniques are best suited are given, both for gaseous and for solid state work. As was the case for photoelectron spectroscopy, solid state work often means surface work, and aspects of the surface sensitivity of the techniques[15,28,29] are thefefore also discussed.

P. Day (ed.), Electronic States of Inorganic Compounds. 425–447. All Rights Reserved.
Copyright © 1975 by D. Reidel Publishing Company, Dordrecht-Holland.

Table 1.   Types of electron spectroscopy

Name	Abbreviations	Basis of the technique	References of leading articles
Photoelectron spectroscopy (ultra-violet excitation)	PES, UPS, or MPS	Energy analysis of electrons ejected from the sample by impact of monoenergetic UV photons	(1)-(6)
Photoelectron spectroscopy (X-ray excitation)	XPS or ESCA	As PES, but soft X-rays used in place of UV photons	(7)-(11)
Auger electron spectroscopy	AES	Energy analysis of Auger electrons (secondary electrons) ejected from the sample following ionization byphoton or electron impact	(12)-(15)
Penning ionization	PI	Energy analysis of electrons ejected from the sample of impact of metastable atoms	(16)-(19)
Electron impact energy loss spectroscopy	ELS	Energy analysis of a monoenergetic beam of electrons inelastically scattered from the sample (no ionization)	(15),(20)-(22)
Ion neutralization spectroscopy	INS	Energy analysis of Auger electrons ejected from the sample following ionization by valence electron transfer from the sample to an impacting ion	(15),(23),(24)

(continued)

Name	Abbreviations	Basis of the technique	References of leading articles
Autoionization electron spectroscopy		Energy analysis of electrons ejected from the sample in an autoionizing decay from super-excited state caused by photon or electron impact	(25)
Resonance electron capture		Elastic scattering cross-section measured as a function of the energy of the impacting monoenergetic electron beam	(26)
Chemi-electron spectroscopy		Energy analysis of electrons produced by collision between atoms or molecules	(27)

2.   BASIC PRINCIPLES

The physical processes which may occur during photon or
particle impact are illustrated in Figure 1.   A fundamental
difference exists between excitation or ionization resulting from
photon and from electron collision.   When a photon is annihilated
during the collision process, all its energy, $h\nu$, is transferred
to the target molecule (except in the special case of Compton
scattering).   Thus excitation of a bound electron to an empty
molecular orbital (bound-bound transition) is only possible if
$h\nu$ is exactly equal to the excitation energy, ΔE, i.e. all photon
absorption spectra involving bound→bound transitions (vibrational,
electronic) are resonance phenomena (Figure 1(a)).   In the case
of ionization (Figure 1(b)), this is not so since the annihilated
photon energy, $h\nu$, is only partly used up in removing a bound
electron from the molecule (bound→free transition), the remainder
being carried off as kinetic energy, K.E. of the electron which
has been freed.   This leads to the well-known equation used by
photoelectron spectroscopists for determining ionization potentials (I.P.'s) of bound electrons:

$$IP = h\nu - K.E.$$

Thus ionization by photons is not a resonance process, $h\nu \geqslant$ I.P.
being the necessary and sufficient condition for it to occur.
The probability of its occurrence, i.e. the photoionization cross-
section, will however vary with variations in $h\nu$.)   In the case
of electron impact the excitation process (Figure 1(c)) is not a
resonance phenomena (Figure 1(c)) because the impacting electron,
E_p is not annihilated but inelastically scattered with lower
energy.   The energy loss of the scattered electron, E_s, given
by $E_p - E_s$ is equal to the excitation energy, ΔE.   Thus in
Electron Impact Energy Loss Spectroscopy (ELS), one measures ΔE
by providing a monoenergetic beam of primary electrons, energy E_p,
allowing them to interact with the target molecule, and measuring
the kinetic energy, E_s, of the scattered electrons. ΔE's corresponding to all allowed bound-bound transitions are therefore
obtainable and ELS is the electron impact equivalent of photon
absorption spectroscopy (Figure 1(a)).   However, there are two
important differences that make it useful as an alternative or
complementary technique to absorption spectroscopy.   First, with
one primary energy E_p one is able to cover the whole frequency
range from vibrational excitation (I.R. photon region), through
to high energy electronic transitions (U.V. and vacuum U.V. photon
region), a situation requiring three or four different optical
spectrometers.   The resolution obtainable depends on how monoenergetic one can make E_p and at what resolution one can analyse
E_s.   Figure 2 represents a typical ELS system where E_p is made
monoenergetic by using a deflection analyser as a monochromator
and the same type of analyser is used in the second stage for E_s.

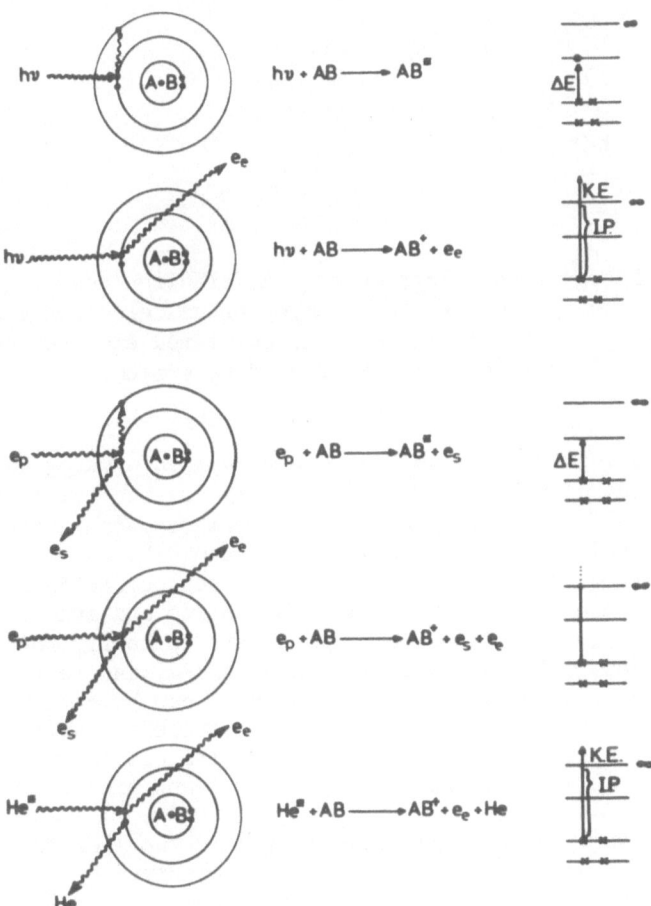

Figure 1.   Schematic representation of processes occurring during photon or particle impact on molecule AB.

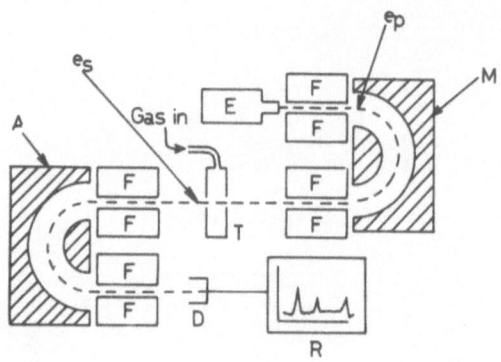

Figure 2.   Energy loss spectrometer: E, electron gun;
F, focusing lens system;   M, 180° hemispherical electron energy
monochromator;   T, target chamber;   A, scattered electron energy
analyser;   D, detector system;   R, recording system.

The best practical resolution that can be obtained using such a
system is about 10 meV (or 80 cm^{-1}), a very low resolution in
the vibrational region compared to I.R., but comparable to
resolutions routinely available at the vacuum U.V. end of the
spectrum.   The second important point is that selection rules
differ from those controlling photon absorption and are in general
less rigorous, so that transitions which are optically forbidden
may be observed.   Variation in the value of E_p, and in the
scattering angle along which E_s is observed, can have marked
effects on the relative probabilities of different transitions,
which may often be related to the types of transition involved,
thus allowing an extra method of assigning transitions.

Ionization by electron impact results in two electrons
leaving the target molecule, the scattered electron E_s and the
ejected electron E_e (Figure 1(d)).   The excess energy E_p – IP
(of the ejected electron, E_e) is partitioned between E_s and E_e
(E_p – IP = E_s + E_e).   Thus for a large number of collisions
there will be a range of E_s values from 0 to E_p – IP, and a range
of E_e values from 0 to E_p – IP.   All that is observed in ELS
when an ionization process occurs, therefore, is a weak step at
the ionization limit E_p – IP, followed by a continuum consisting
of the indistinguishable E_s and E_e electrons of varying energies.
Ionization processes are thus not of great value in ELS.

Collision using a metastable excited atom is phrnomenologic-
ally equivalent to photon impact since the excited atom gives up

all its energy and returns to its ground state.   Bound→bound
transitions in the target molecule will only occur therefore if
the energy E* of the metastable matches the transition energy,
ΔE – i.e. it is a resonance phenomenon.   Ionization can occur
provided E* ⩾ IP (cf. ℏʋ ⩾ IP), as illustrated in Figure 1(e),
thus giving the relationship:

$$I.P. = E* - K.E. \text{ (cf. photoelectron spectroscopy).}$$

The process is known as Penning Ionization (P.I.), and the most
common excited atom to use is He, which has two long-lived metastable states (He*), 2^1S and 2^3S at E* values of 20.61 eV and
19.82 eV.   The use of He* enables I.P.'s of up to 20.61 eV to be
determined although, unless one of the two metastable states is
suppressed (they are formed together by electron bombardment of
He), two superimposed spectra separated by 20.61 – 19.82 eV will
be observed.   The experimental arrangement for Penning Ionization
spectroscopy is the same as for UV photoelectron spectroscopy
except that a source of He* must replace the HeI photon source.
A suitable way of achieving this is shown in Figure 3.   The
distinguishing feature of Penning Ionization compared to photoelectron spectroscopy is an increased contact time between He*
and the target molecule, owing to a decreased force-field, which
may result in metastable collision complexes being generated.
The internal energy of the collision complex may then be distributed differently in the eventual dissociation products of the
complex compared to that found in these species produced directly
by photon impact.

    The remaining two techniques to be considered, Auger
Spectroscopy (AES)[30] and Ion Neutralization Spectroscopy (INS)
are both secondary electron processes.   Since INS is used only

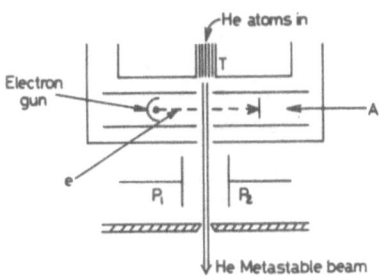

Figure 3.   Helium metastable source for Penning Ionization.
T, capillary plug; A, anode; e, electron beam; P_1P_2, electrostatic plates for removing charged species.

for surface studies, and is really a special form of the Auger
process, consideration will be delayed to section 4.   The Auger
process is one of the relaxation processes (the other being X-ray
fluorescence) which may follow the production of a hole in a
core-level, and is quite independent of the means of production
of that hole (photon, electron, or high energy ion collision).
Figure 4 illustrates the process.   An electron drops into the
vacancy in the core-level W from a higher level X (transition 1,
Figure 4) releasing sufficient energy ($E_W - E_X$) for a second
electron (the Auger electron) to be ejected from level X

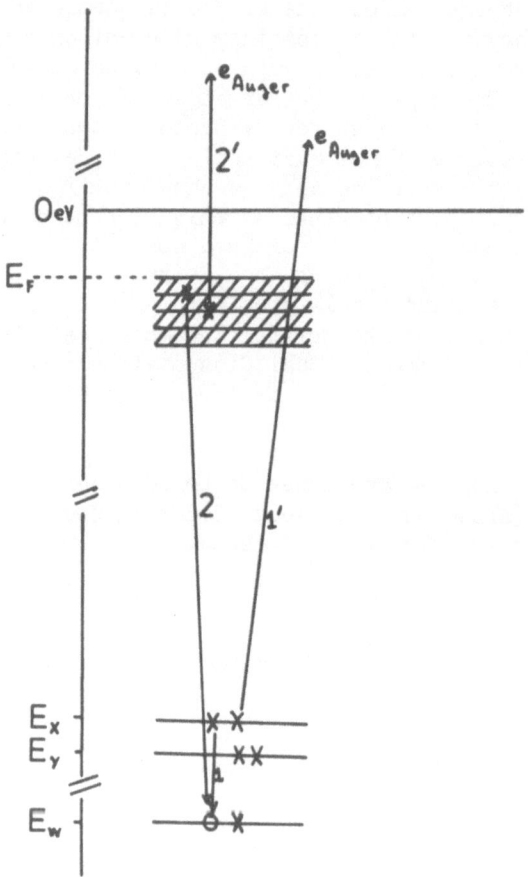

Figure 4.   Schematic representation of the Auger process showing
a case where core-levels only are involved (1,1'), and a case
where valence levels are involved (2,2').

(transition 1', Figure 4) or some other level Y, leaving a doubly charged ion. The amount of kinetic energy possessed by the Auger electron is given by $E_W - 2E_X$ or $E_W - E_X - E_y$. If all the levels involved are atomic core-levels, then the Auger electron energy will be characteristic of the atom concerned, providing an atomic identification (cf. XPS), and chemical shifts in the core-levels will be reflected by chemical shifts in the Auger energies, providing information on the electronic environment of the atom (cf. XPS). Auger spectroscopy therefore has the potential of elemental and chemical analysis. E_X or E_y, or both, may be valence levels, however (transitions 2, 2' in Figure 4), which means that though the Auger kinetic energy is still characteristic of the atom (E_W being the dominant term), any chemical shifts observed for different compounds of that atom are the result of 'proper' chemical shifts in E_W, plus changes in the valence levels. Thus a mixture of XPS and UPS type behaviour results, which may be difficult to separate out, but which offers a wealth of information if an analysis is possible.

     The Auger process is the major decay mechanism unless the energy difference between initial and final states is rather large. At a difference of about 10 KeV X-ray fluorescence is of comparable importance and becomes dominant at higher energies. In practice, this means light elements ($Z \leqslant 20$) have low fluorescence yields and high Auger cross-sections, and heavy atoms have high fluorescence yields and low, but usable, Auger cross-sections.

     As the Auger process is independent of the process originally generating the core hole it is not necessary to use monochromatic X-rays, or monoenergetic electrons. Since it is cheaper and more convenient to produce a beam of high energy electrons than X-rays, and because much higher beam intensities can be achieved for electrons, it is common practice to perform Auger spectroscopy using electron impact. This can have disadvantages when studying surfaces, as discussed in section 4, but advantages if one wishes to irradiate only a small area, since electron beams are easily focussed. The advantage in using X-rays is that if one is primarily interested in XPS, the X-ray induced Auger spectrum is obtained free along with the XPS spectrum.

3.   EXAMPLES OF USE IN GAS PHASE STUDIES

3.1   ELS

     Figure 5 shows part of the electron spectrum obtained from the impact of 34 eV energy electrons on gaseous helium at a scattering angle of 25%.[22]   Each peak represents a promotion process from the filled He(1s) orbital to an empty one.   The

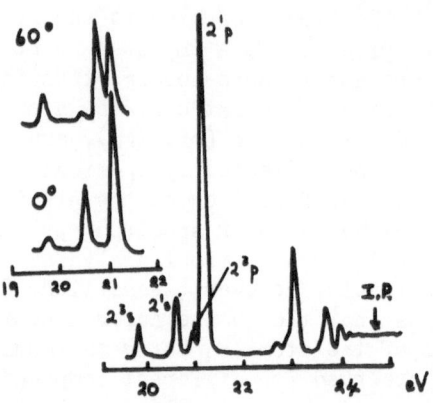

Figure 5.   Energy loss spectrum of helium taken using E_p = 34 eV, and at a scattering angle of 25°.   The section 19 eV to 22 eV energy loss is also recorded at 0° and 60°.[22]

two He* states used for Penning ionization, 2^3S and 2^1S, corresponding to the configuration He(1s)' (2s)' are present at the energy loss values of 19.82 and 20.61 eV.   The singlet-triplet transition resulting inthe 2^3S state is of course forbidden in optical excitation (both spin and symmetry forbidden), yet is quite strong in this spectrum.   Inset into Figure 6 is the same region of the spectrum taken at angles of 0° and 60°, to illustrate the way in which the relative intensities of the transitions change with angle.   The review by Trajmar et al.[22] considers in detail the variations in differential cross-sections with scattering angle and impacting electron energy for helium and a number of small molecules.   For molecules, vibrational structure is superimposed on the electronic transitions in exactly the same way that it is for UPS.   In addition, many transitions are possible, the number increasing with increasing electronic complexity of the molecule.   Figure 6, an ELS of SO_2, shows a large number of bands, with some resolved fine structure. The ELS spectrum of a molecule is generally much more complex than its UPS equivalent because there are many more bound-bound transitions possible than ionization potentials.   The transitions can sometimes be arranged in Rydberg series and such an analysis used to determine electronic structures, usually in conjunction with UPS and optical absorption studies.   The advantage over the optical spectrum is the ease with which it can be obtained for a wide energy loss range (several different continuous sources and a vacuum UV monochromator are required to cover the same range by classical vacuum UV absorption spectroscopy) and the flexibility of angular and impact energy variations.   The disadvantage of a complex spectrum is similar in both cases, but made worse

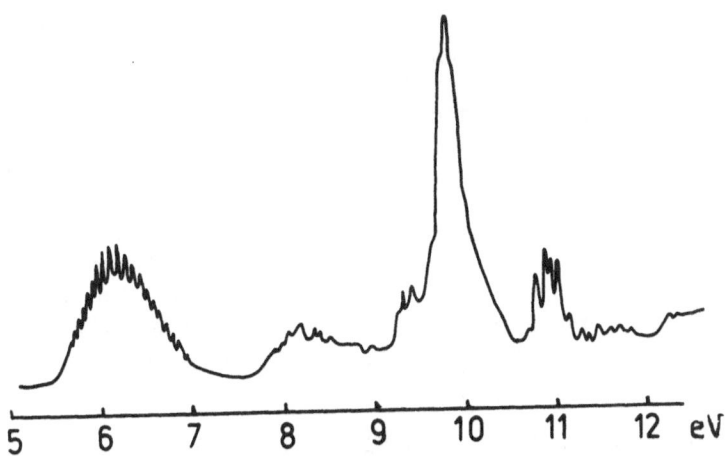

Figure 6.   An ELS spectrum of SO_2.

for ELS because of poorer resolution.   There have been suggestions
that the technique has analytical potential for studying gaseous
mixtures, but this only seems reasonable for very simple molecules
(with simple spectra).

3.2 Penning ionization

     Though PI can be used for the determination of molecular
electronic structure, as explained in section 2, it is generally
easier to perform the UPS measurement to the same end.   It can
be used in reverse to measure the energy of metastable atoms
generated by electron impact by using them as ionizing sources for
a target gas of known ionization potentials.   One of the
previously unknown higher lying states of N_2^* was identified in
this fashion.[31]

     However, the main application of Penning ionization is to
study the basic mechanisms involved in ionization caused by
excited atoms, and their differences from photon impact.   Such
processes are important in gaseous discharges, photolysis,
radiolysis and plasmas.   When the excited atom A* collides with
the target molecule X-Y, a short-lived collision complex, (A-X-Y)*
is formed before an electron is ejected.   The 'normal' result
of the eventual ejection of e is to give A + X-Y$^+$ + e, the energy
of e being given by the appropriate energy balance equation.

There are alternative processes depending on the lifetime of the collision complex, and its energetics. A discussion of these mechanisms is beyond the scope of the present chapter, but a brief description of their effects upon the electron spectrum is presented. First, the population of the vibrational levels in the ionized states might be expected to be different from that covered by direct ionization to $X-Y^+$. This does not seem to be the case for molecules studied so far, since the vibrational populations in PI (when resolved) are very similar to those for UPS of the same molecule.

A common occurrence in PI is that the IP's deduced using a straightforward energy balance equation are displaced by ΔE from their correct positions as found by UPS, due to two processes. In the first ΔE represents translational energy which can be given either to the final ion $X-Y^*$ or to the ejected electron, e. An example of the latter, increasing the apparent IP, is the process

$$He\ 2^3S + H_2 \rightarrow He - H_2 \rightarrow H_2^+ + e + He1'S$$

where $\Delta E \sim 0.1$ eV.[32]

In the second process, which can result in a shift ΔE, the product ion is different from that resulting from the 'normal' Penning process. An <u>associative</u> ionization can take place:

$$(A-X-Y)^*\qquad A-X-Y^+ + e\ or\ AX^+ + Y + e.$$

Depending on the relative positions and shapes of the $(A-X-Y)^*$ potential well, ΔE can be positive or negative. PI therefore offers a means of studying energy conversion processes and interatomic potentials.

3.3 Auger electron spectroscopy (AES)

The 'normal' Auger process described in section 2 can be accompanied by several variations, all of which will create extra lines in an Auger spectrum.[33] Some of these are schematically represented in Figure 7. Those that involve excitations of valence electrons to other unfilled valence orbitals are obviously going to produce Auger electron energies differing from the 'normal' energies by amounts related to the valence level transitions. Auger spectroscopy therefore affords an alternative means of studying such transitions. Even if we consider only the normal process, the spectra produced can become quite complex. Consider Figure 8(a), which represents the normal Auger transition possible for an isolated C atom. There is only one core level, C(1s) at about -290 eV plus the 2p and 2s valence levels. Three different Auger processes are possible, leaving C^{++} final state

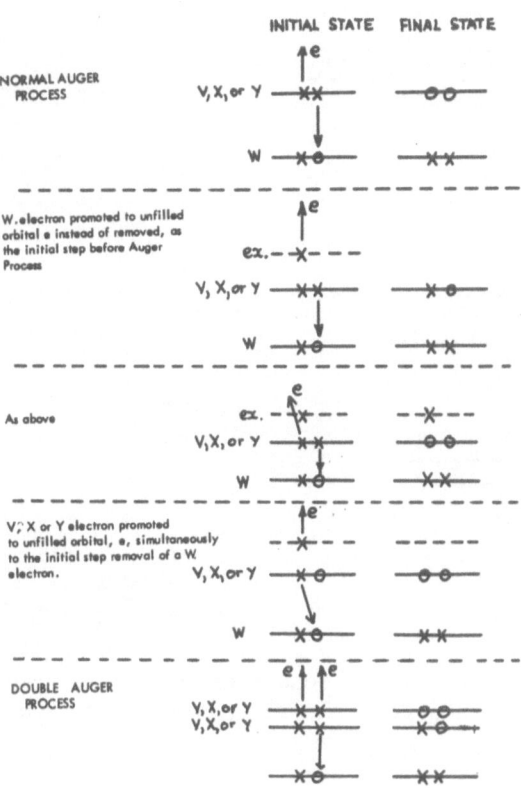

Figure 7. Processes which may compete with the 'normal' Auger processes.[33]

configurations of $(2s)^0 (2p)^2$, $(2s)^1 (2p)^1$, and $(2s)^2 (2p)^0$. However, when one considers the various L/S couplings possible in the final states, it becomes apparent that nine distinct final states are theoretically possible, leading to a maximum of nine lines in the Auger spectrum (seven have been observed). Figure 8(b) illustrates the situation for a carbon atom as part of the CO molecule. The molecule has four valence levels and the C(1s) and O(1s) core-levels. The valence levels can be sub-divided into three low IP levels, V, and the deeper lying σ level, V .

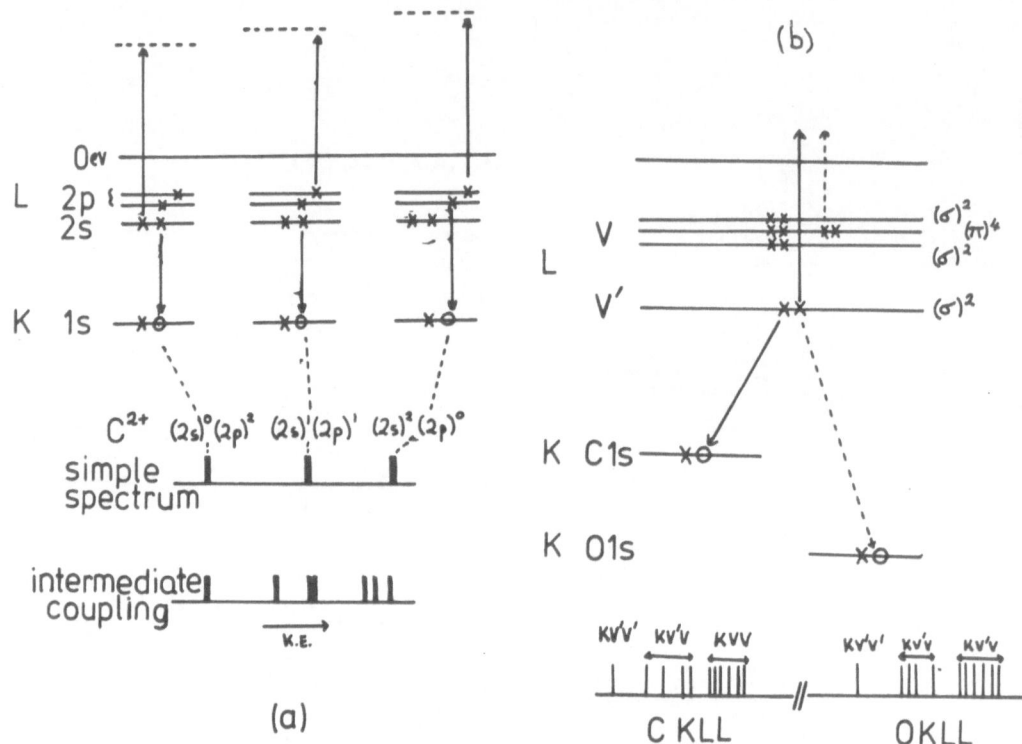

Figure 8. (a) Auger transitions possible for a free carbon atom;
(b) Auger transitions possible for a free CO molecule.

Both the carbon KLL and oxygen KLL Auger processes now have
a multiple line structure which depends intimately on both the
energies of the valence levels (defining the relative positions
of the multiple lines) and their atomic orbital parentage
(defining the intensities of the lines; the strength of a KLL
transition depending on the carbon 2p and 2s contribution to the
L levels concerned). One thus has an atomic identification of
carbon and oxygen, a molecular identification of CO from the
multiple line pattern, and a means of studying the parentages of
the valence orbitals. Figure 9 shows the O KLL region of the
Auger spectra of CO and H_2O[33] to illustrate the fingerprinting
aspect of the multiple line pattern.

4.   SOLID-STATE AND SURFACE APPLICATIONS[15,28,29]

Electrons originating from deep within a solid lose energy
through inelastic collisions on the way out, and therefore do not
appear at their expected energies in the electron spectrum, which

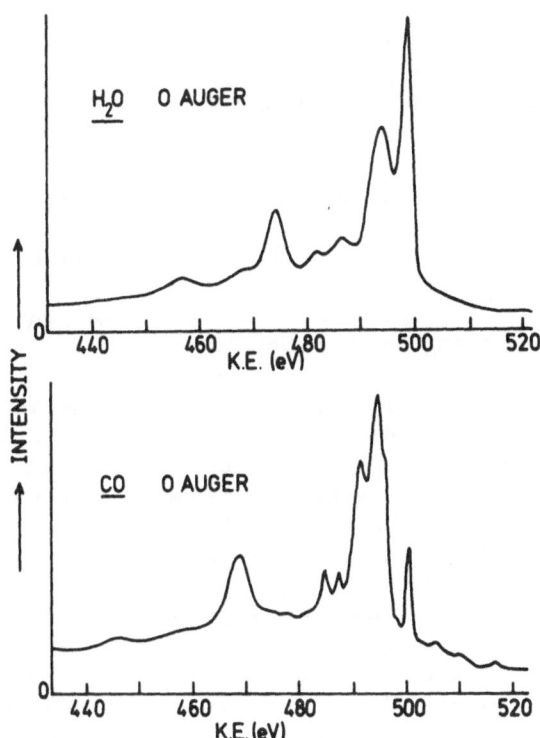

Figure 9.   O KLL Auger spectrum at high resolution for gaseous H_2O and CO.[33]

is therefore characteristic only of a surface slice.   The relationship between electron escape depth and K.E. of the electron was discussed in the chapter on surface effects in photoelectron spectroscopy, in which it was seen that the shortest values were obtained for energies between 20 and a few hundred eV.   The surface sensitivity of ELS and AES (PI is not used in solid-state studies) therefore depends on the K.E. of the ejected electrons in a similar fashion to UPS or XPS.   If the impacting primary electrons are of low energy also the surface sensitivity of the techniques will be enhanced further.

4.1   ELS

The transitions in solids analogous to those of gases, i.e. promotion of electrons from filled energy bands to empty or partially filled ones (band to band transitions) have been used to help interpret the electronic structures of solids, particularly semiconductors.   The surface sensitivity of the technique means that for the impacting energies commonly used (10's, to 100's of eV) a surface slice only is being examined.   Electronic states may be present which are entirely characteristic of the clean surface (surface states).   These will be observed in the ELS spectrum and the way in which they change on reaction at the surface (adsorption) can be followed, as can the appearance of features in the density of states due to the presence of the adsorbate.   Recently a considerable amount of work has been done on silicon surfaces along these lines.[34]

Two other types of energy loss process are important in solid state and surface studies: plasmon and phonon excitations. A plasmon excitation may be thought of as a collective oscillation of the entire Fermi electron gas, with respect to the solid lattice. An impacting electron causing such a vibration will be scattered with energy decreased by the plasmon energy, typically between 2 and 30 eV.   Multiple plasmon excitations can also sometimes be observed.   Plasmon excitations are quite well understood theoretically, at least for free electron-like materials, and their frequencies are calculable.   It is therefore feasible to use them for chemical studies, e.g. the composition and properties of a range of binary alloys.   Theoretically, surface plasmons (two-dimensional collective oscillations) should have $\sqrt{2}$ of the bulk plasmon value for an ideal surface and $\sqrt{3}$ for a spherical grain surface.   The presence, absence or modifications of such features in a spectrum can be used to characterize the condition of a surface during adsorption processes.

The phonon process represents the collective oscillations of the lattice ions (again in three or two dimensions).   Phonon energies are small, in the meV range, but multiple phonons of up to 20 components are possible.   A very high resolution spectrometer is needed to separate phonon excitations from the elastically scattered peak.   When adsorption occurs at a surface it becomes possible to observe the vibrational spectrum of the adsorbed species by examining the first 2 eV energy loss region, and therefore to find out whether the molecule is dissociated.[35] The technique is directly equivalent to I.R. spectroscopy when used in this fashion.

4.2 AES

Electron impact Auger spectroscopy has become a routine
technique for studying the elemental composition of surfaces
owing to the following factors: (a) the transition probabilities
and impacting electron beam flux are sufficiently high that spectra
can be obtained rapidly (oscilloscope scans if necessary); (b)
the atomic identification properties of the techniques (cf. XPS);
(c) the large number of Auger energies which fall within the
20-1000 eV range - i.e. short escape depths, and (d) the focussing
properties of an electron beam, which allows very fine spatial
resolution and the possibility of scanning a surface.   It is to
be expected that Auger/Scanning electron microscope combinations
will become commonplace within a few years.   The serious disadvantage of the technique is the surface damage by desorption,
cracking, induced adsorption, etc. produced by the probing
electron beam, although it can be reduced by decreasing the beam
intensity, with concommittent decrease in sensitivity.

Figure 10 shows typical Auger spectra.[36]   The normal mode
of recording, to take the first differential of the N(E) versus
E distribution as shown in the Figure, has the effect of
flattening out what can be a tremendously steep background (most
of the electrons in the spectrum are scattered background, upon
which sit the small Auger signals), and if phase-sensitive
detection is used (i.e. the differentiation is achieved by
modulating the analyser potentials) sensitivity is improved.
The two traces in Figure 10 are from the surfaces of fractured
steel samples after two different heat treatments.   The sample
showing the presence of antimony at the fracture grain boundary
was a sample embrittled by the heat treatment.

As explained in section 3, chemical shifts arise from two
sources, one in which only core levels are involved, and one in
which valence levels are involved.   Since the valence levels for
a solid are broadened into a band or bands, the discrete multiple
line structure observed in gaseous Auger spectra[33] involving
valence levels is no longer observed, and is replaced by a band
envelope, within which the features may bear a relationship to the
density of states (DOS) of the valence band of the solid.   Several
attempts have been made to extract DOS information from Auger
spectra, but with little success so far, probably mainly because
of the lack of understanding of the variation of transition
probabilities involved.   Well-resolved structure within the band
envelope is often observed, however, and it may be used in a
finger-printing fashion (cf. gaseous studies) to identify the
chemical nature of the surface.   Figure 11(a) shows silicon LVV
Auger spectra of a silicon surface before and after oxygen
treatment.[22]   Oxygen is present on the surface after the treatment,
as can be confirmed by the 0 KLL Auger region.   The Si LVV Auger

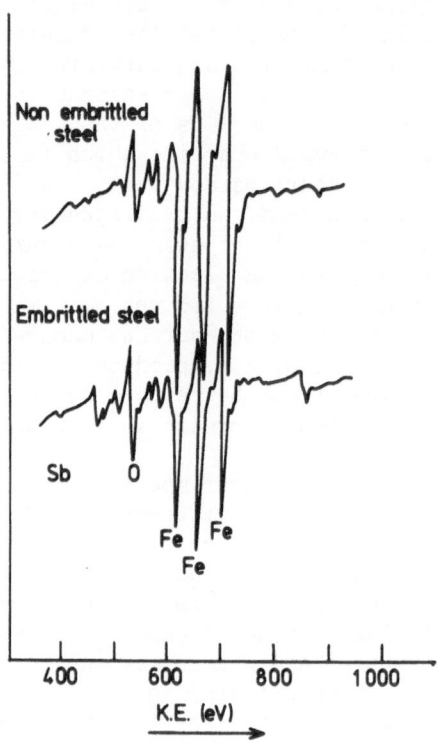

Figure 10.   Auger electron spectrum of stainless steel surfaces
found by fracturing two different samples within the vacuum
systems.[36]

spectrum indicates that it is in the form of SiO_2 because the
band envelope has changed from that characteristic of clean Si to
one characteristic of SiO_2.   Carbon KLL Auger band shapes have
been used empirically in the same fashion to distinguish between
carbon present in adsorbed molecular form, graphite, and carbided
surface.[37]   In XPS studies the X-ray excited Auger spectra
sometimes show chemical shifts when the equivalent shift in the
XPS spectrum is small or non-existent.   The Cu/CuO system is a
good example.   The Cu2p XPS line of Cu metal suffers a zero
chemical shift when oxidised to Cu_2O, but the Cu Auger line shifts
considerably (Figure 11(b)), thus allowing an easy distinction
between Cu^0 and Cu^I.[38]   Discrepancies between XPS and Auger
chemical shifts which are quite common, are apparently traceable
to the different amounts of relaxation involved in the singly and
doubly charged final states for the metal and oxide.[39]

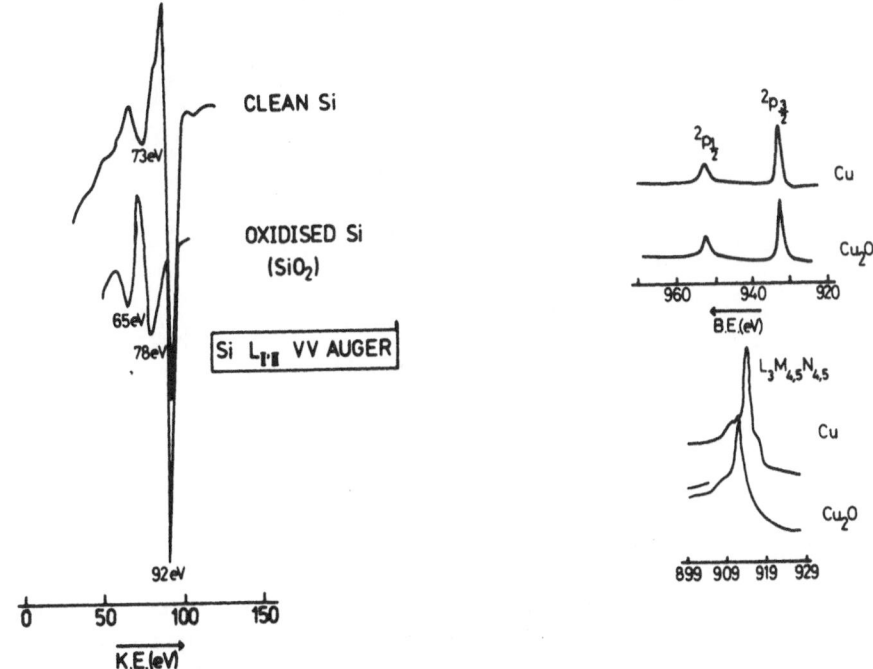

Figure 11. (a) LVV Auger spectra (electron impact involved) of a silicon surface before and after oxygen treatment;[23] (b) Cu Auger spectra (X-ray induced) of Cu and Cu_2O; XPS Cu 2p spectra of Cu and Cu_2O.[38]

4.3 INS

The INS process, illustrated schematically in Figure 12, is a special case of the Auger process (cf. Figure 4). It is a technique used entirely for the study of surfaces. A slow beam of He+ ions is allowed to impinge on the surface of the sample, and an electron transfers from the sample to the ion (see Figure 12), neutralizing it and releasing sufficient energy to eject a second electron. Since the available hole of He+ is only ca. 24 eV below the vacuum level, the Auger transitions are restricted entirely to the valence region. The major feature distinguishing INS from a normal WVV Auger transition is that the slow He+ ion never penetrates the surface, so the density of states sampled by the INS process represents that on, or even just outside, the surface. Hagstrum has demonstrated the extreme surface sensitivity of the technique by comparing INS and UPS spectra of several adsorption systems and showing that INS observes only what is present right at the surface, whereas UPS is affected by the presence of adsorbate in the sub-surface region.[23] Figure 13 shows the INS spectrum for the adsorption of Se on a Ni(100)

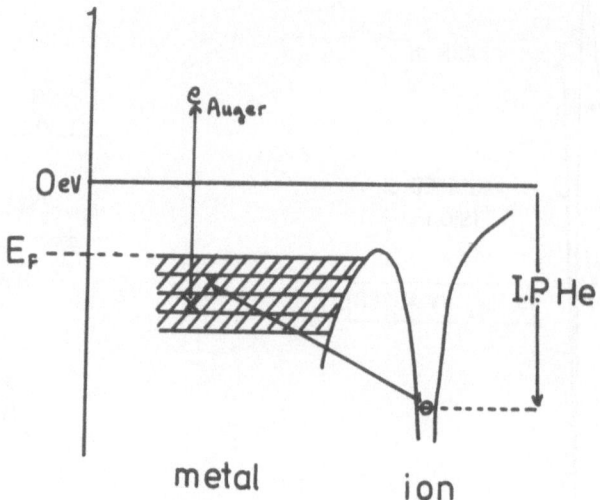

Figure 12.   Schematic representation of the ion neutralization
processes showing the neutralization of He+ as it approaches the.
surface by the tunneling of .a valence electron from the sample,
and the subsequent release of an Auger electron.

surface.[24]   The raw data has been mathematically unfolded to
remove the effect of the involvement of two electrons from the
valence levels, so that at this point it could be compared to the
equivalent UPS spectrum.   The UPS spectrum of gaseous H_2Se,
shown in the same Figure, reveals that the adsorbate features of
the Se/Ni(100) spectrum are rather similar to the MO's of H_2Se
recorded by UPS.   One interpretation is that they represent a
pseudo-orbital structure of Ni_2Se surface molecules with a similar
geometry and electronic structure to H_2Se.

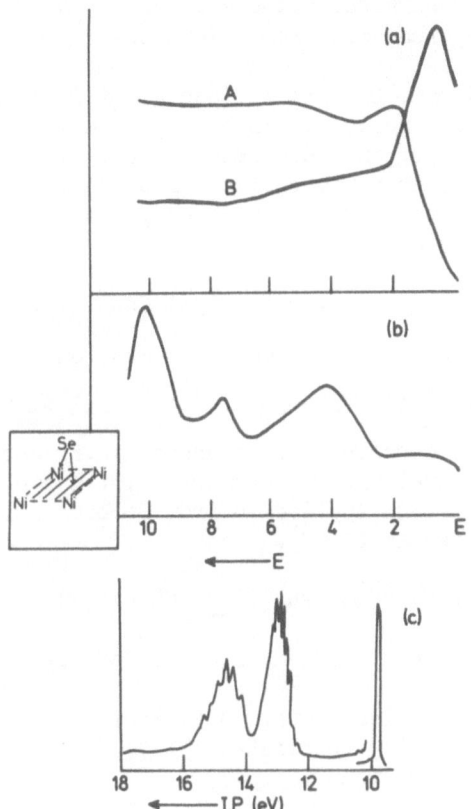

Figure 13. (a) INS spectrum of clean Ni(100), A raw data, B
unfolded; (b) unfolded spectrum of Se/Ni(100) system; (c) UPS
spectrum of H₂Se for comparison.

REFERENCES

1.   D.W. Turner, C. Baker, A.D. Baker and C.R. Brundle,
       'Molecular Photoelectron Spectroscopy', Wiley, London
       (1970).

2.   A.D. Baker and D. Betteridge, 'Photoelectron Spectroscopy -
       Chemical and Analytical Aspects', Pergamon, Oxford (1972).

3.   J.H.D. Eland, 'Photoelectron Spectroscopy', Halstead Press,
       1974 (N.Y.); Butterworths, London (1974).

4.   A.D. Baker, C.R. Brundle and M. Thompson, Chem. Soc. Rev.
       355 (1972).

5. C.R. Brundle and M.B. Robin in 'Determination of Organic
    Structures by Physical Methods', eds. F. Nachod and
    G. Zuckerman, Academic Press, New York.

6. A. Hamnett and A.F. Orchard in Specialist Periodical Reports
    of the Chemical Society: 'Electron Structure and
    Magnetism of Inorganic Compounds', vol. 3, 218 (1974).
    This is a comprehensive collection of data on both UV
    and X-ray photoelectron spectroscopy research in the
    period 1972-73.

7. D. Hercules and J.C. Carver, Analyt. Chem. $\underline{46}$, 133R (1974).

8. C.J. Allen and K. Siegbahn, MTP Int. Rev. Sci. Phys. Chem.
    Ser. One, $\underline{12}$, 1 (1973).

9. K. Siegbahn, C. Nordling, G. Johansson, P.F. Heden,
    K. Hamrin, U. Gelius, T. Bergmark, L.O. Werme, R. Manne
    and Y. Baer, 'ESCA Applied to Free Molecules', North
    Holland, Amsterdam (1969).

10. D.A. Shirley, Adv. Chem. Phys. $\underline{23}$, 85 (1973).

11. D.T. Clark, Ann. Rept. (Chem. Soc. London) $\underline{69}$, 66 (1972).

12. W. Mehlhorn, Proc. Ins. Conf. Inner Shell Ioniz. Phenomena
    Future Appl. (1972) (published 1973), U.S. Atomic Energy
    Comm., Oak Ridge, Tenn., p. 437.

13. C.C. Chang, Surface Science, $\underline{25}$, 53 (1971).

14. L.A. Harris, Analyt. Chem. $\underline{40}$, 24A (1968).

15. C.R. Brundle, in 'Defect and Surface Properties of Solids',
    Specialist Periodical Report of the Chemical Society,
    vol. 1, ed. J.M. Thomas and M.W. Roberts.

16. V. Cermak, J. Chem. Phys. $\underline{44}$, 3781 (1966).

17. H. Hotop, in 'Advances in Mass Spectrometry', ed. A. Quayle,
    vol. 5, Institute of Petroleum, London (1971) p. 116.

18. R.S. Berry, Ann. Rev. Phys. Chem. $\underline{20}$, 357 (1969).

19. C.E. Brion, C.A. McDowell and W.B. Stewart, J. Electron
    Spectrosc. $\underline{1}$, 113 (1972).

20. A. Skerbele and E.N. Lassettre, J. Chem. Phys. $\underline{40}$, 1232
    (1965).

21. G.R. Wright and C.E. Brion, J. Electron Spectrosc. 4,
     313, 327, 335, 347 (1974) and references therein.

22. S. Trajmar, J.K. Rice and A. Kuppermann, Advances in
     Chemical Physics, 18, 15 (1970).

23. H.D. Hagstrum and G.E. Becker, Proc. Roy. Soc. A 331, 395
     (1972).

24. H.D. Hagstrum and G.E. Becker, Phys. Rev. Lett. 22, 1054
     (1969); J. Chem. Phys. 54, 1015 (1971).

25. T. Baer, W.B. Peatman and E.W. Schlag, Chem. Phys. Lett.
     4, 243 (1969).

26. G.J. Shulz, in 'Atomic Physics, Proceedings International
     Conference, 1968', Bederson and Benjamin, published
     Plenum Press, New York (1969).

27. V.I. Oginstov and Yu. F. Bydin, Zh. Eksp. Terr. Fiz. 57,
     1908 (1969).

28. C.R. Brundle, J. Electron Spectrosc. 5, 291 (1974).

29. C.R. Brundle, J. Vac. Sci. Tech. 11, 212 (1974).

30. P. Auger, J.Phys. Radium, 6, 205 (1925).

31. V Cermak, Chem. Phys. Lett. 4, 515 (1970).

32. H. Hotop and A. Niehus, Z. Physik. 228, 68 (1969).

33. W.E. Moddeman, T.A. Carlson, M.O. Krause, B.P. Pullen,
     W.E. Bull and G.K. Schweitzer, J. Chem. Phys. 55, 231
     (1971).

34. J.E. Rowe and H. Ibach, Phys. Rev. Lett. 32, 421 (1974).

35. F.M. Propst and T.C. Piper, J. Vac. Sci. Tech. 4, 53 (1966).

36. R.E. Weber, Solid State Techn. Dec. 1970, p. 49.

37. T.W. Hass, J.T. Grant and G.J. Dooley III, J. Appl. Phys.
     48, 1853 (1972).

38. P.E. Larson, J. Electron Spectra, 4, 213 (1974).

39. C.D. Wagner and P. Biloen, Surface Science, 35, 82 (1973).

X-RAY SPECTROSCOPY

D.S. Urch

Department of Chemistry, Queen Mary College,
London, E.1

1.   INTRODUCTION

The purpose of this chapter is to present in outline the
technique of X-ray spectroscopy and to discuss the importance and
relevance of the results which can be obtained especially from
high resolution emission spectra.   No attempt will be made to
give a fundamental discussion of the physical processes involved
and neither a comprehensive description of all types of spectrometer, nor a complete review of all experimental work will be
contemplated.   Rather an attempt will be made to introduce the
subject at an elementary level and then to discuss selected
spectra: emphasising the importance of related spectroscopies
such as XPS (X-ray photoelectron spectroscopy), UPS (ultraviolet
photoelectron spectroscopy) and Auger, and demonstrating the use
of the results in understanding the chemical bond, especially in
inorganic molecules.

When matter is irradiated with high energy electromagnetic
radiation (e.g. X-rays or gamma rays) or bombarded with energetic
electrons or heavier particles, electrons are ejected.   If the
electrons come from inner orbitals then a highly excited ion is
formed which can relax either by the emission of an X-ray (or
X-rays) or by the non-radiative ejection of an electron (Auger
process).   These processes are all summarised in Figure 1.   Of
course electrons can only be ejected if their binding energies are
less than the energy of the incident radiation or particle.
When monochromatic radiation or electrons of a unique energy are
used the kinetic energies of the ejected electrons can be measured
and thus their original binding energy inferred, this is the basis
of photoelectron spectroscopy (see chapters by Orchard, Lloyd and
Brundle).

P. Day (ed.), Electronic States of Inorganic Compounds. 449–493. *All Rights Reserved.*
Copyright © 1975 by D. Reidel Publishing Company, Dordrecht-Holland.

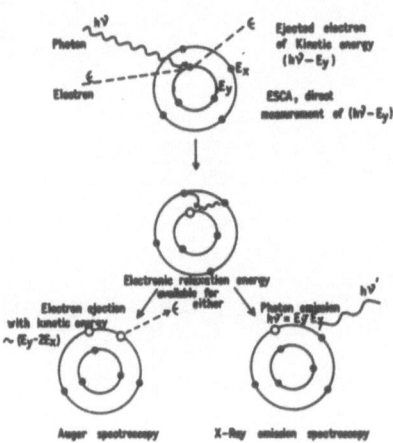

Figure 1.　Basic processes in X-ray and electron spectroscopy.

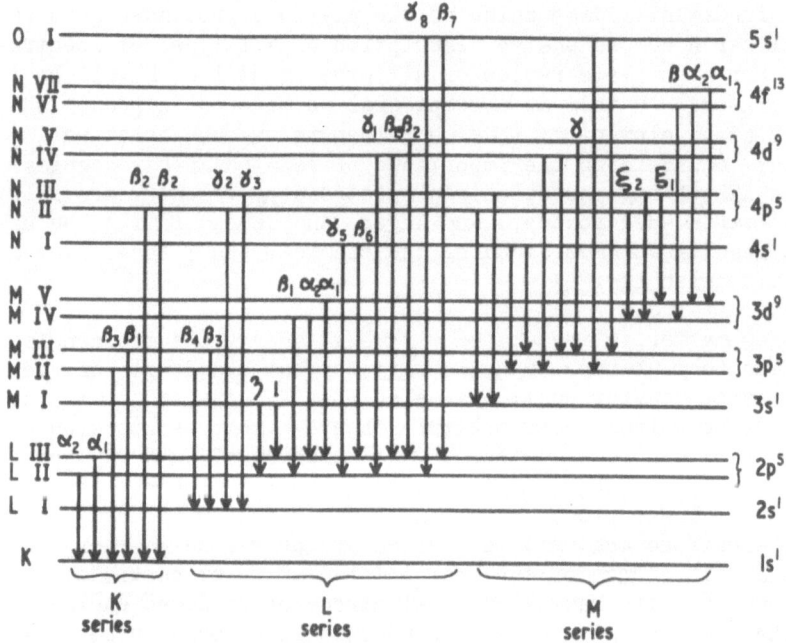

Figure 2.　X-ray emission: nomenclature for 'diagram-lines' permitted by the dipole selectionrule.　Electron configuration and spectroscopic states are indicated on the far right for shells with but one vacancy.

X-rays were discovered by Roentgen[1] in 1895 and were studied
extensively in the ensuing years.   It was quickly recognized
that X-rays of different energies were emitted by different
elements and also that any one atom could (usually) emit many
different X-rays.   The number of different lines and the general
complexity of the X-ray spectrum increases with atomic number.
X-rays were classified as $K\alpha$, L etc., and in 1913 Moseley[2]
demonstrated the elegant and simple relationship between atomic
number (Z) and $K\alpha$ wavelength:

$$\lambda^{-1} = C(Z-\sigma)^2$$

(λ is X-ray wavelength, Z is the atomic number and C and σ are
constants.)

1.1   Nomenclature

The variation in wavelength with Z fitted in well with the
Bohr theory of the atom and the general physics of X-rays.   The
nomenclature which is still used to describe various X-ray lines
reflects the physics of half a century ago.   The basic classification depends upon the primary vacancy: thus K spectra result
from the relaxation of an atom with a vacancy in an orbital with
principal quantum number 1, L spectra;   principal quantum number
2, M spectra;   3, N;   4, 0,5 etc.   It was also established that
a j-j rather than an L-S coupling scheme was the most convenient
for describing the state that arose from the creation of a vacancy
in an otherwise closed shell inner orbital.   The different
possible sub-shells due to different orbital angular momenta are
denoted by Roman subscript numerals, thus: 2s, L_I;   2p, L_{II} and
L_{III};   3d, M_{IV} and M_V etc. L_{II} and L_{III} correspond to the two
possible j-j states that arise from the p^5 configuration $^2P_{\frac{1}{2}}$ and
$^2P_{3/2}$.   (Arabic numerals are used by some authors, e.g. L_2 and
L_3.)

An inner vacancy can of course be filled by transitions from
a variety of less tightly bound orbitals.   These are denoted by
Greek letters further differentiated by Arabic numerals;   thus a
2p→1s transition could rise to two possible X-rays of similar
energy depending upon the final electronic state of the ion $P_{\frac{1}{2}}$ or
$P_{3/2}$.   These two X-rays are $K\alpha_1$ and $K\alpha_2$.   When the two peaks are
not resolved the subscripts are run together.   Thus for the
3p→1s relaxation, we have $K\beta_{1,3}$ while 4p→1s is known as $K\beta_2$.
Apart from the use of K, L etc. for the initial vacancy there is
no obvious logic in the names used for X-rays.   The reason of
course is historical, as the α's and the β's were discovered first
and then it was shown that the α peak could be resolved into two
components, α_1 and α_2, etc.   Unfortunately this system of nomenclature is now well established.   The symbols used for the
principal X-ray emission lines are summarized in Figure 2.

1.2 Selection rules

The lines which have been included in the Figure are the so-called diagram lines, lines which represent X-ray which conform to the atomic dipole selection rule $\Delta l = \pm 1$ and also $\Delta j = 0, \pm 1$. As Z increases so it is found that other transitions are also (weakly) observed, e.g. $K_{\beta_5}(3d \rightarrow 1s)$ in which $\Delta l = \pm 2$. For the most part however the observed X-ray emission lines of light and medium weight elements $(Z < 40)$ are very well described by the diagram – lines.

1.3 Auger emission

As the atomic number of an atom decreases it is found that the probability that the excited ion with an inner orbital vacancy will relax by X-ray emission decreases and the probability that an Auger electron will be ejected increases. Thus more than 99% of carbon atoms with 1s holes decay by the Auger process whereas for $Z > 30$ decay is almost exclusively by X-ray emission. This makes a study of X-ray quanta emitted by light elements particularly difficult even though Auger electrons are of course worthy of study in their own right. However, Auger spectroscopy is much more complicated than photoelectron or X-ray emission spectroscopy because the first state in the Auger process is a doubly ionised atom, which can give rise to a variety of electronic states of different energies. Therefore even simple Auger spectra will be more complex than corresponding XE (X-ray emission) or PE (photoelectron) spectra. The nomenclature for Auger spectra follows that used for X-rays, the initial vacancy is given first followed by the locations of the final two holes, e.g. $K_I L_{II} L_{III}$.

1.4 Lifetimes of excited states: widths of emission lines

The probability of an electronic transition is given by the Einstein equation, which shows a dependence upon frequency. A relaxation process which involves the emission of a highly energetic quantum will proceed more rapidly than the emission of a less energetic quantum. It is therefore to be expected that the lifetimes of excited ions with inner shell vacancies will be very short, much shorter than those associated with ultraviolet excitations, an expectation which is borne out well in practice. Typical lifetimes are of the order of 10^{-17}-10^{-14} seconds. The ion that results after X-ray emission will also have a characteristic lifetime but this will be relatively longer, since lower energy processes are involved in its decay. A consideration of lifetimes is of importance for two reasons. The first is the relationship between the uncertainty in the energy of the emitted X-ray (natural line width E) and the lifetime of the shortest lived excited state (Δt). The Heisenberg equation shows that $\Delta E. \Delta t = \hbar = 6.5 \times 10^{-16}$ eV. secs. Some typical line widths are:

S,K$_\beta$ ~1.3 eV, Al Kα ~0.8 eV, NKα < 0.2 eV.   Such figures are
important as they give an indication of the ultimate experimental
resolution that can be achieved.   The second reason for paying
some attention to the lifetime of the excited ion state is to
determine whether other relaxation processes might take place
before its decay, in particular whether vibrations can take place.
The simple answer would seem to be that the duration of the first
state is too short for any appreciable nuclear motion to take
place although in some cases line broadening may result.
However, the final relaxation of the inner hole state may well
involve a transition from a valence shell orbital and the vibrational structure of the resulting ion would then be observed in
the emission spectrum.

1.5 Emission probability

The lifetime of a particular state is determined by all the
processes which lead to its decay, both radiative and non-
radiative.[3]   Thus a weak peak which might be thought to result
from a transition with a low probability (and therefore a
relatively long lifetime for the initial state), might nevertheless be quite broad because of other processes leading to the
removal of the primary ion.   For light elements Auger processes
are a particularly common source of peak broadening.   Competing
radiative processes can also give rise to short-lived ions and
thus broad peaks.   The probability for radiative emission depends
upon the spatial overlap between he two orbitals involved.   Thus
whilst a 1s vacancy can be filled by transition from 2p, 3p, 4p,
etc. orbitals the intensities of the peaks are Kα > K$\beta_{1,3}$ > Kβ_2.
The relative intensity will of course vary with atomic number as
the 1s, 2p : 1s, 3p : 1s, 4p overlap varies.

1.6 X-ray absorption

Closely related to X-ray emission is X-ray absorption, in
which X-ray quanta promote electrons from inner atomic orbitals
to outer vacant or virtual orbitals.   Unless the atom has a
partially filled shell, and provided it is in its ground electronic
state, the maximum energy for X-ray emission should be less than
the corresponding energy for absorption (i.e. $E(3p \rightarrow 1s) < E(1s \rightarrow 4p)$).
This situation is not common for atoms or metals but is usually
achieved in stable molecules.   Thus X-ray emission and absorption
spectra for metals will usually show regions of overlap whilst
those for compounds do not.   The existence of such overlapping
regions can cause problems in X-ray emission spectroscopy due to
self-absorption which can give rise to spurious features in the
emission spectrum[4] (Figure 3).

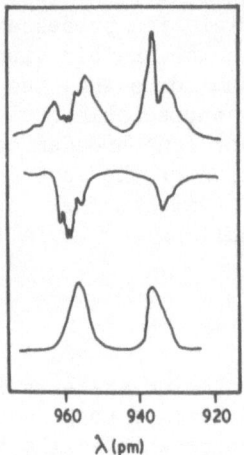

960   940   920
λ (pm)

Figure 3.   The effect of self-absorption on X-ray emission
spectra.   M$_{\alpha\beta}$ emission spectrum of dysprosium under normal
conditions (top);   corresponding absorption spectrum (middle);
M$_{\alpha\beta}$ emission spectrum of dysprosium under conditions of negligible
self-absorption (bottom).

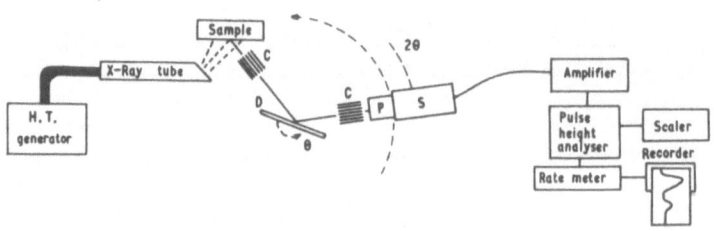

C = collimator,   D = diffracting crystal,   P = proportional counter,

S = scintillation counter.

Figure 4.   X-ray emission spectrometer – diagrammatic.

1.7 X-ray spectroscopy: photoelectron spectroscopy

In X-ray emission spectroscopy (XES) a transition is observed
between two electronic states of an ion. The energy required to
generate both the initial and final ionic states can be determined
directly using photoelectron spectroscopy (usually with X-ray
excitation, XPS or ESCA). A direct correlation between XES and
XPS spectra is therefore possible, and indeed desirable.
However, XPS does not give a well resolved indication of the
molecular orbitals in the valence bond region; this is the
province of photoelectron spectroscopy which uses a much lower
energy excitation source, e.g. He-I (UPS). If X-ray emissions
which involve transitions from the valence bond can be observed,
and if the ionisation energy of the inner orbital can be determined or calculated, then the energies of the molecular orbitals
can be found. In this way a variety of different X-ray emission
spectra can be correlated with each other and with photoelectron
spectra. X-ray absorption spectra can also be incorporated into
the picture; since they probe the corresponding antibonding
orbitals they should give information complementary to that of
the emission spectra.

2.   EXPERIMENTAL TECHNIQUES IN X-RAY EMISSION SPECTROSCOPY

High resolution X-ray spectrometers use either crystals or
gratings to disperse the X-rays. At the present time crystals
are widely used at wavelengths below 25 Å and gratings for longer
wavelengths (some crystals and soap film pseudo-crystals have been
used up to about 160 Å). The simplest type of spectrometer which
is readily available commercially (Philips, Siemans, etc.) uses
flat crystals and is shown diagramatically in Figure 4. The
sample is irradiated with X-rays from a suitable X-ray tube
(common anodes are Cr, Cu, Mo, W, Au). Whilst such a tube will
emit the X-ray lines characteristic of the anode a very great
deal of the emitted X-radiation is also found over a very wide
wavelength range, 'white radiation'. Since the efficiency with
which electrons are ejected from the various orbitals of the
different atoms that may be present in the sample varies considerably for a particular irradiating photon energy, it is expedient
to have a wide range of exciting energies available. A parallel
beam of X-rays emitted by the sample is selected by means of the
Soller collimator slits. The degree of divergence of X-rays
passing through this collimator is one of the factors which limits
instrumental resolution. The X-rays are diffracted by a suitable
crystal of known 2d spacing according to the Bragg equation

$$n\lambda = 2d \sin\theta$$

where n = order of diffraction, λ = wavelength of diffracted ray,
d = characteristic spacing of crystal and θ = angle of incidence
and emergence for diffracted ray.   A list of commonly used
crystals and wavelength ranges for which they are useful is given
in Table 1.

Those X-rays which satisfy the Bragg requirements pass through
a second stage collimation and then fall upon a proportional
counter behind which is mounted a scintillation counter.   Soft
X-rays are detected in the former, hard X-rays pass through and
are detected in the scintillator.   The output from either device
is a pulse of the order of a few millivolts or less which is
proportional to the energy of the initial X-ray.   After amplific-
ation the pulses pass through a pulse height analyser which is
used to reject interfering signals of the wrong order.   (For a
given crystal and θ setting the wavelengths λ, $\lambda/2$, $\lambda/3$ etc. are
all diffracted and could all be counted.   However, the energies
of the photons corresponding to the different orders also vary,
e.g. E, 2E, 3E, etc.   A judicious setting of the pulse height
analyser thus enables the required order to be selected.)   The
output of the pulse height analyser is fed to a scaler and a
ratemeter.   Spectra are obtained by rotating the crystal through
θ° and simultaneously rotating the detector assembly through $2\theta^\circ$,
which is equivalent to scanning a range of wavelengths.   The
ratemeter is usually connected to a chart recorder equipped with
an angle marker so that a graphical record of the spectra (counts
vs.θ) is obtained.   A more accurate method of obtaining spectra
is to count for fixed periods of time at pre-set angular positions.
This step-scanning procedure can readily be automated and the
digital record lends itself to computer analysis.

X-ray emission spectrometers of this general type usually
work with solid samples but can be adapted to take liquids and
even gases.   The main factors which limit resolution are the
quality of the diffracting crystal and the degree of collimation.
Increased resolution may be obtained by adding a second stage of
diffraction (2-crystal spectrometer), or by the use of a spectrometer of fundamentally different design – the curved or bent
crystal spectrometer.   If the crystal can be bent to a suitable
radius then X-rays from a point source can be brought a focus
after diffraction.

2.1   Soft X-rays

For wavelengths longer than 26 $\overset{\circ}{A}$, grazing-incidence grating
spectrometers are often used.   The efficiency of such instruments
is quite low so that signal intensity is a problem.   Spectrometers of this type must operate under vacuum conditions ($< 10^{-4}$ torr)

because of the low energies of the photons being studied
($\lambda > 25$ Å, E < 500 eV).   A vacuum (10^{-2}–10^{-4} torr) within conventional crystal spectrometers is also necessary to work at for wavelengths of greater than about 5 Å.

Table 1 (Crystals)

Crystal	(Diffracting plane)	(Abbreviation)	2d(pm)
Lithium fluoride	420	(LiF 420)	198
Lithium fluoride	220	(LiF 220)	285
Lithium fluoride	200	(LiF 200)	403
Sodium chloride	200	(NaCl)	564
Silicon	111	Si	626
Germanium	111	Ge	655
Pentaerythritol		PE	873
Ethylene diamine tartrate		EDDT	881
Ammonium dihydrogen phosphate		ADP	1065
Gypsum		Gyp	1515
Mica		–	1980
Thallium acid phthalate		TAP	2590
Rubidium acid phthalate		RbAP	2621
Potassium acid phthalate		KAP	2664
Chlinoclore		–	2840
Octadecyl hydrogen maleate		OHM	6350
			(nm)
Lead myristate		PbMy	7.90
Dioctadecyl adipate		OAO	9.38
Lead stearate		PbSt	10.06
Lead lignocerate		PbLi	13.04
Lead melissate		PbMe	16.0

2.2 Sample irradiation

As described above, X-ray emission from a sample can be
stimulated by irradiation with X-rays (secondary excitation:
X-ray fluorescence) but much more intense emission can be produced
by irradiating the sample directly with electrons (primary
excitation).   This has advantages for the study of very soft
X-rays, in association with a grating spectrometer, but a disadvantage is that direct electron bombardment, by depositing its
energy in a much smaller volume of sample than the corresponding
X radiation, causes more chemical decomposition of the sample.[5]
Experimentally samples for direct electron bombardment are most
easily mounted on the anode of a conventional X-ray tube.   The
Telesec betaprobe is an example of a commercial X-ray spectrometer which utilises direct electron bombardment of the sample to
induce emission.

2.3 X-ray absorption

X-ray absorption may be studied in a conventional spectrometer
with only a minimum of alterations.   A very thin sample is placed
in the beam of diffracted X-rays just before the detector assembly.
The position occupied by the sample in fluorescence studies is
filled with a suitable reflecting material.   Spectra can then be
recorded with the absorbing sample present, then absent, and the
absorption spectrum itself is obtained by difference.

2.4 Other types of spectrometer

Crystal and grating spectrometers have been discussed in
detail above because they are capable of giving the high resolution which is required by chemists.   Other much simpler types of
X-ray spectrometers also exist which are primarily used for
element analysis.   Characteristic X radiation is produced from
the sample by X-ray, electron or nuclear particle bombardment
(the latter technique, using a radioactive source, is the basis
of a very compact portable spectrometer - such as was used in
early non-manned Apollo flights for the analysis of moon rocks).
The various X-rays are differentiated directly in a semi-conductor
detector, or else by means of a scintillation counter whose
resolution can be aided by filters.   Because they disperse with
a crystal or grating, such spectrometers are called non-dispersive.
Resolution in these instruments is limited to about 150 eV.

3.   THE APPLICATION OF X-RAY SPECTROSCOPY TO ELEMENTAL ANALYSIS

X-ray spectrometers are widely used for element analysis
(references B4-7).   For this purpose high resolution is not
usually required and spectrometers and the crystals used in them

are optimised to transmit the greatest intensity of characteristic
radiation emitted by the sample.   Even so, commercial spectrometers can be used to give results of quite high precision
without any modification whatsoever.   An interesting analytical
use of X-ray spectroscopy is in electron microscopy.   A very
small portion of the sample will be in the electron beam at any
one time and the electron bombardment will cause it to emit X-rays
characteristic of the elements present.   X-ray spectrometers are
suitably placed so that these X-rays can be analysed.   As the
electron beam is swept across the sample the variations in the
concentrations of the elements present can be determined.   Many
older 'electron microprobes' are equipped with two or even four
crystal dispersive X-ray spectrometers so that two (or four)
elements can be scanned simultaneously.   The more recent development of the scanning electron microscope has been paralleled by
introduction of non-dispersive spectrometers, which have the
great advantage of being able to analyse for all elements at once.
The limitation from which these detectors suffer, of only being
able to accept a rather low total count rate (e.g. 10^4 – 10^5
pulses sec^{-1}) is here almost an advantage, since the area scanned
at any one time is small and so therefore is the amount of
emitted X radiation.

For more details of the role of X-ray emission spectroscopy
in analysis, the reader is referred to the many text books on the
subject.   The purpose of the present chapter is rather to
describe how changes in X-ray spectra may be correlated with
chemical bonding.

4.   X-RAY EMISSION SPECTROSCOPY AND THE CHEMICAL BOND

With changes in the chemical bond X-ray emission peaks may
exhibit any, or none of the following:

(a) a change in energy, i.e. 'peak shift' or 'chemical
shift';

(b) a change in peak shape;

(c) the formation of neighbouring new peaks – 'satellites'
(which may be of higher or lower energy or both, than the
peak they accompany).

For the purpose of the more detailed discussion to follow,
it will be important to make a distinction between those X-ray
emission peaks which arise (i) solely from transition between
inner atomic orbitals (ii) transitions between valence shell
orbitals (i.e. molecular orbitals) and inner orbitals.

Peaks arising from type (i) transitions usually show only very slight changes in energy and shape. New satellite peaks are not usually found but existing satellite peaks (usually of a higher energy than the parent peak) may vary in their relative intensities from compound to compound. From the chemical point of view it is of course transitions of type (ii) which hold the most direct and obvious interest and changes under (a), (b) and (c) are usually observed as the ligand environment of a particular atom is altered.

Typical examples of such effects are shown by (i) $AlK\alpha_{1,2}$ and (ii) $AlK\beta_{1,3}$ peaks in aluminium metal and alumina[6] (Figure 5). That slight changes in coordination environment can also affect X-ray emission peaks and that these changes are greater for type (ii) peaks[7] than type (i) peaks,[8] is demonstrated in Figure 6. Comparable or similar effects have been found for magnesium,[9] silicon,[9] phosphorus,[10] sulphur,[11] chlorine,[12,13] etc. and also for first row[14] and transition metal elements.[15] Molecules with unpaired electrons (especially transition metal compounds) often show broadening of type (i) peaks. This effect can sometimes be manifest as new and separate satellites (section 6.1). Binding of an atom to a complex ligand is reflected in the type (ii) X-ray emission spectra, as might be expected (Figure 7). Most of the spectra shown above have relatively poor resolution (natural line width limitations do not exceed 1-2 eV in any of Figures 5-7), yet it is quite clear that the structure of type (ii) X-ray emission peaks carries considerable conformation about the molecular orbital array in a molecule, molecular orbital energies and also about the degree of participation of particular atomic orbitals in these molecular orbitals. More highly resolved spectra will be of even greater value. In order that X-ray emission spectra may be discussed in detail and their significance assessed it is necessary to compare such spectra with theoretical calculations of orbital energies and atomic coefficients and estimates of relative X-ray emission intensities. This has been attempted for a few anions and compounds,[16,17] but for the most part, detailed m.o. calculations are lacking and a simple qualitative model must be used.

4.1 Simple molecular orbital interpretation of chemical bond effects in X-ray emission spectra

The application of molecular orbital theory, using the Huckel approximations and with neglect of overlap, to a simple molecular or ionic tetrahedral unit XL_4 (X, central atom, L, ligand atom, T_d, point group) enables the relative energies of molecular orbitals to be calculated and a rough indication of the atomic orbital coefficients in those orbitals to be estimated provided that approximate values of Coulomb integrals are taken from atomic orbital ionization energies.[18,19] Such a one-

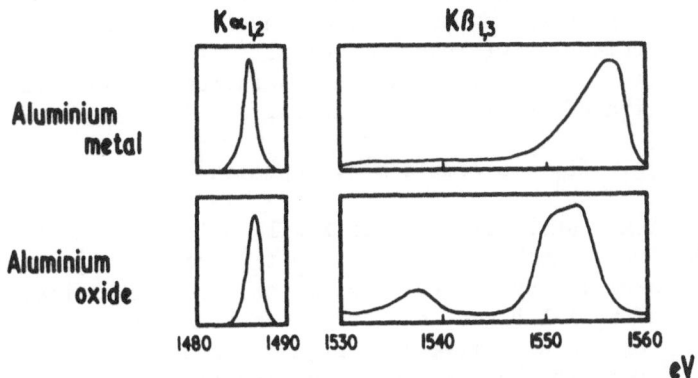

Figure 5.   X-ray emission spectra of aluminium and alumina.

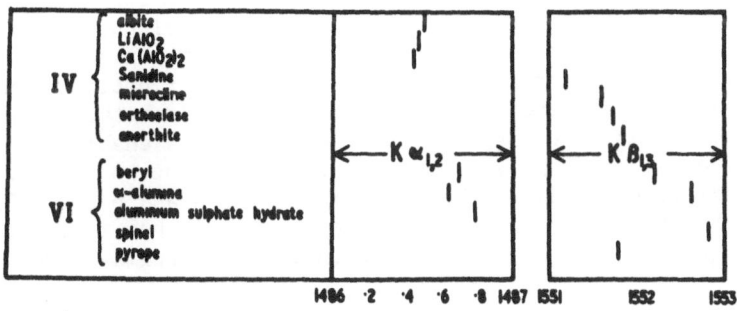

Figure 6.   Energies of $K\alpha_{1,2}$ and $K\beta_{1,3}$ X-ray emission peaks for aluminium in four-fold (IV) and six-fold (VI) coordination.

electron model gives only a qualitative framework within which trends due to variations of X or L may be discussed, but it can also be applied to other high symmetry molecules and also used as a guide to probable atomic orbital involvement in molecular orbitals in molecules of lower symmetry. It should therefore be possible to test the validity and usefulness of this model against spectra of the type shown above.

4.2 Bonding in XL_4: a simple m.o. approach[18]

The valence shell orbitals of X will belong to the following irreducible representations: $\underline{s}(a_1)$, $\underline{p}(t_2)$ and $\underline{d}(e+t_2)$; and for the ligand atoms: four $\underline{s}$ orbitals (a_1+t_2), four p orbitals orientated along X-L bonds, one from each ligand (a_1+t_2), eight remaining p orbitals $(e+t_1+t_2)$. Ligand d orbitals could be included if required but are not necessary to demonstrate the essence of this approach. The classification of ligand and central atom orbitals can be taken a stage further if it is assumed that ligand orbitals of t_2 symmetry that are orientated along the X-L bonds will interact strongly with Xp orbitals but that ligand p orbitals perpendicular to X-L σ bonds will interact more favourably with X d orbitals, (π bonds).[20] The number of atomic orbitals (or symmetry orbitals derived from atomic orbitals) within each irreducible representation is in this way reduced to a maximum of three. The corresponding secular equations can be solved if suitable values are chosen for the various Coulomb and resonance integrals. For the purposes of a general energy level diagram, however, a qualitative approach will suffice, using the general rule that interactions will be greatest for orbitals of comparable energies and least for those orbitals with widely different ionization energies. The corollary which determines atomic orbital coefficients is equally general, coefficients will be of similar magnitude for atomic orbitals of similar energies, but dissimilar in molecular orbitals derived from atomic orbitals of greatly different ionization energies. The tendency will be for the molecular orbitals in this latter case to resemble strongly the atomic orbital which has a similar ionization energy. These general ideas enable a qualitative m.o. energy level diagram such as is shown in Figure 8 to be constructed. It has been assumed that Xs and p orbitals and Lp orbitals have similar ionisation energies, but that the Ls orbitals are somewhat more tightly bound. X d orbitals are assumed to have very low 'ionization' energies.

Before a diagram of this type or even a much more sophistic- ated m.o. calculation can be applied to X-ray emission spectra, the factors which control such emission must first be considered. The intensity will be proportional to $|\int \psi_i . P . \psi_{ii}|^2$. Where ψ_i is the overall electronic wavefunction for the initial state and ψ_{ii} the corresponding function for the final state. For typical

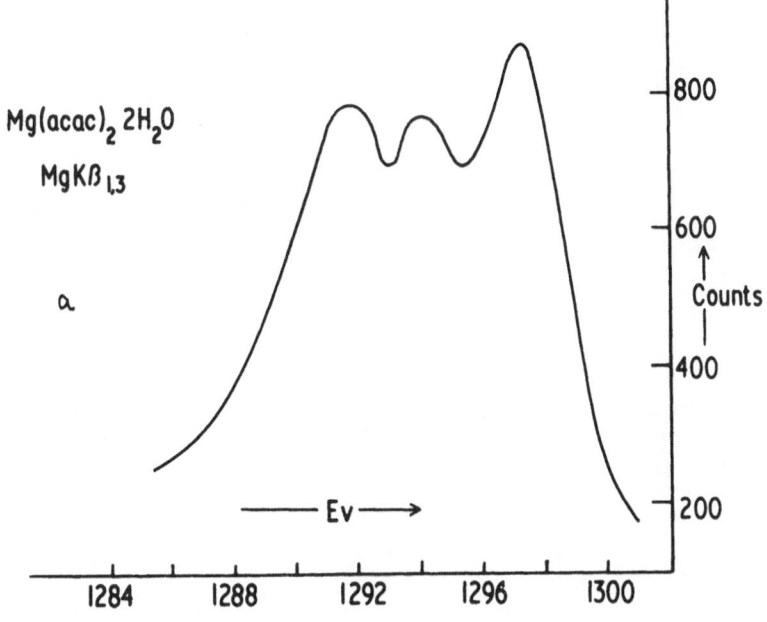

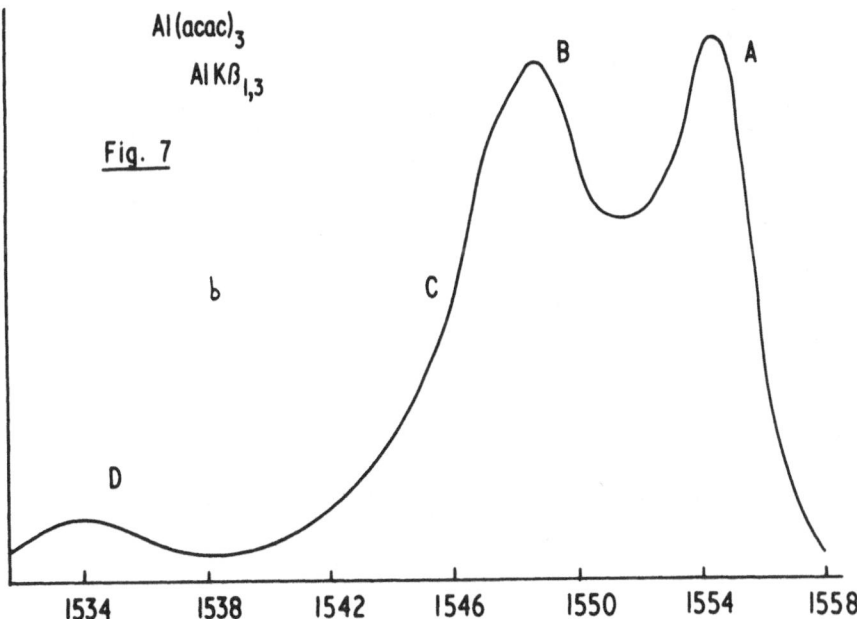

Figure 7. Kβ₁,₃ X-ray emission spectra for acetylacetonate complexes of (a) magnesium and (b) aluminium.

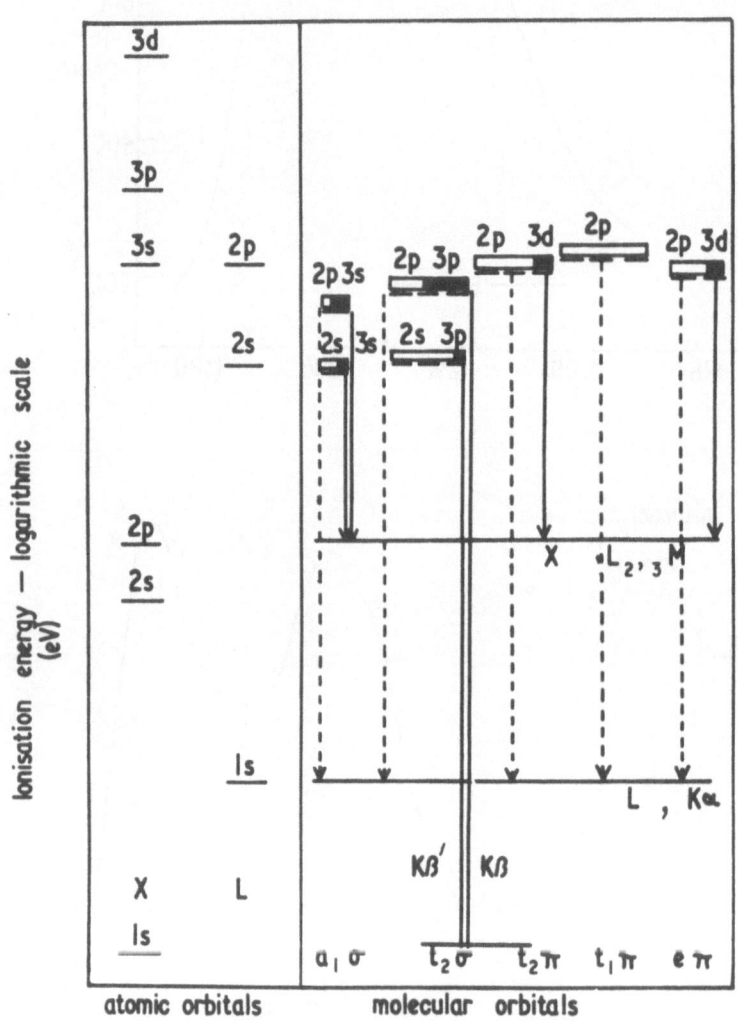

Figure 8.   Qualitative molecular orbital energy level diagram
for a tetrahedral molecule (or ion) XL_4.

X-ray emission ψ_i will have a vacancy in an inner orbital and ψ_{ii} a vacancy in an outer orbital, perhaps even a valence shell molecular orbital; $\underset{\sim}{P}$ is the transition operator. $\underset{\sim}{P}$ represents a collection of specific operators corresponding to dipole, quadrupole, magnetic, etc. terms. The electric dipole term is the one which gives rise to the most intense transitions and is responsible for the selection rule $\Delta l = \pm 1$. As L–S coupling breaks down with increasing atomic number, so also will the rigidity of this rule – thus allowing many 'non-diagram' lines to be observed in the spectra of heavy elements. A simplification of the integral given above may be had by assuming that orbitals which are doubly occupied in both ψ_i and ψ_{ii} may be ignored. By way of example let us consider now the form of the integral for a K emission line of X in XL_4 which involves a transition from a valence shall molecular orbital,

$$\int \underline{\Psi}_{X,1s} \cdot \underset{\sim}{P} \cdot (\underline{a}\,\phi_{X,v} + (1-\underline{a}^2)^{-\frac{1}{2}}\,\phi_{L,v})$$

$$= \underline{a}\int \underline{\Psi}_{X,1s}\,\underset{\sim}{P}\cdot\phi_{X,v} + (1-\underline{a}^2)^{-\frac{1}{2}}\int \underline{\Psi}_{X,1s}\,\underset{\sim}{P}\cdot\phi_{L,v} \quad,$$

$$\qquad\qquad\text{'atomic'}\qquad\qquad\qquad\text{'cross-over'}$$

where subscript v indicates a valence shell orbital and $\underline{a}$ is an atomic coefficient in the LCAO MO equation. The integral breaks down into two parts, one of which is wholly concerned with X orbitals, designated 'atomic' and the other which involves orbitals of both the central atom and the ligand, 'cross-over'. An approximate evaluation of these integrals using S.C.F. orbitals has shown that the 'atomic' term exceeds the 'cross-over' term by a factor of about a hundred.[19] Since it is the square of the integral which determines the intensity of the emitted X-ray it follows that the only significant term will be $\underline{a}^2[\int \underline{\Psi}_{X,1s}\cdot\underset{\sim}{P}\cdot\underline{\Psi}_{Xv}]^2$. Thus it may be concluded that the intensity of X-ray spectra arising from transitions involving molecular orbitals may be regarded as localised and atomic in character, modified only by the square of the coefficient for the specific valence shell atomic orbital in a particular molecular orbital. Within this general level of approximation it should therefore be possible to use the intensities of X-ray emission lines to determine atomic coefficients in LCAO molecular orbitals. Many factors, theoretical and experimental, conspire to making an absolute determination of X-ray emission intensities very difficult indeed. However, relative figures are easily obtained (e.g. $I(K\alpha_{1,2}):I(K\beta_{1,3})$, where I denotes intensity) and can be corrected for changes in frequency. Within a valence band array of molecular orbitals the change in frequency will have very little effect, so intensities can be directly related to the relative contributions of a specific atomic orbital to various m.o.'s. This means that

there should be a very simple and direct relationship between
the structure of the molecular orbitals indicated in Figure 8
and corresponding X-ray emission spectra.

4.3   Application of the m.o. approach to X-ray spectra of SiO_2

Figure 9 shows a series of emission spectra for SiO_2.[21,22]
SiO_2 is chosen, not because its structure is best thought of as
built from isolated XL_4 units, but because of the availability
of the X-ray data.   Indeed SiO_2 is a rather poor example;   the
local environment of each silicon is indeed tetrahedral but each
oxygen ligand is shared with a neighbouring silicon atom.   If
a reasonable correlation between the X-ray spectra and the m.o.
model for XL_4 can be achieved it would then give an indication
of just how local are the bonding factors which influence a
particular atom.

4.3.1   Si $K\beta$ spectrum

The $K_{\beta_1,3}$ emission arises from the 3p→1s transition
(Figure 8) suggesting that silicon 3p character would be found in
two sets of triply degenerate orbitals;   in the least tightly
bound there would be comparable amounts of Si3p and O2p and
perhaps a small contribution from O2s;   on the other more tightly
bound orbital an excess of O2s is to be expected with only a
small amount of Si3p and perhaps some O2p.   The X-ray spectrum
should therefore reflect this disposition of 3p character, a main
peak from the former orbital and a less intense peak at a lower
energy from the latter.   As can be seen from Figure 9 this is
precisely what is observed.

4.3.2   Si L spectrum

Transitions to vacancies in Si2s orbitals should give rise
to spectra very similar to those described above, but of course
at much longer wavelengths.   But relaxation of the vacancy by
non-radiative Auger emission associated with 2p→2s means that
these emissions are very weak.   The $L_{II,III}$ spectrum can be
observed, however, and is shown in Figure 9.   The 2p vacancy
can be filled by either 3s or 3d electrons.   3s character is to
be found in the two molecular orbitals of a_1 symmetry in a very
similar way to 3p character in the t_2 orbitals.   A main peak and
a low energy satellite are therefore anticipated and are indeed
observed.   The spectrum in Figure 9 also shows a further feature
of comparable intensity to the main peak, but at a higher energy.
This could be due to 3d orbital participation in orbitals with
considerable oxygen lone-pair character, but because of the
bridging nature of the ligand atoms in SiO_2 this high energy
feature may be due to 3s character in the more complex m.o. array
that must be present in the macro-molecule or lattice.   Quite

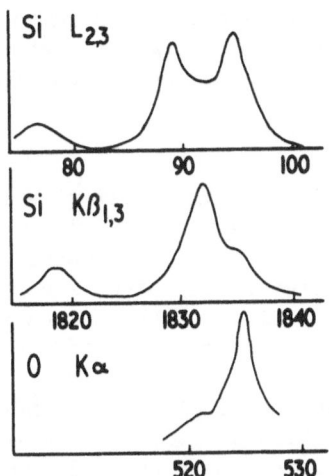

Figure 9.   X-ray emission spectra from silica.

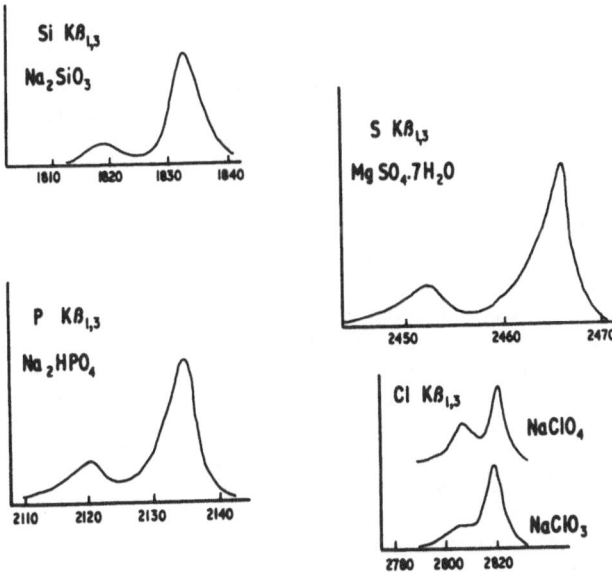

Figure 10.   Typical $K\beta_{1,3}$ emission spectra from some oxy-anions.

similar $L_{II,III}$ spectra are observed for phosphate, sulphate and
perchlorate, where with increasing confidence one might suggest
that the anion is to be regarded as an isolated XL_4 unit and
perhaps in these cases the high energy peak should be regarded as
evidence for 3d orbital participation in bond formation.[23]

4.3.3 O $K\alpha$ spectrum

The transition $2p \rightarrow 1s$ in oxygen will give rise to $K\alpha$ emission.
As can be seen from Figure 8, a wide variety of molecular orbitals
of quite similar energies would all have considerable 2p character.
A broad X-ray peak is therefore expected, associated with the e,
t_1 and t_2 'lone-pair' orbitals.   Weaker peaks at slightly lower
energies should arise from the 2p character present in the a_1 and
t_2 orbitals associated with σ bond formation (interaction with
3s and 3p orbitals).   This is in fact the general form of the
oxygen emission spectrum from silica, a broad intense peak and a
less intense feature at lower energies.   If better resolved it
would be possible to measure this energy difference which is the
ionisation energy difference between electrons in o bonds and in
lone-pairs and which should therefore be related to bond strength.

Bearing in mind that the SiO_2 structure does not contain
isolated XL_4 units, the simple qualitative m.o. model has great
success in explaining the principal features of all the X-ray
emission spectra, their relative intensities and energies.   The
same general ideas can be applied to a wide range of other
molecules and ions based on XL_4 and analogous models can be
developed for other symmetry situations.   In all cases it is
found that the m.o. approach provides an elegant explanation of
the spectra and gives considerable insight into the nature of
chemical bonds.

Despite this overall general success, the limitations of
this approach must also be emphasised.   It is essentially a one-
electron model and effects due to electron correlation, relaxation,
repulsion, etc., etc., are explicitly ignored.   It is also most
important to remember that X-ray emissions arise as a result of
electronic transitions in a moiety which carries one more positive
charge than the species for which a diagram such as Figure 8 might
be constructed.   That one is justified in equating calculated
orbital energies with actual ionisation energies is suggested by
Koopmans theorem[24] but it must be borne in mind that the assumptions inherent in this theorem might not always be justified,
so that the actual ordering of emission spectra peaks might not
correspond to the ordering of molecular orbitals in the original
species.   In most cases it would appear that errors of not more
than 3 or 4 eV can be ascribed to deviations from Koopmans theorem,
but in some specific cases it is just this order of magnitude of
energy which is important.[25]   In the rest of this chapter it will

be assumed that it is proper to attempt to correlate spectra
directly with molecular orbital energy level diagrams, but even
so, the warning given above should not be forgotten.

4.4 Low-energy satellite peaks[18]

The $K\beta_{1,3}$ emission spectra of the elements magnesium to
chlorine lie in an easily accessible part of the X-ray spectrum
and have therefore been much studied. The molecular orbital
arguments that follow from Figure 8 can be generalised to state
that when an element X is in combination with a ligand L we should
expect to find Xp character in two main types of orbital; those
with ligand p character and those with ligand s character, and
that Xp character will be greater in the former than the latter.
A general feature of $K\beta_{1,3}$ spectrum for the elements Mg-Cl should
therefore be the presence of a low energy satellite peak. This
is shown for a few typical cases in Figure 10. The satellite is
called $K\beta'$. It follows also from Figure 8 that there should be
a rough correlation between the ligand s-p ionisation energy
difference and the separation energy (Δ) of $K\beta_{1,3}$ and $K\beta'$, and
this also is found to be so, as is shown in Figure 11. Average
values of Δ are: F, 20 eV; O, 14 eV; N, 11 eV; C, 8 eV (all
$\pm$ 2 eV). The changes in can in fact be correlated with changes
in atomic number. On going to the right in the periodic table
Δ decreases and the relative intensity $I (K\beta') : I (K\beta_{1,3})$
increases, both trends which are easily rationalised by m.o.
theory. The presence of these low-energy satellite peaks can be
turned to analytical advantage, for they are indicative of a bond
to a particular ligand and therefore the chemical presence of that
ligand atom in the sample. Gross contamination, hydrolysis, etc.
can sometimes be detected in this way.[26] If two ligand atoms are
bound to one atom this is revealed as two separate characteristic
$K\beta'$ peaks, e.g. topaz (Figure 12) in which the aluminium atom is
coordinated by four oxygen and two fluorine atoms.

Exactly comparable features are observed in the X-ray emission
spectra of transition metal complexes. Here, both $K\alpha_{1,2}$ and
$K\beta_{1,3}$ are to be regarded as arising from inner orbital transitions
and it is the $K\beta_{2,5}$ peak which arises from valence bond orbitals.
(The $K\alpha_{1,2}$ and especially $K\beta_{1,3}$ peaks do show 'chemical effects'
but these are associated with the magnetic properties of the
complex and will be considered later.) The $K\beta_{2,5}$ peak is often
referred to as $K\beta_5$, implying that it arises solely from the
quadrupole transition 3d→1s. This designation is due to Idei[27]
who argued with admirable logic that the 4p orbitals of transition
metal ions would be empty. However, it is worth noting that the
X-ray emission spectra of isolated transition metal ions have
never been studied. What have been studied are the spectra of
such ions in a wide variety of chemical states and in the presence
of many different ligands. In all cases it is supposed that 4p

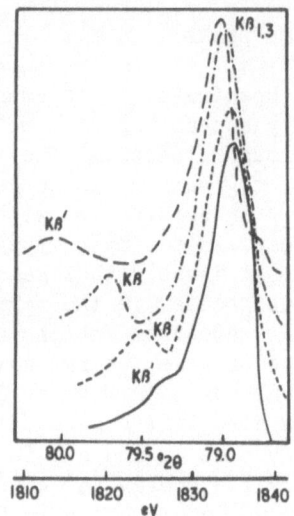

Figure 11.   Silicon $K\beta_{1,3}$ X-ray emission spectra: — — — Na_2SiF_6,
-.-.- SiO_2, ---- Si_3N_4, —SiC.   Spectra obtained using an
ammonium dihydrogen phosphate (101) crystal; $°2\theta$ scale refers
to diffraction angle.

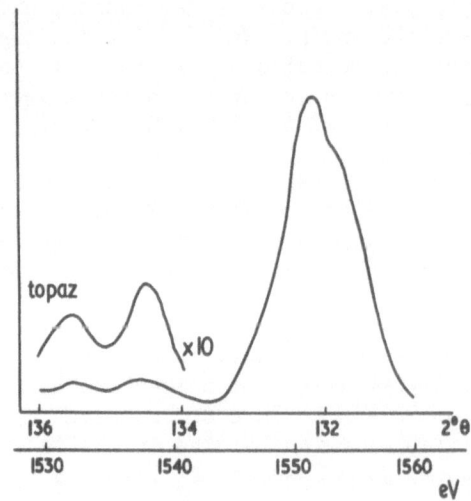

Figure 12. $AlK\beta_{1,3}$ X-ray emission spectrum from topaz.

orbitals are to some extent involved in bonding.   There would
therefore seem no good reason to deny that occupied molecular
orbitals have some admixture of 4p character.   In elements
where it is possible to observe true K_{β_2} and K_{β_5} emissions the
K_{β_5} is invariably much weaker than K_{β_2} (by a factor of about ten).
Finally, the $K_{\beta_2,5}$ feature is prominent in the spectra of ions
which would, on the wholly ionic view, have not only no 4p
electrons but no 3d electrons either!   There would therefore seem
to be no reason to retain K_{β_5} as the name for this peak inthe
X-ray emission spectra of transition metal complexes; K_{β_2} more
accurately describes its electronic origin.

Some typical vanadium spectra are shown in Figure 13;   the
similarities with the main group spectra discussed above are
obvious.   One reason[28] why the low energy satellite peak (here
called $K\beta''$) is relatively so intense is the admixture of a very
little 3p character into the predominantly ligand 2s orbital.

4.5   The structure of valence band K emission peaks (type (ii) peaks)

We have shown above that simple molecular orbital theory can
provide a rationalisation for the gross features of the X-ray
emission spectra of compounds and, in particular, features due
to bond formation involving ligand s and p orbitals.   These
usually give rise to distinct peaks because of the fairly large
energy separation between them.   But it is the ligand p orbitals
which usually play the main role in bond formation and in
situations less constrained by symmetry than XL_4 the interaction
between X s and p orbitals and L p orbitals will often be quite
complex, giving a range of molecular orbitals of different ionization energies and with different contributions from Xs, Xp, Lp,
etc. orbitals.   If the X,K emission peak that results from the
valence bond→1s transition is studied, its structure will reveal
the degree to which X valence shell p orbitals are being used in
the various molecular orbitals.   Some examples of the complex
peak shapes which can be found are shown in Figures 7, 14 and 15.
It will be of interest to see just how much information can be
obtained from single X-ray spectra of the type shown in these
Figures (as opposed to the much more complete picture that results
from the superposition of many XRE spectra, in combination with
XPE spectra - see section 5).

4.5.1   Aluminium tris(acetylacetonate)[29]

The aluminium $K\beta$ spectrum shows four features, a high energy
peak (A), the main peak (B), a shoulder (C) andthe low energy
satellite (D).   Only D can be easily understood, as due to the
presence of a little Al 3p character in molecular orbitals that
are mostly oxygen 2s.   To explain the origin of A, B and C it is

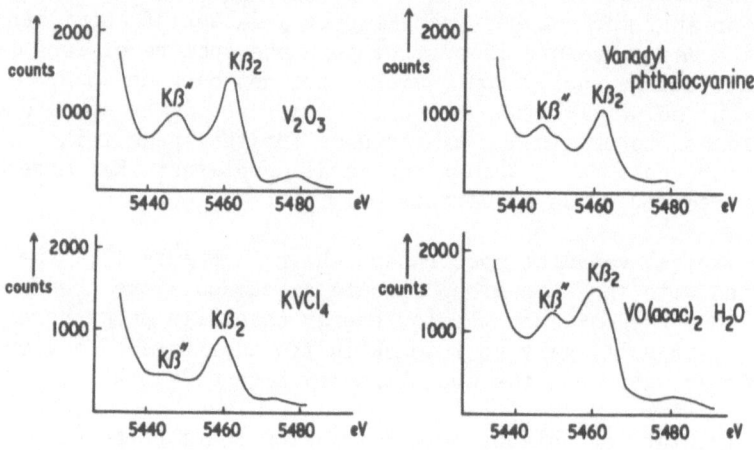

Figure 13.   Vanadium K$\beta_{2,5}$ spectra.

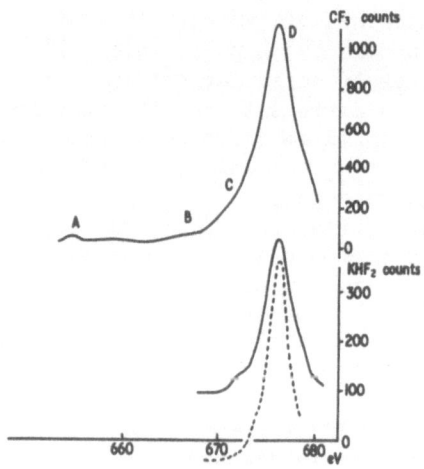

Figure 14.   Fluorine Kα emission spectra: from – CF$_3$ (top) and
FHF$^-$ (bottom; dashed curve is for F$^-$ in sodium fluoride).

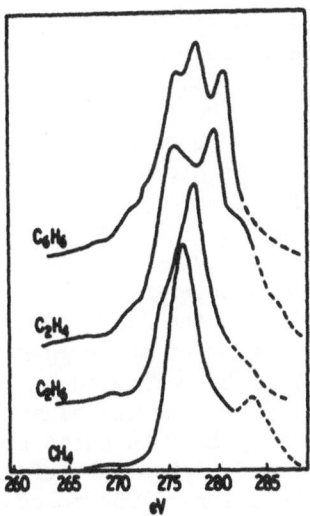

Figure 15.   C Kα emission spectra from hydrocarbon gases.
Dotted region is due to transitions in doubly ionised species.

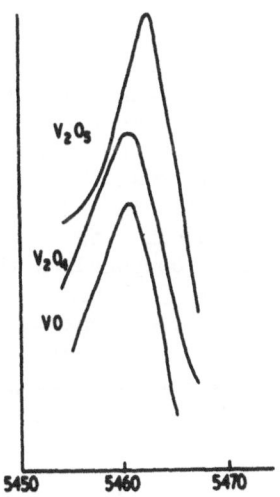

Figure 16.   Position of main V, Kβ2,5 peak for a series of
vanadium oxides.

necessary to consider the bonding between aluminium and the
acetylacetonate ligand.   The overall symmetry of the complex is
D_3 but the local atomic environment of the aluminium is nearly
O_h.   The bonding characteristics of the oxygen atoms 2p orbitals
are (a) a σ bond to the neighbouring carbon atom;   (b) participation in the delocalised π orbitals of the anion;   (c) a lone
pair.   Orbitals of type (c) are thus the most easily available
for σ bonding to aluminium.   To obtain orbitals directly
oriented towards the aluminium atoms it is necessary to involve
orbitals of type (a).   The simple picture in which six ligand
atoms could be so arranged that their σ oriented orbitals would
transform as $a_{1g} + t_{1u} + e_g$, giving rise to a single degenerate
trio of orbitals which would interact with aluminium 3p orbitals
to give a single sharp peak in the AlKβ spectrum, cannot therefore be used.   The involvement of the oxygen atoms with the
remainder of the ligand will be reflected in the energies of the
ligand orbitals that are offered to the aluminium for o bond
formation.   Hence the main peak B is quite broad;   this broadening may also encompass peak C.   Orbitals of type (b) participate
in   bond formation in the anion.   The molecular orbitals that
are formed are such that only π_1 and π_3 (the most and the least
tightly bound of the occupied orbitals) have the correct symmetry
to interact with aluminium p orbitals and thus to influence the
AlKβ spectrum. π_3, the least tightly bound, is also the orbital
which will have the greatest charge density on the oxygen atoms
and will therefore be the orbital most able to participate in
bonding with aluminium. π-bonding is less strong than σ bonding
and so the   orbitals with aluminium 3p character due to this
interaction will be the least tightly bound, and give rise to
peak A in the spectrum. π_1 may also form π bonds with aluminium,
but these would be more tightly bound;   this may be the origin of
peak C.   Whilst the allocation of peaks A and B to π and σ bonds
respectively is reasonable, it is possible to confirm the assignment in the following way.   The oxygen 2s orbitals only form o
bonds with the aluminium.   Peak D can therefore be used as a
reference level for the peak in the main K$\beta_{1,3}$ spectrum which is
due to Al3p participation in σ bonds with oxygen 2p orbitals,
since the value of Δ for simpler compounds with only σ bonds is
known.   The peak which is about 15 eV higher in energy than D
is the required Al3p–O2p σ peak, i.e. peak B.   An examination
of this one spectrum thus sheds considerable light on the way in
which the aluminium atom is bound in this complex, but understandably the illumination of the overall bonding situation is
incomplete.

4.5.2 Fluorine Kα spectra (HF$_2$ and –CF$_3$)

Fluorine provides a simple example of a typical ligand atom.
Figure 14 shows the fluorine Kα emission spectra from the bifluoride anion[30] and the trifluoromethyl group.[31]   These ligand

spectra show considerable structure which can be related to the
nature of the chemical bond in a manner complementary to X-ray
spectra from central atoms.   In both spectra some structure is
present on the high energy side of the most intense peak, probably
due to transitions in doubly ionised atoms (see section 6.2).
In the case of the bifluoride anion four of the six fluorine 2p
orbitals will be lone pairs and, of the two that are involved in
bonding with the hydrogen atom, symmetry determines that one of
the molecular orbitals must be non-bonding and localised on the
fluorine atoms.   A total of ten electrons is thus consigned to
a non-bonding role and only two are involved in the F-H-F bond.
It is therefore to be expected that the fluorine $K\alpha$ emission peak
should show an intense peak and a lower energy satellite peak
with an intensity of less than 20% of the main peak.   This
qualitative picture is clearly borne out of comparison with
Figure 14.   A more detailed estimation of peak heights should
enable the disposition of electric charge in the bonding orbital
between fluorines and hydrogen to be determined.   A comparison
can also be made with the detailed molecular orbital calculations
of Clementi and McLean.[32]   This shows that the minimum basis set
gives incorrect values for the molecular orbital energies of the
FHF anion, but that the results of the extended basis (double γ)
calculations are in very good agreement both for orbital energies
and charge distribution.

The fluorine $K\alpha$ emission spectrum from fluorine bound in a
trifluoromethyl group can be used to examine the nature of the
C-F bond.   Four features can be distinguished: the rather broad
main peak D, shoulders C and B and, at lower energy, a very weak
satellite peak A.   A very simple molecular orbital model would
suggest that from the three fluorine atoms six 2p orbitals would
be non-bonding and three (one from each atom orientated along the
C-F bonds) would be involved in bonding to the carbon.   Because
of their proximity to each other, some mutual repulsion between
the lone pair orbitals is to be expected, giving rise to peak
broadening (for comparison the narrower F $K\alpha$ emission peak from
sodium fluorine is also shown).   This splitting of non-bonding
lone pair orbitals can be clearly seen under the higher resolution
of photoelectron spectra as in, for example, CF_4 where the lone
pair e, t_1 and t_2 orbitals have ionisation energies of 19, 16,
17.7 eV.[33]   The three 2p orbitals directed towards the carbon
will transform under the local C_{3v} symmetry as e and a_1; the
former can interact only with carbon 2p orbitals, whilst under
the latter representation interaction with carbon 2s and $2p_z$
orbitals is possible.   This will give rise to two distinct
molecular orbitals with F 2p character, the first doubly degenerate and less tightly bound, the other more tightly bound.   It
seems reasonable to allocate features C and B to these two sets
of orbitals.   With only slightly better resolution it would be
possible to use the sizes of the peaks to estimate the amount of

fluorine 2p character in each of these bonding molecular orbitals,
using peak D as a reference level (six electrons). The very
weak peak A, about 20 eV below the main peak D, corresponds to a
transition from orbitals which would have considerable fluorine
2s character. Since these F2s orbitals belong to irreducible
representations a_1 and e interaction with other orbitals of the
same symmetry is possible. The presence of A shows that such
interaction does take place and that to a very limited extent
fluorine 2p is mixed with fluorine 2s. The relative intensity
of A can be used to measure the degree of this interaction
quantitatively.

4.5.3 Carbon Kα spectra

Carbon Kα spectra have been measured from a few simple
gaseous hydrocarbon molecules; typical examples are shown in
Figure 15.[34] As in the case of the fluorine spectra, some high
energy peaks are present: these are probably due to transitions
in doubly ionised atoms (section 6.2). The main spectra increase
in structure, reflecting the increased complexity of the molecular
orbital array as the molecules become larger. Carbon 2p orbitals
in methane will be present in a trio of degenerate t_2 orbitals
making C–H bonds, so only one feature is expected in the spectrum.
Ethane at once presents a more complicated situation with 2p
orbitals being involved in both C–H and C–C bonds; p_x and p_y as
a degenerate pair and p_z in combination with the 2s orbital.
In ethylene complexity is further increased since C2p character
enters both σ and π bonds. Even more detailed structure is
apparent in the benzene spectrum. The value of such spectra is
greatly enhanced when considered in combination with photoelectron
spectra (cf. section 5) and also when the orbital energies and
amount of 2p character are compared with calculations.[35]

The purpose of considering these three examples has been to
show the value of individual X-ray emission spectra, whether of
central or ligand atoms, in giving information concerning the
chemical bond. It must be apparent however that a much more
complete picture will emerge if all available spectra for all
atoms in a molecule are considered together, an approach which
will be considered in section 5. One more aspect of individual
emission peaks remains to be considered, however: the so-called
'chemical shift' effect, i.e. changes in characteristic X-ray
wavelengths associated with different types of chemical bond.

4.6 'Chemical shift' of X-ray emission peaks

Examples of changes in X-ray emission energy associated with
different chemical environments are to be seen in Figures 5, 6,
7, 9 and 16. Such changes can be brought about by an alteration
of oxidation state, by variation in the nature of ligand atoms

and by differences in coordination number and bond lengths. Both inner-inner orbital transitions (type (i) in section 4) and valence inner orbital transitions (type (ii) in section 4) are affected, but the former usually much less than the latter.

As discussed in much more detail in the chapters concerning X-ray photoelectron spectroscopy, the ionisation energies of inner atomic orbitals are modified by chemical bond effects in an indirect way: withdrawal of valence bond electrons by whatever means (i.e. an increase in the actual positive charge at an atom) will reduce electron repulsion terms with inner orbitals, thus enhancing their ionisation energies, whilst conversely donation of charge to an atom results in inner orbital ionisation energies being decreased. It is curious that the size of ionisation energy perturbation is just about the same for all inner orbitals. This in turn means that the corresponding changes in X-ray emission energies will be very small indeed. They do exist, however, and can with diligence be found.

As a rough rule of thumb it can be expected that changes in X-ray emission energy will be about one-tenth of changes in inner orbital ionisation energies. Even so, these small shifts can be of significant use, as in the determination of the coordination number of aluminium[8] by studying the exact position of the $K\alpha_{1,2}$ peaks (Figure 6). Extensive studies have been made of many transition metal complexes in various formal oxidation states, and from the observed shifts in $K\alpha_{1,2}$ and $K\beta_{1,3}$ peaks it is possible to calculate the effective charge on the metal atom.[36] Experimentally it is often easier to study the shifts of these type (i) peaks than those involving valence bond orbitals because, although the shifts are smaller the peaks are usually narrower and more clearly defined. Type (ii) peaks will sometimes exhibit bigger shifts but because they arise from many molecular orbitals, maybe even a broad band of such orbitals, they are usually broader and may be split into two or more peaks: which peak then is the 'main' peak and which has suffered a 'shift'? Even so, correlations with coordination number and bond length, comparable to those for $AlK\alpha_{1,2}$, have been demonstrated for both aluminium (Figure 6) and silicon $K\beta_{1,3}$ peaks in aluminosilicates.[7]

When the formal oxidation state of an atom increases, quite large type (ii) shifts can sometimes be observed. Thus in Figure 16 the $K\beta_{2,5}$ peak of vanadium moves to higher energies in going from V^{3+} to V^{5+} even though the ligand environment remains oxygen.[28] Since the V-O bonds are rather ionic, the ionisation energy of the valence orbitals will be dominated by the effect of the oxygen and so will remain relatively unaffected by changes in the vanadium. On the other hand the inner orbital ionisation energies increase as the vanadium is oxidised, i.e. V1s is more tightly bound in V^{5+} than in V^{3+}, which causes the V4p→V1s transition to take place at higher energies in pentavalent than trivalent compounds, as observed. An opposite effect can be seen in Figure 11, in which large changes in the ligand environment have only the very slightest effect upon the Si $K\beta_{1,3}$ energy. The reason for this apparent indifference to the ligand is easy to understand.[37] When surrounded by ligands of increasing electronegativity the inner orbitals of an atom become increasingly tightly bound. However, the valence band molecular orbitals are characterised by ionisation energies related to the ionisation energies of the ligands, which also increase with increasing electronegativity. The energy difference between the inner orbitals and the valence molecular orbitals therefore remains almost constant. A situation in which type (ii) peaks might be expected to shift occurs when the bond ionicity changes while the ligand remains the same. If the central atom increases its effective positive charge and the ligands become increasingly negative, the core orbitals on the central atom become more tightly bound, whilst valence bond molecular orbitals, which are ligand dominated, become less tightly bound. The valence orbital - inner orbital transition in the central atom - therefore increases in energy. This is probably the explanation for the correlation that has been observed for $AlK\beta_{1,3}$ and silicon $K\beta_{1,3}$ emission peak shifts and Al-O and Si-O bond lengths (Figure 6).

5. COMBINATION OF X-RAY EMISSION AND PHOTOELECTRON DATA

It is clear from the preceding discussion that a study of X-ray emission spectra can reveal considerable information about the chemical bond. However, the data are very specific and refer to the bonding role of just one type of atomic orbital. For a more complete overall picture of bonding it is necessary to assemble all the relevant X-ray emission spectra in such a way that the composition of particular molecular orbitals can be understood. For spectra from the same element this can often be done with the aid of other X-ray emission spectra (e.g. $K\beta_{1,3}$ and $K\beta_{2,5}$ spectra of transition metal atoms can be placed on a common energy scale with $L_{II,III}$ spectra, provided the $K\alpha_{1,2}$ emission energy is known). When the spectra of different elements are to be compared the ionisation energies of some inner orbitals of the atoms concerned must be known. In many cases it will not be possible to measure the ionisation energy of the required inner orbital directly (e.g. a 1s level for Z > 12 using $AlK\alpha$ radiation). The required level can be found easily by measuring an accessible level (e.g. 2p) and calculating the required ionisation energy in combination with the $K\alpha$ X-ray emission energy. This determination of inner orbital ionisation energies using X-ray photoelectron spectroscopy must, of course, be repeated for each compound studied, since it is precisely the chemical

perturbations of these ionisation energies that must be determined
quantitatively.   It is important to note that charging effects,
which so bedevil attempts to determine absolute ionisation
energies, do not affect the usefulness of X-ray photoelectron
spectroscopy in this application.   What is required to be known
is the position, in the same energy scale, of the ionisation
energy of an inner orbital from each of the atoms present in a
molecule.   All energies may be subject to error, but provided
it is the same error the only effect will be that the valence
shell molecular orbital ionisation energies computed for all the
spectra (XPS and XES) will also be subject to the same error.
The relative positions will be correct and the X-ray spectra will
be correctly related one to another and to the XPS valence band
spectrum.   In some cases it will also be of great interest to
try to line up HeI photoelectron spectra with the XPS-XES array.
Since it is extremely difficult to estimate work functions of
non-conducting solids and to be sure of the absolute calibration
of XPS data, the quantitative inclusion of the UPS spectra is
well-nigh impossible.   Nevertheless it is often possible to
identify characteristic parts of XPS or XES with the UPS spectra
(e.g. fluorine 2p lone pairs, oxygen 2s level etc.) thus permitt-
ing a correlation between the different classes of spectra.

    Once this has been achieved it should be possible to identify
particular ionisations in the UPS or XPS valence band spectra and
to determine by reference to the various XES spectra the atomic
composition of the molecular orbitals.   It is in this dissection
of molecular orbitals into their atomic components, revealing
the structure and the nature of the bonds between atoms, that the
great potential of X-ray emission spectroscopy lies.   Up to now
only a very few attempts have been made to assemble spectra in the
way described above (e.g. SF_6);[38]   first, two examples will be
briefly discussed below to give some idea of the power of the
technique.

5.1 Magnesium oxide

    The combined Mg $K\beta_{1,3}$ Mg $L_{2,3}$ O $K\alpha_{1,2}$ emission spectra
together with the XPS spectrum for the valence bond region are
shown in Figure 17.[39,40]   Uncorrected XPS values for the Mg 1s,
2s,2p and oxygen 1s levels are also included.   In none of the
spectra does the relative intensity of the peaks have any significance except within a particular spectrum.   The resolution of
the various spectra is unfortunately very different, and once
this factor has been taken into account it can be seen that all
the X-ray emission spectra are split into two peaks separated by
about 2 eV.   In order to understand the significance of these
spectra and the bonding in magnesium oxide, it is necessary to
devise some molecular orbital model. MgO_6^{10-} suggests itself
but whilst it easily explains the low energy satellite peak $K\beta$,

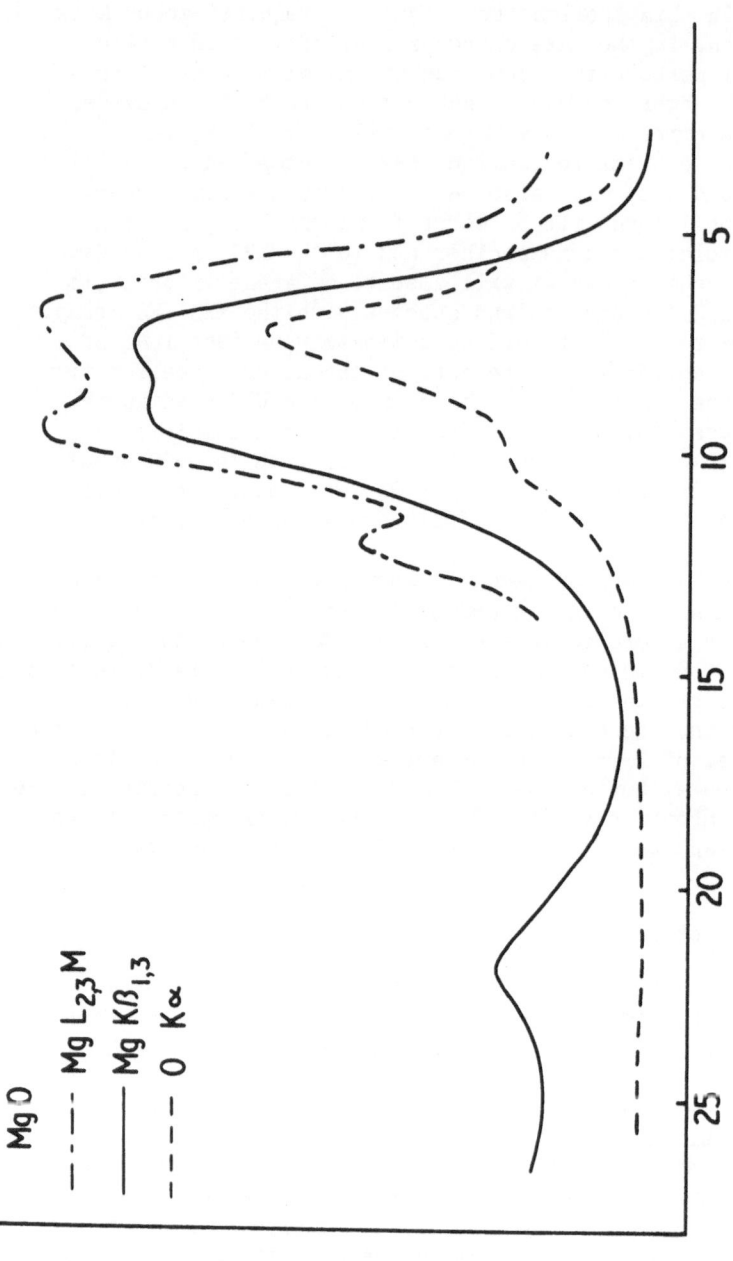

Figure 17. Magnesium oxide X-ray emission spectra. Peak at about '13 eV' in L2,3M spectrum is in fact due to the $L_{2,3} \rightarrow L_1$ transition; structure at about 5 eV on the 0 Kα spectrum is due to transition in doubly ionised atoms.

MgO

– · – Mg L_{23}M

—— Mg $K\beta_{1,3}$

– – – 0 Kα

it is not easy to generalise from thus unit to the MgO lattice,
even in combination with the complementary cation $O\ Mg_6^{10+}$!   A
much more realistic model starts with an Mg_4O_4 cube with magnesium and oxygen atoms at alternate corners, i.e. two interlocking
tetrahedra, one of magnesium, the other of oxygen atoms.   The
ionic partition Mg^{++} and O^{2-} is a convenient starting point for
allocating the electrons.   It is unlikely that oxygen 2s orbitals
will enter significantly into the bonding interactions between
magnesium and oxygen and so they will be ignored (the 2s orbitals
do of course interact to a very slight extent with magnesium 3s
and 3p orbitals, as is shown by the presence of low energy
satellite peaks in the Mg emission spectra, which line up well
with the O2s feature in the XP spectrum;   the relative intensities
and degree of energy separation show that this is not an important
bonding interaction).   Magnesium 3s and 3p orbitals have very
similar ionisation energies and must both therefore be considered
in the interactions with oxygen.   For convenience the atoms can
be sp hybridised, one hybrid from each magnesium points to the
centre of the tetrahedron.   Their mutual interaction will give
rise to a bonding a_1 and a trio of antibonding orbitals, t_2.
Similar interactions will take place between magnesium and oxygen
a_1 and magnesium and oxygen t_2 orbitals.   Figure   shows that
the former interaction will be much greater than the latter and
that one strongly bonding and three weakly bonding orbitals will
result, into which the eight electrons may be placed.   Furthermore, the contribution of the hybrid Mg(3p,3s) orbital to the a_1
orbital will, if these contributions are small, be three times
the contribution to each of the t_2 orbitals: i.e. equal contributions in toto.   This is the origin of the split in the $MgK\beta_{1,3}$
and $L_{II,III}$ peaks.

     The lower energy part lines up well with the low energy
bonding feature of $OK\alpha$, the higher energy part with the main $OK\alpha$
peak.   Similar calculations involving the peripheral octets of
2p orbitals on oxygen and 3p orbitals on magnesium (octets
transforming as $e+t_1+t_2$) show that much smaller interactions are
to be expected.   Thus no more charge is transferred from oxygen
to magnesium and so the $MgK\beta_{1,3}$ peak is not affected.   The four
2p electrons on each oxygen atom can be regarded to a first
approximation as being lone pairs, somewhat split in energy by
mutual interactions.   In this splitting the orbitals will have
decreasing ionisation energy $e > t_2 > t_1$, and this will in turn
affect the shape of the $O\ K\alpha$ emission peak, broadening it and
giving it asymmetry on the low energy side.   In the oxygen X-ray
emission spectra the low energy feature is not well resolved,
but if it is assumed the Gaussian shaped peaks are involved the
intensity ratio of the low energy peak to the main peak is about
2:10, whereas the model so far developed would suggest 1:11 or
1:3.8 depending upon the relative energies of the low pair peak
and the 't_2 bonding' peak.   As suggested above, however, it is

probable that the peaks are <u>not</u> gaussian but rather distorted,
which would have the effect of lowering the relative intensity
of the low energy peak.

When the model is expanded to $Mg_{32}O_{32}$ while considering a
cube to be joined at the site of each atom of the original Mg_4O_4
unit, the degree of broadening of the molecular orbital levels
can be estimated.   It can be shown that resonance integrals are
reduced to one quarter when expansion to the larger unit takes
place.   This enables the shape and energy spread of the band
structure of MgO to be estimated: it is somewhat broader than
for Mg_4O_4 but not fundamentally different.   In the larger unit
the relative intensity of the low energy structure falls to about
1:11 in keeping with the predictions of the simple model.   The
splitting of the Mg 3p and 3s levels is due to interactions with
oxygen orbitals - to the presence of 'covalent character' - in
the MgO linkage.

Whilst this discussion ebables some of the structure of the
bonding in MgO to be unravelled there are tantalising omissions.
Because it is <u>not</u> possible to relate the intensities of the
magnesium spectra and the oxygen spectrum to each other it is
not possible to estimate the degree of covalency in the MgO
bonds, i.e. it is not possible to estimate the values of the
atomic orbital coefficients in LCAO molecular orbitals; a few
ratios can be determined but no absolute values.

5.2 Aluminium tris(hexafluoroacetylacetonate)[26]

Figure 18 shows $AlK\beta_{1,3}$ $OK\alpha$ and $FK\alpha$ X-ray emission spectra,
the X-ray photoelectron spectrum and helium photoelectron spectrum
for the tris-hexafluoroacetylacetonate complex of aluminium.
Some points of immediate interest are the correlation of the XPS,
O2s level with the $AlK\beta'$ peak and the F2s level with peak A
(Figure 14) in the $FK\alpha$ X-ray spectrum, as suggested above
(4.5.2).   The main $FK\alpha$ peak correlates well with features
believed to be due to fluorine lone pairs in the helium photoelectron spectrum.   It is clear that the low energy peaks in
this spectrum are derived from aluminium-oxygen bonds.   This
general correlation enables the arguments of 4.5.1 to be extended,
but the extension is somewhat restricted at present, due to the
poor quality of the oxygen emission spectrum.   However, the
role of the X-ray emission spectra in helping to identify and
characterise photoelectron peaks is clearly demonstrated.

6.   MANY-BODY EFFECTS IN X-RAY EMISSION SPECTRA

It is not proposed in this section to give a detailed
account of either the origin or occurrence of all possible many-

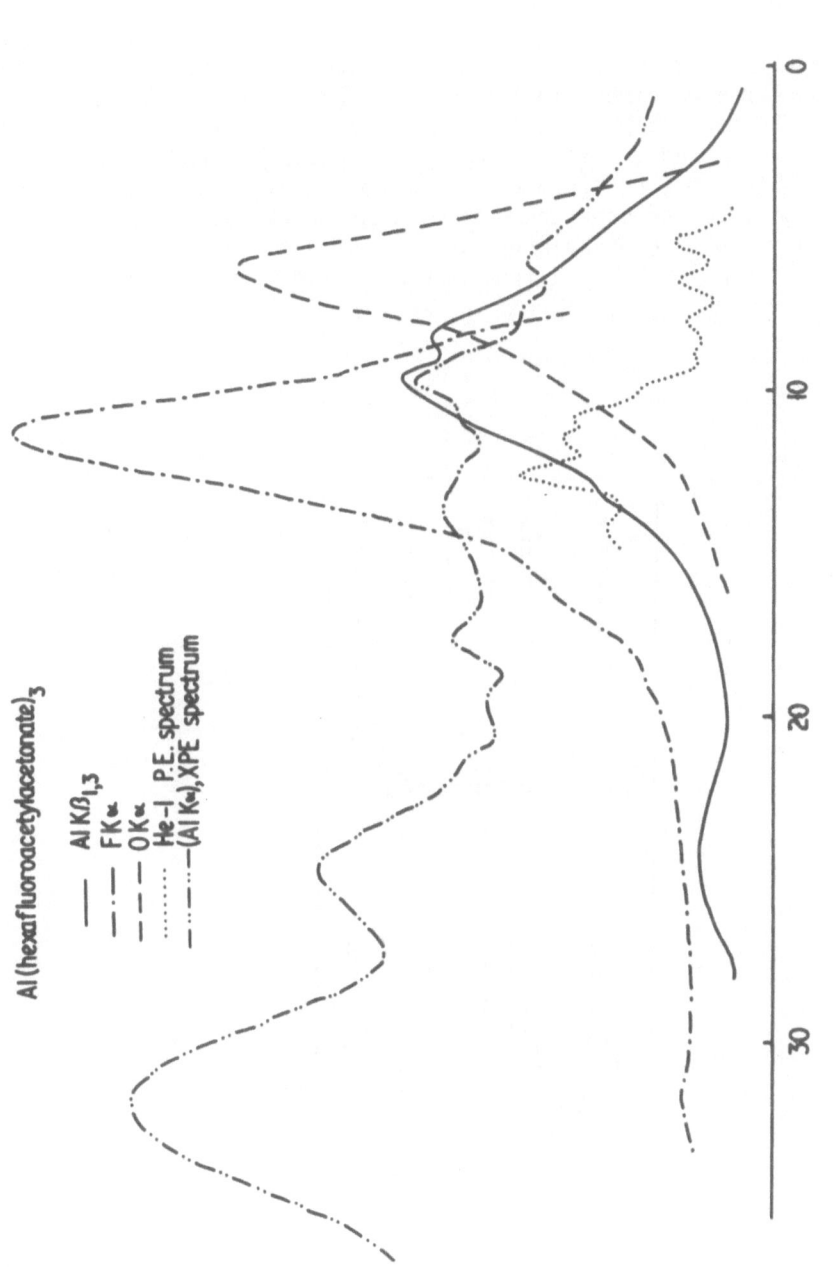

Al(hexafluoroacetylacetonate)$_3$

—— Al K$\beta_{1,3}$
—·— F Kα
—·— O Kα
········· He-I P.E. spectrum
—··— (Al Kα),XPE spectrum

Figure 18.   X-ray emission spectra from aluminium tris(hexafluoroacetylacetonate).
Dotted curve at bottom is taken from S. Evans, A. Hamnett, A.F. Orchard and D.R. Lloyd,
Faraday Disc. <u>54</u>, 227 (1972).

body effects in X-ray spectra.   Rather, what we shall do is to
indicate two regions in which the one-electron model can no
longer be used, and discuss observations of significance in
inorganic chemistry.

6.1 Exchange and spin-orbit effects

None of the examples which have been considered up to now
has had unpaired electrons in its ground state.   All orbitals
have been electronically closed shells save the one from which
ionisation took place.   It is found experimentally that when
compounds with unpaired electrons are examined by means of X-ray
emission spectroscopy some peaks are broadened whilst others are
split: new satellite peaks are formed.   Some typical examples
are shown in Figure 19.[41]   Iron compounds have been chosen

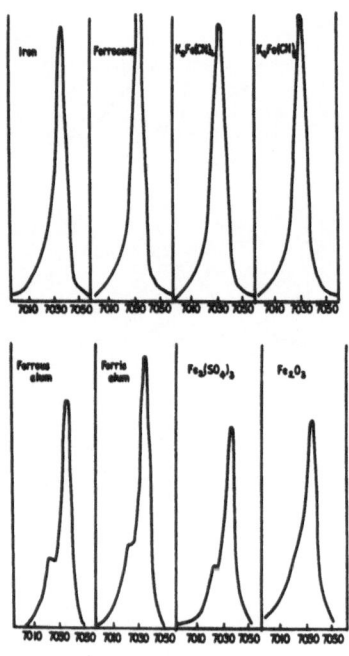

Figure 19. $K\beta_{1,3}$ and $K\beta'$ from a series of iron compounds.

because variation of the ligand enables high spin and low spin
compounds to be obtained easily.   The K$\beta_{1,3}$ peak, which relates
to an orbital not directly involved in bond formation, is seen
to be split in high spin complexes while remaining a single peak
in low spin complexes.   The observation has been repeated for a
wide variety of other transition metal compounds and so would
appear to be quite general.[42,43]   A related observation is the
corresponding broadening of the K$\alpha_{1,2}$ peak, but no new satellite
peaks have been observed as distinct features.[44]   Correlation
of the satellite-to-main peak intensity ratio with the number of
unpaired electrons is fairly exact and relatively weak Kβ' peaks
are observed in compounds with only one or two unpaired electrons:
intense Kβ' features are associated with compounds containing
four or five unpaired electrons per ion.   Indeed this correlation
might almost be suggested as a way of determining oxidation state
(Figure 20).

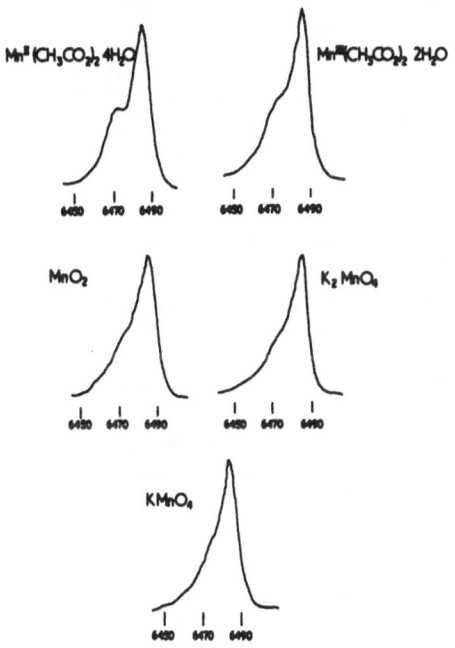

Figure 20.   K$\beta_{1,3}$ and Kβ' from a series of manganese compounds
showing the dependence upon valence state (and also the number
of unpaired electrons).

It is tempting to try and interpret this splitting of the $K\beta_{1,3}$ as simply due to exchange but the relative intensities $K\beta':K\beta_{1,3}$ forecast by this effect are much too large.[45] It is not sufficient to consider the spin associated with the 3p hole state and the spin state of the 3p electrons, because orbital effects must also be taken into account. The analogous problem is important in photoelectron spectroscopy where, of course, final state is created identical to that in XES. (In one case an electron is ejected from an orbital (XPS); in the other case it relaxes from one orbital to an inner orbital (XES).) Even when spin orbit effects are included in the calculations, the correlation with experiment is poor. Chemical bond formation has been suggested as a solution but would not seem to be either the whole or the correct answer: small changes are observed in $K\beta'$ and $K\beta_{1,3}$ both in relative energies and intensities, but not large enough to make the calculations agree with experiment. The most promising suggestion is that made in Wertheim's chapter that configuration interaction of excited states must be included in the calculations.

Since the correlation between peak splitting and the presence of unpaired electrons seems so well established, at least experimentally, it might be possible to use XES to differentiate between those features in photoelectron spectra due to similar spin coupling and to shake-up effects. Shake-up peaks result from the relaxation of an ion, generated suddenly by the ejection of a fast electron, to an excited electronic state. Whilst it has been suggested that this might be the origin of the $K\beta'$, $K\beta_{1,3}$ structure in transition metal ion compounds,[46] such an explanation cannot rationalise the strong correlation with the number of unpaired electrons. It therefore seems reasonable to reject this proposal and to suggest the opposite: that shake-up does not give rise to low energy satellites in X-ray emission spectra. Shake-up may be associated with the formation of a vacancy but not with its relaxation. A direct correlation of satellite structure in XPS and XES thus points to spin coupling effects, whilst peaks unique to XPS probably result from shake-up. The wide occurrence of shake-up structure in X-ray photoelectron spectra suggests that many such ions must also be created by the X-ray irradiation of samples needed to observe X-ray emission spectra. If this is so, the relaxation of such ions should be observable and should give rise to X-ray emission peaks at higher energies than normally anticipated. This may be the origin of peaks such as $K\beta'''$ (Figure 13).

6.2 High-energy satellite peaks

Many X-ray emission peaks have associated high energy structure which cannot be explained as due to simple electronic transitions. The origin of the additional structure can be rationalised, however, as due to transitions in doubly ionised atoms.   A typical array of high energy satellite peaks of this type associated with $AlK\alpha_{1,2}$ is shown in Figure 21.   The initial doubly ionised atom may have a hole configuration $1s^1 2s^1$ or $1s^1 2p^1$ and final hole states may be $2s^2$, $2s^1 2p^1$ or $2p^2$.   Associated with these configurations there will be a variety of spectroscopic states (both singlet and triplet) each with its own energy.   Many possible transitions are therefore possible: the relative intensities and energies can be calculated[47] and good agreement with experiment has been established.   The presence of a vacancy in an outer shell will result in the 1s ionisation energy being somewhat enhanced: the $2p \rightarrow 1s$ relaxation will therefore take place at higher energies in the doubly ionised than in single ionised atoms.   The relative intensity of the high energy satellite peaks to the main peak varies with the bombarding particles or radiation and with their energy.   The use of heavy ions to excite X-ray emission, for example, gives rise to excessive satellite emission from multiply ionised atoms and only a very little normal X-ray intensity.[48]   It is also found that the relative intensity of high energy satellites varies even if electrons or photons are used.   When the energy of the irradiation is just sufficient to cause the primary vacancy no high energy satellites are observed.   As the energy of excitation is increased the relative intensity of the high energy satellite peaks increases.

High energy satellites often show quite pronounced chemical effects: relative intensities of peaks in the high energy satellite region can be changed by as much as fifty percent;   the energies of particular peaks sometimes show slight alterations.[6] This can be clearly seen in the aluminium spectra shown in Figure 21.   In other cases the resolution and energy spread of

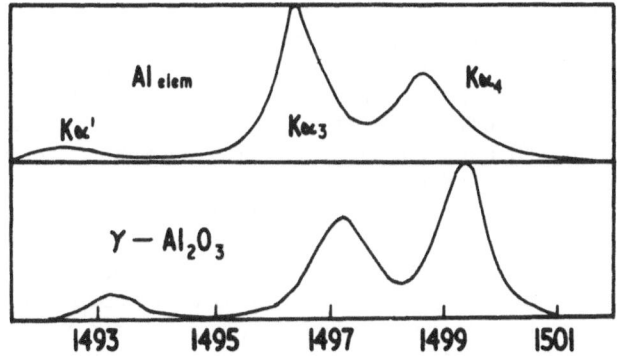

Figure 21.   Some high energy satellites of aluminium and alumina (from K. Lauger, J. Phys. Chem. Solids, 33, 1343 (1972).

the high energy satellites may not be so great and they will then
appear as shoulders on the high energy side of the main peaks,
e.g. Figures 14, 15 and 17.

The origin of the doubly ionised atom which gives rise to
these satellites has been the subject of some discussion.   Two
mechanisms have been suggested and indeed both may be operative.
The simplest idea is that a single photon or electron can cause
the simultaneous ejection of two electrons;[48] the other proposal
is that a second electron is ejected as the atom reacts to the
shock of the ionisation of the first electron ('shake-off').[49]

7.   CONCLUSIONS

X-ray emission spectra which arise from electronic transitions from the valence shell to inner atomic orbitals carry
considerable information about molecular orbital energy levels
and the relative contributions which particular types of atomic
orbital make to different molecular orbitals.   X-ray emission
spectroscopy allows the chemist to dissect molecular orbitals
into their atomic constituents.   Whilst considerable information
can be obtained from a single X-ray spectrum, a much more detailed
picture of bonding in a molecule is to be had by combining all
possible emission spectra from all the elements in the molecule.
These spectra may be aligned on a common energy scale if inner
orbital ionisation energies are determined directly using X-ray
photoelectron spectroscopy.   The array of valence shell molecular
orbitals revealed by UPS, and also by XPS, can then be analysed
in terms of their atomic components.   Since the initial and
final ionic states in XES are precisely those which are generated
in XPS it should be possible to align spectra exactly by this
method.

Two words of warning which must be injected at this stage
both concern the way in which the electrons of the molecule relax
in response to the creation of the initial electron vacancy.
First, the process of relaxation is affected by the speed with
which the electron leaves the atom;   for an exact correspondence
between XPS and XES spectra the ionic states which are being
considered should be formed in similar ways.   The second point
which must be borne in mind is that all the spectra discussed in
this chapter derive from transitions in ions.   It is very
convenient to assume Koopmans theorem and to equate the experimental ionisation energy of an electron with a molecular orbital
energy level in the ground electronic state of the unionised
molecule.   Such a procedure has been implicit throughout this
chapter.   But deviations from Koopmans theorem can and do occur,
different orderings of, and spacing between, molecular orbitals
being found for the ground state and the ionised state.[50]   Thus

whilst XPS and XES spectra may, with care, be used in conjunction to build up a detailed picture of the structure of molecular orbitals, caution must always be exercised in assuming that this picture, directly and correctly, shows the situation in the un-ionised molecule. The only really safe guide is a detailed molecular orbital study of both the ground state and the various possible ions.

Despite these warnings, it seems reasonable to assume that X-ray emission spectroscopy will play an important role in the future in enhancing our knowledge of the chemical bond. The most important developments that will take place in the experimental field in the near future will be the enhancement of resolution, so that detailed spectra may be obtained as a matter of routine and the development of spectrometers which will enable very soft X-ray spectra (25 up to 500 A) to be obtained with ease. On the theoretical side, more detailed and sophisticated m.o. calculations (e.g. by the ab initio method) will be carried out on a wide range of inorganic materials. The predictions of these calculations can be checked by using XES and XPS spectra so that a much more detailed understanding of chemical bonding will result.

8.   ACKNOWLEDGEMENTS

I should like to thank Drs. P. Day and A.F. Orchard for their kind invitation to participate in this NATO symposium, and to express my gratitude to the Royal Society, the Central Research Fund of the University of London, and the Science Research Council for their generous financial support for the purchase of apparatus. I should also like to acknowledge the help and cooperation of my research students, Dr. C.J. Nicholls, J.B. Jones, E.I. Esmail, P. Wood and M. Webber.

REFERENCES

1.   W.C. Roentgen, Ann. Physik u. Chem. 64, 1 (1898).

2.   H.G.J. Moseley, Phil. Mag. 26, 1024 (1913).

3.   See Bibliography, B3, p. 187.

4.   D.W. Fischer and W.L. Baun, J. Appl. Phys. 38, 4830 (1967).

5.   D.W. Fischer and W.L. Baun, Anal. Chem. 37, 902 (1965).

6.   D.W. Fischer and W.L. Baun, J. Appl. Phys. 36, 534 (1965).

7.   E.W. White and G.V. Gibbs, Amer. Mineralogist, 52, 985
     (1967); 54, 931 (1969).

8.   D.E. Day, Nature, 200, 649 (1963).

9.   D.W. Fischer, Adv. X-ray Analysis, 13, 159 (1970).

10. E. Schnell, Monatsh. Chem. 93, 1383 (1962).

11. D.W. Wilbur and J.W. Gofman, Adv. X-ray analysis, 9, 354
     (1966).

12. E. Schnell, Monatsh. Chem. 94, 703 (1963).

13. G. Gilberg, Z. Physik. 236, 21 (1970).

14. D.W. Fischer and W.L. Baun, Adv. X-ray Analysis, 9, 329
     (1966).

15. S.A. Nemnonov, A.Z. Menshikov, K.M. Kurmayev and V.A.
     Trapeznikovs, Trans. of the metallurgical Soc. AIME,
     245, 1191 (1969).

16. J.A. Connor, I.H. Hillier, V.R. Saunders, M.H. Wood and
     M. Barber, Mol. Phys. 23, 81 and 24, 497 (1972).

17. D.W. Clack, J. Chem. Soc. Faraday Trans. II, 68, 1672 (1972).

18. D.S. Urch, Adv. X-ray Analysis, 14, 250 (1971).

19. D.S. Urch, J. Physics C: Solid State Physics, 3, 1275 (1970).

20. M. Wolfsberg and L. Helmholz, J. Chem. Phys. 20, 837 (1952).

21. G. Wiech, 'Soft X-ray Bond Spectra and the Electronic
     Structure of Metals and Materials', ed. D.J. Fabian,
     pub. Academic Press, p. 59 (1968).

22. D.W. Fischer, J. Chem.Phys. 42, 3814 (1965).

23. D.S. Urch, J. Chem. Soc. A, 3026 (1969).

24. T. Koopmans, Physica L, 104 (1934).

25. W.G. Richards, Int. J. Mass Spec. and Ion Phys. 2, 419
     (1969).

26. C.J. Nicholls and D.S. Urch, J. Chem. Soc. Dalton Trans.
     901 (1974).

27.   S. Idei, Nature, 123, 643 (1929).

28.   J.B. Jones and D.S. Urch, J. Chem. Soc. Dalton Trans.,
      sub. for pub. (1975).

29.   C.J. Nicholls, Ph.D Thesis, London University (1974).

30.   E.I. Esmail and D.S. Urch, J. Chem. Soc. Chem. Comm. 213
      (1974).

31.   E.I. Esmail, C.J. Nicholls and D.S. Urch, J. Chem. Soc.
      Chem. Comm. 39 (1974).

32.   E. Clementi and A.D. McLean, J. Chem. Phys. 36, 745 (1962).

33.   W.C. Price, A.W. Potts and D.G. Streets, 'Electron
      Spectroscopy', ed. D.A. Shirley, pub. North Holland,
      Amsterdam, p. 187 (1972).

34.   R.A. Mattson and R.C. Ehlert, J. Chem. Phys. 48, 5465
      (1968).

35.   V.I. Nefedov, J. Struct. Chem. USSR (Eng. trans.), 5, 605
      (1964).

36.   R. Manne, J. Chem. Phys. 52, 5733 (1970).

37.   D.S. Urch, 'Chemical Bonding Effects in X-ray Emission
      Spectroscopy', Analytic Equipment Bulletin published by
      Philips, Eindhoven, Netherlands (1971).

38.   R.E. LaVille, J. Chem. Phys. 57, 899 (1972).

39.   V.A. Fomichev, T.M. Zimkina and I.I. Zhakova, Sov. Phys.
      - Solid State (Eng. trans.) 10, 2421 (1969).

40.   H.U. Chun and D. Hendel, Zeit Naturforschung A22, 1401
      (1967).

41.   R.A. Slater and D.S. Urch, J. Chem. Soc. Chem. Comm. 564
      (1972).

42.   P.R. Wood and D.S. Urch, Proc. 8th Conf. on X-ray Analytical
      Methods (Birmingham) in press, pub. Philips, Eindhoven,
      Netherlands (1975).

43.   K. Tsutsumi and H. Nakamori, J. Phys. Soc. Japan, 25, 1418
      (1968).

44.   J.F. Priest, J. Appl. Phys. 42, 4750 (1971).

45.   G. Ekstig, E. Kallne, E. Noreland and R. Manne, Physical
       Scripta 2, 38 (1970).

46.   L.G. Parratt, Rev. Mod. Phys. 31, 616 (1959).

47.   M. Sawada, K. Taniguchi and H. Nakamura, 'X-ray Spectra
       and Electronic Structure of Matter', 2, 122 (1969),
       pub. Inst. Metal Phys. Acad. Sci. Ukranian SSR, Kiev,
       USSR.

48.   A.R. Knudson, P.G. Burkhalter, D.J. Nagel and K.L. Dunning,
       Phys. Rev. Lett. 26, 1149 (1971).

49.   G. Graeff, J. Siivola, J. Utrianinen, M. Linkvaho and
       T. Aberg, Phys. Lett. A29, 464 (1969).

50.   P.E. Cade, K.D. Sales and A.C. Wahl, J. Chem. Phys. 44,
       1973 (1966).

BIBLIOGRAPHY

General introduction

at an elementary level   B1 R. Jenkins, 'An Introduction
                              to X-ray Spectrometry', pub.
                              Heyden, Rhaine/Westf., Germany
                              (1974).

at a more advanced level   B2 A.H. Compton and S.K. Allinson,
                              'X-ray in Theory and Experiment',
                              Van Nostrand, New York (1935).

a detailed account of   B3 Editor L.V. Azaroff, 'X-ray
current research topics   Spectroscopy', McGraw-Hill, New
                              York (1974).

Applications in analysis

B4 L.S. Birks, 'X-ray Spectrochemical Analysis', Wiley-
Interscience, New York (1969).

B5 R. Jenkins and J.L. de Vries, 'Practical X-ray Spectrometry', MacMillan, London (1970).

B6 H.A. Liebhafsky, H.G. Pfeiffer, E.H. Winslow and P.D.
Zemany, 'X-ray Absorption and Emission in Analytical
Chemistry', Wiley, New York (1960).

Chemical bonding effects in X-ray spectra reviews

    B7   D.S. Urch, Quart Rev. $\underline{25}$, 343 (1971).

    B8   D.J. Fabian, L.M. Watson and C.A.W. Marshall, Rep.
    Prog. Phys. $\underline{34}$, 601 (1971).

Conference reports

    B9   'Rontgenspektren und Chemische Bindung', pub. Phys–Chem
    Inst. der Karl Marx Univ., Leipzig DDR (1967).

    B10 Ed: V.V. Nemoshkalenko, 'X-ray Spectra and Electronic
    Structure of Matter', pub. Inst. Metal Phys., Acad. Sci.
    Ukrainian SSR, Kiev, USSR.

    B11 J. Phys. (Paris), Suppl. 10 (1971) (Colloque CNRS,
    No. 196).

    B12 Ed: A. Faessler, 'Proc. Int. Symp. on X-ray Spectra
    and Electronic Structure of Matter', pub. Dept. Physics
    U. of Munich (1973).

    B13 Phys. Fennica $\underline{9}$, Suppl. S1 (1974).

    B14 Ed: D.J. Fabian, 'Soft X-ray Bond Spectra and the
    Electronic Structure of Metals and Materials', Academic
    Press, London (1968).

    B15 Ed: D.J. Fabian and L.M. Watson, 'Bond Structure
    Spectroscopy of Metals and Alloys', Academic Press, London
    (1973).

THE USE OF INELASTIC NEUTRON SCATTERING TO DETERMINE THE ELECTRONIC STATES OF INORGANIC MATERIALS

M.T. Hutchings

Materials Physics Division, A.E.R.E., Harwell, Didcot, Oxon OX11 ORA.

1. INTRODUCTION

Neutron scattering is rapidly becoming a familiar tool used by chemists for studying the physical properties of materials. So far this use has been confined primarily to structure analysis and the study of defects using elastic scattering of neutrons in the same way as X-rays are used, the study of molecular vibrations in solids, liquids and glasses, and the study of diffusive motion in solids and solutions.[1] The subject of this article, the use of magnetic inelastic scattering to study magnons and excitons, has up to now been largely explored by physicists, although magnetic diffraction has recently been exploited by chemists to study covalency effects. Not only physicists and chemists, but an increasing number of biophysicists, metallurgists, polymer scientists, and scientists in other fields are now using neutron techniques.

One should also say at the outset that there are now clear administrative channels in several countries allowing access to neutron beams by University, Industrial, and Research Institute workers. The great costs of the spectrometers are shared by governments and by research councils and agencies, so that the individual, if he has an experiment of sufficient merit, can use the technique without financial concern. In Britain, for example, university scientists have access to the medium flux reactors and the LINAC pulsed neutron source at A.E.R.E., Harwell, the HERALD reactor at Aldermaston, and also the high-flux reactor at the Institut Laue-Langevin (I.L.L.), Grenoble. At such installations scientists of different disciplines can come together to share a common technique to solve their particular problems, and

P. Day (ed.), Electronic States of Inorganic Compounds. 495–541. *All Rights Reserved.*
Copyright © 1975 by D. Reidel Publishing Company, Dordrecht-Holland.

as a consequence the technique of neutron scattering has
benefited greatly.

The use of neutrons for the study of excited electronic
states has strong advantages and some limitations, and the
information gained may overlap that obtained from other techniques.
It is clearly important only to use neutrons when the area of
information is unique.   What are these areas?

The basic advantage of neutrons as a radiation probe stems
from their energy-momentum relation, their electric neutrality
and generally low absorption, and in our case the fact that they
possess a spin of $\frac{1}{2}$ and a magnetic moment ~ -1.91 nuclear magnetons,
or -9.65×10^{-27} JT^{-1}.   The neutrons from the reactor sources
we shall be concerned with are thermally moderated with energies
in the range used of ~ 4 meV to ~ 400 meV.   By using a pulsed
neutron source with suitable moderator the upper limit may be
extended into the epithermal range of a few volts energy (1 meV =
8.067 cm^{-1} = 11.6 K $\cong$ 0.242 THz).

The energy E, momentum $\underline{p}$, velocity $\underline{v}$, wavelength λ, and
wavevector $\underline{k}$ of the neutrons are related by

$$\underline{p} = m\underline{v} = \hbar\underline{k}, \quad E = \frac{\hbar^2 k^2}{2m},$$

where $k = 2\pi/\lambda$ and m is the neutron mass.   If E is expressed in
meV, λ in Å, k in $Å^{-1}$, and v in m/μsec, they are related by:

$$E = 81.8/\lambda^2 = 2.072\ k^2 = 5.23 \times 10^6\ v^2.$$

We immediately see that a neutron of wavelength of 1.1Å,
corresponding to the peak flux from a 'thermal' reactor moderator
at 40°C has an energy of ~ 68-meV or ~ 545 cm^{-1}, or ~ 784 K.   This
wavelength is of the order of atomic spacings in crystals, and
the energy is of the order of that of the lowest excitations:
magnons, phonons and excitons.   Consequently coherent elastic
diffraction effects may be observed, and inelastic scattering may
occur in which the neutron excites or de-excites one of these
excitations and undergoes a change in energy which is comparable
to its original energy and is therefore easily measurable.
Furthermore as the wavelength of the neutron is of the same order
as that of the excitation the energy dispersion of the latter may
be investigated.   We can contrast this with X-rays, where the
energy of a photon of wavelength 1.1Å is 10 keV so that for these
low energy excitations the fractional change in energy is very
difficult to measure, and to light scattering or absorption where
the energies can be determined accurately but only very small
wavevectors can be sampled.

There are, however, several limitations to the use of the

technique for the types of problem with which most of the lectures
at this Institute are concerned.   The range of excitations
measurable by neutron techniques, up to ~500 meV maximum and
generally below 100 meV, is much smaller than that of most other
techniques to be discussed.   Furthermore, as we will see, it is
the magnetic interaction between the neutron and excitation which
is involved in the scattering process, so that only electronic
states with non-zero matrix elements of the magnetic moment operator with the initial state can be observed.   For insulators
this restricts the technique to the study of compounds containing
unpaired spins, such as the ions of the 3d, 4d or 5d transition
series, or the rare earths or actinides with unfilled 4f or 5f
shells.   Neutron sources are relatively weak, and large (~few
c.c.) samples are generally required for inelastic scattering
experiments.

2.   SCATTERING THEORY

In this section we shall outline the basic results of
scattering theory.[2]   For perspective we shall briefly include
nuclear scattering and magnetic elastic scattering, before devot-
ing the rest of the talks to the main subject matter.

We consider a neutron incident upon a target with energy E_0,
wavevector $\underline{k}_0$ and polarisation state δ_0, which is scattered to
values E', $\underline{k}'$, δ' respectively whilst the target changes state
from J_0 to $\overline{J}'$ (see Figure 1).   The cross section (the ratio of
the probability that the scattered neutron has direction within
a solid angle $d\Omega'$ about $\underline{k}'$ and final energy between E' and dE' to
the incident flux multiplied by $d\Omega'$ dE') is found in the first
Born approximation using Fermi's golden rule and the appropriate
density of final states.   It is

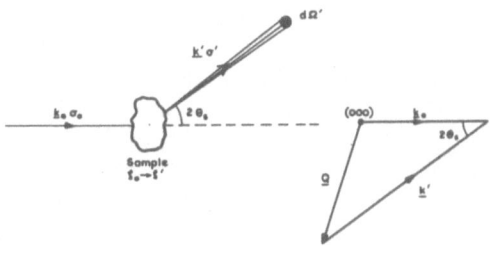

Figure 1.   Neutron scattering process and definition of
scattering vector $\underline{Q}$.

$$\frac{d^2\delta}{d\Omega'dE'} = \frac{k'}{k_0} \left(\frac{m}{2\pi\hbar^2}\right)^2 \sum_{\delta_0 s_0} p_{\delta_0} p_{s_0} \sum_{\delta's'} \left|\langle \underline{k}' \delta's'|\hat{V}|\underline{k}_0\delta_0 s_0\rangle\right|^2 \times$$

$$\delta(\hbar w + E_{s_0} - E_{s'}), \tag{1}$$

where p_{s_0} and p_{δ_0} are the probabilities of occupation of the initial

target state and neutron polarisation, and $\hat{V}$ is the interaction potential between the neutron and target atoms.   In this talk we shall consider only unpolarised neutron beams, so that $p_\uparrow = p_\downarrow = \frac{1}{2}$.

We define the energy and momentum change of the neutron by

$$\hbar w = E_0 - E' = \frac{\hbar^2}{2m} (k_0^2 - k'^2), \quad \hbar\underline{Q} = \hbar\underline{k}_0 - \hbar k' , \tag{2}$$

where $\underline{Q}$ is called the scattering vector.   For elastic scattering $|\underline{k}_0| = |\underline{k}'|$ and $\hbar w = 0$.   For inelastic scattering $\hbar w$ must equal the change of energy of the target.

There are two major contributions to $\hat{V}$ which we need consider, the nuclear-force interaction with the target nuclei, and the interaction of the neutron's magnetic moment with the moment distribution of the electrons in the target ions.

2.1 Nuclear interaction

The exact form of nuclear forces is unknown, but they are short range ($\sim 1.5 \times 10^{-15}$m) and as the wavelength is much larger than this range, and the size of the nucleus, the scattering is isotropic and $\hat{V}$ may be written

$$\hat{V}_{nucl} = \frac{2\pi\hbar^2}{m} b \, \delta(\underline{r} - \underline{R}) \tag{3}$$

for a given nucleus at $\underline{R}$ and neutron at $\underline{r}$.   b is called the bound scattering length of the nucleus and is determined by experiment. Scattering lengths vary erratically from nuclei to nuclei, so that unlike the X-ray scattering length from the ions' electrons, which increases monotonically with atomic number, neighbouring nuclei in the periodic table may have very different scattering lengths and even light ions may have strongly scattering nuclei.

The scattering length depends on both the particular isotope scattering, and its spin state relative to that of the neutron. This leads to the fact that nuclear scattering can be split into two parts.   That from the average scattering length with cross section $\sim \bar{b}^2$, is called 'coherent scattering'.   Elastic nuclear

coherent scattering gives rise to Bragg diffraction peaks when
the scattering vector $\underline{Q} = \underline{\tau}$, a reciprocal lattice vector of the
crystal.   The scattering angles, $2\theta_s$, at which these peaks occur
give the unit cell lattice dimensions and space group of the
target sample, and obey the familiar relations,

$$2d \sin \theta_s = \lambda, \text{ or } 2k_o \sin \theta_s = |\underline{Q}| = |\underline{\tau}| . \tag{4}$$

The intensities of the diffraction peaks give information on the
positions of ions in the unit cell of the sample, and on their
average thermal motion through the Debye-Waller factor, just as
in the case of X-ray diffraction.   One major difference from X-
ray diffraction, however, is that there is no form factor involved, since the Fourier transform of the scattering potential is
a constant.   As absorption is usually much less than for X-rays,
neutrons examine the whole bulk of a sample.   However the
diffraction cross sections, few barns, are somewhat smaller[3] as
are the intensity of neutron sources.   The neutron nuclear
diffraction pattern is proportional to the square of the Fourier
transform of the nuclear density in the sample, whereas the X-ray
diffraction pattern is proportional to the square of the Fourier
transform of the charge density in the sample.

The second part of the nuclear scattering is that from the
mean square deviation of the scattering lengths.   This is called
'incoherent scattering' and the cross section is $\propto (b - \bar{b})^2 =$
$(\bar{b^2} - \bar{b}^2)$.   The elastic nuclear incoherent scattering from
crystals is isotropic, apart from the Debye-Waller factor.

Turning now to inelastic nuclear scattering, inelastic nuclear
coherent scattering yields direct information on phonon modes or
molecular vibrational energies and their energy dispersion in a
crystal, in a similar way to that which we shall discuss for
magnons.   Inelastic nuclear incoherent scattering enables a
weighted density of phonon states to be determined directly, and
has been used extensively to investigate the energies of hydrogen
motion in molecular crystals since hydrogen has a large incoherent
cross section (82 barns).[1]

2.2 Magnetic interaction

When the target sample contains unpaired electrons magnetic
scattering may also occur.   Unlike the nuclear scattering, the
scattering of neutrons by the magnetic moment of the electrons on
ions in a sample may be written in a known form.   The interaction
potential is just that of the moment of the neutron $\underline{\mu}_n$ in the
magnetic field $\underline{H}$ of the moment distribution in the sample:

$$V = -\underline{\mu}_n \cdot \underline{H} \tag{5}$$

There are contributions to $\underline{H}$ from both the spin and the orbital or band motion of each electron in the sample, and it can be shown[4] that the cross section involves matrix elements of an operator $\underline{A}_\perp$ defined by

$$\underline{A}_\perp = \sum_i \exp\left(i\,\underline{Q}\cdot\underline{r}_i\right)\left\{\hat{\underline{Q}}_\wedge(\underline{s}_i{}_\wedge\hat{\underline{Q}}) - \frac{i}{\hbar}\frac{\hat{\underline{Q}}_\wedge\underline{p}_i}{|Q|}\right\},\tag{6}$$

where $\underline{r}_i$, $\underline{p}_i$ and $\underline{s}_i$ are the position, momentum, and spin operators of the ith electron, and $\hat{\underline{Q}} = \underline{Q}/|Q|$. $\underline{A}_\perp$ is the component perpendicular to $\hat{\underline{Q}}$ of an operator $\underline{A}$, i.e. $\underline{A}_\perp = \hat{\underline{Q}}_\wedge(\underline{A}_\wedge\hat{\underline{Q}})$. In terms of the operator $\underline{A}_\perp$ the general magnetic cross section is given by

$$\frac{d^2\sigma}{d\Omega' dE'} = \left(\frac{\gamma e^2}{m_e c^2}\right)^2 \frac{k'}{k_0} \sum_{J_0,J'} P_{J_0}\langle J_0|\underline{A}_\perp^+|J'\rangle\langle J'|\underline{A}_\perp|J_0\rangle\delta(\hbar\omega + E_{J_0} - E_{J'}),\tag{7}$$

where $\gamma = -1.91$ is the gyromagnetic ratio of the neutron, and e and m_e are the electron charge and mass.

Using the identity $\underline{A}_\perp^+ \cdot \underline{A}_\perp = \sum_{\alpha,\beta}(\delta_{\alpha\beta} - \hat{Q}_\alpha\hat{Q}_\beta) A_\alpha^+ A_\beta$ the cross section may be written

$$\frac{d^2\sigma}{d\Omega' dE'} = \left(\frac{\gamma e^2}{m_e c^2}\right)^2 \frac{k'}{k_0} \sum_{\alpha,\beta}(\delta_{\alpha\beta} - \hat{Q}_\alpha\hat{Q}_\beta) \sum_{J_0,J'} P_{J_0}\langle J_0|A_\alpha^+|J'\rangle\langle J'|$$
$$A_\beta|J_0\rangle\,\delta(\hbar\omega + E_{J_0} - E_{J'})\tag{8}$$

The evaluation of $\langle\underline{A}_\perp\rangle$ is very complicated in the general case – when the wavefunctions of the electrons on the ions involve both spin and orbital contributions. It can be shown that $\underline{A}_\perp$ may also be written in the form[5]

$$\underline{A}_\perp = \frac{-1}{2\mu_B} \int_{sample} d\underline{r}\, \exp\left(i\,\underline{Q}\cdot\underline{r}\right)\hat{\underline{Q}}_\wedge\{\underline{M}(\underline{r})\wedge\hat{\underline{Q}}\},\tag{9}$$

where μ_B is the Bohr magneton and $\underline{M}(\underline{r}) = \underline{M}_L(\underline{r}) + \underline{M}_S(\underline{r})$ is the magnetic moment density operator for the sample, $\underline{M}_L(\underline{r})$ and $\underline{M}_S(r)$ being orbital and spin contributions. In general cases $\langle\underline{M}(\underline{r})\rangle$ is very complicated. However, the expression for $\underline{A}_\perp$ may be simplified if the orbital content of the wavefunctions is low or if it is strongly coupled to the spin as in the case of rare earth ions.

The first approximation one can make, and one which we shall use throughout, is the so-called 'dipole approximation', which is valid when the mean radius of the unpaired electrons is much less than $|Q|^{-1}$. We shall assume a Heitler–London model throughout this article. Then grouping the electrons as belonging to given ions d, in the unit cell l,

$$\underline{A} \sim \underline{A}^{(D)} = \sum_{1,d} \exp\left(i \ \underline{Q} \cdot \underline{R}_{1d}\right) \left\{ j_o \ (Q) \ \underline{S}_{1d} + {}^1\!/2\left\{j_o(Q) + j_2 \right.\right.$$

$$(Q)\left.\left.\right\}\underline{L}_{1d}\right]. \tag{10}$$

$\underline{R}_{1,d}$ is the position of ion $(1,d)$ with total spin operator $\underline{S}_{1d}$ and orbital operator $\underline{L}_{1,d}$, and the $j_K(Q)$ are radial integrals defined by

$$j_K(Q) = \int_c^\infty dr \ r^2 |f(r)|^2 j_K(Qr) \tag{11}$$

where $f(r)$ is the radial part of the ion's wavefunction (assumed identical), and $j_K(Qr)$ is a spherical Bessel function of order K. If $Q = 0$, $\underline{A}^D \sim \sum_{1,d}(\underline{L}_{1d} + 2\underline{S}_{1d})$, and so the contribution to the scattering amplitude from each ion is proportional to its total moment.

For the two special cases mentioned above, which are of frequent occurrence, $\underline{A}$ can be expressed more simply:-

(a) Nearly quenched orbital angular momentum, e.g. Ni^{2+} in an intermediate crystal field, when $\underline{L} = (g - 2)\underline{S}$. Then

$$\underline{A}^D = \tfrac{1}{2}g \ F \ (Q) \sum_{1,d} \exp\left(i \ \underline{Q} \cdot \underline{R}_{1d}\right) \underline{S}_{1d} \ , \tag{12}$$

where $F(Q) = j_o(Q) + \left(\frac{g - 2}{g}\right) j_2 \ (Q)$.

(b) When $\underline{S}$ and $\underline{L}$ are strongly coupled to give a total angular momentum $\underline{J}$, as in the rare earth ion case. Then within states of given $\underline{J}$,

$$\underline{A}^D = \tfrac{1}{2}g_J \ F \ (Q) \sum_{1,d} \exp\left(i \ \underline{Q} \cdot \underline{R}_{1d}\right) \underline{J}_{1d}, \tag{13}$$

where $F(Q) = j_o(Q) \ \frac{g_S}{g_J} + \left\{j_o(Q) + j_2(Q)\right\}\frac{g_L}{g_J}$

and g_J is the Landé g factor, $2\underline{S} = g_S\underline{J}$, and $\underline{L} = g_L\underline{J}$.

$F(\underline{Q})$ is called the 'form factor' of the ion, and is related to the Fourier transform of the normalised effective spin density on the ion

$$\tilde{F}(\underline{Q}) = \int_{ion} \exp\left(i \ \underline{Q} \cdot \underline{r}\right) \ \tilde{S}(\underline{r})d\underline{r}\Big/ \int_{ion} \tilde{S}(\underline{r})d\underline{r}. \tag{14}$$

We use $\tilde{S}$ to stand for $\underline{S}$ or $\underline{J}$ and $\tilde{g}$ for g or g_J in the particular cases, $\tilde{F}(\underline{Q})$ being taken appropriately. It is analogous to the X-ray form factor, but as the unpaired spins are usually associated

with outer orbitals it will fall off faster with Q than in the
X-ray case, when the charge density concerned includes that of
the inner orbitals.   A typical form factor, that for chromium,
is shown in Figure 2.   In the general case when the expression
for the scattering is more complex it is not convenient to define
a form factor explicitly.[5]

On substituting into equation (8) we obtain the general
expression for the magnetic cross section in the form

$$\frac{d^2\sigma}{d\Omega' dE'} = \left(\frac{\gamma e^2}{m_e c^2}\right)^2 \frac{k'}{k_0} \left\{\frac{1}{2} \tilde{g} \tilde{F}(Q)\right\}^2 \sum_{\alpha,\beta} (\delta_{\alpha\beta} - \hat{Q}_\alpha \hat{Q}_\beta) \sum_{J_0 J' J_c} P_{J_0} \sum_{ld} \sum_{l'd'}$$

$$\langle J_0 | \exp(-i\, \underline{Q}.\underline{R}_{ld})\, \tilde{S}^{\alpha}_{ld} | J' \rangle \langle J' | \exp(i\, \underline{Q}.\underline{R}_{l'd'})\, \tilde{S}^{\beta}_{l'd'} | J_0 \rangle$$

$$\delta(\hbar\omega + E_{J_0} - E_{J'}) \, . \tag{15}$$

We have here assumed all the ions to be magnetically equivalent, the extension to inequivalent ions necessitating $F_d(Q)$ to be
taken inside the summation.

The cross sections for magnetic scattering are typically of
the same order as those for nuclear scattering,

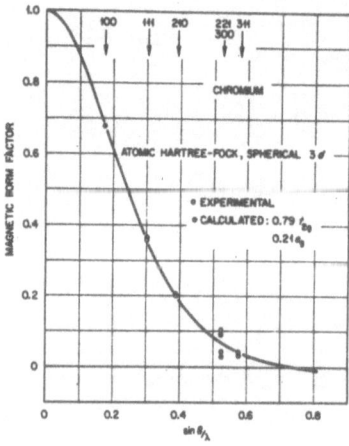

Figure 2.   Form factor of chromium (Ref. 6).

$$\left(\frac{\gamma e^2}{m_e c^2}\right)^2 \sim 0.29 \times 10^{-28} \text{ m}^2 = 0.29 \text{ barns.}$$ It should be remembered

that it is always the components of $\underline{S}$ or $\underline{J}$ perpendicular to the scattering vector $\underline{Q}$, whose matrix elements give rise to the scattering.

2.2.1 Elastic magnetic scattering

At high temperatures a system of weakly interacting moments will give rise to quasielastic scattering called paramagnetic scattering which depends on $\underline{Q}$ only through $|F(Q)|^2$ and the Debye-Waller term.

At temperatures below that at which the moments order, the magnetic scattering will contain an elastic coherent component due to the ordered ferromagnetic, antiferromagnetic, or ferri-magnetic arrangement of moments. This gives rise to diffraction patterns which are proportional to the square of the Fourier transform of the moment density in the crystal. The scattering angles at which the peaks occur give the dimensions and space group of the magnetic unit cell. The peak intensities give the moment and the form factor of the magnetic ions, and frequently information on the spin direction although in practice magnetic domain formation may prohibit its unique identification. The magnetic diffraction peaks from a ferromagnet will coincide with the nuclear peaks, a fact which may be exploited to measure mag-netic scattering amplitude by use of polarised neutrons, whereas those from an antiferromagnet may not or may coincide depending on whether the magnetic unit cell is larger than the chemical (nuclear) cell or not.

As an example we give the magnetic cross section for diffraction from a simple two-sublattice antiferromagnet, the most common type of ordering found in ionic compounds. For a sample of N unit cells of volume v_0, with identical sublattices, and spins $\underline{\tilde{S}}$ parallel $\sigma_d = +1$, or antiparallel $\sigma_d = -1$, to a given direction $\underline{\eta}$ this is given by

$$\frac{d\sigma}{d\Omega'} = \left(\frac{\gamma e^2}{m_e c^2}\right)^2 \frac{N}{v_o} \ (2\pi)^3 \sum_{\underline{\tau}} \delta(\underline{Q} - \underline{\tau}) \ \exp\left\{-2W(\underline{Q})\right\} |F_M(\underline{\tau})|^2 \times$$

$$\left\{1 - (\hat{\underline{\tau}} \cdot \hat{\underline{\eta}})^2\right\}_{Av} , \tag{16}$$

where $F_M(\underline{\tau}) = \frac{1}{2} \, \tilde{g} <\underline{\tilde{S}}> \tilde{F}(\underline{\tau}) \sum_{d}^{cell} \sigma_d \exp\left(i \, \underline{\tau} \cdot \underline{d}\right)$

is called the magnetic unit cell structure factor. $\langle \ \rangle$ denotes
the thermal average

$$\langle S \rangle = \sum_{\mathfrak{z}_0} p_{\mathfrak{z}_0} \langle \mathfrak{z}_0 | S | \mathfrak{z}_0 \rangle, \text{ where } p_{\mathfrak{z}_0} = \exp(-E_{\mathfrak{z}_0}/kT)/\sum_{\mathfrak{z}_0}$$

$$\exp(-E_{\mathfrak{z}_0}/kT).$$

By measuring intensities at $\underline{Q} = \underline{\tau}$, the mean spin value and the
ionic form factor $F(\underline{\tau})$ may thus be found. These quantities may
be used to deduce information on the ground state of the ion, as
they depend intimately on its wavefunction as we have seen. At
higher temperatures averages over low-lying states may be
involved. This type of measurement may therefore shed light on
the energy levels of the magnetic ion, and although they require
high accuracy have been used to deduce the configuration of ions,
the wavefunction, and the amount of covalency.[7]

2.2.2 Inelastic magnetic scattering

As we shall discuss the inelastic magnetic cross sections
more fully in sections 5-8 in connection with the investigation
of excited electronic states, we shall only mention here a useful
form in which the basic inelastic cross section, equation (15),
may be re-written in terms of the Van Hove scattering function
$\mathcal{J}(\underline{Q},\omega)$. This is

$$\frac{d^2\sigma}{d\Omega'dE'} = \mathcal{A}(\underline{k}_0,\underline{k}') \sum_{\alpha,\beta} (\delta_{\alpha\beta} - \hat{Q}_\alpha \hat{Q}_\beta) \mathcal{J}^{\alpha\beta}(\underline{Q},\omega), \text{ where}$$

$$\mathcal{A}(\underline{k}_0\underline{k}') = \left(\frac{\gamma e^2}{m_e c^2}\right)^2 \frac{k'}{k_0} \{{}^1\!/2 \, \tilde{g} \, \tilde{F}(Q)\}^2 \frac{N}{\hbar}\{\exp -2W(\underline{Q})\}, \text{ and}$$

$$\mathcal{J}^{\alpha\beta}(\underline{Q},\omega) = \frac{1}{2\pi} \sum_{\underline{r}} \int_{-\infty}^{\infty} dt \, e^{i(\underline{Q}\cdot\underline{r} - \omega t)} \langle \tilde{S}_0^\alpha(o) \, \tilde{S}_{\underline{r}}^\beta(t)\rangle. \quad (17)$$

$\mathcal{J}^{\alpha\beta}(\underline{Q}\omega)$ is thus the Fourier Transform of the two spin, unequal time, spin-spin correlation function. It may be related
to the generalised susceptibility $\chi^{\alpha\beta}(\underline{Q},\omega)$ by the Fluctuation
Dissipation Theorem.[8] This expression for the cross section in
terms of the Fourier transform of a correlation function is useful as the latter can be calculated using Green function theory
or more sophisticated techniques of statistical mechanics. It
should be mentioned that it is also possible to relate the nuclear
inelastic cross section to the Fourier transform of the appropriate
density correlation functions.

In general inelastic cross sections are three to four orders of magnitude less than elastic cross sections. We should note that besides giving rise to the Debye-Waller factor the fact that the ions move in the lattice can give rise to 'magneto vibrational scattering' or scattering from phonons via the magnetic interaction, which is sometimes difficult to distinguish from the true magnetic scattering.

Magnetic inelastic scattering may be distinguished from nuclear scattering by its temperature dependence in the case of ordered materials, by the orientation factor in the cross section, and by its dependence on Q of the form $|F(Q)|^2$. It may thus be distinguished from scattering from phonons which varies as Q^2, but it is often difficult to distinguish it from magneto vibrational scattering unless some theoretical knowledge is involved. The $|F(Q)|^2$ dependence of the magnetic cross section usually means that every effort must be made to observe the inelastic scattering at as low a momentum transfer as possible.

3. NEUTRON SOURCES[9]

3.1 Steady state reactors

By far the most commonly used neutron source at present is the steady state reactor shown schematically in Figure 3. The neutrons are produced by fission reactions in the core fuel elements immersed in light or heavy water which acts as a thermal moderator, and the whole is surrounded by a reflector of graphite, water, or beryllium, and a large biological shield. The shield

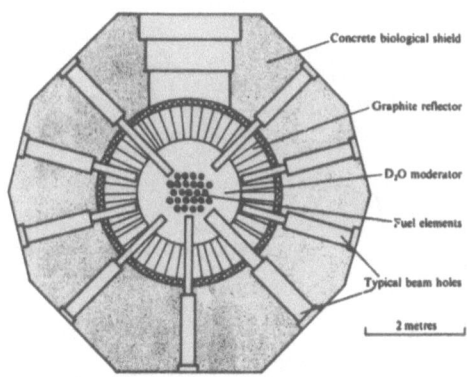

Figure 3.   Schematic layout of a beam reactor, based on Harwell DIDO (Ref. 10).

is pierced by a number of beam tubes which end in the moderator,
preferably in an area of high thermal flux, allowing a fraction
of the moderated neutrons to pass down the collimator tube to the
neutron spectrometer.   The thermal neutrons will have a
Maxwellian-like distribution, as shown in Figure 4, but this will
be accompanied by a fast-neutron tail.   By use of sources of
'hot' graphite or beryllium at 1000-2000 K, or 'cold' liquid
hydrogen at ~20 K, the peak in the flux spectrum can be shifted
to ~300 meV or ~5 meV respectively.

Typically the thermal flux in the core of the Harwell reactors is ~1.4 x 10^{14} n/cm^2/sec, whilst at the I.L.L. high flux
source is ~1.5 x 10^{15} n/cm^2/sec.   Collimation and monochromatisation cut these down to ~10^5 - 10^7 n/cm^2/sec at the sample
position.

3.2   Pulsed neutron sources

Reactors such as that at the I.L.L. are closely approaching
the maximum flux which can be obtained from continuous sources.
In order to achieve higher neutron flux at the sample, nearly
always a desirable requirement for inelastic scattering, pulsed
sources with high peak powers for short duration are the best
answer.   The velocities of thermal neutrons, typically 0.36 cm/
 sec for 68 meV neutrons, enable time-of-flight techniques to be
used to determine their energy, and pulsed sources are mainly
used with such techniques.   Present pulsed sources are of much
lower than maximum power - but even so they have some advantages.
In principle there are two types:

(a)   Fast neutrons are produced by pulsing a subcritical fast
reactor either by reactivity pulsing, or by injection pulsing
from an accelerator source (below), when the reactor acts as a
booster.

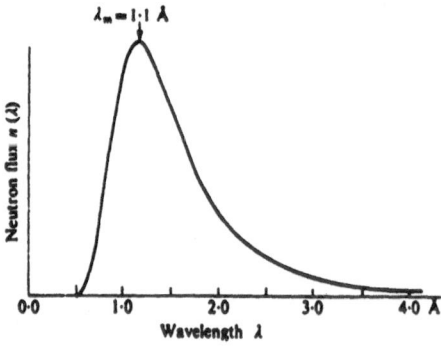

Figure 4.   Typical flux distribution from a moderator at 40°C.

(b) A linear accelerator is used to provide charged particles
which then produce neutrons by reaction with a target.   For
example, at the Harwell 45 MeV LINAC the pulsed electrons produce
Bremsstrahlung γ-rays from a gold target, which then yield neutrons
by (γ,n) or (γ, fission) reactions from a natural uranium secondary
target.

The fast neutrons produced by such sources are then moderated
by materials: paraffin, water, etc., of carefully chosen size to
produce the desired pulse shape of a thermal spectrum.   The flux
spectrum produced by a 3 cm polythene moderator at 293K from the
Harwell LINAC[11] is shown in Figure 5.   The main feature is the
high energy, 'epithermal', tail where there are many more neutrons
in this energy range than from a thermal or even a hot reactor
source.   Although they represent a potential major advantage,
the use of these high energy neutrons for inelastic scattering
has not yet been fully exploited.   Typical thermal flux at the
Harwell LINAC source is 6×10^9 n/cm^2/sec and 3×10^5 n/cm^2/sec,
both total (over all thermal energies), at the sample.

4.   INELASTIC SCATTERING SPECTROMETERS

We shall only have time here to outline the main experimental
features of the inelastic neutron scattering spectrometers, giving
two examples.   The neutron beam from the reactor is first monochromated by selection of a narrow range of energies from the

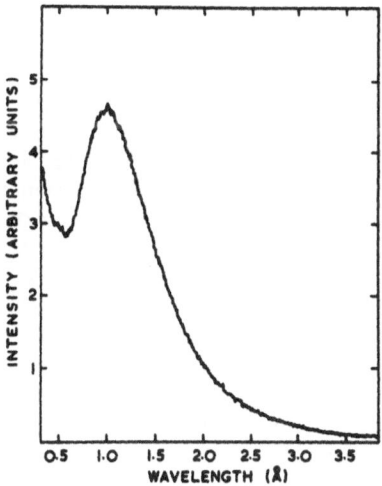

Figure 5.   Typical intensity distribution from the moderated
pulsed source at the Harwell LINAC (R. Sinclair private communication).   The intensity is ∝ flux n(λ).

Maxwellian distribution (Figure 4). This may be done either by
Bragg reflection from a large monochromator crystal, velocity
selection using rotating slits in an absorbing material or Bragg
reflection from a rotating crystal, or by filter techniques. The
same methods may be used to determine the energy of the scattered
neutrons, the second involving time-of-flight measurement. The
neutrons are detected by detection of secondary charged particles
using a proportional counter or solid-state scintillation techniques. The reactions most commonly used are $^{10}B(n,\alpha)^7Li$, ^3He
$(n,p)^3H$, or $^6Li(n,\alpha)^3H$. Beam sizes are usually up to $\sim$25 cm^2,
and directions are defined by apertures of slow-neutron absorbing
materials (B, Cd or Gd), or by soller slit collimators coated
with these materials. Often the vertical angular resolution is
not as tightly defined as that in the horizontal plane in order
to increase intensities, as the scattering usually takes place in
a symmetry plane; as in most techniques one has always to com-
promise between intensity and good resolution. Except for highly
absorbing samples it is usually an advantage to use as large a
sample as possible for inelastic scattering - unlike elastic
scattering when extinction effects must be avoided. The detector
and neutron flight paths must be carefully shielded to reduce
high background from fast neutrons, stray thermal neutrons, and
γ-rays, and discrimination techniques are used in the detection
system to isolate as far as possible only the neutron capture
events.

In all cases the observed intensity $I(\underline{Q}_0,\omega)$, when the
instrument is set to observe a particular energy transfer ω_0 and
scattering vector $\underline{Q}_0$, is a convolution of the cross section $\sigma(\underline{Q},\omega)$
and the resolution function of the instrument $R(\underline{Q} - \underline{Q}_0, \omega-\omega_0)$.

$$I(\underline{Q}_0,\omega_0) = \iint \sigma(\underline{Q},\omega) \, R(\underline{Q} - \underline{Q}_0, \omega-\omega_0) \, d\omega \, d\underline{Q}. \qquad (18)$$

The resolution function R gives the probability of detection of a
process of $(\underline{Q},\omega)$ when the instrument is set for $(\underline{Q}_0,\omega_0)$ and may
often be approximated by a 4-dimensional Gaussian function
$R = R_0 e^{-\underline{x} \underline{M} \underline{x}}$, where x is a 4-vector with components Q_x, Q_y, Q_z
and ω, and R_0 is a normalisation factor. In order to obtain
$\sigma(\underline{Q},\omega)$ some form of deconvolution must therefore be carried out.
This is particularly necessary if intensity or width measurement
is involved. For certain cases it can be shown that R will not
shift the peak intensity position.

Inelastic scattering processes may be observed both where
the neutron loses or gains energy to excite or de-excite an energy
level in the sample. Clearly the former requires $E_0 > (E_{\gamma} - E_{\gamma_0})$,
and for the latter the temperature of the sample must be high
enough so that the level is populated. Scattering may be
investigated from both powders and single crystals depending on
the information desired, but the information from the former is

usually limited, particularly regarding the wavevector dependence of the excitation. It is possible to vary the temperature of samples between 1 K-1500 K with the use of cryostats and furnaces, and magnetic fields of up to 10T are available at some establishments. The application of electric fields and pressures to the sample is also possible.

We now outline two types of spectrometer.

4.1 Time-of-flight spectrometer

As an example we shall describe the A.E.R.E. Harwell PLUTO 7H1R twin-rotor spectrometer[12] shown schematically in Figure 6. Each rotor is made of Mg-Cd alloy which absorbs neutrons, and has curved slots cut in it. The neutrons are monochromated by the successive opening to the beam of the narrow apertures of the slots at the two extreme positions. Bursts of monochromatic neutrons at ∼200-450/sec thus impinge on the sample with energy usually <20 meV defined by the speed of rotation, slot width, and phasing of the two rotors. There is no higher order of the incident beam, as occurs for crystal monochromators, but frame overlap effects may pose problems. The neutrons detected by the three low-efficiency monitor counters in the incident beam, and the ∼30 counters in the main bank, are fed into a multi time-channel analyser. E_0 and k_0 are measured by determining v_0 from the time of flight between monitors, and E' and k' from v', or the time after hitting the sample at which they are detected. For each counter a given time channel corresponds to a certain ω, and $\underline{Q}$, and if these coincide with

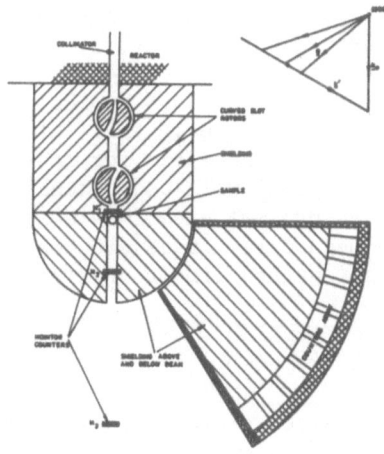

Figure 6. Schematic layout of Harwell PLUTO time-of-flight spectrometer with scattering diagram.

$(E_{J'} - E_{J_0})$ and the appropriate wavevector of an excitation in the
sample the number of neutrons in that channel will increase over
those in neighbouring channels as the time of the experiment
elapses. One can carry on the experiment for sufficient time to
give the desired accuracy of peak determination, and of course
gather much information over (Q,ω) space from the whole bank of
counters. As in all neutron experiments, counts are usually
normalised to a certain number of incident neutrons measured by
the first monitor counter. A typical spectrum is shown later in
Figure 18. Since E_0 is low, the instrument is often used to
observe energy-gained neutrons. However, the time channels are
evenly spaced so that the energy resolution of E' gets worse for
high energy transfers, $\Delta E' \sim E'^{3/2}$, with corresponding worsening
resolution in Q.[13]

Time-of-flight instruments are ideally suited to measuring
low-lying excitons, or crystal field levels, with little dispersion when powdered samples may be used. They may also be used
to observe exciton dispersion from single crystals when the data
cover an area of (Q,ω) space. However only one counter at a
time can be set so that all its time channels correspond to the
locus of Q (which lies along k', see Figure 6) being parallel to
a given direction in the crystal. In general one must work back
from the time channel in which the peak intensity occurs to
determine what value of (Q,ω) an excitation has. Heavy data
analysis using a computer usually accompanies such experiments,
but as on-line computers become more common this should not present
many problems.

There are many modifications of the arrangements described
above, such as the use of a rotating crystal to provide the pulsed
incident beam, which we cannot go into now. However, one point
should be made regarding pulsed sources, as these by necessity
must involve a combination of time-of-flight and crystal diffraction techniques in order to observe inelastic scattering. The
use of the 'high epithermal energy tail' of the neutron distribution of the pulsed source spectrum for the observation of high
energy excitations at low Q will undoubtedly be exploited more
in the future, though present experiments are relatively few.[14]
There has recently been a lot of development of more sophisticated
use of time-of-flight methods incorporating correlation
techniques.[15]

4.2 Triple-axis spectrometer

The triple axis spectrometer[16] is the most widely used
instrument for detailed measurement of both the ω and Q dependence
of the inelastic scattering cross section, usually from single
crystal samples. A schematic diagram of the A.E.R.E. Harwell
PLUTO 7H2L instrument[17] is shown in Figure 7. The heavy shielding

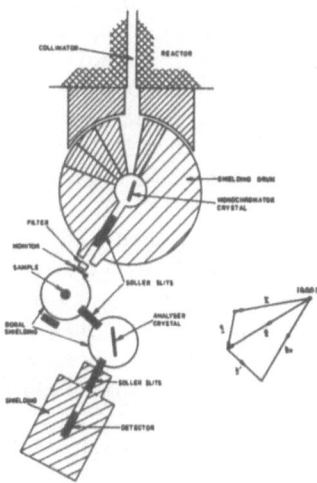

Figure 7.   Schematic layout of Harwell PLUTO triple axis
spectrometer.

around the counter is moved on air pads - a technique which now
is becoming very popular.

     Both the incident and final energies are defined by Bragg
reflection from crystals of known lattice spacing set at the
reflection angle.   The portion of the reactor spectrum selected
depends on the mosaic width of the monochromator crystal, which
should match the angle of the monochromator-sample soller.   The
monochromator and analyser crystals are selected for their size,
mosaic, and reflectivity, and Cu, Zn, Al, Be, Ge, and most
recently pyrolytic graphite, are used.   The last has a very
high reflectivity.[18]   If the monochromator is set to reflect a
beam of wavelength λ_0 from planes of spacing d, then the Bragg
condition will also hold for λ_0/n from planes of spacing d/n.
This 'high order' contamination of the incident beam may be
reduced by (a) choosing λ_0 so that the $\lambda_0/2$, $\lambda_0/3$, etc. components
are from a low intensity part of the spectrum, (b) use of a guide[19]
or filter[20] to scatter these high orders out of the beam, or (c)
use of a crystal plane, e.g. Ge(111), with a forbidden second
order reflection.   The observation of high order diffraction
from the analyser must also be considered.

     The triple axis spectrometer takes its name from the three
rotation axes of the monochromator, sample, and analyser crystals.
There are also three more angles defining the scattered beams, and
all six angles may be computer controlled so that a given setting

will sample a particular ω and $\underline{Q}$.   One major advantage of the
3-axis spectrometer is that one may calculate a set of positions
to scan ω or $\underline{Q}$ independently.[16]   The two most common modes are
a constant-$\underline{Q}$, ω scan, or a constant-ω, Q-scan.   The first of
these may be achieved by varying either E_0 which has advantages
of allowing for the normalisation of the resolution function and
keeping the analyser efficiency constant, or by varying E' which
allows for a filter to be set in the incident beam to filter out
$\lambda_0/2$.   Typical data are shown in Figure 12.

The resolution function of this instrument has been the
object of extensive discussion.[21]   It involves all collimation
angles, mosaics, the scattering sense at each axis, as well as
E_0, E' and $\underline{Q}$.   The function may be approximated by considering
the half maximum intensity contours as an elongated ellipsoid in
$\underline{Q}$-ω space, and the orientation of this ellipsoid relative to the
cross section 'dispersion-surface' gives rise to 'focussing'
effects which must be carefully considered by the experimentalist.[22]

There are two variations of the 3-axis instrument we should
mention.   One is the use of a cooled finely-divided beryllium
filter detector system instead of the analyser-detector system
described above.[23]   This is shown schematically in Figure 8.
The beryllium acts as a band pass filter for E' in the range 0-5
meV, and scatters neutrons with higher energies into absorbing
material.   When the energy resolution is fully treated one finds
that effectively a narrower range of E' is passed, centred on
~3.3 meV.[24]   The wide solid angle of acceptance gives rather
poor $\underline{Q}$ resolution which makes the filter most suitable for determination of excitations with little energy dispersion using powder

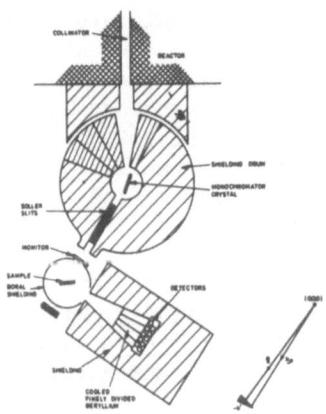

Figure 8.   Schematic layout of Harwell beryllium filter detection mode of the PLUTO triple axis spectrometer.

samples.   In an experiment one scans E_O, and obtains a peak in
the detected neutrons when $E_O - 3.3 = \hbar\omega = (E_{\gamma'} - E_{\gamma_L})$meV, the
energy of the excitation.   Use for magnetic excitations is limited
to small energies as $\underline{Q} \sim \underline{k}_O$ and the form factor reduces the high
energy peak intensities.

The other variation is the MARC, or multi-angle reflecting
crystal, spectrometer developed at A.E.K. Risφ.[25]   The analysing
system consists of a large crystal close to the sample with no
intervening soller collimation, and a large position sensitive
detector, as shown in Figure 9, the neutron pulses being fed to
a multi-position channel analyser.   Each position on the detector
corresponds to a different E' and k', so that one setting of the
axes corresponds to a scan in $(\underline{Q},\omega)$ space.   The scan obtained is
similar to that of the time-of-flight instruments but both its
centre point, set as in a conventional 3-axis, and its direction
may be chosen with flexibility.   The direction of $\underline{k}'$ varies over
the scan, so that the locus of $\underline{Q}$, in the direction of the perpendicular bisector to the analyser plane reciprocal lattice vector
τ_A (see Figure 9), may be set to lie along a required direction
in the crystal.   The MARC spectrometer has been used to determine
energy levels of rare earth ions in crystals.[25,55b]

5.   MAGNETIC EXCITATIONS IN SOLIDS

In the next few sections we shall give a number of examples
illustrating different cases where neutrons have been used to
investigate magnons and excitons in solids.   Before discussing

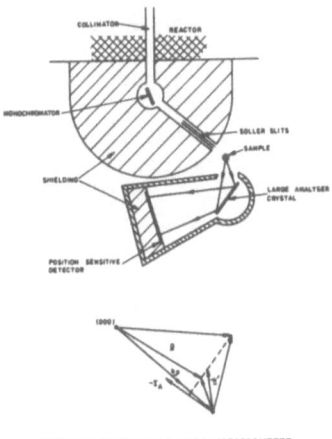

SCHEMATIC DIAGRAM OF A MARC SPECTROMETER

Figure 9.   Schematic layout of a multi-angle reflecting crystal
spectrometer.

each case we shall briefly review the theories so far developed
to explain the wavevector dependence of these magnetic excitations.

The true excited states of magnetic crystals are excitons,
or wavelike excitations in which the excitation is shared by all
the magnetic ions in the crystal, the amplitude of the excitation
being the same for all ions at equivalent positions in all the
unit cells.   The excitons are Frenkel excitons as the excitation
involves electron transfer from one state to another on the same
ion.   These single ion excitations propagate as a result of an
interaction between the ions, such as exchange, and the individual
ions' degenerate energy levels are spread out into a band of
energy states characterised, as is any excitation in a periodic
structure, by their wavevector q.   Viewed in this way there is
no formal distinction between a magnon or an exciton, the former
name being usually given to the lowest lying exciton in an
ordered material.   For S-state ions the magnons are usually very
much lower than the exciton levels at a few eV, and there is no
difficulty in assignment, nor is there when the orbital levels
are nearly quenched.   In both these cases of a magnon the ion's
electronic excitation involves unit change in spin, or effective
spin, component along the quantisation axis.   When the orbital
moment is not quenched, or the interaction energy is comparable
to the low-lying individual ion levels, such as in the case of
Co^{2+} or rare earth ions, the situation is more complex to treat.

Loudon[26], from whom some of the above definitions are taken,
has considered the symmetry properties of excitons in detail with
particular reference to the case of the rutile structure, and has
shown how they can give the conditions for magnetic Davidov
splittings of the otherwise twofold-degenerate exciton modes.
He also derives expressions for the exciton dispersion in the
rutile case.   The symmetry properties of magnons in ferro-,
antiferro- and ferrimagnetically ordered crystals have been
discussed by Brinkman and Elliott.[27]   As neutron scattering
transitions, in the dipole approximation, involve the component
of the effective spin operator perpendicular to the scattering
vector Q, it is the symmetry of the magnetic dipole operator
which must be considered when discussing the group theory of
magnetic excitons.

Excitons may be observed using neutron scattering at all
temperatures at which their lifetime is sufficiently long,
although their energies may depend on the state of magnetic order
of the sample.   Indeed excitons may be intimately involved in
magnetic transitions by exhibiting soft mode behaviour at the
ordering temperatures.   Magnons, being excitations from a state
of order, require long range order or highly correlated spins
with very well developed short-range order for their existence
and observation.

We shall first illustrate magnon excitations in ferro- and antiferromagnetically ordered crystals with quenched orbital moment where the exchange interaction dominates the Hamiltonian. Then we shall discuss the case of paramagnetic rare earth compounds with dominant crystal field interactions and relatively weak exchange.   Finally we consider cases where the inter-ion interactions and single ion terms in the Hamiltonian are comparable in magnitude.

6.   MAGNONS

We first consider the excitations in the case where the lowest lying ionic states are well separated from the higher states.   In this case we can describe the magnetic system by a spin Hamiltonian[28] involving either the real or effective spin S. As an example we take

$$\mathcal{H} = - 2 \sum_{i>j} J_{ij} \underline{S}_i \cdot \underline{S}_j - D \sum_i (S_i^z)^2 \, , \qquad (19)$$

where J_{ij} represents an isotropic Heisenberg exchange interaction between spins on sites i and j, and $D>0$ represents a single-ion axial anisotropy.   We shall only consider the magnon energies at zero temperature.

6.1   Ferromagnetic magnons[30]

When $J_{ij} > 0$ the lowest state of $\mathcal{H}$ will be a parallel ordered ferromagnetic alignment of the spins, which will precess together about the easy direction Oz with $\langle S_i^z \rangle = S$.   The lowest lying excited states are those in which the spins are no longer quite parallel, but precess with a phase angle between adjacent spins of $\phi = qa$, where a is the lattice spacing, as shown classically in Figure 10.   The two components of $\underline{S}$ perpendicular to Oz

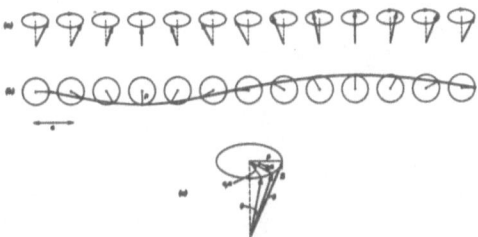

Figure 10.(a) ferromagnetic spin wave on a classical linear chain with wavevector $\underline{q}$, (b) view down Oz, (c) angular relationship of three adjacent spins (after Ref. 29).

therefore have a sinusoidal spatial variation at a given time with wavelength $\lambda_s = 2\pi/q$. The energy of these states may be seen to increase as λ decreases, since the angle between adjacent spins increases, and clearly depends on the exchange interaction between the spins. It is easily seen classically that for small $q, E_q \sim Jq^2a^2$. From periodic boundary conditions, the values of q equal $0, \frac{2\pi}{Na}, \frac{4\pi}{Na} \ldots \frac{2\pi}{Na}(N-1)$. Each value corresponds to a certain 'spin wave' state whose excitation can be likened to the creation of a pseudo-particle called a magnon.

To determine the magnon energies generally, we must solve $\mathcal{H}$ above. This can be done by use of the equations of motion of the spin operators $\dot{S}_\alpha = \frac{i}{\hbar}[\mathcal{H}, S_\alpha]$. They may be linearized by replacing S_z by its average value, $\langle S_z \rangle = S$ at low temperatures, using the so-called Random Phase Approximation (R.P.A.). One may then solve by transforming to reciprocal space operators via a Fourier transform.

An alternative method is to use the Holstein-Primakoff transformation.[31] The steps are (a) write $\mathcal{H}$ in terms of S^+, S^- and S_z; (b) transform $\mathcal{H}$ to site, spin-deviation, boson creation and annihilation operators a_j, a_j^+, where $\left[a_j, a_j^+\right] = \delta_{ij}$,

$$S_i^z = S - a_i^+ a_i$$

$$S_i^+ = \sqrt{2S}\, a_i \qquad\qquad\qquad (20a)$$

$$S_i^- = \sqrt{2S}\, a_i^+$$

(c) diagonalise by transforming to reciprocal space, or magnon, variables

$$a_j = \frac{1}{\sqrt{N}} \sum_q e^{-i\underline{q}\cdot\underline{R}_j} b_{\underline{q}} \;,\quad a_j^+ = \frac{1}{\sqrt{N}} \sum_q e^{i\underline{q}\cdot\underline{R}_j} b_{\underline{q}}^+$$

$b_{\underline{q}}^+$ is an operator which creates a magnon of wavevector $\underline{q}$, and $b_{\underline{q}}$ annihilates a magnon of wavevector $\underline{q}$. Denoting a state of $n_{\underline{q}}$ magnons excited with wavevector $\underline{q}$ as $|n_{\underline{q}}\rangle$, then

$$b_{\underline{q}}^+ |n_{\underline{q}}\rangle = \sqrt{(n_{\underline{q}} + 1)}\, |n_{\underline{q}} + 1\rangle$$

$$b_{\underline{q}} |n_{\underline{q}}\rangle = \sqrt{n_{\underline{q}}}\, |n_{\underline{q}} - 1\rangle \qquad\qquad (20b)$$

$$b_{\underline{q}}^+ b_{\underline{q}} |n_{\underline{q}}\rangle = n_{\underline{q}} |n_{\underline{q}}\rangle.$$

We then find that the part of $\mathcal{H}$ correctly bilinear in these variables is

$$\mathcal{H}_0 = \sum_q \left\{ 2S(\gamma_o - \gamma_{\underline{q}}) + (2S - 1)D \right\} b_{\underline{q}}^+ b_{\underline{q}} \;, \qquad (21)$$

where $\quad \gamma_{\underline{q}} = \sum\limits_{j'} J_{jj'} \; e^{i\underline{q}\cdot(\underline{R}_j - \underline{R}_{j'})}$ $\qquad\qquad\qquad$ (22)

is the Fourier transform of the exchange interaction between the ions. Usually in ionic compounds the exchange falls off sufficiently quickly with distance that only the nearest few neighbouring ions are involved in the summation.

The operator $b_{\underline{q}}^{+}b_{\underline{q}}$ is the occupation number operator for each magnon state $\underline{q}$, so that $\mathcal{H}_0 = \sum\limits_{\underline{q}} \hbar\omega_{\underline{q}} n_{\underline{q}}$, and the magnon energy is given by

$$E_{\underline{q}} = \hbar\omega_{\underline{q}} = \left\{ 2S(\gamma_0 - \gamma_{\underline{q}}) + (2S - 1)D \right\}. \qquad (23)$$

Each magnon state excited corresponds to a reduction in total spin component along Oz of 1.

We therefore see that measurement of $E_{\underline{q}}$ can give very directly the exchange interaction and anisotropy constants in $\mathcal{H}$. If these are between effective spins they can in principle be used to calculate the interaction between real spins – and thence the effect of interactions on the higher levels to give their energies and wavefunction. If the interactions are large, however, use of a spin Hamiltonian for the lowest levels may be invalid because of mixing with the upper states, so that all the levels must then be treated together – we shall consider this case later.

The magnon energy dispersion may be measured using the magnetic interaction between the neutron probe and the sample to excite or de-excite the spin wave states. The cross section for magnon creation or annihilation may be derived from equation (15) or equation (17), by calculation of the appropriate matrix elements, or correlation functions, respectively. In the former case the spin operators are transformed as above, and their matrix elements evaluated between states $|n_{\underline{q}}\rangle$ and $|n_{\underline{q}} \pm 1\rangle$.

The result for creation (+) or annihilation (−) of spin wave of wavevector $\underline{q}$ is

$$\frac{d^2\sigma_{(\pm)}}{d\Omega dE'} = \left(\frac{\gamma e^2}{m_e c^2}\right)^2 \left\{ \tfrac{1}{2} \, \widetilde{g} \, \widetilde{F}(Q) \right\}^2 \frac{k'}{k_0} \, (1 + \hat{Q}_z^2) \, \exp\left\{ -2W(\underline{Q}) \right\} \frac{S}{2} \times$$

$$\frac{(2\pi)^3}{V_0} \sum\limits_{\underline{q},\underline{\tau}} (\langle n_{\underline{q}} \rangle + \tfrac{1}{2} \pm \tfrac{1}{2}) \, \delta(\hbar\omega \mp \hbar\omega_{\underline{q}}) \, \delta(\underline{Q} \mp \underline{q} - \underline{\tau}). \qquad (24)$$

$\langle n_{\underline{q}} \rangle = \left\{ \exp(\hbar\omega_{\underline{q}}/k_B T) - 1 \right\}^{-1}$ is the mean Bose-Einstein occupation number for magnons of wavevector $\underline{q}$.

The two delta functions in this formula show that both energy and 'crystal' momentum must be conserved to observe scattering by magnon creation or annihilation (see Figure 14). The excitations repeat themselves in each Brillouin zone and their dispersion may be measured about any reciprocal lattice point since there is no dynamic structure factor involved. When $T < \hbar\omega_{\underline{q}}$ the creation process will always have the larger cross section, but at high temperatures the cross sections $\sim \dfrac{k'}{k_o} \dfrac{k_B T}{\hbar\omega_{\underline{q}}}$

so that the annihilation process has larger intensity.

Spin waves in metallic ferromagnets were among the first excitations to be investigated with neutrons, but as there are few known non-metallic ferromagnets only $CrBr_3$, EuO, and EuS have been studied.   In the former case both acoustic and optical magnons are observed as there are two inequivalent Cr^{3+} ions per cell, a complication not discussed above.   In the latter two cases polycrystalline samples were used, and measurements taken with $\underline{q}$ referred to $\underline{\tau} = o$, i.e. the straight through beam.[33]   In this way an average dispersion $E_{\underline{q}}$ over all directions of $\underline{q}$ was measured.   The EuO data are shown in Figure 11, where the q^2 dependence of $E_{\underline{q}}$ at low q is clearly seen for this case of negligible anisotropy.   Values of nearest and next nearest neighbour interaction constants, J_1 and J_2, were determined from fits to the data.

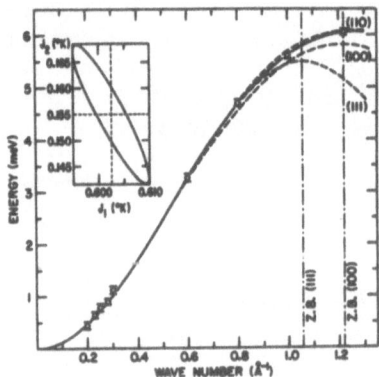

Figure 11.   Spin wave dispersion in polycrystalline EuO.   The solid line is the best fit to the data yielding values of J_1 and J_2 given in the insert (Ref. 33).

6.2 Antiferromagnetic magnons

The ground state of the antiferromagnet is not so simple as that of the ferromagnet as the classical up-down Néel state is not an eigenstate of $\mathcal{H}$. However this Néel state is taken as the initial state from which the spin waves are excited.

Both the equation of motion and the Holstein-Primakoff transformation methods may be used to solve the Hamiltonian equation (19) for the case when $J < 0$. In the simple two sub-lattice case we now have to define spin deviation operators, and spin wave creation operators, for each sublattice. The Fourier transform to the spin wave of operators for the two sublattices, c_q^+, c_q, d_q^+, d_q, does not now diagonalise $\mathcal{H}$, and we have to make a further transformation, often called the Bogoliubov transformation, to α_q, β_q given by

$$\alpha_q = u_q c_q - v_q d_q^+ \qquad \alpha_q^+ = u_q c_q^+ - v_q d_q$$

$$\beta_q = u_q d_q - v_q c_q^+ \qquad \beta_q^+ = u_q d_q^+ - v_q c_q$$

where $u_q^2 - v_q^2 = 1$.

The wavevector dependent part of $\mathcal{H}$ becomes

$$\mathcal{H} = \sum_q \mathcal{H} \omega_q \left(\alpha_q^+ \alpha_q + \beta_q^+ \beta_q + 1 \right), \tag{25}$$

with the energy dispersion given by

$$E_q = \mathcal{H} \omega_q = 2S \left\{ \omega_{2q}^2 - \omega_{1q}^2 \right\}^{\frac{1}{2}}, \tag{26}$$

$$\omega_{2q} = (1 - \tfrac{1}{2}S) D + (\gamma_{ao} - \gamma_{po} + \gamma_{pq})$$

$$\omega_{1q} = \gamma_{aq}.$$

Here γ_{aq} and γ_{pq} are the Fourier transforms of the exchange interactions between antiparallel and parallel spins respectively, defined as in equation (22). u_q and v_q are given by

$$u_q^2 + v_q^2 = 2S\omega_{2q}/E_q, \text{ and } 2u_q v_q = -2S\omega_{1q}/E_q. \tag{27}$$

α_q^+, α_q and β_q^+, β_q are associated with two magnons, which in the case considered are degenerate. The two magnons each involve a combination of spin waves as shown in Figure 10 on each sublattice; they each change the total spin component $\Sigma_i S_i^z$ by +1 or −1. The extent to which each change is shared by each sublattice is given by u_q and v_q.

When higher order terms in the deviation operators are taken into consideration it is found that right hand side of the expression for E_q in equation (26) must be increased by a few percent by a correction factor, R, which decreases as S increases. This has been discussed by Oguchi[34] and others. If the two sublattices are inequivalent or there is a directional anisotropy such as dipolar interactions between sublattices, the modes may split.

The neutron cross section for magnon creation or annihilation may be found in the same manner as for a ferromagnet, by calculating the matrix elements in terms of the spin wave operators, and one finds[35] for each mode in the degenerate case

$$\frac{d^2\sigma^{(\pm)}}{d\Omega dE'} = \left(\frac{\gamma e^2}{m_e c^2}\right)^2 \left\{\tfrac{1}{2} \tilde{g}\, \tilde{F}(Q)\right\}^2 \frac{k'}{k_0}\, \frac{S}{2}\, (1 + \hat{Q}_z^2)\{\exp[-2W(Q)]\}_x$$

$$\frac{(2\pi)^3}{v_0}\sum_{q,\tau} (\langle n_q\rangle + \tfrac{1}{2} \pm \tfrac{1}{2})\, \delta(\hbar\omega \mp \hbar\omega_q)\, \delta(Q \mp q - \tau)\{u_q^2 + v_q^2 + 2u_q v_q \times$$

$$\cos(\rho.\tau)\}, \tag{28}$$

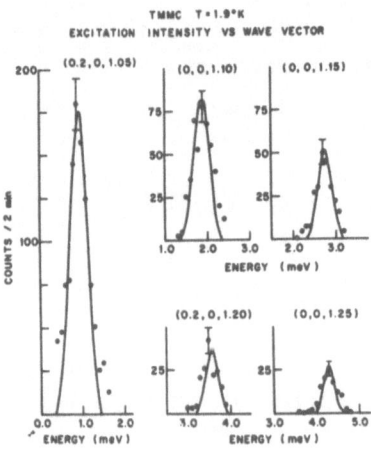

TMMC T = 1.9°K
EXCITATION INTENSITY VS WAVE VECTOR

Figure 12.   Neutron groups observed in TMMC at 1.9 K using constant −Q method. The value of Q for each scan is given in parentheses (Ref. 42). The solid lines are the calculated intensity.

where $\underline{p}$ is the lattice translation from an ion on one sublattice to its nearest neighbour on the other.

The formula is therefore very similar to that for the ferromagnet except that we now have a q dependent dynamic structure factor - the term in the last bracket. Although the excitations repeat within each Brillouin zone of the antiferromagnetic reciprocal lattice the intensity of the neutron scattering is larger in some zones than others; in the simple two-sublattice case, in zones where the elastic magnetic structure factor is large.

We therefore see that the intensity of the magnon scattering from antiferromagnets as well as the energy at which it occurs depends on the interaction constants in $\mathcal{H}$, and it may help to distinguish between different sets of parameters which fit the energies equally well. This can be particularly useful in more complicated antiferromagnets. The derivation of E_q and cross sections for more general $\mathcal{H}$, or when there are more than two sublattices, has been discussed in detail by a number of authors,[36,37] and as an example the application to the corundum structure of Cr_2O_3 and $\alpha - Fe_2O_3$ by Samuelsen[38] may be given.

The treatment of different terms in the anisotropy, such as $E(S_i^{x2} - S_i^{y2})$, to the correct order in the sublattice spin wave operators is not simple.[39] A more rigorous method used recently for NiO[40] is to first diagonalise the molecular field Hamiltonian involving the anisotropy terms and mean field interaction term, and then to consider excitations between these levels. We will discuss this approach in more detail when we consider exciton dispersion.

Magnons may be also investigated using electromagnetic radiation which can excite $\underline{Q} = 0$ excitations. The $q = 0$ antiferromagnetic resonance mode E_0 may be observed in first order by infrared absorption, or Raman light scattering. Excitations involving two magnons of equal and opposite q may also be observed in second order processes by infrared absorption or Raman scattering, and one magnon sidebands may be observed on optical transitions.[41] In all these cases a density of one or two magnon states weighted by a trignometrical function of q which emphasises the contributions from certain parts of the Brillouin zone is observed. The density of magnon states $\rho(E)$ calculated from the parameters in $\mathcal{H}$ found from neutron scattering has been used in these cases to test the theories of the optical processes.

We now give three examples from the many cases studied of neutron scattering from antiferromagnets.

(a) $(CD_3)_4 NMnCl_3$

Tetramethyl ammonium manganese trichloride (TMMC) may for
the purposes of this talk be considered to be an essentially
ordered one-dimensional antiferromagnet,[42] the $MnCl_3$ ions forming
linear chains with two Mn^{2+} ions per chemical cell separated by
$c/2 \sim 3$ Å.   In fact the spins do not show long range magnetic
order unless $T < T_N \sim 0.8K$, when the interchain interactions become
effective, but we may use the highly correlated low temperature
states at 1.9K to illustrate the simplest type of antiferromagnet
– a linear array of alternating plus and minus spins with effectively no anisotropy.   The Mn^{2+} ions are in a $(3d)^5$ ^6S state.

For nearest-neighbour only interactions and no anisotropy,
we would expect

$$\gamma_{\underline{q}} = 2J \cos(q_z c/2)$$

and $E_{\underline{q}} = 4JS \sin (q_z c/2)$, with $S = 5/2$.   (29)

The observed neutron 'groups' in the plots of intensity against
energy transfer for constant-Q scans are shown in Figure 12.
The reciprocal lattice of a chain of spins is a series of planes
perpendicular to the unique direction, and we can use any point
on these planes as $\underline{\tau}$ in equation (28).   The energy dispersion of
the intensity peaks is shown in Figure 13 where we see a simple
sine curve is in fact observed, with linear slope at low q characteristic of an antiferromagnet.   There is no measurable

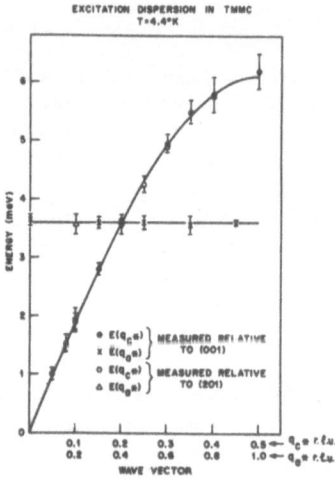

Figure 13.   Energy dispersion of spin-wave-like excitations in
TMMC along and perpendicular to the c-axis.   The solid lines
are the best fit to the data of equation (29) (Ref. 42).

dispersion perpendicular to the chain indicating that interaction
between chains is too weak to be observed.   The solid curve in
Figure 13 is the best fit of equation (29) to the data and gives
J_{nn} = -7.07K = 1.71 meV.   This value actually includes the
correction factor R, which since the system is disordered is
relatively large,   1.07, so that the true exchange constant is
J_{nn} = 6.6($\pm$0.15)K.   The solid lines in Figure 12 are the calculated profiles from the convolution of the measured resolution
function and the cross section of equation (28).   There is one
normalisation factor and we see there is excellent agreement with
experiment.   (We shall not ask here why the agreement is so good
in this special case of no long range order!)[43]

(b) NiF_2

     As a second example we take rutile structured NiF_2, T_N =
73.2 K.[44]   The $(3d)^8$ ^3F level of the free Ni^{2+} ion is split to
give an orbital singlet ground state, and we treat this using
the spin Hamiltonian,

$$\mathcal{H} = \sum_{i\uparrow}\left\{ DS_{i\uparrow}^{z^2} + E(S_{i\uparrow}^{x^2} - S_{i\uparrow}^{y^2})\right\} + \sum_{i\downarrow}\left\{ DS_{i\downarrow}^{z^2} - E(S_{i\downarrow}^{z^2} - S_{i\downarrow}^{y^2})\right\} -$$

$$2\sum_{i>j} J_{ij}\underline{S}_i\underline{S}_j + \mathcal{H}_{dd} , \tag{30}$$

where $\mathcal{H}_{dd}$ is the dipolar interaction, and i$\uparrow$ and i$\downarrow$ refer to the
two sublattices.   The spins in this case lie in the $\underline{a}$ - $\underline{b}$ plane
and are slightly canted.

     The reciprocal lattice is shown in Figure 14 where a typical
vector diagram for conservation of crystal momentum for a constant
-Q scan is illustrated.   The energy dispersion relative to the
(100) point is shown in Figure 15.   The two modes are split at
q = 0 due to the nature of the single ion anisotropy and come
close together at the zone boundary if dipolar-interactions are
small.   The best fit to the dispersion using the theoretical
expressions of equation (26) gives values for E, D, and the
exchange to the three nearest neighbours.   In Figure 16 we show
the agreement obtained between the observed two-magnon Raman
spectrum and that calculated from the neutron results including
the effect of magnon-magnon interaction in the scattering process.[45]

(c) Fe_2O_3

     Finally we illustrate in Figure 17 the magnons in hematite,
α-Fe_2O_3, which has the corundum structure, as an example of an
antiferromagnet with four sublattices, T_N=960K.[46]   In this case
a twofold degenerate optical branch at $\sim$100 meV is observed as

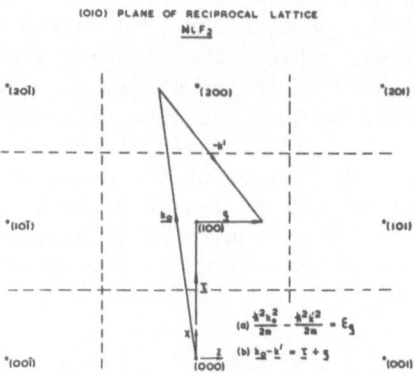

Figure 14.   Scattering diagram and reciprocal lattice for the
(010) plane of NiF₂ illustrating observation of magnons in the
[00ζ] direction about (100).

well as the usual acoustic magnons.   The authors used a comparison of calculated and observed intensity of the scattering from
the two modes to distinguish between two possible fits to the
energy dispersion, and values of exchange out to fifth nearest
neighbour were determined.

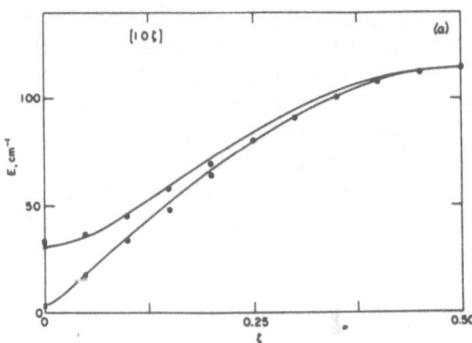

Figure 15.   Spin wave energy dispersion in [00ζ] in NiF₂.   The
solid lines represent the best fit of theoretical expressions to
the observed points at 4.2K obtained by adjusting the parameters
in the spin Hamiltonian (Ref. 44).

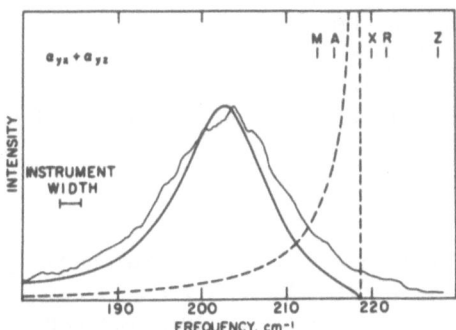

Figure 16.   Experimental line shape for Raman scattering from
NiF_2 compared with theory (heavy line) calculated using the
neutron scattering results and allowing for magnon-magnon interaction effects.   The dotted line is theory without the interaction effects (Ref. 44).

7.   CRYSTAL FIELD LEVELS

     In this section we shall discuss the observation of ionic
energy levels in crystals in the absence of strong interactions
between the ions and long range order of the effective spins, i.e.
in paramagnetic crystals with only weak exciton dispersion.   In
the truly paramagnetic state (no short range order) there will be
no magnon scattering;   the transitions within the ground or
excited states give inelastic paramagnetic neutron scattering
centred on zero energy transfer with an energy width which gives
a measure of the interactions between ions.[47]   We take this

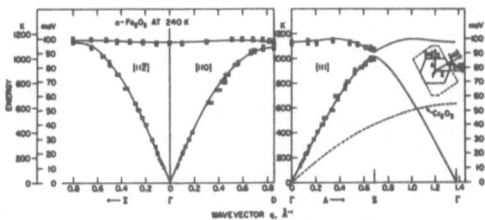

Figure 17.   Spin wave energy dispersion in α-Fe$_2$O$_3$.   The solid
lines are the best theoretical fit to the data points for both
acoustic and optical magnons (Ref. 46).

width to be small in our considerations here. The group of
compounds which fall readily into this category are rare earth
compounds with low magnetic ordering temperatures. The predominant interaction is then between the rare earth ion and the
surrounding crystal field. This interaction is usually smaller
than the spin orbit coupling and may be expressed to a good
approximation by a Hamiltonian acting within a given J state of
the form

$$\mathcal{H} = \sum_{i,n,m} B_n^m O_n^m (J_i), \tag{31}$$

where the B_n^m are parameters and the O_n^m are operators involving
powers of J_x, J_y and J_z transforming like tesseral harmonics.[48]
For a cubic crystal field there are only two parameters B_4^0 and
B_6^0 and

$$\mathcal{H}_{cubic}^i = B_4^0 \left[O_4^0(J_i) + 5\, O_4^4\, (J_i) \right] + B_6^0 \left[O_6^0\, (J_i) - 21\, O_6^4(J_i) \right]. \tag{32}$$

When there is little or no energy dispersion powder samples
may be used to investigate the excitons, and time-of-flight
techniques prove particularly advantageous as the energy spectrum
is scanned at one setting. Levels in Ho_2O_3 and Er_2O_3 were the
first to be observed,[49] but the first positive identification was
of the $\Gamma_7 - \Gamma_8$ transition of Ce^{3+} in the cubic field in metallic
$CeAs$[50] (T_N = 16 K). The data are shown in Figure 18 where the

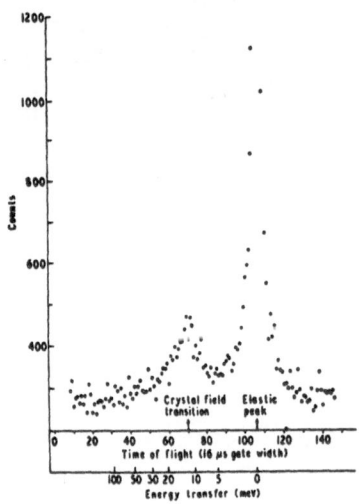

Figure 18.   Time-of-flight distribution of neutrons scattered
from CeAs powder at 293K.   Note inelastic peak at 12 meV due to
de-excitation of Γ_8 level (Ref. 50).

exciton is clearly seen by neutron energy gain scattering at room
temperature.   The level is found at 12 meV or 140 $\pm$ 10 K.

The cross section is given[51] in the dipolar approximation
by equation (15) with $\underline{A}_\perp = \frac{1}{2}$ g_J $F(Q)$ $\underline{J}_\perp$, the states of the sample
being taken as those of the single ion.   There is no correlation
between different ions so that the only Q dependence is from the
form factor and Debye-Waller factor,

$$\frac{d^2\sigma}{d\Omega' dE'} = \left(\frac{\gamma e^2}{m_e c^2}\right)^2 N \exp\{-2W(\underline{Q})\} \left\{\tfrac{1}{2} g_J F(\underline{Q})\right\}^2 \frac{k'}{k_o} \sum_{\mathfrak{J}_o, \mathfrak{J}'} p_{\mathfrak{J}_o}$$

$$|\langle \mathfrak{J}' | \underline{J}_\perp | \mathfrak{J}_o \rangle|^2 \delta(\hbar\omega + E_{\mathfrak{J}_o} - E_{\mathfrak{J}'}). \qquad\qquad (33)$$

For low symmetries use of a single crystal can be an advantage,
as the angle $\underline{Q}$ makes with the crystal axes and therefore $\underline{J}_\perp$ may
be varied in order to identify transitions.   For a powder the
cross section must be averaged over all directions relative to $\underline{Q}$.

At low Q the point symmetry of the ions may be used to
determine allowable transitions.   For example in the case of
cubic crystals the states $|\mathfrak{J}\rangle$ may be labelled by their irreduc-
ible representation $|\Gamma_n v\rangle$, where v denotes their degeneracy.
In cubic symmetry $\underline{J}$ transforms as Γ_4, and the finite transition
probabilities can easily be determined by seeing which product
representations Γ_n x Γ_4 x $\Gamma_{n'}$ contain Γ_1.   The cross section in
cubic symmetry may be calculated in a straightforward manner
from the matrix elements

$$|\langle \mathfrak{J}' | J_\perp | \mathfrak{J}_o \rangle|^2 = 2|\langle \mathfrak{J}' | J_z | \mathfrak{J}_o \rangle|^2. \qquad\qquad (34)$$

These matrix elements have been tabulated by Birgeneau.[52]

Identification of excitations as excitons may be made using
the $|F(Q)|^2$ variation of the intensity, and also their temperature
variation which from equation (33) should vary as $p_{\mathfrak{J}_o}$ and so obey
Boltzmann statistics in contrast to the Bose-Einstein variation
of $\langle n_q \rangle$ for phonons.

As an example of a complex-level system we give the results
found for Tm^{3+}, $(4f)^{12}$ $^3H_6$, in a cubic crystal field in the paramagnetic metallic compound TmSb which has the rock-salt structure.[53]   The variation of the energy levels of the Tm^{3+} ion in
cubic crystal fields is shown in Figure 19;   they are characterised by the two parameters x and W where

$$B_4^o = Wx/ F(4) \text{ and } B_6^o = W(1 - |x|)/F(6) \qquad\qquad (35)$$

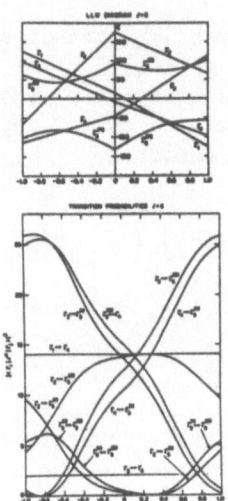

Figure 19.   Upper-eigenvalues of equations (32,35) for J = 6 in
units of W as a function of x.   Lower-values of the matrix
elements equation (34) as a function of x (Ref. 53).

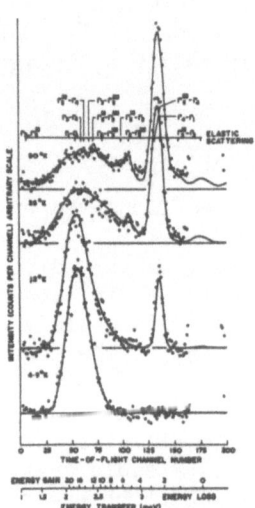

Figure 20.   Time-of-flight neutron spectra for TmSb as a function
of temperature.   The smooth curves represent the best fit to
the observed intensities using the theoretical transition probabilities and instrumental resolution function (Ref. 53).

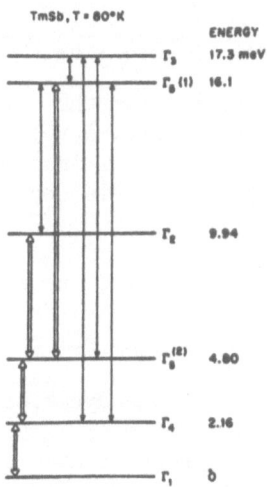

Figure 21.   Energy level diagram for TmSb at 80K deduced from
the 80K neutron spectrum.   Arrows indicate allowed transitions,
with those seen clearly at 80K marked with double bars (Ref. 53).

and F(4) and F(6) are numerical factors which for Tm^{3+} are respectively 60 and 7560.   Also shown is the variation of the matrix
elements equation (34).   It is clear from Figure 19 that both
the levels and intensities are strongly dependent on x, for which
a point charge model suggests a value of -0.96.   Neutron scattering time-of-flight spectra are shown in Figure 20.   Due to frame
overlap the energy gain and energy loss spectra are superimposed,
for example, at 12K a transition at ~2.2 meV, which is identified
as $\Gamma_1 - \Gamma_4$, is seen in both gain and loss.   Other identifications
are labelled.   The energy levels at 80 K deduced by the authors
are shown in Figure 21:   they find x = -0.79 ($\pm$ 0.02) and W = -
0.086 ($\pm$ 0.002) meV.

     The Brookhaven workers[54,55] have investigated a number of
other conducting rare earth compounds, notably the praseodymium
monopnictides and monochalcogenides where identification of the
levels and determination of B_4^o and B_6^o leads to values which are
in accord with a simple point charge model.

     It is clear from these examples that neutron scattering can
play a major role in characterising the energy levels of weakly
interacting ions both in conducting compounds, which are difficult
to probe with electromagnetic radiation, and in cases where the
orientation factor can assist in the identification of levels in

a single crystal.[56]

8.   EXCITONS IN STRONGLY INTERACTING COMPOUNDS

We now turn our attention to the more general case when there
are a number of single ion energy levels spaced by energies of the
order of the interaction energy.   In some cases it is possible
to treat only the lowest few levels separately, in others all have
to be considered together.

The usual simple treatment of the static magnetic properties
is to use the molecular field model, in which the interactions of
an ion with its neighbours are approximated by the interaction
with their average spin value.[57]   This is the starting point for
the pseudo-boson approach to the dynamical properties first given
by Grover.[58]   This is a low temperature model in that it assumes
all the ions are in their ground state, and treats only the excitations from the ground state.   The Hamiltonian is first split
into two parts

$$\mathcal{H} = \mathcal{H}_{si} + \mathcal{H}_{int} = \mathcal{H}_1 + \mathcal{H}_2 \; ,$$

where $\mathcal{H}_1 = \mathcal{H}_{si} + \mathcal{H}_{MF}$, the single ion Hamiltonian plus the interaction Hamiltonian in the molecular field approximation.   For an
exchange with z antiparallel neighbours, for example,

$$\mathcal{H}_{int} = -2 \sum_{i>j} J_{nn} \underline{S}_i \cdot \underline{S}_j \; , \; \text{and}$$

$$\mathcal{H}_{MF} = -2 J_{nn} \; z < S_j^z > \sum_i S_i^z \; ,$$

Oz being taken to lie along the mean spin direction. $\mathcal{H}_2$ is then
given by

$$\mathcal{H}_2 = \mathcal{H}_{int} - \mathcal{H}_{MF} \; ,$$

and it is this term which connects the excitations on different
ions and gives rise to propagation of the mode and dispersion of
the exciton.   The exciton modes thus correspond to transitions
to the molecular field single-ion levels in the same way as magnon
modes correspond to transitions to an ion's excited spin levels in
the ground state of an ordered system.   Clearly the exciton modes
may be also present in the paramagnetic state, when $\mathcal{H}_{MF} = 0$.   The
exciton modes may be characterised as either longitudinal or
transverse depending on whether it is the zz, or the xx or yy,
components of $\mathcal{H}_2$ which have matrix elements from the ground state
to the excited level.

If we label the eigenstates of $\mathcal{H}_1$ as p = 0, 1,2 ... we can define pseudo-boson operators for ions on each sublattice, a_p^+, a_p which excite, or de-excite, an ion from, or to, the ground (p=o) state to, or from, the pth state.   The spin operators on each site are then written as a linear combination of the pseudo-boson operators such that the correct matrix elements between the molecular field states are reproduced.   For example, for a given site,

$$S^+ = \sum_p \left\{ \langle p|S^+|0\rangle a_p^+ + \langle 0|S^+|p\rangle a_p \right\}$$

$$S^z = \langle 0|S^z|0\rangle + \sum_p \langle p|S^z|0\rangle \left\{ a_p^+ + a_p \right\} + \sum_p \left\{ \langle p|S^z|p\rangle - \langle 0|S^z \right.$$

$$\left. |0\rangle \right\} a_p^+ a_p \; .$$

The Hamiltonian $\mathcal{H}$ is transformed using these expressions to bi-linear form in the pseudo-boson operators for each sublattice. The operators are then transformed to q-space operators by a Fourier transform and $\mathcal{H}$ is then diagonalised by a more generalised form of Bogoliubov transformation such as given by Walker.[30b] In the many-level case this diagonalisation must be carried out on a computer.   We now consider a few examples.

(a)   KCoF$_3$

Exciton dispersion was first observed in CoF$_2$[59] and CoO.[60] We shall use the nearly cubic perovskite KCoF$_3$,[61] T_N = 114K, as an example as this was the first case in which the theory was fully fitted to the observed data.   Co^{2+} has a $(3d)^7$ ^{4}F term lowest and in a cubic field this is split into two triplets and a singlet.   The lowest $^4T_1$ triplet may be considered as having an effective orbital angular momentum l = 1, and spin orbit coupling splits this into three j multiplets where j = l + S. The magnon mode may be fitted well to a model of effective spins s' = j = $\frac{1}{2}$.   However, to give a complete analysis of the spectrum the pseudo-boson theory was used.   The single ion levels as a function of molecular field are shown in Figure 22, and the best fit of the theory to the data at T < 22K for the two modes is shown in Figure 23.   The fit shows the exchange between real spins to be very predominantly isotropic and between nearest neighbours, J_{nn} = -1.28 ($\pm$ 0.01) meV.   Values were also obtained for the spin-orbit coupling constant and crystal field parameter.   The calculated dispersion of all the excitations is shown in Figure 24.   Intensities calculated from equation (15) with the appropriate transformation to exciton creation operators gave good agreement with experiment.

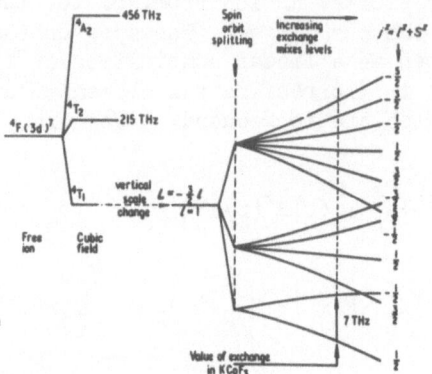

Figure 22.   Energy levels of $\mathcal{K}_1$ for Co^{2+} in $KCoF_3$ as a function
of molecular field energy (Ref. 61b).

(b)   Pr,Pr₃Tl

The light rare earth metals provide another example where
the exchange interaction is of the same order as the crystal field
levels.   Praseodymium has attracted a great deal of attention
from physicists because of its singlet ground state and induced
magnetism.[62]   A comprehensive study of d.h.c.p. Pr metal has been
carried out at Risø[63] using a small single crystal sample which
gave no indication of magnetic order (polycrystalline samples have
shown apparent antiferromagnetic order with $T_N \sim 25K$).   There are
two types of sites in this structure and the dispersion of the

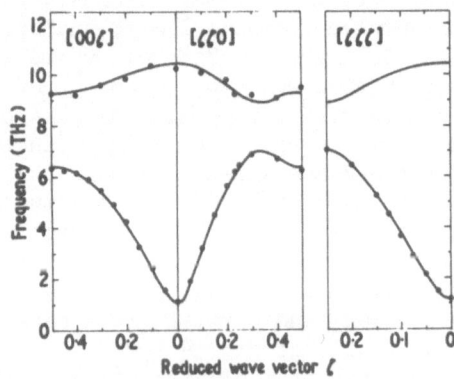

Figure 23.   Energy dispersion curves of lowest two excitations
in $KCoF_3$ (T<22K).   The solid line is the best fit to the experimental points using an isotropic exchange interaction (Ref. 61b).

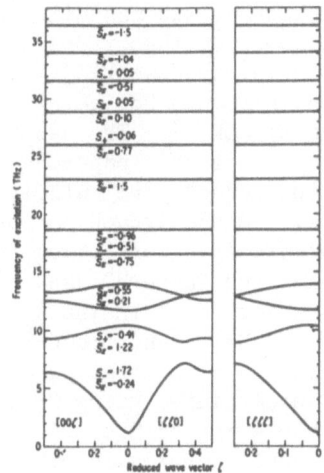

Figure 24.   Calculated dispersion of all distinct excitations
from the ground state to $^4T_1$ levels in $KCoF_3$ using the parameters
giving the best fit to the data on the lowest two modes.   Values
of $\langle p|S_x|0\rangle$ and $\bar{S}_z = \langle p|S_z|p\rangle$ are given for each excitation
(Ref. 61b).

exciton modes from hexagonal sites is shown in Figure 25, from
which the authors deduced the form of the Fourier transform of
the interaction between these sites, Y_q, using the pseudo-boson
theory.

    Polycrystalline samples of the cubic form of Pr, which order
ferromagnetically at $T_c \sim 20K$, and of the ordered alloy Pr_3Tl with
$T_c \sim 11.6K$, have also been investigated with neutron scattering.[64]

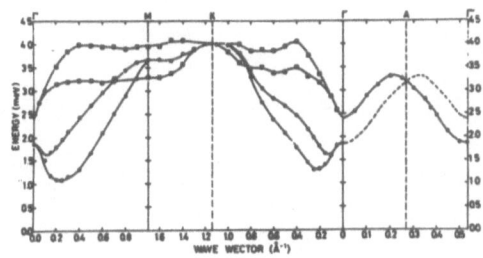

Figure 25.   Energy dispersion of excitation on the hexagonal
sites of Pr at 6.4K (Ref. 63).   (The solid lines are guides to
the eye.)

The lattice site of the $Pr^{3+}(4f)^2 \, ^3H_4$ ions has cubic symmetry, and the main interest has centred on the $\Gamma_1 - \Gamma_4$ singlet-triplet exciton mode and its behaviour at the magnetic transition. Early theories based on the pseudo-boson theory predicted a softening of these modes at T_C, but none was observed, the spherically averaged dispersion, shown in Figure 26, being found to be almost temperature independent up to 60K.[64] The behaviour of this system, treated originally in Grover's paper,[58] has greatly stimulated the theory of exciton modes, particularly those between singlet and triplet states. One approach has been to use a pseudo-spin model when the standard techniques for treating spin dynamics can be used. Wang and Cooper[65] used a $S' = \frac{1}{2}$ model to treat the singlet-singlet system both in the R.P.A. and including short range correlations, and Hsieh and Blume[66] have used a model of combinations of two spins of $\frac{1}{2}$ to treat singlet-triplet exciton modes at zero and finite temperatures using Green function theory in the R.P.A. Cooper[67] has extended Grover's pseudo-boson theory of Pr^{3+} at T = 0 to include all the levels of the Pr^{3+} ion, and Holden and Buyers[68] have treated all the excitons in Pr_3Tl, including those between excited levels, at different temperatures using Green function theory to calculate $\chi(\underline{Q},\omega)$ and hence the neutron scattering cross section. However none of these theories gives a completely satisfactory explanation of why the $\Gamma_1 - \Gamma_4$ mode does not soften at T_C, and it appears that the correct answer may lie in a suggestion by Smith[69] that the true 'soft mode' is an

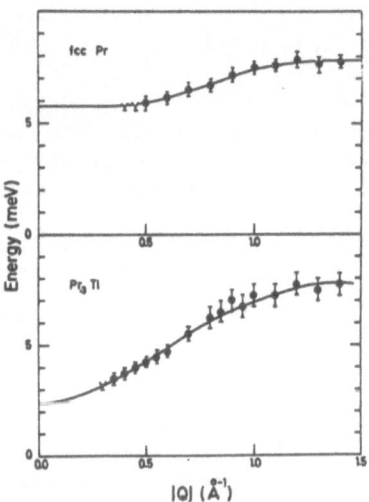

Figure 26.   Spherically averaged $\Gamma_1 - \Gamma_4$ exciton dispersion relation for T<60K in f.c.c. Pr and Pr_3Tl. The solid lines are the best fit of theory to the data yielding values for the crystal field parameter and interaction parameters (Ref. 64).

intra-triplet mode of zero frequency throughout the transition.
This suggestion has recently received support from Lines[70] using
a correlated effective field approach, and from more general considerations by Blume and Birgeneau.[71]

(c) TbSb

A further system with a singlet ground state recently studied
and explained in terms of the pseudo-boson approach is terbium
antimonide.[72] This compound is antiferromagnetic below T_N = 15K,
and again no soft mode behaviour of the excitons is observed.

(d) TbVO$_4$

Finally we mention TbVO$_4$ near its cooperative Jahn-Teller
phase transition at $T_c \sim 33K$[73] as an example of how exciton levels
may be deduced by their interaction with other modes. Above T_c,
there is a singlet B$_1$ level at 0.69 THz which has very small matrix
elements of $\underline{J}$ with the ground state and so is unobservable by
neutron scattering. At T_c however it interacts strongly with
the acoustic phonon causing a large splitting in this mode as
shown in Figure 27.[74] Another example of such an interaction is
observed in PrAlO$_3$.[75]

9. SUMMARY

With these examples it is hoped we have illustrated the
scope of neutron scattering as a technique for investigating

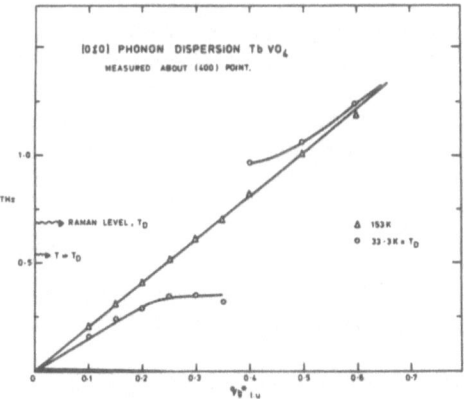

Figure 27.   Excitations in the [0ʝ0] direction in TbVO$_4$.   The
acoustic phonon mode shows normal behaviour at 153K, but is split
by the strong interaction with an exciton level near T_c.   Strong
scattering near ω = o is also observed near T_c.

electronic excitations in inorganic compounds. The principal
advantages are the ability to observe the wavevector dependence
of the excitations, and the ability to probe conducting materials.
The different selection rules and orientational dependence of the
scattering also make the technique a useful supplement to the use
of electromagnetic radiation. From the intensity of the scattering and the energy at which it occurs much can be learnt about
the dynamics of exciton levels and the nature of the interaction
between ions. There is no doubt that the technique will be used
increasingly in the future, stimulating further theory, and we
expect the range of accessible energies probed to increase with
the availability of new stronger sources of epithermal neutrons.

ACKNOWLEDGEMENTS

I would like to thank my colleagues at Harwell for useful
discussions, and Drs. R.D. Lowde, M.W. Stringfellow, B.C. Tofield
and B.T.M. Willis for helpful comments on the manuscript.

REFERENCES

1. See 'Chemical Applications of Thermal Neutron Scattering',
     ed. B.T.M. Willis (Oxford University Press, 1973).

2. We shall generally follow the theory given in W. Marshall
     and S.W. Lovesey, 'Theory of Thermal Neutron Scattering'
     (Oxford, 1971) to which the reader is referred for
     rigorous derivation of formulae. We have made some
     small changes in notation.

3. Comparative lists of neutron and X-ray cross sections and
     absorption are given by G.E. Bacon in 'Neutron Diffraction'
     (Oxford, 1962) to which the reader is also referred for
     an elementary introduction to diffraction. More recent
     values of scattering lengths are given by G.E. Bacon,
     Acta Cryst. A$\underline{28}$, 357 (1972).

4. Ref. 2, Chapter 5.

5. Ref. 2, Chapter 6, S.W. Lovesey and D.E. Rimmer, Rep. Prog.
     Phys. $\underline{32}$, 333 (1969).

6. R.M. Moon, W.C. Koehler and A.L. Trego, J. Appl. Phys. $\underline{37}$,
     1036 (1966).

7. See, for example, A.J. Jacobson, Ref. 1, Chapter 12, and
     B.C. Tofield, Structure and Bonding $\underline{21}$ (to be published
     1975).

8. W. Marshall and R.D. Lowde, Repts. Prog. Phys. 31, 705 (1968).

9. See S.J. Cocking and F.J. Webb, 'Thermal Neutron Scattering', ed. P.A. Egelstaff (Academic Press, London and New York, 1965), Chapter 4. This volume contains a good survey of all aspects of neutron scattering.

10. See G.C. Stirling, Ref. 1, Chapter 2.

11. D.H. Day and R.H. Sinclair, J. Chem. Phys. 55, 2807 (1971).

12. R.D. Lowde, J. Nucl. Energy, A11, 69 (1960); R.F. Dyer and G.G. Low, 'Inelastic Scattering of Neutrons in Solids and Liquids' (IAEA, Vienna, 1961), p. 179.

13. S. Komura and M.J. Cooper, Jap. J. Appl. Phys. 9, 866 (1970).

14. See Ref. 11 for examples of measurement of energy transfers $\hbar\omega$ up to 0.5 eV.

15. See, for example, F. Hossfeld, A. Amadori and R. Scherm, 'Instrumentation for Neutron Inelastic Scattering', (IAEA, Vienna, 1970), p. 117.

16. See B.N. Brockhouse, 'Inelastic Scattering of Neutrons in Solids and Liquids' (IAEA, Vienna, 1961), p. 113.

17. L. Chumbley, R.F. Dyer and D.E. Wallis, J. Phys. E1, 528 (1968).

18. T. Riste, 'Instrumentation for Neutron Inelastic Scattering' (IAEA, Vienna, 1970), p. 91.

19. B. Jacrot, 'Instrumentation for Neutron Inelastic Scattering' (IAEA, Vienna, 1970), p. 225.

20. J. Bergsma and C. Van Dijk, Nucl. Inst. and Meth. 51, 121 (1967); G. Shirane and V.J. Minkiewicz, Nucl. Inst. and Meth. 89, 109 (1970).

21. See, for example, M.J. Cooper and R. Nathans, Acta Cryst. 23, 357 (1967); B. Dorner, Acta Cryst. A28, 319 (1972).

22. See, for example, B.N. Brockhouse, 'Phonons', ed. R.W.H. Stevenson (Oliver and Boyd, Edinburgh and London, 1966), p. 110.

23. B.C. Haywood and M.F. Collins, J. Phys. C4, 1299 (1971).

24.   P.H. Gamlen, N.F. Hall and A.D. Taylor, Report USS/P29,
       Materials Physics Division, Harwell, England.

25.   J.K. Kjems and P.A. Reynolds, 'Proc. Symposium on Neutron
       Inelastic Scattering' (IAEA, Vienna, 1972), p. 733.

26.   R. Loudon, Advances in Physics $\underline{17}$, 243 (1968).

27.   W. Brinkman and R.J. Elliott, J. Appl. Phys. $\underline{37}$, 1457
       (1966) and Proc. Roy. Soc. A$\underline{294}$, 343 (1966).

28.   See, for example, K.W.H. Stevens in 'Magnetism', eds. G.T.
       Rado and H. Suhl (Academic Press, New York and London,
       1963), Vol. 1, p. 1.

29.   A.M. Morrish, 'The Physical Principles of Magnetism' (Wiley,
       New York, 1965), p. 296.

30.   For complete discussions of spin waves, see (a) C. Kittel
       'Quantum Theory of Solids' (Wiley, New York, 1963), p. 49;
       (b) L.R. Walker in 'Magnetism', eds. G.T. Rado and H. Suhl
       (Academic Press, New York and London, 1963), Vol. 1, p.
       299; (c) F. Keffer, 'Encyclopedia of Physics', ed. S.
       Flugge (Springer Verlag, Berlin, 1966), Vol. 18/2, p. 1;
       (d) R.D. Lowde, 'Advanced Course in Magnetic Exchange
       Interactions', Report KR93, Institutt for Atomenergi,
       Kjeller, Norway (1965), p. 35; (e) Ref. 2, Chapter 9.

31.   T. Holstein and M. Primakoff, Phys. Rev. $\underline{58}$, 1098 (1940).

32.   E.J. Samuelsen, R. Silberglitt, G. Shirane and J.P. Remeika,
       Phys. Rev. B$\underline{3}$, 157 (1971).

33.   L. Passell, O.W. Dietrich and J. Als-Nielsen, 'Proc. 17th
       Conf. on Magnetism and Magnetic Materials', A.I.P. Conf.
       Proc. No. 5, p. 1251 (1972).

34.   T. Oguchi, Phys. Rev. $\underline{117}$, 117 (1960).

35.   O. Nagai and A. Yoshimori, Prog. Theor. Phys. $\underline{25}$, 595 (1961).

36.   A.W. Saenz, Phys. Rev. $\underline{125}$, 1940 (1962).

37.   D.C. Wallace, Phys. Rev. $\underline{128}$, 1614 (1962).

38.   E.J. Samuelsen, Physica $\underline{43}$, 353 (1969).

39.   See, for example, M.E. Lines, Phys. Rev. $\underline{156}$, 543 (1967).

40.   R. Loudon (private communication); A. Stevens, J. Phys.
      C5, 1859 (1972).

41.   See Ref. 26, and P.A. Fleury and R. Loudon, Phys. Rev. 166,
      514 (1968).

42.   M.T. Hutchings, G. Shirane, R.J. Birgeneau and S.L. Holt,
      Phys. Rev. B5, 1999 (1972).

43.   See F.B. McLean and M. Blume, Phys. Rev. B7, 1149 (1973);
      S.W. Lovesey, J. Phys. C7, 2008 (1974); and K. Tomita
      and H. Mashiyama (to be published).

44.   M.T. Hutchings, M.F. Thorpe, R.J. Birgeneau, P.A. Fleury
      and H.J. Guggenheim, Phys. Rev. B2, 1362 (1970).

45.   R.J. Elliott, M.F. Thorpe, G.F. Imbusch, R. Loudon and
      J.B. Parkinson, Phys. Rev. Letters 21, 147 (1968).

46.   E.J. Samuelsen and G. Shirane, Phys. Stat. Sol. 42, 241
      (1970).

47.   P.G. de Gennes, J. Chem. Phys. Solids 4, 223 (1958).

48.   See, for example, M.T. Hutchings, 'Solid State Physics',
      ed. F. Seitz and D. Turnbull (Academic Press, New York
      1964), Vol. 16, p. 227.

49.   D. Cribier and B. Jacrot, C.R. Acad. Sci. Paris 250, 2871
      (1960).

50.   B. Rainford, K.C. Turberfield, G. Busch and O. Vogt, J.
      Phys. C1, 679 (1968).

51.   P.G. de Gennes,'Magnetism', ed. G.T. Rado and H. Suhl
      (Academic Press, New York, 1963), Vol. 3, p. 115.

52.   R.J. Birgeneau, J. Phys. Chem. Solids 33, 59 (1972).

53.   R.J. Birgeneau, E. Bucher, L. Passell and K.C. Turberfield,
      Phys. Rev. B4, 718 (1971).

54.(a) K.C. Turberfield, L. Passell, R.J. Birgeneau and E. Bucher,
      Phys. Rev. Letters 25, 752 (1970), and J. Appl. Phys. 42,
      1746 (1971); (b) R.J. Birgeneau, E. Bucher, L. Passell,
      D.L. Price, and K.C. Turberfield, J. Appl. Phys. 41, 900
      (1970); (c) R.J. Birgeneau, E. Bucher, J.P. Maita, L.
      Passell and K.C. Turberfield, Phys. Rev. B8, 5345 (1973).

55.   See also: (a) A. Fürrer, W. Bührer, H. Heer, W. Hälg,
      J. Benes and O. Vogt, 'Proc. Symposium on Neutron
      Inelastic Scattering' (IAEA, Vienna, 1972), p. 563;
      (b) A. Fürrer, J. Kjems and O. Vogt, J. Phys. C5, 2246
      (1972); (c) H.L. Davis and H.A. Mook, 'Proc. 18th Conf.
      on Magnetism and Magnetic Materials', A.I.P. Conf. Proc.
      No. 10, p. 1548 (1973).

56.   See, for example, W. Holmes, H.J. Guggenheim and J. Als-
      Nielsen, 'Proc. International Conf. on Magnetism', Moscow
      1973 (to be published).

57.   The dynamical extension of this theory is discussed by R.
      Alben, Phys. Rev. 184, 495 (1969).

58.   B. Grover, Phys. Rev. 140, A1944 (1965).

59.   R.A. Cowley, P. Martel and R.W.H. Stevenson, Phys. Rev.
      Letters 18, 162 (1967).

60.   J. Sakurai, W.J.L. Buyers, R.A. Cowley and G. Dolling, Phys.
      Rev. 167, 510 (1968).

61.(a) T.M. Holden, W.J.L. Buyers, E.C. Svensson, R.A. Cowley,
      M.T. Hutchings, D. Hukin and R.W.H. Stevenson, J. Phys.
      C4, 2127 (1971); (b) W.J.L. Buyers, T.M. Holden, E.C.
      Svensson, R.A. Cowley and M.T. Hutchings, J. Phys. C4,
      2139 (1971).

62.   R.J. Birgeneau, 'Proc. 18th Conf. on Magnetism and Magnetic
      Materials', A.I.P. Conf. Proc. No. 10, p. 1664 (1973).

63.   B.D. Rainford and J. Gylden Houmann, Phys. Rev. Letters 26,
      1254 (1971), and private communication.

64.   R.J. Birgeneau, J. Als-Nielsen and E. Bucher, Phys. Rev.
      Letters 27, 1530 (1971), Phys. Rev. B6, 2724 (1972), and
      'Proc. Symposium on Neutron Inelastic Scattering' (IAEA,
      Vienna, 1972), p. 543.

65.   Y-L. Wang and B.R. Cooper, Phys. Rev. 172, 539 (1968).

66.   Y.Y. Hsieh and M. Blume, Phys. Rev. B6, 2684 (1972).

67.   B.R. Cooper, Phys. Rev. B6, 2730 (1972).

68.   T.M. Holden and W.J.L. Buyers, Phys. Rev. B9, 3797 (1974).

69.   S.R.P. Smith, J. Phys. C5, L157 (1972).

70.   M.E. Lines, J. Phys. C7, L287 (1974), Phys. Rev. 9, 950 and
       3927 (1974).

71.   M. Blume and R.J. Birgeneau, J. Phys. C7, L282 (1974).

72.   T.M. Holden, E.C. Svensson, W.J.L. Buyers and O. Vogt,
       'Proc. Symposium on Neutron Inelastic Scattering' (IAEA,
       Vienna, 1972), p. 553.

73.   R.J. Elliott, R.T. Harley, W. Hayes and S.R.P. Smith,
       Proc. Roy. Soc. A328, 217 (1972).

74.   M.T. Hutchings, R. Scherm, S.H. Smith and S.R.P. Smith
       (private communication).

75.   J.K. Kjems, G. Shirane, R.J. Birgeneau and L.G. Van Uitert,
       Phys. Rev. Letters 31, 1300 (1973).